Current Topics in Behavioral Neurosciences

Volume 17

Series editors

About this Series

Current Topics in Behavioral Neurosciences provides critical and comprehensive discussions of the most significant areas of behavioral neuroscience research, written by leading international authorities. Each volume offers an informative and contemporary account of its subject, making it an unrivalled reference source. Titles in this series are available in both print and electronic formats.

With the development of new methodologies for brain imaging, genetic and genomic analyses, molecular engineering of mutant animals, novel routes for drug delivery, and sophisticated cross-species behavioral assessments, it is now possible to study behavior relevant to psychiatric and neurological diseases and disorders on the physiological level. The *Behavioral Neurosciences* series focuses on "translational medicine" and cutting-edge technologies. Preclinical and clinical trials for the development of new diagnostics and therapeutics as well as prevention efforts are covered whenever possible.

More information about this series at http://www.springer.com/series/7854

Klaus A. Miczek · Andreas Meyer-Lindenberg
Editors

Neuroscience of Aggression

Springer

Editors
Klaus A. Miczek
Moses Hunt Professor of Psychology,
Psychiatry, Pharmacology
and Neuroscience
Tufts University
Medford, MA
USA

Andreas Meyer-Lindenberg
Department of Psychiatry
and Psychotherapy
Central Institute of Mental Health
Mannheim
Germany

ISSN 1866-3370 ISSN 1866-3389 (electronic)
ISBN 978-3-662-44280-7 ISBN 978-3-662-44281-4 (eBook)
DOI 10.1007/978-3-662-44281-4

Library of Congress Control Number: 2014949316

Springer Heidelberg New York Dordrecht London

Printed on acid-free paper

Springer is part of Springer Science+Business Media (www.springer.com)

Preface

Despite its relevance to science as well as society, aggression remains an understudied topic in the basic neurosciences and in the psychopharmacological management and treatment of patients. The present volume is intended to appeal to both clinical practitioners and preclinical researchers who seek to enhance their understanding of basic molecular and cellular processes, neural circuitry, and behavioral mechanisms that are critical for the motivation and expression of aggressive behavior. The current contributions advance our insights into neurobiological mechanisms and begin to identify genes and gene networks which—in interaction with environmental triggers—render aggressive behavior more or less likely to occur. It is evident that the current research proceeds in a more rational, mechanistic way relative to its serendipitous beginnings. For example, long before the sites and mechanisms of action of lithium or benzodiazepines were understood, it was discovered that they calm agitated and aggressive individuals (Cade 1949; Randall et al. 1960).

From a psychiatric perspective, research on aggression has been impeded by inadequate recognition of these behavioral domains in consecutive versions of the Diagnostic and Statistical Manual, discouraging the development of compounds and interventions with specific anti-aggressive effects. One cardinal criterion for novel medications in this area should be their degree of specificity for reducing aggressive behavior relative to undesirable adverse effect profiles. Quantitative ethological methods enable the comparison of reducing aggressive acts with effects on other elements in the behavioral repertoire to learn about the behavioral specificity of novel compounds.

A fundamental challenge in aggression research is the distinction of the neurobiological mechanisms mediating escalated *pathological* aggressive behavior relative to those for species-typical patterns of behavior that are necessary for the survival of the organism (Miczek et al. 2013). In humans, of course, this distinction is further compounded by legal, ethical, and societal restriction on aggression no matter the biological concomitants. The neural circuits of different kinds of mammalian aggression have begun to be delineated by immunohistochemistry of immediate-early gene expression, intracranial microinjection, in vivo microdialysis, optogenetics techniques in experimental animals, and high resolution functional imaging techniques. It is now feasible to conceptualize neurocircuits for different kinds of aggressive behavior to encompass nuclei in the mesencephalon projecting to hypothalamic, amygdaloid, septal, and hippocampal sub-nuclei, loops between striatum and thalamus with frontal and prefrontal structures and important feedback to limbic and mesencephalic nuclei (chapters by Kruk; Gobrogge; Bedrosian and Nelson; Barr and Driscoll; Takahashi and Miczek; Haller; Morrison and Melloni; de Almeida et al.). Identifying not only the neural structures and projections, but also the precise molecular processes within these critical cell groups that mediate escalated aggressive behavior remains an urgent task.

A most instructive example is our emerging understanding of the role of serotonin, the most intensively investigated neurotransmitter system that has been linked to aggression. It is evident that the classic serotonin deficiency hypothesis that associates defects in synthesis, release, receptor activation, or metabolism to a heightened propensity to engage in aggressive behavior has been replaced with a framework that incorporates a much more nuanced set of modulatory and regulatory mechanisms. The current contributions (Barr and Driscoll; Takahashi and Miczek; Morrison and Melloni; Bedrosian and Nelson; Gobrogge; Beck et al.; Buckholtz et al.) illustrate how serotonergic activity is modulated by a cascade of direct and indirect systems that range from nitric oxide synthase, neuropeptides and neurosteroids to excitatory and inhibitory amino acids and thereby reduce or intensify aggressive behavior.

We invite you to be inspired by the current contributions which we hope will renew interest in the neurobiology of aggressive behavior and reactivate this largely neglected area of research.

References

Cade JFJ (1949) Lithium salt in the treatment of psychotic excitement. Med J Aust 2:349–352

Miczek KA, de Boer SF, Haller J (2013) Excessive aggression as model of violence: a critical evaluation of current preclinical methods. Psychopharmacology 226:445-458

Randall LO, Schallek W, Heise FA et al (1960) The psychosedative properties of methaminodiazepoxide. J Pharmacol Exp Ther 129:163–171

Medford — Klaus A. Miczek
Mannheim — Andreas Meyer-Lindenberg

Contents

Part I
Preclinical Research

Neurogenetics of Aggressive Behavior: Studies in Rodents

Aki Takahashi and Klaus A. Miczek

Abstract Aggressive behavior is observed in many animal species, such as insects, fish, lizards, frogs, and most mammals including humans. This wide range of conservation underscores the importance of aggressive behavior in the animals' survival and fitness, and the likely heritability of this behavior. Although typical patterns of aggressive behavior differ between species, there are several concordances in the neurobiology of aggression among rodents, primates, and humans. Studies with rodent models may eventually help us to understand the neurogenetic architecture of aggression in humans. However, it is important to recognize the difference between the ecological and ethological significance of aggressive behavior (species-typical aggression) and maladaptive violence (escalated aggression) when applying the findings of aggression research using animal models to human or veterinary medicine. Well-studied rodent models for aggressive behavior in the laboratory setting include the mouse (*Mus musculus*), rat (*Rattus norvegicus*), hamster (*Mesocricetus auratus*), and prairie vole (*Microtus ochrogaster*). The neural circuits of rodent aggression have been gradually elucidated by several techniques, e.g., immunohistochemistry of immediate-early gene (c-Fos) expression, intracranial drug microinjection, in vivo microdialysis, and optogenetics techniques. Also, evidence accumulated from the analysis of gene-knockout mice shows the involvement of several genes in aggression. Here, we review the brain circuits that have been implicated in aggression, such as the hypothalamus, prefrontal cortex (PFC), dorsal raphe nucleus (DRN), nucleus accumbens (NAc), and olfactory system. We then discuss the roles of glutamate and γ-aminobutyric acid (GABA), excitatory and inhibitory amino acids in the brain, as well as their receptors, in controlling aggressive behavior, focusing mainly on recent findings. At the end of this chapter,

A. Takahashi (✉)
Mouse Genomics Resource Laboratory, National Institute of Genetics, (NIG), 1111 Yata, Mishima, Shizuoka 411-8540, Japan
e-mail: aktakaha@nig.ac.jp

K. A. Miczek
Departments of Psychology, Psychiatry, Pharmacology and Neuroscience, Tufts University, Medford and Boston, MA, USA

Curr Topics Behav Neurosci (2014) 17: 3–44
DOI: 10.1007/7854_2013_263

Published Online: 7 December 2013

we discuss how genes can be identified that underlie individual differences in aggression, using the so-called forward genetics approach.

Keywords Aggression · Rodents · Brain Circuit · Glutamate · GABA · Neurogenetics

Contents

1 Introduction

Epigenetic studies have begun to reveal how salient life experiences during critical periods of development determine the probability of subsequently engaging in aggressive confrontations (Caspi et al. 2002; Veenema 2009). Control over the breeding history and each facet of early development make rodents the most intensively studied subjects for behavioral and molecular genetic analysis of aggressive behavior (Crawley et al. 1997). Since 2002, when the mapping of the mouse genome was completed (Waterston et al. 2002), mice have been the focus of most rodent aggression studies.

During the past five decades, neurogenetic research of aggressive behavior in rodents has progressed from "top-down" to "bottom-up" to epigenetic studies. Early strain comparisons and domestication studies initiated "top-down" genetics where the genetic basis for an aggressive trait was investigated chiefly via selective breeding (Cairns et al. 1983; Lagerspetz 1964; Popova et al. 1991; van Oortmerssen and Bakker 1981); "bottom-up" genetics focuses on a gene for a specific candidate receptor or transporter molecule and manipulates the expression of this gene (Cases et al. 1995; Nelson et al. 1995; Saudou et al. 1994). Given the polygenic nature of genetic influences on aggressive behavior, it is likely that future studies in rodents will uncover gene networks for each type of aggressive behavior.

The most intensively investigated neurochemical system for the control of adaptive and pathological forms of aggressive behavior involves all aspects of serotonin—which was early on labeled the "civilizing neurohumor" (Nelson and Chiavegatto 2001; Takahashi et al. 2011). Every facet of the synthetic and metabolic pathways, uptake and storage processes as well as somatodendritic, pre and postsynaptic receptor mechanisms of serotonin has been explored in terms of its relevance to the neural control of aggressive behavior (de Boer and Koolhaas 2005; Barr and Driscoll, this volume; Bedrosian and Nelson, this volume). Several major themes have emerged from this considerable data base. For example, depletion studies have highlighted the importance of tonic levels of serotonin in the likelihood of impulsive outbursts. By contrast, inhibition of 5-HT impulse flow due to somatodendritic autoreceptor stimulation in the dorsal raphe nucleus reduces several types of species-specific and maladaptive aggressive behavior. However, activation of 5-HT1A receptors in prefrontal cortical regions can increase aggressive behavior, pointing to functionally separate receptor pools. Differences in the alleles of genes that encode for specific serotonin receptor subtypes, transporter molecules, synthetic, and metabolic enzymes may contribute to variable outcomes in pharmacotherapeutic treatments. Findings of this nature have led to a re-examination of the seductively simple serotonin deficiency hypothesis of aggressive behavior. Phasic changes in 5-HT emerge during aggressive episodes as illustrated by a sudden decrease in accumbal serotonin at the termination of a confrontation (van Erp and Miczek 2000) and this decrease can be conditioned by repeated aggressive experiences (Ferrari et al. 2003).

The brain areas involved in aggressive behavior have been elucidated using traditional lesion and electrical stimulation studies. Analysis of c-Fos expression has provided novel insights into the neuronal circuits that are involved in specific types of aggressive behaviors. Intracranial microinjections of receptor-selective agonists and antagonists have demonstrated the roles of specific receptor populations in particular areas of the brain that mediate aggressive behaviors. Furthermore, recently developed optogenetic techniques have allowed in vivo manipulation of the activity of specific neurons in the desired brain area in milliseconds to study the role of microcircuits in aggressive behavior.

In this chapter, we will first discuss the brain circuits that have been implicated in aggression, such as the hypothalamus, prefrontal cortex (PFC), dorsal raphe nucleus (DRN), nucleus accumbens (NAc), and olfactory system. We will then discuss the roles of glutamate and γ-aminobutyric acid (GABA), major inhibitory and excitatory amino acids in the brain, as well as their receptors, in controlling aggressive behavior, focusing mainly on recent findings. The relationships between other neurotransmitters, neuropeptides, and neuromodulators that have been implicated in aggression, such as serotonin (5-HT), dopamine, vasopressin, oxytocin, testosterone, estrogen, corticotrophin releasing factor (CRF), opioids, neuronal nitric oxide synthase (nNOS), and monoamine oxidase A (MAOA), will be discussed elsewhere in this book. At the end of this chapter, we will discuss attempts to identify genes that underlie individual differences in aggression, using the so-called forward genetics approach.

2 Rodent Models for Aggressive Behavior

2.1 Definition of Aggressive Behavior

Aggression encompasses a range of diverse behavioral patterns and is multidimensional in terms of its origins, motivations, expressions, and functions (Miczek et al. 2007). Definitions of aggression refer to different subtypes of aggression, or define it broadly as "behavior that inflicts harm and injury or threatens to do so" (Berkowitz 1993) or "any form of behavior directed toward the goal of harming or injuring another living being who is motivated to avoid such treatment" (Baron and Richardson 1994). As Scott (1966) originally defined from an ethological perspective, aggressive or agonistic behaviors are "*adaptations* for situations involving physical conflict or contests between members of the same species" (Scott 1966). Thus, research on aggression using animal models has tended to investigate the ethological significance of the behavior, such as its functionality in the survival and reproduction of animal, and its phylogenetic and ontogenetic development. Such aggressive behavior is beneficial to the individual because this behavior can be used to obtain food, water, and other resources, including female mates in the case of males, and to defend its territory, offspring, or social rank. However, given that aggressive behavior carries the associated risks (costs) of injury or even death as well as high energy expenditure, the exercise of aggressive behavior is determined by a *cost-benefit* analysis (Maynard Smith and Price 1973; Haller 1995). When this balance is disrupted and an animal shows an exaggerated level of aggression, the behavior is no longer adaptive, and it could be called maladaptive or pathological aggression. Most research on aggression in both human and veterinary medicine seeks to understand and control pathological aggression (Volavka and Nolan 2008; Siever 2008).

Aggressive behavior in rodent models can be classified as "*offensive*" or "*defensive*" based upon the distal and proximal antecedent conditions that precipitated it, the topography of the behavior, and its consequences (Blanchard and Blanchard 1977; Brain 1979; Adams 2006). Offensive aggression is observed in interactions between mature males, and most of the research on the genetics and neurobiology of aggression studies this offensive behavior (Table 1a). In rodents, offensive aggressive behavior between conspecifics is ritually organized, composed of sideways threats, chasing, tail rattling (especially in mouse), defensive upright postures, and attack bites (Miczek and O'Donnell 1978). The attack is usually directed toward less vulnerable body areas such as the back and flanks of the opponent (Blanchard and Blanchard 1977; Blanchard et al. 1979, 2001).

Table 1 Types of aggressive behavior in rodent models

(a) Species-typical aggression

	Definition	Test situation or experimental variable	Reference
Territorial aggression (intermale aggression)	The behavior of a male in control of an "exclusive territory." A breeding male resident rodent threatens and attacks an adult male that intrudes into the territory	Resident–intruder test in the established territory. In the laboratory, the experimental male (*resident*) is pair housed with a female. A male stimulus animal (*intruder*), which has been group housed with other males, is introduced into resident's cage. Both the frequency and duration of both aggressive behaviors (e.g. attack bites, pursuit, and sideways threat) and nonaggressive behaviors (e.g., walking, rearing, and self-grooming), and the latency to the first bite are recorded by human observation	Crawley et al. (1975), Murphy (1976), Miczek and O'Donnell (1978)
Dominance	The behavior of one male who dominates others. Dominance hierarchy is common in rats but group housed male C57BL/6 mice also show hierarchy	Observation of the agonistic and submissive behaviors in the group housed home cage. Also, in mice, the tube test can be used to evaluate social dominance. In this test, two mice are released into the opposite ends of a narrow tube. The expectation is that the dominant male forces the opponent forward and out of the tube	Popova and Naumenko (1972), Blanchard and Blanchard (1977), Wang et al. (2011)
Maternal aggression	Aggressive behavior expressed by females shortly before gestation, which reaches a peak during the first week postpartum, and declines thereafter	Resident–intruder test in the home cage of lactating females from postpartum day 1–7. Either a male or female intruder is introduced into dam's cage. In addition to the variables observed for territorial aggression in the male, attack bites directed at the snout and the face can be observed	Noirot et al. (1975), Hurst (1987), Haney et al. (1989), Lonstein and Gammie (2002), Sgoifo et al. (2006)
Female aggression	Aggressive behavior expressed by females to compete for reproductive opportunities	Female rats pair housed with a breeding male. Resident–intruder test using sexually matured female as an intruder. In the hamster, females show male-like territorial behavior toward both male and female intruders	Payne and Swanson (1970), DeBold and Miczek (1981), Bowler et al. (2002)

(continued)

Table 1 (continued)

(a) Species-typical aggression

	Definition	Test situation or experimental variable	Reference
Pair-bonding-induced aggression	Monogamous prairie voles show exclusive aggressive behavior toward intruders once they form pair bonds with a female	Within 24 h of mating with a female, the sexually-experienced male prairie vole attacks conspecific intruders of both sexes and exhibits affiliative behaviors only toward his mate	Insel et al. (1995)

(b) Escalated aggression

	Definition	Test situation and experimental design	Reference
Social provocations (social instigation)	The presence of a breeding male provokes the resident into intense and frequent aggressive behavior, when given the opportunity. It is likely that the olfactory, visual, and auditory cues emanating from the intruder induce "aggressive arousal" or "attack readiness" in the resident male	A resident male is exposed to another breeding male (instigator) in his home cage without direct agonistic interaction. The instigator is behind a protective screen, and thus the resident male can see, smell, and hear the instigator male. Immediately after the provocation, an intruder male which is different from the instigator is introduced into the home cage of the resident male	Heiligenberg (1974), Potegal and Tenbrink (1984), Potegal (1991), Fish et al. (1999)
Social isolation (SI)	Social isolation induces higher levels of attack on the background of increased defensiveness	Males isolated for at least 4 weeks prior to resident–intruder encounter	Guidotti et al. (2001), Tóth et al. (2008),
Alcohol-heightened aggression (AHA)	A subset of mice and rats (ca. 30 %) show escalated aggressive behavior after the consumption of alcohol compared to their base-line aggression (aggressive behavior after water consumption)	Animals receive ethanol (1.0 g/kg) orally 15 min before the resident–intruder encounter. Water access is restricted for 21 h before the test. Operant conditioning is used to allow the animal to self-administer a certain amount of ethanol	Blanchard et al. (1987), Miczek et al. (1992, 1998), Miczek and de Almeida (2001)

(continued)

Table 1 (continued)

(b) Escalated aggression

	Definition	Test situation and experimental design	Reference
Hypoglucocorticoid status (ADXr)	Surgery and replacement of a low level of glucocorticoid escalates aggressive behavior in the male rats. ADXr animals provide a model of hypoarousal aggression owing to their low heart rates during aggressive encounters	Animals are adrenalectomized and implanted with a low corticosterone pellet (ADXr). Escalation of aggression is observed 1 week after the surgery	Haller et al. (2001, 2004)
Chronic anabolic–androgenic steroids (AAS)	Exposure to chronic anabolic–androgenic steroids during adolescence predisposes male hamsters and rats to escalate aggressive behavior	Animals are treated with a cocktail of anabolic–androgenic steroids (AAS) by subcutaneous injections daily for 30 consecutive days during puberty	Melloni et al. (1997)
Artificial selection on aggression	Genetic model to examine the heritability of aggressive behavior. Selective breeding has successfully produced highly aggressive lines for intermale aggression, isolation-induced aggression, and maternal aggression in the mouse	Breeding of highly aggressive individuals in the population to produce the next generation. Characterization of aggressive behavior of all offspring is performed every generation, and the highly aggressive individuals are selected each generation	van Oortmerssen and Bakker (1981), Cairns et al. (1983), Sandnabba (1996), Gammie et al. (2006)

Defensive aggression involves attack in defense of the self; it occurs in response to threatening or fear-inducing stimuli, and is often accompanied by escape. In male rodents, specific defensive behaviors include escape, freezing, defensive postures, and threats, with defensive attacks targeted at the snouts of predators or conspecifics (Blanchard and Blanchard 2003).

In most common laboratory rodents, aggressive behavior is restricted almost exclusively to males. However, both male and female golden hamsters show a high level of aggression, with females usually dominating males (Payne and Swanson 1970). Females of many rodent species show intense aggressive behavior during specific periods after giving birth, referred to as maternal aggression, this protects their newborn offspring from male intruders (Palanza et al. 1994). In most rodents, female aggressive behavior becomes more frequent shortly before gestation, reaches a peak at the first week postpartum, and declines thereafter (Noirot et al. 1975; Erskine et al. 1980). Maternal aggression includes both defensive and offensive elements; lactating females engage in defensive attacks toward males and offensive attacks toward female intruders (Parmigiani et al. 1989; Lucion and de Almeida 1996; de Almeida et al. this volume). In some respects, infanticide (Svare et al. 1984), predatory aggression, such as mouse killing behavior by rats (Karli et al. 1972), and play fighting in juvenile rats and hamsters (Pellis and Pellis 1988; Pellis and Iwaniuk 2000) also qualify as aggressive behavior. However, these are considered to be qualitatively different from offensive or defensive aggression. Each type of aggression is most likely regulated by different genes and neurobiological pathways, with some overlap.

2.2 Escalated Aggression

Aggressive behavior in animals becomes increasingly relevant to clinical concerns in humans when it differs from the typical pattern for the species in question. Aggressive behaviors that exceed species-typical levels or patterns are known as *escalated aggression*, or sometimes as maladaptive or pathological aggression that models violent behavior in humans. Several rodent models have been used to investigate the neurobiology and genetics of escalated aggression (Miczek et al. 2013; Table 1b). Escalated aggression is characterized quantitatively by shorter attack latencies (readiness to initiate aggressive behavior) and higher frequencies and longer durations of fighting (inability to terminate aggressive behavior) than normal adaptive aggression (Miczek et al. 2004; de Almeida et al. 2005). There are also qualitative differences. Attack bites aimed with high intensity at vulnerable parts of the opponent's body, such as the head and throat, are considered to be abnormally aggressive, while adaptive aggression consists of less-injurious bites directed at the intruder's back and flanks (Haller and Kruk 2006). Additional qualitative distinctions between adaptive and escalated aggression include context-independent attacks regardless of the sex and nature of the opponent, such as the responsiveness of the target (moving and responsive vs. anaesthetized), or the test environment (home or neutral cage), a failure to respond to appeasement signals, and a lack of ritualistic behaviors, quantified using Attack/Threat (A/T) ratios (Haller et al. 2005a, 2006; de Boer et al. 2009). Therefore, in principle, "violence" in animal models could refer either to quantitatively escalated and to qualitatively abnormal forms of aggression (for review see Natarajan and Caramaschi 2010; Miczek et al. 2013).

3 The Neurobiology of Aggressive Behavior in Rodents

3.1 Brain Circuits Activated by Aggressive Encounters

To understand the neural circuits involved in aggressive behaviors, immediate early gene expression has been analyzed to identify brain areas that are activated by aggressive acts. c-Fos is the protein product of an immediate early gene, *fos*, which is expressed in neurons shortly after their depolarization (activation), and then induces the expression of downstream genes. The expression of c-Fos can be readily visualized by immunohistochemistry staining, and the number of c-Fos positive neurons in each brain area is used to quantify the activation of the area. Investigations of c-Fos activity have implicated several brain areas in aggressive behavior: the PFC; the medial preoptic area (MPOA); lateral septum (LS); some hypothalamic nuclei including the anterior hypothalamus (AH), ventromedial hypothalamus (VMH), and lateral hypothalamus (LH); the paraventricular nucleus (PVN); the medial and central amygdala (MeA and CeA); bed nucleus of the stria terminalis (BNST); the periaqueductal gray (PAG); the locus coeruleus (LC); and the DRN (for review, see Lonstein and Gammie 2002; Nelson and Trainor 2007; Heinz et al. 2011).

Many of these areas are involved in aggressive behavior in nonhuman primates, and even in humans (Nelson and Trainor 2007). Note, however, that these areas are not specific to aggressive behavior but are also involved in a wide range of other social behaviors (Newman 1999). Figure 1 summarizes the brain areas that are activated by intermale aggressive behavior, maternal aggressive behavior, and escalated aggressive behavior, as indicated by c-Fos expression. There is a large overlap among brain areas that are involved in different types of aggression, but also some differentiation. In particular, when the animal is engaging in escalated levels of aggression, activity in some brain areas may be either overactivated or suppressed, depending on the model of escalated aggression, when compared with the species-typical level (Fig. 1). For example, the PAG is involved in both intermale and escalated aggression (and phosphorylation of the cAMP response element binding protein (pCREB) suggests the involvement of the PAG in maternal aggression). The PAG is activated during species-typical aggression, but at the same time, the degree of PAG activation is inhibited in animals that show escalated aggression as a result of selective breeding compared to a line that is selected for low aggression (Kollack-Walker and Newman 1995; Delville et al. 2000; Gammie and Nelson 2001; Haller et al. 2006).

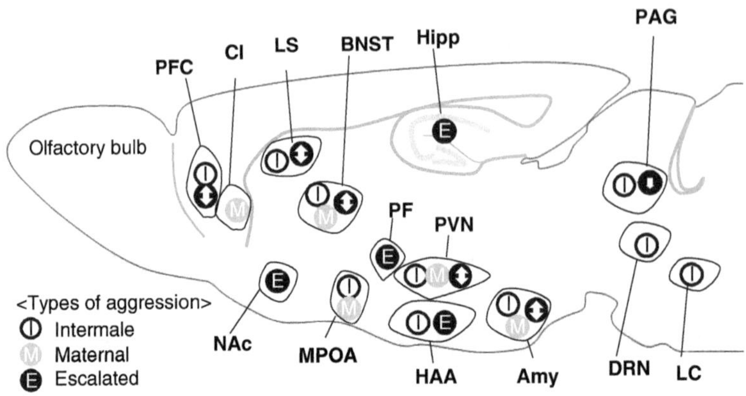

Fig. 1 Schemas of brain areas that are activated during and after an aggressive encounter. Observed by c-Fos immunohistochemistry. *Open circles* indicate intermale aggression; *shaded circles*, maternal aggression; and *black circles*, escalated aggression. An *arrow* in a circle indicates the area where the c-Fos activation was inhibited compared with intermale aggression (*downwards arrow*) or mixed reports, depending on the study (*bidirectional arrow*). Prefrontal cortex (*PFC*), claustrum (*Cl*), lateral septal nucleus (*LS*), bed nucleus of the stria terminals (*BNST*), nucleus accumbens (*NAc*), piriform cortex (*Pir*), medial preoptic area (*MPOA*), paraventricular nucleus (*PVN*), parafacicular nucleus of thalamus (*PF*), hypothalamus attack areas (*HAA*), amygdala (*Amy*), hippocampus (*Hipp*), periaqueductal gray (*PAG*), 5-HT neurons in the dorsal raphe nucleus (*DRN*), and locus coeruleus (*LC*). In this figure, HAA includes the anterior, ventromedial, and lateral hypothalamic nuclei. (Kollack-Walker and Newman 1995; Joppa et al. 1995; Potegal et al. 1996; Wang et al. 1997, 2011; Delville et al. 2000; Gammie and Nelson 2001; Halász et al. 2002, 2006; van der Vegt et al. 2003; Davis and Marler 2004; Haller et al. 2005b, 2006; Veening et al. 2005; Gobrogge et al. 2007; Pan et al. 2010; Nehrenberg et al. 2012; Wall et al. 2012; Konoshenko et al. 2013)

3.2 Hypothalamus and Aggression

Lesion or stimulation studies have confirmed the involvement of each brain area of the aggression circuit in aggressive behavior. Of all of the areas of the brain, the hypothalamic area is one of the best-studied areas in relation to aggression, ever since this area in aggression in cats was identified in the early twentieth century (Hess and Akert 1955). The hypothalamic attack area (HAA) is a region containing several hypothalamic nuclei where electrical stimulation can elicit attack behavior (Kruk et al. 1983; Siegel et al. 1999). Electrical stimulation of parts of the HAA, such as lateral or ventromedial hypothalamus, also induces escalated aggression in the rat, with the resulting behavioral pattern not simply categorized as “offensive” or “defensive” aggression but rather as a mixed combination of several types of aggression (Siegel et al. 1999). This hypothalamically induced attack behavior can be directed against males, females, anesthetized or dead rats, and even mice (Koolhaas 1978; Kruk 1991; Roberts and Nagel 1996). Also, observation that the attack is directed toward vulnerable areas of the body of the opponent suggests that the aggressive behavior induced by stimulation of the hypothalamic area represents abnormal or pathological aggression. One of the neuromodulators in the hypothalamus with a strong link to the level of aggression is arginine vasopression (AVP) (Donaldson and Young 2008). In male hamsters, microinjection of AVP into the anterior or ventrolateral hypothalamus enhances aggressive behavior, whereas antagonists for the AVP receptor V1aR reduces aggressive behavior (Ferris and Potegal 1988; Potegal and Ferris 1989; Ferris et al. 1997; Caldwell and Albers 2004). The effect of intra-hypothalamic AVP on aggression depends on the photoperiod, the basal testosterone level, and the sex of the individual (Delville et al. 1996; Caldwell and Albers 2004; Gutzler et al. 2010). The HAA is further discussed in detail by (Kruk, this volume).

Recently developed optogenetic techniques (Deisseroth et al. 2006) have shown that manipulation of phasic activity of the hypothalamic attack area can elicit aggressive behavior in mice (Lin et al. 2011). In this method, genetic manipulation to artificially express light-sensitive opsin receptors (i.e., channelrhodopsin ChR2) that are derived from other organisms (such as *Chlamydomonas*) on desired mouse neurons enables light stimuli to activate or inhibit target neurons in milliseconds. Temporal activation of the ventrolateral area of the ventromedial hypothalamus (VMHvl) of male mice, using optical stimulation of artificially expressed ChR2, provoked attacks toward either male or female opponents, and even against inanimate objects (Lin et al. 2011). Attacks started and stopped concurrently with the onset and termination of the light stimulus, suggesting that VMHvl is a switch that controls the execution of attack behavior. However, Lin et al. (2011) used a promoter that is active in all neurons within the VMHvl region, so all types of neurons within the VMHvl area expressed ChR2. The key advantage of optogenetics over electrical stimulation is that it can be used to manipulate a specific type of neuron or a specific neural circuit. By using specific promoters, such as the CaMKIIα promoter which is specifically expressed in excitatory neurons, or parvalbumin (PV) promoter which is specifically expressed in GABA neurons, it is possible to express ChR2 or other opsins in a specific type of neuron and to manipulate the activity of these subpopulations of neurons specifically. Thus, optogenetic techniques can be used to dissect the functional microneurocircuit in the hypothalamic area that is associated with aggressive behavior. For example, optogenetics techniques can be applied to determine whether AVP projections from the medial supraoptic nucleus (mSON) or the nucleus circularis (NC) into the anterior hypothalamus facilitate aggressive behavior (Ferris et al. 1997).

3.3 Prefrontal Cortex (PFC) and Aggression

The importance of the PFC in the inhibitory control of aggression has been reported in primates, including humans (Nelson and Trainor 2007). The PFC sends glutamatergic projections to several brain areas that are linked to aggression, such as the hypothalamus, amygdala, and DRN (Peyron et al. 1998; Rosenkranz and Grace 2002; Hoover and Vertes 2011). In rodents, the PFC also has some regulatory roles in aggressive behavior. Among several subareas in the PFC, the medial prefrontal cortex (mPFC) and orbitofrontal cortex (OFC) are activated in rats and mice engaged in inter-male aggression (Haller et al. 2006; Halász et al. 2006; Wall et al. 2012), or by winning the social dominance test in mice (Wang et al. 2011). The activation of the mPFC was blunted in socially isolated rats after 4 weeks of isolation (Wall et al. 2012). In addition, bilateral lesions of the OFC increased inter-male aggression in the rat (De Bruin et al. 1983). Recently, we used the optogenetic technique with the aim to modulate the activity of mPFC excitatory neurons during an aggressive encounter in the male mouse. Our data show that the activation of the mPFC inhibits inter-male aggressive behavior specifically, without affecting any other behaviors (Takahashi et al., submitted). We conclude that the mPFC has an important inhibitory role on aggression so that it is maintained at species-typical level of aggression.

Serotonin (5-HT) receptors seem to have an important role in modulating the activity of the PFC and therefore in inhibiting aggressive behavior. For example, microinjection of the 5-HT_{1B} agonist CP-94,253 into the ventral OFC suppressed inter-male aggression, maternal aggression, and aggression induced by social instigation (De Almeida et al. 2006; Veiga et al. 2007; Centenaro et al. 2008). The activation of 5-HT_{1B} receptors in the mPFC also reduced species-typical territorial aggression (Faccidomo et al. 2012). On the other hand, the same treatment escalated aggressive behavior when it was microinjected into the mPFC, but not into the OFC, under the effect of alcohol (Faccidomo et al. 2008). In vivo microdialysis showed that the extracellular level of 5-HT in the mPFC is reduced during and after a species-typical aggressive encounter in rats (van Erp and Miczek 2000, Fig. 2). By contrast, pharmacological treatment that induced escalated aggression caused a phasic increase of 5-HT in the mPFC (Takahashi et al. 2010b, Fig. 3). Furthermore, the dominant male mouse in a group-housed cage showed higher mPFC activity in terms of glutamatergic synaptic transmission and c-Fos expression, and the genetic manipulation of the gulutamatergic AMPA current in the mPFC caused a change of social rank in the cage (Wang et al. 2011). Therefore, the PFC modulates several types of aggressive behaviors in different ways.

3.4 Dorsal Raphé Nucleus (DRN) and Aggression

Since its discovery in the brain, 5-HT has been considered a key transmitter in the neurocircuitry that controls aggression in several species ranging from invertebrates to humans (Coccaro et al. 1997; Miczek et al. 2002, 2004; Olivier 2004; de Boer and Koolhaas 2005; Takahashi et al. 2011). 5-HT in the mammalian central nervous system originates mainly from the midbrain raphé nuclei. In particular, the DRN contains the largest accumulation of 5-HT neuronal cell bodies in the brain, and it sends afferent projections to several distinct targets including the limbic structures and cortex (Dahlström and Fuxe 1964; Azmitia and Segal 1978;

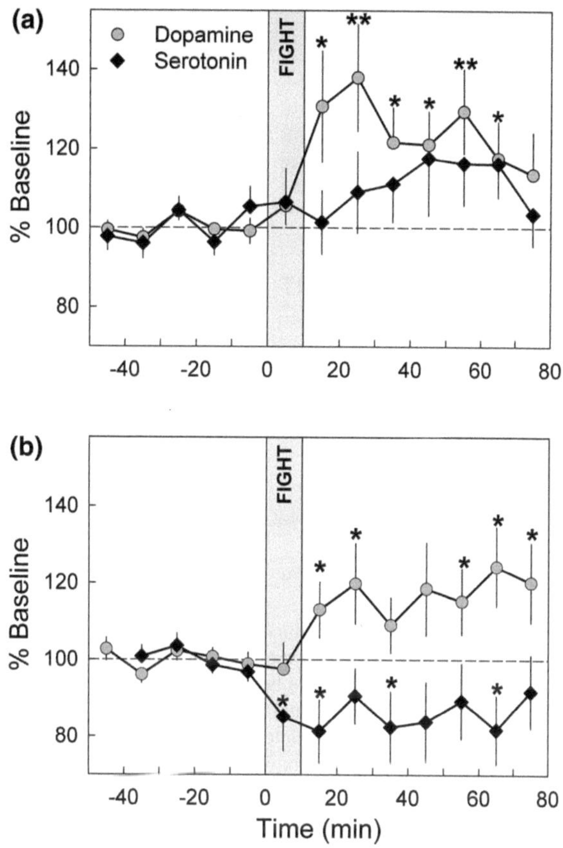

Fig. 2 Changes of dopamine and serotonin in the PFC and NAc during species-typical aggressive behavior in the rat. Measurements of extracellular dopamine and serotonin via in vivo microdialysis in resident male rats before, during, and after a confrontation with an intruder. **a** In the nucleus accumbens (*top panel*), dopamine levels increased after the confrontation, while serotonin levels did not change significantly. **b** In the prefrontal cortex (*bottom panel*), dopamine levels increased after the confrontation, whereas serotonin decreased after the confrontation. The *vertical light gray bar* indicates the occurrence of the 10-min fight. * $p < 0.05$ and ** $p < 0.01$ compared with baseline. Reprinted with permission from van Erp and Miczek (2000)

Michelsen et al. 2007). Firing rates of DRN neurons change in distinct patterns during defensive and offensive fighting in tree shrews (Walletschek and Raab 1982). In rats, c-Fos expression is increased in the 5-HT neurons of the DRN after an aggressive encounter (van der Vegt et al. 2003). The activity of the DRN is modulated by its own released 5-HT via autoreceptors, and also by GABA, glutamate, norepinephrine, and neuropeptides such as CRF (Adell et al. 2002). Activation of 5-HT_{1A} and 5-HT_{1B} autoreceptors in the DRN by microinjection of selective receptor agonists consistently reduced aggressive behavior in rats and mice, but in the case of 5-HT_{1A} receptor agonists with concomitant reduction of motor activity and social interactions (Mos et al. 1993; Bannai et al. 2007; Faccidomo et al. 2008). GABA receptors in the DRN seem to have a more aggression-specific effect. Administering baclofen, a $GABA_B$ receptor agonist, into the DRN escalated intermale aggressive behavior in the mouse (Takahashi et al. 2010b). Interestingly, administration of the $GABA_A$ receptor agonist muscimol into the DRN only heightened aggressive behavior in mice after consuming a modest amount of alcohol (Takahashi et al. 2010a). By contrast, intra-DRN administration

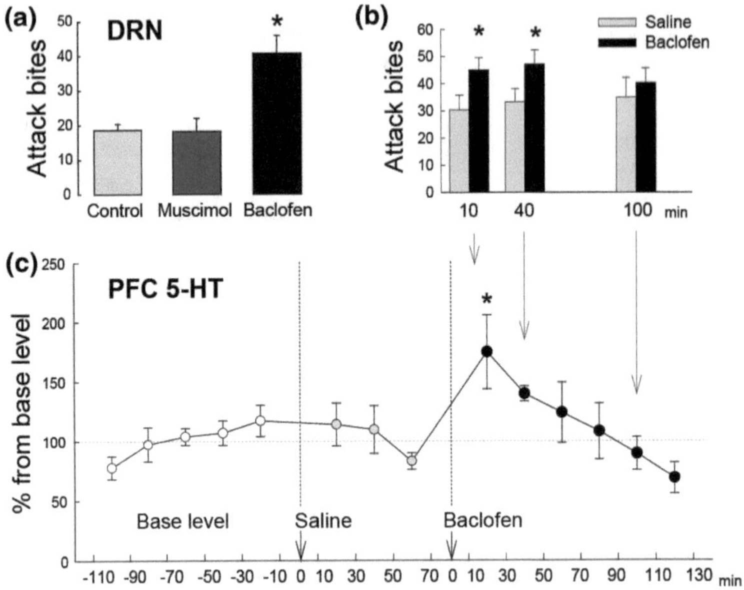

Fig. 3 Modulation of the dorsal raphe nucleus (*DRN*) by GABA receptors and escalated aggression in the mouse. **a** Microinjection of the $GABA_B$ receptor agonist baclofen into the DRN increased intermale aggressive behavior, whereas microinjection of the $GABA_A$ receptor agonist muscimol into the DRN did not have any effect. **b** Temporal change in the effect of 0.06 nmol baclofen on attack bites. Escalated attack bites were observed both 10 and 40 min after the intra-DRN baclofen injection, and return to basal level after 100 min. **c** Extracellular serotonin (5-HT) concentration in the medial prefrontal cortex (mPFC) of mice after intra-DRN baclofen injection. Baclofen increased 5-HT release in the mPFC whereas saline injection did not change the level of 5-HT. Data are expressed as percentage of baseline ($n = 7$). * $p < 0.05$ compared with the vehicle control **a** and **b** or baseline **c**. Adopted from Takahashi et al. (Takahashi et al. 2010b)

of muscimol either reduced aggressive behaviors in rats (van der Vegt et al. 2003) or had no effect on aggression (Takahashi et al. 2010b) in the absence of alcohol. This pattern of effects suggests that both subtypes of GABA receptors are involved in escalated forms of aggressive behavior via different mechanisms. In vivo microdialysis showed that $GABA_B$ activation in the DRN increased the extracellular level of 5-HT in the mPFC (Takahashi et al. 2010b). This suggests that the activation of DRN 5-HT neurons occurs during escalated inter-male aggression in mice (Fig. 3).

By contrast, there is evidence for inhibition of 5-HT neurons during escalated aggression. Audero et al. (2013) used transgenic mice with overexpression of $5\text{-}HT_{1A}$ receptors specifically located on soma of 5-HT neurons after adulthood (>P40). These mice showed suppressed 5-HT neural firing in both dorsal and median raphe nuclei, and also showed enhanced aggressive behavior. Pharmacological inhibition of raphe 5-HT neural firing in these mice also increased attack bites, suggesting that transient inhibition of 5-HT neurons can escalate mouse aggression (Audero et al. 2013). CRF in the DRN also has an important modulatory effect on aggressive behavior. Local administration of the CRF1 receptor antagonists CP-154526 or MTIP into the DRN prevented the escalation of aggression induced by alcohol consumption without producing any motor incoordination, whereas an antagonist for another receptor subtype, CRF2, enhanced inter-male aggressive behavior in male mice (Quadros et al. 2009). The anti-aggressive effect of the CRF1 antagonist required intact 5-HT neural activity in the DRN, suggesting that the activation of 5-HT neurons suppresses heightened aggression specifically when under the influence of alcohol (Quadros et al. 2009). Furthermore, there are some lines of knockout mice that show enhanced aggression with reduced 5-HT content or 5-HT neural activities (e.g., gene knockout of TPH2, Pet1, or α-CaMKII) (Chen et al. 1994; Hendricks et al. 2003; Alenina et al. 2009; Mosienko et al. 2012; Angoa-Pérez et al. 2012).

It still needs to be determined under which conditions the DRN 5-HT neurons are activated or inhibited during which type of escalated aggression. It is likely that the basal level of 5-HT (*trait*) and phasic changes of 5-HT (*state*) may have different roles in the different types of aggression. Direct recording of 5-HT neuron activity during aggressive encounter, or optogenetic activation of the DRN 5-HT neurons will clarify the role of phasic activity of 5-HT neurons in the DRN on aggressive behavior.

3.5 Nucleus Accumbens (NAc) and Aggression

Aggressive behavior has a rewarding property for dominant males, and the opportunity for aggressive encounters reinforces conditioned responses in male mice (Fish et al. 2002, 2005). Microinjection of either a dopamine D1 or D2 receptor antagonist into the NAc reduced this operant response, which suggests that the rewarding property of an aggressive encounter is mediated by dopamine receptors in the NAc (Couppis and Kennedy 2008). The mesocorticolimbic dopamine system, which comprises DA neurons in the ventral tegmental area (VTA) and their projections to the NAc and other brain areas, is known to mediate reward processing in addition to aversive motivation (Kelley et al. 2002). In vivo microdialysis showed that the level of dopamine in the NAc increased during and after an aggressive confrontation in male rats (van Erp and Miczek 2000; Ferrari et al. 2003, Fig. 2). After repeated aggressive encounters at the same time of day, resident rats showed an anticipatory increase in dopamine levels, and a decrease in 5-HT, in the NAc at the scheduled time, even without aggressive behavior (Ferrari et al. 2003). On the other hand, defeated male rats also showed increased dopamine release in the NAc (Tidey and Miczek 1996; Anstrom et al. 2009). Thus, the increase of dopamine in the NAc does not simply indicate the reward signal *per se*, but characterizes both individuals reacting to an aggressive confrontation and anticipating such an event (Miczek et al. 2007).

3.6 Olfactory System and Aggression

For rodents, olfaction is the major sensory input regulating social behaviors (Brennan and Keverne 2004). Olfactory information about an intruder is first processed by the olfactory bulb, and then projected to the amygdala via the lateral olfactory tract. Thus, the olfactory system is the first location for evaluating the quality of the opponent (Guillot and Chapouthier 1996). There are two types of olfactory bulb, the main and accessory olfactory bulbs; whereas the main olfactory bulb receives all of the odorant information, the accessory olfactory bulb specifically receives pheromonal information, which is important for intraspecific communication (Dulac and Torello 2003). Dominant male rodents show territorial scent marking (Desjardins et al. 1973), and their urine contains aggression-promoting pheromones, so-called major urine proteins or MUPs (Chamero et al. 2007). In addition, two volatile compounds, dihydroexobrevicomin and 2-(s-butyl)-dihydrothiazole, have been identified as pheromones in male urine that can elicit inter-male aggression. However, these pheromones need to interact with other components of urine (Novotny et al. 1985). The chemoreceptors for the pheromones reside on the sensory neurons of the vomeronasal organ (VNO), and send their axons to the accessory olfactory bulb. It has been shown that lesions of the VNO produce a wide range of deficits in social and reproductive behaviors including inter-male aggression and maternal aggression (Clancy et al. 1984; Bean and Wysocki 1989). Similarly, deletion of genes that relate to VNO signaling reduce aggressive behaviors and the capacity to discriminate between the sexes (Stowers et al. 2002; Keverne 2002; Norlin et al. 2003; Chamero et al. 2011). TRP2 is a putative cation channel that is exclusively expressed in VNO sensory neurons, and deletion of TRP2 in the mixed background of the C57BL/6J and 129/SvEv mouse (TRP2$^{-/-}$) induced an absence of urine-evoked neuronal activity in the VNO. Male TRP2$^{-/-}$ mice showed complete abolition of aggressive behavior, and they also engaged in sexual behavior toward male intruders (Stowers et al. 2002; Leypold et al. 2002). In addition, lactating female TRP2$^{-/-}$ mice exhibited a reduced level of maternal aggression with either the C57BL/6J and 129/SvEv mixed background (Leypold et al. 2002) or the genetic background of a high maternal aggression line (Hasen and Gammie 2009). Pheromone receptors are encoded by two distinct super families of genes, V1R and V2R, and these receptors are located on a separate layer of the VNO (Dulac and Axel 1995; Tirindelli et al. 1998). Deletion of the V1R family cluster (including sixteen V1R genes) on the 129/SvEv background reduced maternal aggressive behavior in lactating females (Del Punta et al. 2002). However, failure to discern a difference in the initiation of aggressive behavior in male V1Rs mutant mice compared with wild-type mice suggests that the V1R-expressing VNO neurons specifically transmit the signals for female aggression. On the other hand, hypoactivity of V2R-expressing VNO neurons, induced by the conditional knockout of a G protein (Gαo) being regulated by the promoter of the olfactory marker protein (OMP), reduced both inter-male aggression and maternal aggression (Chamero et al.

2011). In that study, the intruder male was castrated and then scrubbed with recombinant MUP protein because it has been shown that the MUPs activate V2Rs (Chamero et al. 2007, 2011). Thus, V2R-expressing VNO neurons recognize urine-containing pheromones and subsequently trigger aggressive behavior in both sexes.

4 Excitatory and Inhibitory Neurotransmission and Aggression

4.1 Glutamate and Aggression

Glutamate is the major excitatory transmitter in the mammalian brain. Several kinds of psychiatric diseases such as anxiety disorders, depression, and also aggression are attributed to an imbalance between glutamatergic excitation and GABAergic inhibition in limbic areas (Herman et al. 2004; Miczek et al. 2007; Garcia-Garcia et al. 2009). In particular, activation of glutamate neurons in the hypothalamic attack area (HAA) seems to be strongly linked to aggression. Direct microinjection of L-glutamate into the HAA induced aggressive responses in cats (Brody et al. 1969). In rats, microinjecting a combination of the glutamate receptor agonist kainate and the $GABA_A$ receptor antagonist bicuculline into the HAA induced an increase in attack behavior (Haller et al. 1998). This pro-aggressive effect depended on previous experience; rats with recent aggressive experiences showed an increase in aggressive behavior, whereas naïve animals displayed grooming in response to the same treatment. Male hamsters treated with anabolic–androgenic steroid (AAS) during adolescence displayed both highly escalated levels of offensive aggression and increased activation of glutamatergic neurons in several brain areas, including the anterior hypothalamus (Fischer et al. 2007; Carrillo et al. 2009, 2011). We also found that direct microinjection of L-glutamate in the DRN escalates aggressive behavior of male mice. Furthermore, in vivo microdialysis showed that glutamate release in the DRN was increased in mice that engaged in aggressive behavior (Takahashi et al. unpublished data).

There are three main types of ionotropic receptor for glutamate: N-methyl-D-asparate (NMDA) receptors, 2-amino-3-(3-hydroxy-5-methyl-isoxazol-4-yl)propanoic acid (AMPA) receptors, and kainate receptors. All of these receptors are cation channels and depolarize the surrounding membrane. The AMPA receptors and kainate receptors are responsible for fast excitatory transmission mediated by the influx of Na^+ ions. In contrast, NMDA receptors have a higher permeability to Ca^{2+} than AMPA receptors but require membrane depolarization in order for Ca^{2+} influx to release Mg^{2+} block from the NMDA cation channel. Accordingly, NMDA receptors open more slowly than AMPA receptors. Glutamate also activates a distinct class of G protein-coupled receptors, called metabotropic glutamate receptors (mGluRs) (Kew and Kemp 2005). As a consequence of this diversity of receptor types for glutamate, it is likely that the roles of glutamate and its receptor subtypes in aggression differ. Most pharmacological studies of glutamate and aggression focused primarily on the NMDA receptor because of the availability of selective pharmacological agents. However, the role of other receptors in aggression has been discovered gradually by using gene knockout mice and novel pharmacological tools.

4.1.1 NMDA Receptors

Pharmacological studies have shown that antagonists of NMDA receptors tend to have biphasic effects on aggression, causing an escalation of aggression at low doses and suppression of aggression at higher doses. However, the nature and the extent of the response are affected by the individual's basal level of aggression and/or housing conditions. Systemic administration of prototypic antagonists of NMDA receptors such as phencyclidine (PCP) or dizocilpine (MK-801) can increase aggressive behavior in the mouse and rat, depending on the individual's age and specific history such as how long the individual has been housed in isolation, the extent of sleep deprivation, or the presence or absence of previous aggressive encounters (Russell et al. 1984; Wilmot et al. 1987; McAllister 1990; Lang et al. 1995; Audet et al. 2009). In contrast, high doses of PCP and dizocilpine reduce aggressive behavior and cause abnormal locomotor activation or ataxia (Tyler and Miczek 1982; Miczek and Haney 1994; Lang et al. 1995; Belozertseva and Bespalov 1999). Memantine, the low-affinity NMDA receptor channel blocker, also reduced aggressive behavior in socially isolated male mice when used at the highest dose which induced ataxia (Belozertseva and Bespalov 1999). In mice that showed increased aggression induced by morphine withdrawal, memantine specifically reduced aggressive behavior without any side effects (Sukhotina and Bespalov 2000). On the other hand, memantine and neramexane, another non-competitive NMDA receptor blocker, enhanced territorial aggressive behavior in male mice after alcohol consumption (Newman et al. 2012, Fig. 4). In each case, control animals that were not treated with morphine or alcohol did not show any aggression-specific effect of memantine (Sukhotina and Bespalov 2000; Newman et al. 2012). Anatomically, discrete analysis is required to identify the sites of action for NMDA receptors that produce escalated aggressive behavior.

The NMDA receptor is a heterotetramer that comprises mainly two NR1 subunits and two NR2 subunits (NR2A–NR2D) (Monyer et al. 1992; Kutsuwada et al. 1992). Geneticists have generated knockout mice to examine the role of the NMDA receptor in behavior. Unfortunately, given that complete ablation of NMDA receptor expression in NR1- or NR2B- knockout mouse lines causes lethality during the prenatal period, adult behaviors cannot be studied in these mice (Li et al. 1994; Forrest et al. 1994; Kutsuwada et al. 1996). However, there is a mutant mouse with reduced NMDA receptor expression but with normal development. A mouse mutant affected in the NR1 subunit ($Nr1neo^{-/-}$) expresses 5–10 % of the normal level of the NR1 subunit of the NMDA receptor; these mice

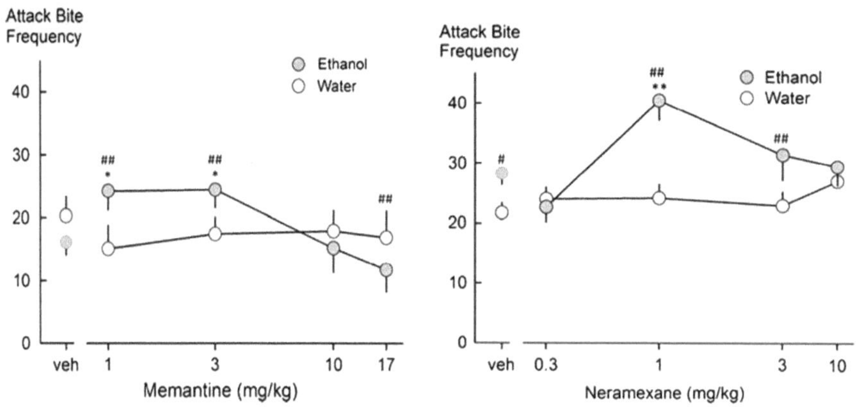

Fig. 4 Systemic administrations of non-competitive NMDA receptor antagonists enhance aggressive behavior specifically when associated with the administration of alcohol. Frequencies of attack bites following self-administration of either water or ethanol (1 g/kg) combined with either memantine (*left panel*) or neramexane (*right panel*) treatment. *$p < 0.05$, **$p < 0.001$ compared with vehicle; #$p < 0.05$, ##$p < 0.001$ ethanol compared with water. Reprinted with permission from Newman et al. (2012)

show strongly reduced social investigation and therefore rarely show aggressive behavior (Mohn et al. 1999; Duncan et al. 2004). This mutant mouse also shows a range of abnormal behaviors, such as hyperactivity, deficits in sensorimotor gating, and reduced motivation to find hidden food (Mohn et al. 1999; Duncan et al. 2004). Those pleiotropic changes in various other phenotypes may contribute to the reduction of aggression in this mutant mouse.

The function of the NMDA receptor in the PFC has an important role in a well-known model for escalated aggression induced by the deficiency of monoamine oxidase A (MAOA; Bedrosian and Nelson, this volume). As in humans, disruption of the MAOA gene in either a C3H/He or 129S6 mouse genetic background escalated aggressive behavior in males such as infliction of skin wounds on cage mates and short attack latency in the resident–intruder test (Cases et al. 1995; Scott et al. 2008). In addition to large increase of 5-HT and norepinephrine, which are normally degraded by MAOA, in the developing brain, MAOA knockout mice showed altered expression of the NMDA receptor subunits in the PFC; higher expression of NR2A and NR2B subunits as well as reduced levels of glycosylated NR1 compared with those in wild-type mice (Bortolato et al. 2012). These alterations reduced NMDA-mediated fluctuations in the amplitude and decay time of excitatory postsynaptic current (EPSC) in the PFC, and a moderate dose of dizocilpine or antagonists for either NR2A or NR2B reduced aggressive behavior in MAOA knockout mice without affecting locomotion (Bortolato et al. 2012). Thus, changes in the composition of the NMDA receptor subunit in the PFC seem to critically contribute to the genesis of pathological aggression in MAOA-deficient mice.

Further evidence points to the involvement of NMDA receptor subtypes in aggression. Rearing in social isolation causes rats to increase their aggressive behavior, and these rats showed increased expression of NR2A and NR2B subunits in the hippocampus, but reduced NR2A expression in the PFC (Zhao et al. 2009). Chronic exposure to AAS, another treatment that facilitates aggression, decreased the abundance of the mRNA that encodes the NR2A subunit in the hippocampus and hypothalamus, but not in the cortex, of male rats (Le Grevès et al. 1997).

4.1.2 AMPA Receptors

Dominant male C57BL/6 mice in group-housed home cages, assessed using the tube test (Table 1), have higher AMPA-mediated current in mPFC pyramidal neurons than otherwise comparable subordinate males (Wang et al. 2011). Artificial overexpression of AMPA receptors in mPFC pyramidal neurons induced by viral infection increased social rank in the home cage, whereas the suppression of AMPA receptor expression in the mPFC reduced the social rank (Wang et al. 2011). Thus, AMPA receptor-mediated activation of the mPFC seems to have an important role in social hierarchy and aggressive behavior. The AMPA receptors are heterotetramers assembled from four subunits, GluR1–GluR4, with GluR1/2 and GluR2/3 combinations mainly form ion channels with distinct functional properties (Hollmann and Heinemann 1994; Derkach et al. 2007). Knockout mice that lack the GluR1 (GluRA) subunit, as well as mutant mice that have a mutation in GluR1 which causes functional reduction of the AMPA receptor, were less aggressive than their wild-type littermates (Vekovischeva et al. 2004). Whereas males of both classes of mutants showed reduced aggressive behavior in their territory after prolonged isolation (30 days) and in a neutral environment, female GluR1 knockout mice did not show altered maternal aggressive behavior (Vekovischeva et al. 2004). GluR1-containing AMPA receptors co-localize with androgen receptors in hypothalamic neurons, and castrated rats exhibited lower levels of GluR1 protein expression in the hypothalamus compared with intact rats or castrated rats treated with either testosterone or estrogen (Diano et al. 1997). In addition, AAS-treated Syrian hamsters showed an increased number of neurons that express GluR1-containing AMPA receptors in the ventrolateral hypothalamus and BNST (Fischer et al. 2007). These results consistently suggest that GluR1 subunit-containing AMPA receptors in several brain areas, including the mPFC, hypothalamus, and BNST, promote aggressive behavior in male rodents. On the other hand, male GluR3 knockout mice with the C57BL/6J background fought more relative to wild-type males (Adamczyk et al. 2012). Overall, the GluR1-containing AMPA receptor and GluR3-containing AMPA receptor have opposite effects on the modulation of aggressive behavior.

In addition to the functional differences among the subtypes of glutamate receptor, differences in location are critical for their effect on aggression. To dissect the effect of each receptor in a specific neuron, a conditional knockout mouse has been established with deletion of the receptor gene in a specific neuron at a specific developmental stage. For example, a conditional knockout of the AMPA receptor GluR2 subunit that is expressed specifically in hypothalamic gonadotropin-releasing hormone (GnRH) neurons, has been established on the mixed genetic background of C57BL/6 and NMRI strain of mice; these mutants lack many kinds of social behavior and thus rarely produce offspring (Shimshek et al. 2006). Male GnRH GluR2 knockout mice showed a lack of interest in females and also displayed no aggressive behavior toward male intruders. Females with this mutation also showed impaired maternal care and no maternal aggression. Conditional knockout and overexpression techniques will allow more detailed dissection of the neuronal pathway that is involved in aggressive behavior.

4.1.3 Kainate Receptor and Delta Receptors

The kainate receptor, which is an ionotropic receptor for glutamate, comprises five subunits, GluR5–GluR7 and KA1–KA2. A human genome-wide association study identified the gene that encodes a subunit of the kainate receptor, GluR6 (GRIK2), as potentially conferring susceptibility to bipolar affective disorder (Hamshere et al. 2009). Deletion of the GluR6 gene subunit in mice induced manic-like behavioral changes, and GluR6 knockout mice are hyperactive, more aggressive toward other males, and exhibited less depressive-like behavior (Shaltiel et al. 2008). These behavioral phenotypes in knockout mice were reversed by chronic treatment with lithium.

The last family of ionotropic glutamate receptors comprises the delta receptors. The function of delta receptors remains poorly understood. However, a recent study has shown that mice with deletion of GluD1 on the mixed background of C57BL/6J and 129/SvEv displayed an increased number of attacks and shorter attack latency compared with characteristically non-aggressive wild-type mice in the resident–intruder test after 3 weeks of isolation (Yadav et al. 2012). These GluD1 mutants showed increased expression of the GluR1 (AMPA receptor) and GluR6 (kainate receptor) subunits in the amygdala. As mentioned above, GluR1-containing AMPA receptors seem to promote aggressive behavior. Accordingly, increased GluR1 levels in the amygdala might account for the escalated aggressive behavior of these mutants compared with their wild-type littermates. However, these mutants showed reduced GluR1 expression in the PFC. It is also possible that activation of the GluD1 receptor itself may have a role in regulating aggression circuits. The emergence of novel and more selective drugs for each of these receptor subtypes promises to elucidate their specific aggression-modulating effects.

4.1.4 Metabotropic Glutamate Receptors (mGluRs)

There are three groups of mGluRs, which are classified as group I (mGluR1, 5), group II (mGluR2, 3), and group III (mGluR4, 6, 7, 8) mGluRs. The group I receptors couple with G protein alpha-q (Gq) to activate the phospholipase C (PLC)-inositol trisphosphate (IP3) cascade. In contrast, group II and III receptors couple with inhibitory Gi/o, which decreases the activity of adenylate cyclase (Kew and Kemp 2005). Regarding aggressive behavior, mGluRs have been studied much less than ionotropic glutamate receptors. A few studies have examined the effects of recently developed mGluR-specific agents on aggression.

Group I (mGluR1, 5) Pharmacological manipulation of group I mGluRs demonstrated their roles in promoting inter-male aggression. A selective antagonist of mGluR1, JNJ16259685, suppressed isolation-induced aggression in male mice without any concurrent motor effects (Navarro et al. 2008). The mGluR5 antagonist 3-[(2-methyl-1,3-thiazol-4-yl)ethynyl]pyridine (MTEP) also reduced aggressive behavior in both socially isolated and pair-housed male mice in a dose-dependent manner (Navarro et al. 2006; Newman et al. 2012). On the other hand, in alcohol-treated animals a low dose of MTEP escalated aggressive behavior (Newman et al. 2012). Therefore, as with NMDA receptor antagonists, the interaction of Group I mGluRs with ethanol to escalate aggressive behaviors remains to be resolved .

Group II (mGluR2,3) In contrast to group I mGluRs, group II mGluRs seem to inhibit aggression. The mGluR2/3 receptor selective agonist LY379268 reduced inter-male aggressive behavior at a high dose, but this dose also reduced locomotor activity (Newman et al. 2012). Another mGluR2/3 agonist, MGS0028, also reduced aggressive behavior in socially isolated (SI) mice. This drug also inhibited hyperactivity in SI mice, causing to return it to the level shown by group-housed males, but had no effect on the group-housed males (Ago et al. 2012).

Group III (mGluR4,6,7,8) A selective agonist for mGluR7 receptors, AMN082, reduced offensive behaviors such as attack and threat without affecting locomotor activity. However, the mGluR8 agonist (S)-3,4-DCPG did not produce significant behavioral changes in male mice (Navarro et al. 2009). It is possible that group III mGluRs have a similar inhibitory effect on aggression to that of group II mGluRs, because both groups of mGluR couple with inhibitory Gi/o. However, their site of action varies due to the different localization of these mGluRs (Shigemoto et al. 1997).

4.2 GABA and Aggression

GABA is the major inhibitory neurotransmitter in the central nervous system. Early reports hypothesized that the loss of GABAergic inhibitory control of the limbic system causes exaggerated aggression (Mandel et al. 1981). Comparisons of GABA contents or the activity of the GABA synthetic enzyme, glutamic acid decarboxylase (GAD), in the postmortem brains of male mice and female hamsters supported this hypothesis; aggressive animals showed a lower GABA level and lower GAD activity in several brain areas such as the olfactory bulb, striatum, and amygdala, when compared with nonaggressive animals (Earley and Leonard 1977;

Simler et al. 1982; Potegal et al. 1982; Clement et al. 1987; Guillot and Chapouthier 1998). On the other hand, gene knockout of a 65-kDa isoform of GAD (GAD65) in mice with a mixed background of C57B/6 and CBA2 induced a decrease in brain GABA content during postnatal development, and these mutant males showed reduced aggressive behavior toward male intruders after 4 weeks of isolation (Stork et al. 2000). Furthermore, in animal models of escalated aggression, increased GAD65 expression and increased GABA neurons were observed in some brain areas of male hamsters given chronic AAS treatment (Grimes et al. 2003; Schwartzer et al. 2009) or repeated cocaine exposure during adolescence (Ricci et al. 2005). One explanation for the divergent findings of the effect of GABA on aggression may be that its actions depend on the brain area involved, the receptor types, and also the specific context of the situation responsible for evoking the aggressive behavior.

4.2.1 $GABA_A$ Receptors

$GABA_A$ receptors are ligand-gated ion channels. The activation of $GABA_A$ receptors by the binding of GABA or corresponding agonists induces Cl^- influx into the cell and thus hyperpolarization of the surrounding membrane. The involvement of $GABA_A$ receptors in aggressive behavior seems to be well established, but the effect (either pro-aggressive or anti-aggressive) is quite intricate, possibly owing to the functional diversity that depends on the heterogeneous composition of receptor subunits.

More than 4 weeks of SI induces increased aggression in most male mice, male rats, and female hamsters (Wise 1974; Simler et al. 1982; Toth et al. 2012). The decrease in $GABA_A$ receptor function caused by SI is possibly the result of reduced levels of the neurosteroid allopregnanolone, a positive allosteric modulator of $GABA_A$ receptors, in the frontal cortex, hippocampus, and basolateral amygdala of male mice (Dong et al. 2001; Pinna et al. 2008; Nelson and Pinna 2011). Fluoxetine treatment prevented decreases in level of allopregnanolone and reduced aggressive behavior in SI mice (Pinna et al. 2003). In addition, SI animals showed reduced sensitivity to $GABA_A$ receptor positive allosteric modulators such as barbiturates, neurosteroids, and benzodiazepines (Matsumoto et al. 1999; Serra et al. 2000; Pinna et al. 2004). These studies consistently illustrate that the down regulation of $GABA_A$ receptors via reduction of allopregnanolone is implicated in the escalated aggression induced by SI. In addition, mice that were selected for highly aggressive behavior showed decreased binding of benzodiazepines and also reduced $GABA_A$-dependent chloride uptake (Weerts et al. 1992). Thus, a reduction in $GABA_A$ receptor function may predispose individuals to initiate aggressive behaviors.

By contrast, agonists and positive allosteric modulators of the $GABA_A$ receptor have a range of effects on aggressive behavior, depending on dose, extending from inhibition to escalation (Puglisi-Allegra and Mandel 1980; Miczek et al. 2003). Allosteric positive modulators such as benzodiazepines, alcohol, barbiturates, and

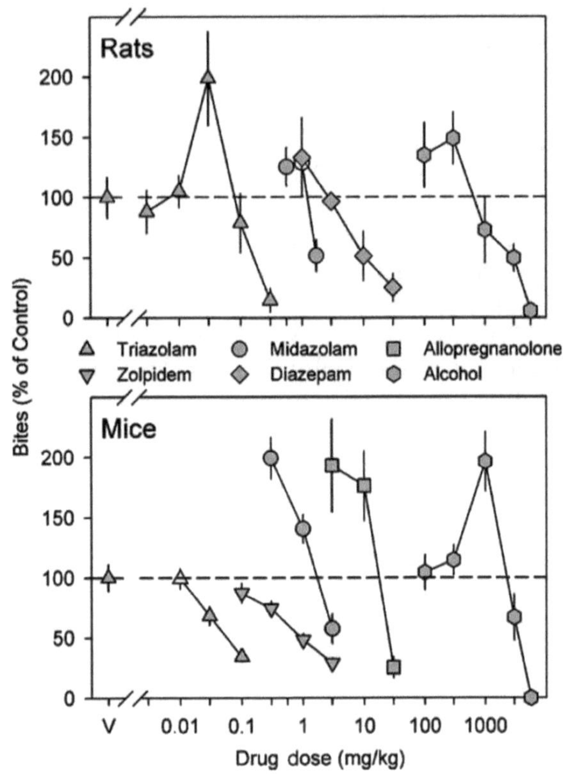

Fig. 5 $GABA_A$-positive allosteric modulators and aggression. Biphasic effects of $GABA_A$ receptor-positive modulators on aggression in rats (*top panel*) and mice (*bottom panel*). Low doses of alcohol, the benzodiazepines diazepam (rats only) and midazolam, and the neurosteroid allopregnanolone (mice only) increase the frequencies of attack bites, expressed as a percentage of vehicle control, whereas higher doses decrease this measure of aggression. Triazolam increases attack bites in rats but not mice. No increase in aggression was seen after treatment with zolpidem, a alpha1 receptor selective agonist (tested for mice only). The *dotted line* represents the baseline at 100 %. Reprinted with permission from Miczek et al. (2007)

neurosteroids enhance the inhibitory transmission of the $GABA_A$ receptor. Clinically, prototypical $GABA_A$-positive modulators have been used widely in the treatment of anxiety, convulsion, muscle tension, and sleep disorders (Shader and Greenblatt 1993). However, these drugs have so-called paradoxical effects, and acute administration of benzodiazepines may increase aggression in human patients, depending on the dose, the context, and the individual's history (Hall and Zisook 1981; Dietch and Jennings 1988; Bond et al. 1995).

Alcohol also acts as a positive allosteric modulator of $GABA_A$ receptors, and alcohol has been linked to violence and aggression in humans more frequently than any other drug (Miczek et al. 2002, 2004). The pro-aggressive effect of $GABA_A$ positive allosteric modulators has been confirmed in many rodent models. Benzodiazepines and some neurosteroids enhanced both inter-male aggression and maternal aggression in mice and rats (Miczek 1974; Miczek and O'Donnell 1980; Mos and Olivier 1989; Ferrari et al. 1997; Fish et al. 2001; Gourley et al. 2005). As shown in Fig. 5, low to moderate doses of benzodiazepines enhanced inter-male aggressive behavior, but high doses of benzodiazepines inhibited aggression. Alcohol intake reliably escalates aggression in approximately 30 % of male mice and rats, whereas the level of aggression in the other 70 % either did not change, or decreased relative to their basal level after an administration of a moderate dose of alcohol (Miczek et al. 1992, 1998; van Erp and Miczek 1997). This individual variation seems to be comparable to the human condition, and differences in the propensity for escalated aggression induced by alcohol may arise from functional and compositional differences in $GABA_A$ receptors. When alcohol is co-administered with allopregnanolone, animals that are vulnerable to the pro-aggressive effect of alcohol (alcohol-heightened aggression, AHA) and animals with alcohol-non-heightened aggression (ANA) respond to the treatment differently (Fish et al. 2001). Furthermore, pharmacological activation of $GABA_A$ receptors in the DRN enhances the pro-aggressive effect of alcohol, and animals with AHA are more sensitive to the $GABA_A$ receptor agonist muscimol than ANA animals (Takahashi et al. 2010a). However, intra-DRN injection of muscimol did not have any effect on aggression in non-drugged mice (Takahashi et al. 2010a), and even reduced inter-male aggression in rats (van der Vegt et al. 2003).

The $GABA_A$ receptor is a pentameric receptor that comprises a combination of five subunits. There are seven main receptor subunit families (α, β, γ, δ, ε, θ, π), and some families have further isoforms (α1–6, β1–3, γ1–3). Benzodiazepines bind to a specific site formed by α and γ subunits; whereas $GABA_A$ receptors that contain either α1, α2, α3, or α5 are sensitive to benzodiazepines, $GABA_A$ receptors that contain α4 or α6 are not. The pharmacological properties of $GABA_A$ receptors vary, depending on the subunit composition (Mehta and Ticku 1999; Tan et al. 2011). It has been proposed that the α2-containing $GABA_A$ receptors are responsible for the anxiolytic effect of benzodiazepine, whereas α1-containing $GABA_A$ receptors are involved in the sedative effect (Rudolph et al. 1999; Löw et al. 2000). The α1-containing $GABA_A$ receptors seem to play a key role in mediating the pro aggressive effect of benzodiazepines. Antagonists such as βCCt and 3-PBC that act preferentially on α1-containing $GABA_A$ receptors reduced the aggression-heightening effects of midazolam or alcohol in mice and rats (De Almeida et al. 2004; Gourley et al. 2005). In addition, 4 weeks of SI alters the expression of $GABA_A$ receptor subunits, and male mice exposed to SI showed reduced α1 subunit expression in the hippocampus and PFC compared with group housed males (Pinna et al. 2006). Moreover, compared with group-housed males, SI males had reduced levels of the α2 and γ2 subunits and increased levels of mRNAs that encode α4 and α5 subunits in the PFC (Pinna et al. 2006). Escalated aggressive behavior after benzodiazepine treatment is eliminated in mice with a point mutation of either the α1 or α2 subunit (Newman et al. unpublished). The attempt to dissect the role of each subunit of the $GABA_A$ receptor in the pro- and anti-aggressive effects of benzodiazepines is just beginning, and further investigation is required. Mice with a point mutation in the benzodiazepine binding site of a specific α subunit and a knockout mouse for each subtype of $GABA_A$ receptor have been established (Rudolph and Knoflach 2011), and these mice will facilitate analysis of the function of each of these subunit in controlling aggression.

4.2.2 $GABA_B$ Receptor

The $GABA_B$ receptor is a metabotropic receptor, which mediates its effect less rapidly than the ionotropic $GABA_A$ receptor through the activation of Giα- or Goα-type G proteins. The $GABA_B$ receptors are located on both pre and postsynaptic terminals, and either suppress neurotransmitter release by inhibiting Ca^{2+} channels (presynaptic) or induce a slow inhibitory postsynaptic current (IPSC) through activation of inwardly rectifying K^+ channels (Bettler et al. 2004). Recently, the $GABA_B$ receptor has received increased attention because of its effect in several psychiatric diseases, such as anxiety, depression, and drug addiction (Cryan and Kaupmann 2005; Cryan and Slattery 2010; Vlachou and Markou 2010), but the role of the $GABA_B$ receptor in aggressive behavior has not yet been examined fully. A clinical study found that oral administration of the $GABA_B$ receptor agonist baclofen inhibited provoked aggressive responses in subjects that had a history of childhood conduct disorder, whereas the same treatment escalated aggressive responses in control subjects in a laboratory setting (Cherek et al. 2002). In rats, baclofen also reduces escalated forms of aggressive-defensive behavior induced by electric shock, ethanol withdrawal, and apomorphine treatment (Rodgers and Depaulis 1982; File et al. 1991; Rudissaar et al. 2000). In contrast, we found recently that systemic administration of baclofen showed an inverse U-shaped dose-effect on territorial aggressive behavior in male mice: low-to-moderate doses of baclofen increased aggressive behavior of male mice, whereas high doses of baclofen reduced aggressive behavior and were accompanied by motor incoordination (Takahashi et al. 2012). The DRN appears to be one of the sites targeted by baclofen, which would explain the aggression-escalating effect of baclofen. Intra-DRN administration of baclofen escalated inter-male aggressive behavior, and this pro-aggressive effect of baclofen extended across different strains of mice, such as C57BL/6J and outbred ICR or CFW strains (Takahashi et al. 2010b, 2012). The selective $GABA_B$ receptor antagonists phaclofen and CGP54626 both blocked this pro-aggressive effect of baclofen. In vivo microdialysis showed that $GABA_B$ receptor activation in the DRN increased the extracellular level of 5-HT in the mPFC (Takahashi et al. 2010b) (Fig. 3). Thus, phasic activation of 5-HT neurons by indirect modulation of $GABA_B$ receptors may promote certain types of escalated aggressive behavior in mice.

5 Forward Genetics of Aggressive Behavior in Rodents

So far, we have seen several examples of gene knockout mice that are targeted at specific receptor subunits or molecules involved in intracellular cascades. These "gene driven" approaches, so-called *reverse* genetics, are useful when examining the effect of a specific gene on behavior. Deletion of any of more than 50 genes can either increase or decrease aggressive behaviors in the mouse (Miczek et al. 2001; Takahashi et al. 2011). On the other hand, most behavioral phenotypes, including aggressive behavior, are known as "quantitative traits," and are attributable to multiple genes that have a subtle effect individually but have strong interactions with the environment as well as epistatic interactions among genes. *Forward* genetics is a top-down method that seeks to identify unknown genetic factors that contribute to individual differences in behavior. This section discusses attempts to identify the genetic basis of aggression using the forward genetics approach.

Inbred mouse strains have been widely used to examine the genetics of aggression (Ginsburg and Allee 1942; Scott 1942). For example, whereas the NZB/B1NJ strain shows extremely high inter-male aggression, the FVB/NtacfBR strain shows aggression toward females, and A/J mice rarely show any aggressive behavior even when they are reared in the same environment as aggressive strains (Canastar and Maxson 2003; Roubertoux et al. 2005). In addition, these inbred strains respond differently to environmental factors such as past experiences of winning, defeat, or SI, suggesting that susceptibility to the environmental challenge is determined genetically (Ginsburg and Allee 1942; Schneider et al. 1992). Selective breeding is another method used to examine the heritability of aggression. Highly aggressive and non-aggressive populations diverge promptly (in fewer than five generations) and, after several generations of artificial selection, these populations stabilize at high and low levels of aggressive behavior. The use of selective breeding to study inter-male aggression (van Oortmerssen and Bakker 1981), isolation-induced aggression (Lagerspetz 1964; Cairns et al. 1983; Sandnabba 1996), and maternal aggression (Gammie et al. 2006) in the mouse has indicated a genetic basis for these types of aggression. These lines selected for aggression have been accepted as good models of escalated aggression (Miczek et al. 2013). For example, the Short Attack Latency line (SAL) and the Long Attack Latency line (LAL) were originally developed from wild house mice that were bidirectionally selected on the basis of their latency to first attack (van Oortmerssen and Bakker 1981). In addition to shorter attack latency, SAL males also showed a "pathological" level of aggression manifested by heightened attack/threat ratios, attacks on vulnerable body parts of the opponent, and aggressive behavior toward females and anesthetized animals (De Boer et al. 2009; Natarajan et al. 2009).

Attempts have been made to identify quantitative trait loci (QTLs) that account for differences in levels of aggression between individuals which have been seen in the comparison of inbred strains and selected lines. In early studies, the involvement of chromosome (Chr) Y and Chr 17 of mice was examined because of the existence of the t haplotype on Chr 17 in the wild population and the comparatively easy establishment of Chr Y consomic strains (Miczek et al. 2001). QTL analysis was undertaken for a genome-wide screen of genetic loci. QTL mapping is a statistical method to identify genetic loci that are linked to the phenotype, using the genetic polymorphisms between strains. To date, there have been only a few attempts to identify QTLs for aggressive behavior, possibly owing to the complexity of the behavior. Four QTLs, on Chr 5, Chr 10, Chr15, and Chr X, were identified by analyzing an F2 intercross between A/J mice and either NZB/B1NJ or

BALB/c mice. Aggression was assessed by dangling an intruder by its tail at the corner of the cage and examining the occurrence of aggressive attack by the resident male (Brodkin et al. 2002; Dow et al. 2011). QTL analysis has also revealed that inter-male aggression and isolation-induced aggression have different, as well as some overlapping, genetic contributions (Roubertoux et al. 2005).

Analysis of consomic strains (or Chromosome substitution strain) is another way to map the chromosomes that determine the level of aggression (Nadeau et al. 2000). Three panels of consomic mouse strains (Nadeau et al. 2000; Gregorová et al. 2008; Takada et al. 2008) and one panel for rats (Malek et al. 2006) are currently available. A consomic strain was made by replacing a single chromosome in a host strain with the corresponding chromosome from a donor strain. Since the mouse has 19 autosomes and X and Y sex chromosomes, a panel of consomic strain involves a total of 22 lines of inbred strains (Nadeau et al. 2000). In our laboratory, we use a consomic strain of the wild-derived mouse strain MSM/Ms and a common laboratory strain C57BL/6J (Takada et al. 2008). Male MSM mice are highly aggressive in the home cage and sometimes kill their littermates, and we found enhanced aggression in the consomic strains that have either Chr 4 or Chr 15 from MSM on the C57BL/6J background (Takahashi et al. 2010c; in preparation).

Identifying the novel genetic factors that contribute to individual differences in aggression is appealing because it will help to elucidate the neurobiological mechanisms responsible for aggression. However, we are far from confirming the actual causal gene within the QTL. Usually, a chromosomal region corresponding to the genetic locus identified by the QTL analysis contains several 100 genes. Also, it is not unusual for the genetic effect on behavior to be diluted, or disappear entirely, after the genetic locus has been narrowed down by production of a congenic mouse (Bryant et al. 2012). The difficulty in identifying the appropriate gene can possibly be attributed to the small effect of each QTL (less than 5 %) as well as to complex epistatic interactions among genes (Flint et al. 2005; Manolio et al. 2009; Eichler et al. 2010).

The complexity of the genetic mechanisms of aggression also has to be considered when analyzing knockout mice. The deletion of one gene can have different effects depending on the genetic background. For example, the spontaneous mouse mutation *fierce* has deletion of the coding region of the Nr2e1 gene (tailless); both male and female *fierce* mice engage in extremely high level of aggressive behavior and are difficult to handle (Monaghan et al. 1997). However, the effect of the *fierce* mutation depends on the genetic background—mice with *fierce* on the C57BL/6J background are highly aggressive, whereas those with *fierce* on a mixed background of C57BL/6J and 129 only show intermediate levels of aggression. This background-dependent effect indicates a salient interaction between the deleted gene and other QTLs or so-called modifiers.

6 Concluding Remarks

- Analysis of c-Fos expression has identified several brain areas that are implicated in inter-male aggression, maternal aggression, and escalated aggression. There are large overlaps among the brain areas that are involved in different types of aggression, but also some differentiation depending on the type of aggression.
- Electrical stimulation studies and optogenetic studies have shown that activation of the hypothalamus (hypothalamic attack area, HAA) induces pathological aggressive behavior in rodents. Arginine vasopression (AVP) modulates the activity of the hypothalamus and enhances aggressive behaviors in male rodents.
- The prefrontal cortex (PFC), especially the medial prefrontal cortex (mPFC) and orbitofrontal cortex (OFC), have an inhibitory effect on aggression. 5-HT modulates the activity of the PFC, but its role in aggression depends on (1) the subtypes of receptors and the region of the brain in which they are expressed, and (2) the types of aggressive behaviors (species-typical aggression vs escalated aggression).
- The activity of the dorsal raphe nucleus (DRN), which sends dense 5-HT neuron projections into the PFC and other regions, is critical in aggression. The basal level of 5-HT (trait) and phasic change of 5-HT (state) have to be considered separately, and it is possible that they have different roles in aggressive behavior.
- Dopamine in the NAc increases during aggressive encounters, both in aggressors and defeated animals. Increased levels of dopamine in the NAc are also observed when an animal is anticipating an aggressive confrontation.
- Pheromonal information that is processed by the accessory olfactory bulb is essential for appropriate social behavior in rodents, including aggressive and sexual behaviors. Disrupting the activation of the vomeronasal organ (VNO) causes males to avoid aggressive behavior and shows sexual behavior toward a male intruder.
- The balance between excitatory glutamate and inhibitory GABA is important in maintaining aggressive behavior at the species-typical (adaptive) level. Pharmacological and genetic studies have shown that almost all subtypes of glutamate (NMDA, AMPA, kainate receptors, and mGluRs) and GABA ($GABA_A$ and $GABA_B$) receptors are involved in aggression. However, the role of each of the receptors may vary depending on the composition of receptor subunits, its localization, and the type of aggressive behavior studied. Combinations of newly developed pharmacological agents that can manipulate a specific subtype of receptors and genetic models, such as conditional knockout or transgenic mice for a receptor subtype on the specific neurons, promise to elucidate the functions of these receptors in aggression.
- The forward genetics approach has successfully demonstrated the genetic contributions (QTLs) to aggression in the mouse. However, a breakthrough in methodology and technology is required to unravel the complicated interactions

among genes and genetic backgrounds, and to understand the entire gene network that is involved in the individual differences in aggression.

References

Adamczyk A, Mejias R, Takamiya K, Yocum J, Krasnova IN, Calderon J, Cadet JL, Huganir RL, Pletnikov MV, Wang T (2012) GluA3-deficiency in mice is associated with increased social and aggressive behavior and elevated dopamine in striatum. Behav Brain Res 229:265–272

Adams DB (2006) Brain mechanisms of aggressive behavior: an updated review. Neurosci Biobehav Rev 30:304–318

Adell A, Celada P, Abellán MT, Artigas F (2002) Origin and functional role of the extracellular serotonin in the midbrain raphe nuclei. Brain Res Brain Res Rev 39:154–180

Ago Y, Araki R, Yano K, Kawasaki T, Chaki S, Nakazato A, Onoe H, Hashimoto H, Baba A, Takuma K, Matsuda T (2012) The selective metabotropic glutamate 2/3 receptor agonist MGS0028 reverses isolation rearing-induced abnormal behaviors in mice. J Pharmacol Sci 118:295–298

Alenina N, Kikic D, Todiras M, Mosienko V, Qadri F, Plehm R, Boyé P, Vilianovitch L, Sohr R, Tenner K, Hörtnagl H, Bader M (2009) Growth retardation and altered autonomic control in mice lacking brain serotonin. Proc Nat Acad Sci USA 106:10332–10337

Angoa-Pérez M, Kane MJ, Briggs DI, Sykes CE, Shah MM, Francescutti DM, Rosenberg DR, Thomas DM, Kuhn DM (2012) Genetic depletion of brain 5HT reveals a common molecular pathway mediating compulsivity and impulsivity. J Neurochem 121:974–984

Anstrom KK, Miczek KA, Budygin EA (2009) Increased phasic dopamine signaling in the mesolimbic pathway during social defeat in rats. Neuroscience 161:3–12

Audero E, Mlinar B, Baccini G, Skachokova ZK, Corradetti R, Gross C (2013) Suppression of serotonin neuron firing increases aggression in mice. J Neurosci 33:8678–8688

Audet MC, Goulet S, Doré FY (2009) Impaired social motivation and increased aggression in rats subchronically exposed to phencyclidine. Physiol Behav 96:394–398

Azmitia EC, Segal M (1978) An autoradiographic analysis of the differential ascending projections of the dorsal and median raphe nuclei in the rat. J Comp Neurol 179:641–667

Bannai M, Fish EW, Faccidomo S, Miczek KA (2007) Anti-aggressive effects of agonists at 5-HT_{1B} receptors in the dorsal raphe nucleus of mice. Psychopharmacology 193:295–304

Baron RA, Richardson DR (1994) Human aggression, 2nd edn. Plenum Press, New York

Bean NJ, Wysocki CJ (1989) Vomeronasal organ removal and female mouse aggression: The role of experience. Physiol Behav 45:875–882

Belozertseva IV, Bespalov AY (1999) Effects of NMDA receptor channel blockade on aggression in isolated male mice. Aggress Behav 25:381–396

Berkowitz L (1993) Aggression: Its causes, consequencecs and control. Temple University Press, Philadelphia

Bettler B, Kaupmann K, Mosbacher J, Gassmann M (2004) Molecular structure and physiological functions of $GABA_B$ receptors. Physiol Rev 84:835–867

Blanchard DC, Blanchard RJ (2003) What can animal aggression research tell us about human aggression? Horm Behav 44:171–177

Blanchard RJ, Blanchard DC (1977) Aggressive behavior in the rat. Behav Biol 21:197–224

Blanchard RJ, Hori K, Blanchard DC, Hall J (1987) Ethanol effects on aggression of rats selected for different levels of aggressiveness. Pharmacol Biochem Behav 27:641–644

Blanchard RJ, O'Donnell V, Caroline Blanchard D (1979) Attack and defensive behaviors in the albino mouse. Aggress Behav 5:341–352

Blanchard RJ, Ohl F, van Kampen M, Blanchard DC, Fuchs E (2001) Attack and defense in conspecific fighting in tree shrews (*Tupaia belangeri*). Aggress Behav 27:139–148

Bond AJ, Curran HV, Bruce MS, O'Sullivan G, Shine P (1995) Behavioural aggression in panic disorder after 8 weeks' treatment with alprazolam. J Affect Disord 35:117–123
Bortolato M, Godar SC, Melis M, Soggiu A, Roncada P, Casu A, Flore G, Chen K, Frau R, Urbani A, Castelli MP, Devoto P, Shih JC (2012) NMDARs mediate the role of monoamine oxidase-A in pathological aggression. J Neurosci 32:8574–8582
Bowler CM, Cushing BS, Carter CS (2002) Social factors regulate female–female aggression and affiliation in prairie voles. Physiol Behav 76:559–566
Brain PF (1979) Differentiating types of attack and defense in rodents. In: Brain PF and Benton D (eds.) Multidiscriplinary approaches to aggression research, Elsevier, Amsterdam, p. 53–77
Brennan PA, Keverne EB (2004) Something in the Air? New Insights into Mammalian Pheromones. Curr Biol 14:R81–R89
Brodkin ES, Goforth SA, Keene AH, Fossella JA, Silver LM (2002) Identification of quantitative trait loci that affect aggressive behavior in mice. J Neurosci 22:1165–1170
Brody JF, DeFeudis PA, DeFeudis FV (1969) Effects of micro-injections of L-glutamate into the hypothalamus on attack and flight behaviour in cats. Nature 224:1330
Bryant CD, Kole LA, Guido MA, Sokoloff G, Palmer AA (2012) Congenic dissection of a major QTL for methamphetamine sensitivity implicates epistasis. Genes, Brain, and Behavior 11:623–632
Cairns RB, MacCombie DJ, Hood KE (1983) A developmental-genetic analysis of aggressive behavior in mice: I. Behavioral outcomes. J Comp Psychol 97:69–89
Caldwell HK, Albers HE (2004) Effect of photoperiod on vasopressin-induced aggression in Syrian hamsters. Horm Behav 46:444–449
Canastar A, Maxson SC (2003) Sexual aggression in mice: effects of male strain and of female estrous state. Behav Genet 33:521–528
Carrillo M, Ricci LA, Melloni RH (2009) Adolescent anabolic androgenic steroids reorganize the glutamatergic neural circuitry in the hypothalamus. Brain Res 1249:118–127
Carrillo M, Ricci LA, Melloni RH (2011) Glutamate and the aggression neural circuit in adolescent anabolic steroid-treated Syrian hamsters (*Mesocricetus auratus*). Behav Neurosci 125:753–763
Cases O, Seif I, Grimsby J, Gaspar P, Chen K, Pournin S, Müller U, Aguet M, Babinet C, Shih JC (1995) Aggressive behavior and altered amounts of brain serotonin and norepinephrine in mice lacking MAOA. Science 268:1763–1766
Caspi A, McClay J, Moffitt TE, Mill J, Martin J, Craig IW, Taylor A, Poulton R (2002) Role of genotype in the cycle of violence in maltreated children. Science 297:851–854
Centenaro LA, Vieira K, Zimmermann N, Miczek KA, Lucion AB, de Almeida RMM (2008) Social instigation and aggressive behavior in mice: role of 5-HT_{1A} and 5-HT_{1B} receptors in the prefrontal cortex. Psychopharmacology 201:237–248
Chamero P, Katsoulidou V, Hendrix P, Bufe B, Roberts R, Matsunami H, Abramowitz J, Birnbaumer L, Zufall F, Leinders-Zufall T (2011) G protein $G\alpha o$ is essential for vomeronasal function and aggressive behavior in mice. Proc Nat Acad Sci USA 108:12898–12903
Chamero P, Marton TF, Logan DW, Flanagan K, Cruz JR, Saghatelian A, Cravatt BF, Stowers L (2007) Identification of protein pheromones that promote aggressive behaviour. Nature 450:899–902
Chen C, Rainnie DG, Greene RW, Tonegawa S (1994) Abnormal fear response and aggressive behavior in mutant mice deficient for α-calcium-calmodulin kinase II. Science 266:291–294
Cherek DR, Lane SD, Pietras CJ, Sharon J, Steinberg JL (2002) Acute effects of baclofen, a γ-aminobutyric acid-B agonist, on laboratory measures of aggressive and escape responses of adult male parolees with and without a history of conduct disorder. Psychopharmacology 164:160–167
Clancy AN, Coquelin A, Macrides F, Gorski RA, Noble EP (1984) Sexual behavior and aggression in male mice: involvement of the vomeronasal system. J Neurosci 4:2222–2229
Clement J, Simler S, Ciesielski L, Mandel P, Cabib S, Puglisi-Allegra S (1987) Age-dependent changes of brain GABA levels, turnover rates and shock-induced aggressive behavior in inbred strains of mice. Pharmacol Biochem Behav 26:83–88

Coccaro EF, Kavoussi RJ, Cooper TB, Hauger RL (1997) Central serotonin activity and aggression: inverse relationship with prolactin response to *d*-fenfluramine, but not CSF 5-HIAA concentration, in human subjects. Am J Psychiatry 154:1430–1435

Couppis MH, Kennedy CH (2008) The rewarding effect of aggression is reduced by nucleus accumbens dopamine receptor antagonism in mice. Psychopharmacology 197:449–456

Crawley JN, Belknap JK, Collins A, Crabbe JC, Frankel W, Henderson N, Hitzemann RJ, Maxson SC, Miner LL, Silva AJ, Wehner JM, Wynshaw-Boris A, Paylor R (1997) Behavioral phenotypes of inbred mouse strains: implications and recommendations for molecular studies. Psychopharmacology 132:107–124

Crawley JN, Schleidt WM, Contrera JF (1975) Does social environment decrease propensity to fight in male mice? Behav Biol 15:73–83

Cryan JF, Kaupmann K (2005) Don't worry "B" happy!: a role for $GABA_B$ receptors in anxiety and depression. Trends Pharmacol Sci 26:36–43

Cryan JF, Slattery DA (2010) $GABA_B$ receptors and depression: current status. Adv Pharmacol 58:427–451

Dahlström A, Fuxe K (1964) Localization of monoamines in the lower brain stem. Experientia 20:398–399

Davis ES, Marler CA (2004) c-fos Changes following an aggressive encounter in female California mice: a synthesis of behavior, hormone changes and neural activity. Neuroscience 127:611–624

de Almeida RMM, Ferrari PF, Parmigiani S, Miczek KA (2005) Escalated aggressive behavior: dopamine, serotonin and GABA. Eur J Pharmacol 526:51–64

de Almeida RMM, Rosa MM, Santos DM, Saft DM, Benini Q, Miczek KA (2006) $5\text{-}HT_{1B}$ receptors, ventral orbitofrontal cortex, and aggressive behavior in mice. Psychopharmacology 185:441–450

de Almeida RMM, Rowlett JK, Cook JM, Yin W, Miczek KA (2004) $GABA_A/\alpha 1$ receptor agonists and antagonists: effects on species-typical and heightened aggressive behavior after alcohol self-administration in mice. Psychopharmacology 172:255–263

de Boer SF, Caramaschi D, Natarajan D, Koolhaas JM (2009) The vicious cycle towards violence: focus on the negative feedback mechanisms of brain serotonin neurotransmission. Front Behav Neurosci 3:52

de Boer SF, Koolhaas JM (2005) $5\text{-}HT_{1A}$ and $5\text{-}HT_{1B}$ receptor agonists and aggression: a pharmacological challenge of the serotonin deficiency hypothesis. Eur J Pharmacol 526:125–139

DeBold JF, Miczek KA (1981) Sexual dimorphism in the hormonal control of aggressive behavior of rats. Pharmacol Biochem Behav 1:89–93

de Bruin JP, van Oyen HG, Van de Poll N (1983) Behavioural changes following lesions of the orbital prefrontal cortex in male rats. Behav Brain Res 10:209–232

Deisseroth K, Feng G, Majewska AK, Miesenböck G, Ting A, Schnitzer MJ (2006) Next-generation optical technologies for illuminating genetically targeted brain circuits. J Neurosci 26:10380–10386

Delville Y, Mansour KM, Ferris CF (1996) Testosterone facilitates aggression by modulating vasopressin receptors in the hypothalamus. Physiol Behav 60:25–29

Delville Y, de Vries G, Ferris C (2000) Neural connections of the anterior hypothalamus and agonistic behavior in golden hamsters. Brain Behav Evol 55:53–76

Derkach VA, Oh MC, Guire ES, Soderling TR (2007) Regulatory mechanisms of AMPA receptors in synaptic plasticity. Nat Rev Neurosci 8:101–113

Desjardins C, Maruniak JA, Bronson FH (1973) Social rank in house mice: differentiation revealed by ultraviolet visualization of urinary marking patterns. Science 182:939–941

Diano S, Naftolin F, Horvath TL (1997) Gonadal steroids target AMPA glutamate receptor-containing neurons in the rat hypothalamus, septum and amygdala: a morphological and biochemical study. Endocrinology 138:778–789

Dietch JT, Jennings RK (1988) Aggressive dyscontrol in patients treated with benzodiazepines. J Clin Psychiatry 49:184–188

Donaldson ZR, Young LJ (2008) Oxytocin, vasopressin, and the neurogenetics of sociality. Science 322:900–904

Dong E, Matsumoto K, Uzunova V, Sugaya I, Takahata H, Nomura H, Watanabe H, Costa E, Guidotti A (2001) Brain 5α-dihydroprogesterone and allopregnanolone synthesis in a mouse model of protracted social isolation. Proc Nat Acad Sci U S A 98:2849–2854

Dow HC, Kreibich AS, Kaercher KA, Sankoorikal GMV, Pauley ED, Lohoff FW, Ferraro TN, Li H, Brodkin ES (2011) Genetic dissection of intermale aggressive behavior in BALB/cJ and A/J mice. Genes Brain Behav 10:57–68

Dulac C, Axel R (1995) A novel family of genes encoding putative pheromone receptors in mammals. Cell 83:195–206

Dulac C, Torello AT (2003) Molecular detection of pheromone signals in mammals: from genes to behaviour. Nat Rev Neurosci 4:551–562

Duncan GE, Moy SS, Perez A, Eddy DM, Zinzow WM, Lieberman JA, Snouwaert JN, Koller BH (2004) Deficits in sensorimotor gating and tests of social behavior in a genetic model of reduced NMDA receptor function. Behav Brain Res 153:507–519

Earley CJ, Leonard BE (1977) The effect of testosterone and cyproterone acetate on the concentration of γ-aminobutyric acid in brain areas of aggressive and non-aggressive mice. Pharmacol Biochem Behav 6:409–413

Eichler EE, Flint J, Gibson G, Kong A, Leal SM, Moore JH, Nadeau JH (2010) Missing heritability and strategies for finding the underlying causes of complex disease. Nat Rev Genet 11:446–450

Erskine MS, Barfield RJ, Goldman BD (1980) Postpartum aggression in rats: II. Dependence on maternal sensitivity to young and effects of experience with pregnancy and parturition. J Comp Physiol Psychol 94:495–505

Faccidomo S, Bannai M, Miczek KA (2008) Escalated aggression after alcohol drinking in male mice: dorsal raphé and prefrontal cortex serotonin and 5-HT_{1B} receptors. Neuropsychopharmacology 33:2888–2899

Faccidomo S, Quadros IMH, Takahashi A, Fish EW, Miczek KA (2012) Infralimbic and dorsal raphé microinjection of the 5-HT(1B) receptor agonist CP-93,129: attenuation of aggressive behavior in CFW male mice. Psychopharmacology 222:117–128

Ferrari PF, van Erp AMM, Tornatzky W, Miczek KA (2003) Accumbal dopamine and serotonin in anticipation of the next aggressive episode in rats. Eur J Neurosci 17:371–378

Ferrari PF, Parmigiani S, Rodgers RJ, Palanza P (1997) Differential effects of chlordiazepoxide on aggressive behavior in male mice: the influence of social factors. Psychopharmacology 134:258–265

Ferris CF, Melloni RH, Koppel G, Perry KW, Fuller RW, Delville Y (1997) Vasopressin/serotonin interactions in the anterior hypothalamus control aggressive behavior in golden hamsters. J Neurosci 17:4331–4340

Ferris CF, Potegal M (1988) Vasopressin receptor blockade in the anterior hypothalamus suppresses aggression in hamsters. Physiol Behav 44:235–239

File SE, Zharkovsky A, Gulati K (1991) Effects of baclofen and nitrendipine on ethanol withdrawal responses in the rat. Neuropharmacology 30:183–190

Fischer SG, Ricci LA, Melloni RH (2007) Repeated anabolic/androgenic steroid exposure during adolescence alters phosphate-activated glutaminase and glutamate receptor 1 (GluR1) subunit immunoreactivity in hamster brain: correlation with offensive aggression. Behav Brain Res 180:77–85

Fish EW, DeBold JF, Miczek KA (2002) Aggressive behavior as a reinforcer in mice: activation by allopregnanolone. Psychopharmacology 163:459–466

Fish EW, DeBold JF, Miczek KA (2005) Escalated aggression as a reward: corticosterone and $GABA_A$ receptor positive modulators in mice. Psychopharmacology 182:116–127

Fish EW, Faccidomo S, DeBold JF, Miczek KA (2001) Alcohol, allopregnanolone and aggression in mice. Psychopharmacology 153:473–483

Fish EW, Faccidomo S, Miczek KA (1999) Aggression heightened by alcohol or social instigation in mice: reduction by the 5-HT_{1B} receptor agonist CP-94,253. Psychopharmacology 146:391–399
Flint J, Valdar W, Shifman S, Mott R (2005) Strategies for mapping and cloning quantitative trait genes in rodents. Nat Rev Genet 6:271–286
Forrest D, Yuzaki M, Soares HD, Ng L, Luk DC, Sheng M, Stewart CL, Morgan JI, Connor JA, Curran T (1994) Targeted disruption of NMDA receptor 1 gene abolishes NMDA response and results in neonatal death. Neuron 13:325–338
Gammie SC, Garland T, Stevenson SA (2006) Artificial selection for increased maternal defense behavior in mice. Behav Genet 36:713–722
Gammie SC, Nelson RJ (2001) cFOS and pCREB activation and maternal aggression in mice. Brain Res 898:232–241
Garcia–Garcia AL, Elizalde N, Matrov D, Harro J, Wojcik SM, Venzala E, Ramírez MJ, Del Rio J, Tordera RM (2009) Increased vulnerability to depressive-like behavior of mice with decreased expression of VGLUT1. Biol Psychiat 66:275–282
Ginsburg B, Allee WC (1942) Some effects of conditioning on social dominance and subordination in inbred strains of mice. Physiol Zool 15:485–506
Gobrogge KL, Liu Y, Jia X, Wang Z (2007) Anterior hypothalamic neural activation and neurochemical associations with aggression in pair-bonded male prairie voles. J Comp Neurol 502:1109–1122
Gourley SL, DeBold JF, Yin W, Cook J, Miczek KA (2005) Benzodiazepines and heightened aggressive behavior in rats: reduction by $GABA_A/\alpha_1$ receptor antagonists. Psychopharmacology 178:232–240
Gregorová S, Divina P, Storchova R, Trachtulec Z, Fotopulosova V, Svenson KL, Donahue LR, Paigen B, Forejt J (2008) Mouse consomic strains: exploiting genetic divergence between Mus m. musculus and Mus m. domesticus subspecies. Genome Res 18:509–515
Grimes JM, Ricci LA, Melloni RH (2003) Glutamic acid decarboxylase (GAD65) immunoreactivity in brains of aggressive, adolescent anabolic steroid-treated hamsters. Horm Behav 44:271–280
Guidotti A, Dong E, Matsumoto K, Pinna G, Rasmusson AM, Costa E (2001) The socially-isolated mouse: a model to study the putative role of allopregnanolone and 5alpha-dihydroprogesterone in psychiatric disorders. Brain Res Brain Res Rev 37:110–115
Guillot PV, Chapouthier G (1996) Olfaction, GABAergic neurotransmission in the olfactory bulb, and intermale aggression in mice: modulation by steroids. Behav Genet 26:497–504
Guillot PV, Chapouthier G (1998) Intermale aggression, GAD activity in the olfactory bulbs and Y chromosome effect in seven inbred mouse strains. Behav Brain Res 90:203–206
Gutzler SJ, Karom M, Erwin WD, Albers HE (2010) Arginine-vasopressin and the regulation of aggression in female Syrian hamsters (*Mesocricetus auratus*). Eur J Neurosci 31:1655–1663
Halász J, Liposits Z, Kruk MR, Haller J (2002) Neural background of glucocorticoid dysfunction-induced abnormal aggression in rats: involvement of fear- and stress-related structures. Eur J Neurosci 15:561–569
Halász J, Tóth M, Kalló I, Liposits Z, Haller J (2006) The activation of prefrontal cortical neurons in aggression–a double labeling study. Behav Brain Res 175:166–175
Hall RC, Zisook S (1981) Paradoxical reactions to benzodiazepines. Brit J Clin Pharmacol 11:99S–104S
Haller J (1995) Biochemical background for an analysis of cost-benefit interrelations in aggression. Neurosci Biobehav Rev 19:599–604
Haller J, Abrahám I, Zelena D, Juhász G, Makara GB, Kruk MR (1998) Aggressive experience affects the sensitivity of neurons towards pharmacological treatment in the hypothalamic attack area. Behav Pharmacol 9:469–475
Haller J, Halász J, Mikics E, Kruk MR (2004) Chronic glucocorticoid deficiency-induced abnormal aggression, autonomic hypoarousal, and social deficit in rats. J Neuroendocrinol 16:550–557

Haller J, Kruk MR (2006) Normal and abnormal aggression: human disorders and novel laboratory models. Neurosci Biobehav Rev 30:292–303
Haller J, Mikics E, Halász J, Tóth M (2005a) Mechanisms differentiating normal from abnormal aggression: glucocorticoids and serotonin. Eur J Pharmacol 526:89–100
Haller J, van de Schraaf J, Kruk MR (2001) Deviant forms of aggression in glucocorticoid hyporeactive rats: a model for "pathological" aggression? J Neuroendocrinol 13:102–107
Haller J, Tóth M, Halász J (2005b) The activation of raphe serotonergic neurons in normal and hypoarousal-driven aggression: a double labeling study in rats. Behav Brain Res 161:88–94
Haller J, Tóth M, Halász J, de Boer SF (2006) Patterns of violent aggression-induced brain c-fos expression in male mice selected for aggressiveness. Physiol Behav 88:173–182
Hamshere ML, Green EK, Jones IR, Jones L, Moskvina V, Kirov G, Grozeva D, Nikolov I, Vukcevic D, Caesar S, Gordon-Smith K, Fraser C, Russell E, Breen G, St Clair D, Collier DA, Young AH, Ferrier IN, Farmer A, McGuffin P; Wellcome Trust Case Control Consortium, Holmans PA, Owen MJ, O'Donovan MC, Craddock N (2009) Genetic utility of broadly defined bipolar schizoaffective disorder as a diagnostic concept. Brit J Psychiat 195:23–29
Haney M, DeBold JF, Miczek KA (1989) Maternal aggression in mice and rats towards male and female conspecifics. Aggress Behav 15:443–453
Hasen NS, Gammie SC (2009) Trpc2 gene impacts on maternal aggression, accessory olfactory bulb anatomy and brain activity. Genes Brain Behav 8:639–649
Heiligenberg W (1974) Processes Governing Behavioral States of Readiness. Adv Study Behav 5:173–200
Heinz AJ, Beck A, Meyer-Lindenberg A, Sterzer P, Heinz A (2011) Cognitive and neurobiological mechanisms of alcohol-related aggression. Nat Rev Neurosci 12:400–413
Hendricks TJ, Fyodorov D V, Wegman LJ, Lelutiu NB, Pehek E a, Yamamoto B, Silver J, Weeber EJ, Sweatt JD, Deneris ES (2003) Pet-1 ETS gene plays a critical role in 5-HT neuron development and is required for normal anxiety-like and aggressive behavior. Neuron 37:233–247
Herman JP, Mueller NK, Figueiredo H (2004) Role of GABA and glutamate circuitry in hypothalamo-pituitary-adrenocortical stress integration. Ann N Y Acad Sci 1018:35–45
Hess WR, Akert K (1955) Experimental data on role of hypothalamus in mechanism of emotional behavior. AMA Arch Neurol Psychiatry 73:127–129
Hollmann M, Heinemann S (1994) Cloned glutamate receptors. Ann. Rev Neurosci 17:31–108
Hoover WB, Vertes RP (2011) Projections of the medial orbital and ventral orbital cortex in the rat. J Comp Neurol 519:3766–3801
Hurst JL (1987) Behavioural variation in wild house mice *Mus domesticus* Rutty: A quantitative assessment of female social organization. Anim Behav 35:1846–1857
Insel TR, Preston S, Winslow JT (1995) Mating in the monogamous male: behavioral consequences. Physiol Behav 57:615–627
Joppa MA, Meisel RL, Garber MA (1995) c-Fos expression in female hamster brain following sexual and aggressive behaviors. Neuroscience 68:783–792
Karli P, Vergnes M, Eclancher F, Schmitt P, Chaurand J (1972) Role of the amygdala in the control of "mouse killing" behaviour in the rat. In: Eleftheriou B (ed) ed. Plenum Press, The Neurobiology of the Amygdala New York
Kelley AE & Berridge KC (2002) The neuroscience of natural rewards: relevance to addictive drugs. J Neurosci 22:3306–3311
Keverne EB (2002) Mammalian pheromones: from genes to behaviour. Curr Biol 12:R807–R809
Kew JN, Kemp JA (2005) Ionotropic and metabotropic glutamate receptor structure and pharmacology. Psychopharmacology 179:4–29
Kollack-Walker S, Newman SW (1995) Mating and agonistic behavior produce different patterns of Fos immunolabeling in the male Syrian hamster brain. Neuroscience 66:721–736
Konoshenko MY, Timoshenko TV, Plyusnina IZ (2013) c-Fos activation and intermale aggression in rats selected for behavior toward humans. Behav Brain Res 237:103–106
Koolhaas JM (1978) Hypothalamically induced intraspecific aggressive behaviour in the rat. Exp Brain Res 32:365–375

Kruk MR (1991) Ethology and pharmacology of hypothalamic aggression in the rat. Neurosci Biobehav Rev 15:527–538

Kruk MR, Van der Poel AM, Meelis W, Hermans J, Mostert PG, Mos J, Lohman AH (1983) Discriminant analysis of the localization of aggression-inducing electrode placements in the hypothalamus of male rats. Brain Res 260:61–79

Kutsuwada T, Kashiwabuchi N, Mori H, Sakimura K, Kushiya E, Araki K, Meguro H, Masaki H, Kumanishi T, Arakawa M (1992) Molecular diversity of the NMDA receptor channel. Nature 358:36–41

Kutsuwada T, Sakimura K, Manabe T, Takayama C, Katakura N, Kushiya E, Natsume R, Watanabe M, Inoue Y, Yagi T, Aizawa S, Arakawa M, Takahashi T, Nakamura Y, Mori H, Mishina M (1996) Impairment of suckling response, trigeminal neuronal pattern formation, and hippocampal LTD in NMDA receptor epsilon 2 subunit mutant mice. Neuron 16:333–344

Lagerspetz K (1964) Studies on the aggressive behaviour of mice. Ann Acad Sci Fenn B 131:1–13

Lang A, Harro J, Soosaar A, Kõks S, Volke V, Oreland L, Bourin M, Vasar E, Bradwejn J, Männistö PT (1995) Role of N-methyl-D-aspartic acid and cholecystokinin receptors in apomorphine-induced aggressive behaviour in rats. Naunyn Schmiedebergs Arch Pharmacol 351:363–370

Le Grevès P, Huang W, Johansson P, Thörnwall M, Zhou Q, Nyberg F (1997) Effects of an anabolic-androgenic steroid on the regulation of the NMDA receptor NR1, NR2A and NR2B subunit mRNAs in brain regions of the male rat. Neurosci Lett 226:61–64

Leypold BG, Yu CR, Leinders-Zufall T, Kim MM, Zufall F, Axel R (2002) Altered sexual and social behaviors in trp2 mutant mice. Proc Nat Acad Sci USA 99:6376–6381

Li Y, Erzurumlu RS, Chen C, Jhaveri S, Tonegawa S (1994) Whisker-related neuronal patterns fail to develop in the trigeminal brainstem nuclei of NMDAR1 knockout mice. Cell 76:427–437

Lin D, Boyle MP, Dollar P, Lee H, Lein ES, Perona P, Anderson DJ (2011) Functional identification of an aggression locus in the mouse hypothalamus. Nature 470:221–226

Lonstein JS, Gammie SC (2002) Sensory, hormonal, and neural control of maternal aggression in laboratory rodents. Neurosci Biobehav Rev 26:869–888

Löw K, Crestani F, Keist R, Benke D, Brünig I, Benson JA, Fritschy JM, Rülicke T, Bluethmann H, Möhler H, Rudolph U (2000) Molecular and neuronal substrate for the selective attenuation of anxiety. Science 290:131–134

Lucion AB, de Almeida RMM (1996) On the dual nature of maternal aggression in rats. Aggress Behav 22:365–373

Mandel P, Ciesielski L, Maitre M, Simler S, Kempf E, Mack G (1981) Inhibitory amino acids, aggressiveness, and convulsions. Adv Biochem Psychopharmacol 29:1–9

Manolio TA, Collins FS, Cox NJ, Goldstein DB, Hindorff LA, Hunter DJ, McCarthy MI, Ramos EM, Cardon LR, Chakravarti A, Cho JH, Guttmacher AE, Kong A, Kruglyak L, Mardis E, Rotimi CN, Slatkin M, Valle D, Whittemore AS, Boehnke M, Clark AG, Eichler EE, Gibson G, Haines JL, Mackay TF, McCarroll SA, Visscher PM (2009) Finding the missing heritability of complex diseases. Nature 461:747–753

Malek RL, Wang HY, Kwitek AE, Greene AS, Bhagabati N, Borchardt G, Cahill L, Currier T, Frank B, Fu X, Hasinoff M, Howe E, Letwin N, Luu TV, Saeed A, Sajadi H, Salzberg SL, Sultana R, Thiagarajan M, Tsai J, Veratti K, White J, Quackenbush J, Jacob HJ, Lee NH (2006) Physiogenomic resources for rat models of heart, lung and blood disorders. Nat Genet 38:234–239

Matsumoto K, Uzunova V, Pinna G, Taki K, Uzunov DP, Watanabe H, Mienville JM, Guidotti A, Costa E (1999) Permissive role of brain allopregnanolone content in the regulation of pentobarbital-induced righting reflex loss. Neuropharmacology 38:955–963

Maynard Smith J, Price GR (1973) The logic of animal conflict. Nature 246:15–18

McAllister KH (1990) Ethological analysis of the effects of MK-801 upon aggressive male mice: Similarity to chlordiazepoxide. Pharmacol Biochem Behav 37:101–106

Mehta AK, Ticku MK (1999) An update on $GABA_A$ receptors. Brain Res Rev 29:196–217

Melloni RH, Connor DF, Hang PT, Harrison RJ, Ferris CF (1997) Anabolic-androgenic steroid exposure during adolescence and aggressive behavior in golden hamsters. Physiol Behav 61:359–364

Michelsen KA, Schmitz C, Steinbusch HWM (2007) The dorsal raphe nucleus–from silver stainings to a role in depression. Brain Res Rev 55:329–342

Miczek KA (1974) Intraspecies aggression in rats: effects of *d*-amphetamine and chlordiazepoxide. Psychopharmacologia 39:275–301

Miczek KA, de Almeida RMM (2001) Oral drug self-administration in the home cage of mice: alcohol-heightened aggression and inhibition by the 5-HT_{1B} agonist anpirtoline. Psychopharmacology 157:421–429

Miczek KA, Barros HM, Sakoda L, Weerts EM (1998) Alcohol and heightened aggression in individual mice. Alcohol Clin Exp Res 22:1698–1705

Miczek KA, de Boer SF, Haller J (2013) Excessive aggression as model of violence: a critical evaluation of current preclinical methods. Psychopharmacology 226:445–458

Miczek KA, Faccidomo S, de Almeida RMM, Bannai M, Fish EW, DeBold JF (2004) Escalated aggressive behavior: new pharmacotherapeutic approaches and opportunities. Ann N Y Acad Sci 1036:336–355

Miczek KA, Faccidomo S, Fish EW, DeBold JF (2007) Neurochemistry and Molecular Neurobiology of Aggressive Behavior. In: Blaustein J (ed) Behavioral Neurochemistry, Neuroendocrinology and Molecular Neurobiology, 3rd edn. Springer, New York, pp 285–336

Miczek KA, Fish EW, DeBold JF (2003) Neurosteroids, $GABA_A$ receptors, and escalated aggressive behavior. Horm Behav 44:242–257

Miczek KA, Fish EW, DeBold JF, de Almeida RMM (2002) Social and neural determinants of aggressive behavior: pharmacotherapeutic targets at serotonin, dopamine and γ-aminobutyric acid systems. Psychopharmacology 163:434–458

Miczek KA, Haney M (1994) Psychomotor stimulant effects of *d*-amphetamine, MDMA and PCP: aggressive and schedule-controlled behavior in mice. Psychopharmacology 115:358–365

Miczek KA, Maxson SC, Fish EW, Faccidomo S (2001) Aggressive behavioral phenotypes in mice. Behav Brain Res 125:167–181

Miczek KA, O'Donnell JM (1978) Intruder-evoked aggression in isolated and nonisolated mice: effects of psychomotor stimulants and L-dopa. Psychopharmacology 57:47–55

Miczek KA, O'Donnell JM (1980) Alcohol and chlordiazepoxide increase suppressed aggression in mice. Psychopharmacology 69:39–44

Miczek KA, Weerts EM, Tornatzky W, DeBold JF, Vatne TM (1992) Alcohol and "bursts" of aggressive behavior: ethological analysis of individual differences in rats. Psychopharmacology 107:551–563

Mohn AR, Gainetdinov RR, Caron MG, Koller BH (1999) Mice with reduced NMDA receptor expression display behaviors related to schizophrenia. Cell 98:427–436

Monaghan AP, Bock D, Gass P, Schwäger A, Wolfer DP, Lipp HP, Schütz G (1997) Defective limbic system in mice lacking the tailless gene. Nature 390:515–517

Monyer H, Sprengel R, Schoepfer R, Herb A, Higuchi M, Lomeli H, Burnashev N, Sakmann B, Seeburg PH (1992) Heteromeric NMDA receptors: molecular and functional distinction of subtypes. Science 256:1217–1221

Mos J, Olivier B (1989) Quantitative and comparative analyses of pro-aggressive actions of benzodiazepines in maternal aggression of rats. Psychopharmacology 97:152–153

Mos J, Olivier B, Poth M, Van Oorschot R, Van Aken H (1993) The effects of dorsal raphe administration of eltoprazine, TFMPP and 8-OH-DPAT on resident intruder aggression in the rat. Eur J Pharmacol 238:411–415

Mosienko V, Bert B, Beis D, Matthes S, Fink H, Bader M, Alenina N (2012) Exaggerated aggression and decreased anxiety in mice deficient in brain serotonin. Transl Psychiatry 2:e122

Murphy MR (1976) Olfactory stimulation and olfactory bulb removal: Effects on territorial aggression in male Syrian golden hamsters. Brain Res 113:95–110

Nadeau JH, Singer JB, Matin A, Lander ES (2000) Analysing complex genetic traits with chromosome substitution strains. Nat Genet 24:221–225
Natarajan D, Caramaschi D (2010) Animal violence demystified. Front. Behav Neurosci 4:9
Natarajan D, de Vries H, Saaltink DJ, de Boer SF, Koolhaas JM (2009) Delineation of violence from functional aggression in mice: an ethological approach. Behav Genet 39:73–90
Navarro JF, De Castro V, Martín-López M (2008) JNJ16259685, a selective mGlu1 antagonist, suppresses isolation-induced aggression in male mice. Eur J Pharmacol 586:217–220
Navarro JF, de Castro V, Martín-López M (2009) Behavioural profile of selective ligands for mGlu7 and mGlu8 glutamate receptors in agonistic encounters between mice. Psicothema 21:475–479
Navarro JF, Postigo D, Martín-López M, Burón E (2006) Antiaggressive effects of MPEP, a selective antagonist of mGlu5 receptors, in agonistic interactions between male mice. Eur J Pharmacol 551:67–70
Nehrenberg DL, Sheikh A, Ghashghaei HT (2012) Identification of neuronal loci involved with displays of affective aggression in NC900 mice. Brain Struct Funct 218:1033–1049
Nelson M, Pinna G (2011) S-norfluoxetine microinfused into the basolateral amygdala increases allopregnanolone levels and reduces aggression in socially isolated mice. Neuropharmacology 60:1154–1159
Nelson RJ, Chiavegatto S (2001) Molecular basis of aggression. Trends Neurosci 24:713–719
Nelson RJ, Demas GE, Huang PL, Fishman MC, Dawson VL, Dawson TM, Snyder SH (1995) Behavioural abnormalities in male mice lacking neuronal nitric oxide synthase. Nature 378:383–386
Nelson RJ, Trainor BC (2007) Neural mechanisms of aggression. Nat Rev Neurosci 8:536–546
Newman EL, Chu A, Bahamón B, Takahashi A, DeBold JF, Miczek KA (2012) NMDA receptor antagonism: escalation of aggressive behavior in alcohol-drinking mice. Psychopharmacology 224:167–177
Newman SW (1999) The medial extended amygdala in male reproductive behavior. A node in the mammalian social behavior network. Ann N Y Acad Sci 877:242–257
Noirot E, Goyens J, Buhot MC (1975) Aggressive behavior of pregnant mice toward males. Horm Behav 6:9–17
Norlin EM, Gussing F, Berghard A (2003) Vomeronasal Phenotype and Behavioral Alterations in Gαi2 Mutant Mice. Curr Biol 13:1214–1219
Novotny M, Harvey S, Jemiolo B, Alberts J (1985) Synthetic pheromones that promote inter-male aggression in mice. Proc Nat Acad Sci USA 82:2059–2061
Olivier B (2004) Serotonin and aggression. Ann N Y Acad Sci 1036:382–392
Palanza P, Parmiglani S, vom Saal FS (1994) Maternal aggression toward infanticidal males of different social status in wild house mice (Mus musculus domesticus). Aggress Behav 20:267–274
Pan Y, Xu L, Young KA, Wang Z, Zhang Z (2010) Agonistic encounters and brain activation in dominant and subordinate male greater long-tailed hamsters. Horm Behav 58:478–484
Parmigiani S, Rodgers RJ, Palanza P, Mainardi M, Brain PF (1989) The inhibitory effects of fluprazine on parental aggression in female mice are dependent upon intruder sex. Physiol Behav 46:455–459
Payne AP, Swanson HH (1970) Agonistic behaviour between pairs of hamsters of the same and opposite sex in a neutral observation area. Behaviour 36:260–269
Pellis SM, Iwaniuk AN (2000) Comparative analyses of the role of postnatal development on the expression of play fighting. Dev Psychobiol 36:136–147
Pellis SM, Pellis VC (1988) Play-fighting in the Syrian golden hamster *Mesocricetus auratus* Waterhouse, and its relationship to serious fighting during postweaning development. Dev Psychobiol 21:323–337
Peyron C, Petit JM, Rampon C, Jouvet M, Luppi PH (1998) Forebrain afferents to the rat dorsal raphe nucleus demonstrated by retrograde and anterograde tracing methods. Neuroscience 82:443–468

Pinna G, Agis-Balboa RC, Pibiri F, Nelson M, Guidotti A, Costa E (2008) Neurosteroid biosynthesis regulates sexually dimorphic fear and aggressive behavior in mice. Neurochem Res 33:1990–2007
Pinna G, Agis-Balboa RC, Zhubi A, Matsumoto K, Grayson DR, Costa E, Guidotti A (2006) Imidazenil and diazepam increase locomotor activity in mice exposed to protracted social isolation. Proc Nat Acad Sci U S A 103:4275–4280
Pinna G, Costa E, Guidotti A (2004) Fluoxetine and norfluoxetine stereospecifically facilitate pentobarbital sedation by increasing neurosteroids. Proc Nat Acad Sci U S A 101:6222–6225
Pinna G, Dong E, Matsumoto K, Costa E, Guidotti A (2003) In socially isolated mice, the reversal of brain allopregnanolone down-regulation mediates the anti-aggressive action of fluoxetine. Proc Nat Acad Sci U S A 100:2035–2040
Popova NK, Kulikov AV, Nikulina EM, Kozlachkova EY, Maslova GB (1991) Serotonin metabolism and serotonergic receptors in Norway rats selected for low aggressiveness to man. Aggress Behav 17:207–213
Popova NK, Naumenko EV (1972) Dominance relations and the pituitary-adrenal system in rats. Anim Behav 20:108–111
Potegal M (1991) Attack priming and satiation in female golden hamsters: Tests of some alternatives to the aggression arousal interpretation. Aggress Behav 17:327–335
Potegal M, Ferris CF (1989) Intraspecific aggression in male hamsters is inhibited by intrahypothalamic vasopressin-receptor antagonist. Aggress Behav 15:311–320
Potegal M, Ferris CF, Hebert M, Meyerhoff J, Skaredoff L (1996) Attack priming in female Syrian golden hamsters is associated with a c-fos-coupled process within the corticomedial amygdala. Neuroscience 75:869–880
Potegal M, Perumal AS, Barkai AI, Cannova GE, Blau AD (1982) GABA binding in the brains of aggressive and non-aggressive female hamsters. Brain Res 247:315–324
Potegal M, Tenbrink L (1984) Behavior of attack-primed and attack-satiated female golden hamsters (*Mesocricetus auratus*). J Comp Psychol 98:66–75
Puglisi-Allegra S, Mandel P (1980) Effects of sodium n-dipropylacetate, muscimol hydrobromide and (R, S) nipecotic acid amide on isolation-induced aggressive behavior in mice. Psychopharmacology 70:287–290
Del Punta K, Leinders-Zufall T, Rodriguez I, Jukam D, Wysocki CJ, Ogawa S, Zufall F, Mombaerts P (2002) Deficient pheromone responses in mice lacking a cluster of vomeronasal receptor genes. Nature 419:70–74
Quadros IM, Miguel T, DeBold JF, Miczek KA (2009) Opposing action of CRF1 vs. CRF2 receptors in the dorsal raphé: modulation of alcohol-heightened aggression. 2009 neuroscience meeting planner. Society for Neuroscience, Chicago, Program No.445.5/T8
Ricci LA, Grimes JM, Knyshevski I, Melloni RH (2005) Repeated cocaine exposure during adolescence alters glutamic acid decarboxylase-65 (GAD65) immunoreactivity in hamster brain: correlation with offensive aggression. Brain Res 1035:131–138
Roberts WW, Nagel J (1996) First-order projections activated by stimulation of hypothalamic sites eliciting attack and flight in rats. Behav Neurosci 110:509–527
Rodgers RJ, Depaulis A (1982) GABAergic influences on defensive fighting in rats. Pharmacol Biochem Behav 17:451–456
Rosenkranz JA, Grace AA (2002) Cellular mechanisms of infralimbic and prelimbic prefrontal cortical inhibition and dopaminergic modulation of basolateral amygdala neurons in vivo. J Neurosci 22:324–337
Roubertoux PL, Guillot P-V, Mortaud S, Pratte M, Jamon M, Cohen-Salmon C, Tordjman S (2005) Attack behaviors in mice: from factorial structure to quantitative trait loci mapping. Eur J Pharmacol 526:172–185
Rudissaar R, Pruus K, Skrebuhhova-Malmros T, Allikmets L, Matto V (2000) Involvement of GABAergic neurotransmission in the neurobiology of the apomorphine-induced aggressive behavior paradigm, a model of psychotic behavior in rats. Methods Find Exp Clin Pharmacol 22:637–640

Rudolph U, Crestani F, Benke D, Brünig I, Benson JA, Fritschy JM, Martin JR, Bluethmann H, Möhler H (1999) Benzodiazepine actions mediated by specific γ-aminobutyric acid$_A$ receptor subtypes. Nature 401:796–800
Rudolph U, Knoflach F (2011) Beyond classical benzodiazepines: novel therapeutic potential of $GABA_A$ receptor subtypes. Nat Rev Drug Discov 10:685–697
Russell JW, Greenberg BD, Segal DS (1984) The effects of phencyclidine on spontaneous aggressive behavior in the rat. Biol Psychiat 19:195–202
Sandnabba NK (1996) Selective breeding for isolation-induced intermale aggression in mice: associated responses and environmental influences. Behav Genet 26:477–488
Saudou F, Amara DA, Dierich A, Lemeur M, Ramboz S, Segu L, Buhot MC, Hen R (1994) Enhanced aggressive behavior in mice lacking 5-HT_{1B} receptor. Science 265:1875–1878
Schneider R, Hoffmann HJ, Schicknick H, Moutier R (1992) Genetic analysis of isolation-induced aggression. I. Comparison between closely related inbred mouse strains. Behav Neural Biol 57:198–204
Schwartzer JJ, Ricci LA, Melloni RH (2009) Interactions between the dopaminergic and GABAergic neural systems in the lateral anterior hypothalamus of aggressive AAS-treated hamsters. Behav Brain Res 203:15–22
Scott AL, Bortolato M, Chen K, Shih JC (2008) Novel monoamine oxidase A knock out mice with human-like spontaneous mutation. NeuroReport 19:739–743
Scott JP (1942) Genetic differences in the social behavior of inbred strains of mice. J Heredity 33:11–15
Scott JP (1966) Agonistic behavior of mice and rats: a review. Am Zool 6:683–701
Serra M, Pisu MG, Littera M, Papi G, Sanna E, Tuveri F, Usala L, Purdy RH, Biggio G (2000) Social isolation-induced decreases in both the abundance of neuroactive steroids and $GABA_A$ receptor function in rat brain. J Neurochem 75:732–740
Sgoifo A, Stilli D, Musso E, Mainardi D, Parmigiani S (2006) Offensive and defensive bite-target topographies in attacks by lactating rats. Aggress Behav 18:47–52
Shader RI, Greenblatt DJ (1993) Use of benzodiazepines in anxiety disorders. N Engl J Med 328:1398–1405
Shaltiel G, Maeng S, Malkesman O, Pearson B, Schloesser RJ, Tragon T, Rogawski M, Gasior M, Luckenbaugh D, Chen G, Manji HK (2008) Evidence for the involvement of the kainate receptor subunit GluR6 (GRIK2) in mediating behavioral displays related to behavioral symptoms of mania. Mol Psychiatry 13:858–872
Shigemoto R, Kinoshita A, Wada E, Nomura S, Ohishi H, Takada M, Flor PJ, Neki A, Abe T, Nakanishi S, Mizuno N (1997) Differential presynaptic localization of metabotropic glutamate receptor subtypes in the rat hippocampus. J Neurosci 17:7503–7522
Shimshek DR, Bus T, Grinevich V, Single FN, Mack V, Sprengel R, Spergel DJ, Seeburg PH (2006) Impaired reproductive behavior by lack of GluR-B containing AMPA receptors but not of NMDA receptors in hypothalamic and septal neurons. Mol Endocrinol 20:219–231
Siegel A, Roeling TA, Gregg TR, Kruk MR (1999) Neuropharmacology of brain-stimulation-evoked aggression. Neurosci Biobehav Rev 23:359–389
Siever LJ (2008) Neurobiology of aggression and violence. Am J Psychiatry 165:429–442
Simler S, Puglisi-Allegra S, Mandel P (1982) γ-Aminobutyric acid in brain areas of isolated aggressive or non-aggressive inbred strains of mice. Pharmacol Biochem Behav 16:57–61
Stork O, Ji FY, Kaneko K, Stork S, Yoshinobu Y, Moriya T, Shibata S, Obata K (2000) Postnatal development of a GABA deficit and disturbance of neural functions in mice lacking GAD65. Brain Res 865:45–58
Stowers L, Holy TE, Meister M, Dulac C, Koentges G (2002) Loss of sex discrimination and male–male aggression in mice deficient for TRP2. Science 295:1493–1500
Sukhotina IA, Bespalov AY (2000) Effects of the NMDA receptor channel blockers memantine and MRZ 2/579 on morphine withdrawal-facilitated aggression in mice. Psychopharmacology 149:345–350
Svare B, Kinsley CH, Mann MA, Broida J (1984) Infanticide: accounting for genetic variation in mice. Physiol Behav 33:137–152

Takada T, Mita A, Maeno A, Sakai T, Shitara H, Kikkawa Y, Moriwaki K, Yonekawa H, Shiroishi T (2008) Mouse inter-subspecific consomic strains for genetic dissection of quantitative complex traits. Genome Res 18:500–508

Takahashi A, Kwa C, DeBold JF, Miczek KA (2010a) $GABA_A$ receptors in the dorsal raphé nucleus of mice: escalation of aggression after alcohol consumption. Psychopharmacology 211:467–477

Takahashi A, Quadros IM, de Almeida RMM, Miczek KA (2011) Brain serotonin receptors and transporters: initiation vs. termination of escalated aggression. Psychopharmacology 213:183–212

Takahashi A, Schilit AN, Kim J, DeBold JF, Koide T, Miczek KA (2012) Behavioral characterization of escalated aggression induced by $GABA_B$ receptor activation in the dorsal raphe nucleus. Psychopharmacology 224:155–166

Takahashi A, Shimamoto A, Boyson CO, DeBold JF, Miczek KA (2010b) $GABA_B$ receptor modulation of serotonin neurons in the dorsal raphé nucleus and escalation of aggression in mice. J Neurosci 30:11771–11780

Takahashi A, Tomihara K, Shiroishi T, Koide T (2010c) Genetic mapping of social interaction behavior in B6/MSM consomic mouse strains. Behav Genet 40:366–376

Tan KR, Rudolph U, Lüscher C (2011) Hooked on benzodiazepines: $GABA_A$ receptor subtypes and addiction. Trends Neurosci 34:188–197

Tidey JW, Miczek KA (1996) Social defeat stress selectively alters mesocorticolimbic dopamine release: an in vivo microdialysis study. Brain Res 721:140–149

Tirindelli R, Mucignat-Caretta C, Ryba NJ (1998) Molecular aspects of pheromonal communication via the vomeronasal organ of mammals. Trends Neurosci 21:482–486

Tóth M, Halász J, Mikics E, Barsy B, Haller J (2008) Early social deprivation induces disturbed social communication and violent aggression in adulthood. Behav Neurosci 122:849–854

Toth M, Tulogdi A, Biro L, Soros P, Mikics E, Haller J (2012) The neural background of hyper-emotional aggression induced by post-weaning social isolation. Behav Brain Res 233:120–129

Tyler CB, Miczek KA (1982) Effects of phencyclidine on aggressive behavior in mice. Pharmacol Biochem Behav 17:503–510

van der Vegt BJ, Lieuwes N, van de Wall EH, Kato K, Moya-Albiol L, Martínez-Sanchis S, de Boer SF, Koolhaas JM (2003) Activation of serotonergic neurotransmission during the performance of aggressive behavior in rats. Behav Neurosci 117:667–674

van Erp AMM, Miczek KA (1997) Increased aggression after ethanol self-administration in male resident rats. Psychopharmacology 131:287–295

van Erp AMM, Miczek KA (2000) Aggressive behavior, increased accumbal dopamine, and decreased cortical serotonin in rats. J Neurosci 20:9320–9325

van Oortmerssen GA, Bakker TC (1981) Artificial selection for short and long attack latencies in wild *Mus musculus domesticus*. Behav Genet 11:115–126

Veenema AH (2009) Early life stress, the development of aggression and neuroendocrine and neurobiological correlates: what can we learn from animal models? Front Neuroendocrinol 30:497–518

Veening JG, Coolen LM, de Jong TR, Joosten HW, de Boer SF de, Koolhaas JM, Olivier B (2005) Do similar neural systems subserve aggressive and sexual behaviour in male rats? Insights from c-Fos and pharmacological studies. Eur J Pharmacol 526:226–239

Veiga CP, Miczek KA, Lucion AB, de Almeida RM (2007) Effect of $5\text{-}HT_{1B}$ receptor agonists injected into the prefrontal cortex on maternal aggression in rats. Braz J Med Biol Res 40:825–830

Vekovischeva OY, Aitta-Aho T, Echenko O, Kankaanpää A, Seppälä T, Honkanen A, Sprengel R, Korpi ER (2004) Reduced aggression in AMPA-type glutamate receptor GluR-A subunit-deficient mice. Genes Brain Behav 3:253–265

Vlachou S, Markou A (2010) $GABA_B$ receptors in reward processes. Adv Pharmacol 58:315–371

Volavka J, Nolan KA (2008) Methodological structure for aggression research. Psychiatr Q 79:293–300

Wall VL, Fischer EK, Bland ST (2012) Isolation rearing attenuates social interaction-induced expression of immediate early gene protein products in the medial prefrontal cortex of male and female rats. Physiol Behav 107:440–450
Walletschek H, Raab A (1982) Spontaneous activity of dorsal raphe neurons during defensive and offensive encounters in the tree-shrew. Physiol Behav 28:697–705
Wang F, Zhu J, Zhu H, Zhang Q, Lin Z, Hu H (2011) Bidirectional control of social hierarchy by synaptic efficacy in medial prefrontal cortex. Science 334:693–697
Wang Z, Hulihan TJ, Insel TR (1997) Sexual and social experience is associated with different patterns of behavior and neural activation in male prairie voles. Brain Res 767:321–332
Waterston RH, Lindblad-Toh K, Birney E, Rogers J, Abril JF et al (2002) Initial sequencing and comparative analysis of the mouse genome. Nature 420:520–562
Weerts EM, Miller LG, Hood KE, Miczek KA (1992) Increased $GABA_A$-dependent chloride uptake in mice selectively bred for low aggressive behavior. Psychopharmacology 108:196–204
Wilmot CA, Vanderwende C, Spoerlein MT (1987) The effects of phencyclidine on fighting in differentially housed mice. Pharmacol Biochem Behav 28:341–346
Wise DA (1974) Aggression in the female golden hamster: effects of reproductive state and social isolation. Horm Behav 5:235–250
Yadav R, Gupta SC, Hillman BG, Bhatt JM, Stairs DJ, Dravid SM (2012) Deletion of glutamate delta-1 receptor in mouse leads to aberrant emotional and social behaviors. PLoS ONE 7:e32969
Zhao X, Sun L, Jia H, Meng Q, Wu S, Li N, He S (2009) Isolation rearing induces social and emotional function abnormalities and alters glutamate and neurodevelopment-related gene expression in rats. Prog Neuropsychopharmacol Biol Psychiatry 33:1173–1177

Neurogenetics of Aggressive Behavior: Studies in Primates

Christina S. Barr and Carlos Driscoll

Abstract Aggressive behavior can have adaptive value in certain environmental contexts, but when extreme or executed inappropriately, can also lead to maladaptive outcomes. Neurogenetic studies performed in nonhuman primates have shown that genetic variation that impacts reward sensitivity, impulsivity, and anxiety can contribute to individual differences in aggressive behavior. Genetic polymorphisms in the coding or promoter regions of the Mu-Opioid Receptor (*OPRM1*), Corticotropin Releasing Hormone (*CRH*), Monoamine Oxidase A (*MAOA*), Dopamine D4 Receptor (*DRD4*), and Serotonin Transporter (*SLC6A4*) genes have been shown to be functionally similar in humans and rhesus macaques and have been demonstrated to contribute to individual differences in aggression. This body of literature suggests mechanisms by which genetic variation that promotes aggressivity could simultaneously increase evolutionary success while making modern humans more vulnerable to psychopathology.

Keywords primate · aggression · genetic · G x E · macaque

Contents

C. S. Barr (✉) · C. Driscoll
Section of Comparative Behavioral Genomics, Laboratory of Neurogenetics, NIH/NIAAA, Rockville, MD, USA
e-mail: cbarr@mail.nih.gov

Curr Topics Behav Neurosci (2014) 17: 45–71
DOI: 10.1007/7854_2013_267

Published Online: 25 December 2013

Aggression is a behavior that can have adaptive value in certain environmental contexts. However, when aggressive behaviors are extreme or if executed out of impulse or in inappropriate contexts, they can also lead to maladaptive outcomes. This chapter aims to present the study of aggressive behavior in nonhuman primates when considered in an evolutionary context, to present the current understanding of neurogenetic systems that contribute to high aggression phenotypes, to discuss G × E interactions, and to translate findings obtained in nonhuman primates to the human condition.

1 Animal Aggression

In many different species, aggression is important for the protection of self and offspring and in the defense and/or acquisition of rank, territory, or resources. It can also be exhibited in response to fear or pain, or in order to execute control over other individuals. Individual and species differences in aggression are observed in both wild and domesticated animals. While aggression in many wild animals, including primates such as the rhesus macaque, commonly plays an adaptive role, among domestic animals aggressive behavior correlates with "wildness" and behavioral problems. Still, even among relatively aggressive species of monkeys, such as rhesus macaques (pictured in Fig. 1), there is a high degree of variation in the expression of aggressive behavior, and these behaviors can be either advantageous or maladaptive, depending on the environmental context. Although human aggressive behavior may originally have been adaptive, excessive and inappropriate aggression is now a feature of many psychiatric disorders, such as borderline personality disorder, antisocial personality disorder, post-traumatic stress disorder, depression, and psychopathy.

In her book *Animals in Translation*, Temple Grandin (2005) describes the different types of aggression exhibited by animals. She argues that there are two types of aggression exhibited across species: predatory aggression and affective (emotional) aggression. Predatory aggression is reward based and involves hard-wired, fixed action patterns. All species of animals have these predatory brain circuits, but some don't engage them. Though the act of stalking and chasing prey engages the reward circuits in the brain, the behavior for the sake of reward rather

Fig. 1 Rhesus macaques *(Macaca mulatta)* in their social group on the island of Cayo Santiago. Rhesus macaques have been shown to be useful for the study of mother–infant behavior, stress reactivity, and, notably, the neurogenetics of behavior. These studies have had translational value for understanding the genetics of neuropsychiatric disorders and aggression. Photo taken by C. Barr

than for food must be inhibited among wild animals because the act of chasing can be energetically costly. Among domesticated animals, there is no need to inhibit the drive to engage in this type of rewarding behavior because they are generally provisioned; the cat can chase the laser pointer beam, and the dog can chase his own tail. Unlike predatory aggression, affective/emotional aggression is reactive rather than rewarding in nature. Examples of emotional aggression include assertive aggression, fear aggression, maternal aggression, irritable aggression, and inter-male aggression. When excessive or inappropriate, both predatory and affective aggression can potentially become pathological.

As stated above, Dr. Grandin describes predatory aggression as being a behavior that involves fixed action patterns that are hardwired. However, it is known that animals are also capable of learned aggression. The more complex the prefrontal cortices are among individuals of a given species, the more likely they are to exhibit learned aggressive behavior. In other words, more complex animals may be able to imitate aggressive behavior or learn that aggression can lead to a desired outcome in a given situation. While a lizard might only exhibit fixed action pattern-based aggression, animals such as canids, felids, and equids might be more capable of learned aggressive behavior, and may also show an increased incidence of planned or organized aggression. Adolescent male killer whales have been known to kill for sport, and dolphins will engage in gang-rapes and will mass murder porpoises despite the fact that they are nonthreatening and do not compete for resources. Goodall (1968) described what she referred to as "mini wars" among the chimpanzees she followed in the field. In effect, the more complex the brain, the more severe and organized aggressive behavior becomes. How this pertains to the human condition will be discussed in more detail at the end of this chapter.

2 Genetics of Aggression in Primates

Studies performed in rodents and primates have shown that a number of neurotransmitter systems are consistently involved in the expression of aggressive behavior. Among these are the serotonin, corticotropin-releasing hormone, oxytocin, neuropeptide Y, and endogenous opioid systems (Takahashi et al. 2012). Of relevance to the human condition, there may be functional genetic variations modulating these systems to promote aggression and violence. The aggression-related traits discussed in the previous section have been shown to be heritable in nature. Studies have shown that both anxious and impulsive behaviors are heritable in vervet monkeys (Fairbanks 2004), and others show anxiety to be heritable in rhesus monkeys (Williamson et al. 2003). Aggressive tendencies have also been shown to be both highly heritable and predictive of reproductive output in rhesus macaques, suggesting that the genetic factors that underlie aggression are likely to impact fitness (Brent 2013). Selection of genetic variants that predict aggression is likely to have had a role in human evolution as well.

Genetic factors that impact aggressivity in rhesus macaques and the mechanisms by which they may promote or moderate aggressive behavior may be predictive of the human condition. What follows is a discussion of some of the key neurogenetic studies that have been performed in the rhesus macaque, showing how genetic factors that may be adaptive in certain contexts can also render individuals more or less sensitive to particular environmental variables, ultimately driving potentially maladaptive responses such as excessive aggression. The chapter will close with a review of the literature presented in terms of how certain genetic variants could simultaneously promote evolutionary success while making modern humans more vulnerable to psychopathology. Interactions between genetic and environmental factors will also be discussed.

2.1 Reward Sensitivity and OPRM1 Genotype: Responses to Both Natural and Artificial Rewards

The reward systems are critical to survival and reproduction because they are involved in driving ingestion of food, social interactions, and sexual activity, to name a few. When resources are scarce, these are particularly important, because they also relate to resource acquisition and control, as behaviors that can increase the likelihood of gaining access to limited resources (i.e., aggression) also involve the reward pathways.

One system that is activated in response to both natural and artificial rewards is the endogenous opioid system. In multiple primate species, there are nonsynonymous single-nucleotide polymorphisms (SNPs) in the first exon of the *OPRM1* gene that produce amino acid changes in the ligand-binding domain of the receptor (Miller et al. 2004; Bond et al. 1998). Among these are the A118G SNP in humans and the C77G SNP in rhesus macaques; early in vitro work showed that both appeared to increase the affinity of the receptor for its endogenous ligand (ß-Endorphin) by approximately threefold. Studies examining intermediate phenotypes likely to be under the control of this receptor (for example, HPA axis activity) also suggest gain-of-function roles for these polymorphisms (Ray and Hutchison 2004; Chong et al. 2006; Barr et al. 2007; Schwandt et al. 2011).

The *OPRM1* SNPs present in humans and rhesus macaques not only predict responses to artificial rewards, such as alcohol, but to natural rewards as well. In rhesus, the 77G allele predicts increased attachment of an infant to its mother, and this is particularly true following repeated periods of maternal separation (Barr et al. 2008a). This finding has also been replicated in a study of human children aged 9–13, in which various measures of attachment to the caregiver were increased as a function of parental inconsistency or unavailability. *OPRM1* genotype predicts individual differences in aggressive behavior in rhesus macaques as well (Miller et al. 2004).

2.2 Impulsivity in a High-Risk Environment—CRH and MAOA

Impulsivity is a trait that is observed in many psychopathological conditions, from the personality disorders to attention deficit hyperactivity disorder (ADHD) to the addictions. It is a coping mechanism that, at certain life history stages and in certain environmental contexts, might be an adaptive one. The corticotropin-releasing hormone (CRH) system is critical to behavioral responses to stress, and studies in which CRH activity is experimentally manipulated suggest that naturally occurring *CRH* gene variation may mediate individual variability in the behavioral traits that determine an individual's coping style. One of the most consistent behavioral correlates of CRH system activity is the way in which an organism approaches novel stimuli and unfamiliar conspecifics (Kalin et al. 2000; Korte et al. 2005).

The *CRH* haplotype has been shown to predict behavioral inhibition in children (Smoller et al. 2003), and studies in rhesus macaques suggest that *CRH* variation in humans may moderate the risk of alcohol use disorders, perhaps through the pathway of behavioral inhibition. The presence of a rhesus polymorphism (−2232 C/G) that has similar in vitro functional effects to some *CRH* haplotypes reported in humans (Wagner et al. 2006) predicts decreased CSF levels of CRH, an intermediate phenotype found in individuals or strains characterized as being particularly extroverted, aggressive, or bold. Infant macaques carrying the G allele are characterized as being more exploratory and bold (Barr et al. 2008b), and, following adolescence, males that are G allele carriers exhibit a more bold and active response to an unfamiliar conspecific.

Variation at the monoamine oxidase A (*MAOA*) gene is also linked to impulsivity and impulsive aggression in both animal models and humans. Monoamine oxidase A degrades the monoamine transmitters (dopamine, norepinephrine, and serotonin), and, therefore can influence synaptic concentrations of these neurotransmitters. A variable number of tandem repeats (VNTR) polymorphism in the transcriptional control region for the human *MAOA* gene (MAOA-LPR) have been shown to produce differential transcriptional activity in vitro (Sabol et al. 1998). Rhesus and other primates also show variation at this locus (Gibbons 2004). This polymorphism is present across macaque species, but differs substantially in its frequency (Wendland et al. 2006). The low activity allele of rhMAOA-LPR contributes additively with peer rearing to increase impulsivity and aggression (Barr et al. 2004a; Newman et al. 2005). These studies in rhesus macaques support the hypothesis that *MAOA* gene promoter variation may specifically increase the risk for conduct or personality disorders characterized by impaired impulse control and aggressivity, particularly among those exposed to early stress or lack of parental influence. The effect of this genetic variant in human populations will be discussed at the end of this chapter.

2.3 *Novelty Seeking: DRD4*

Dopamine neurotransmission underlies many reward-dependent and reinforcing processes. Tandem repeats in the third exon of the dopamine receptor D4 gene (DRD4) exist across a variety of species, including humans, vervets, dogs, dolphins, bears, raccoons, horses, and chimpanzees (see Larsen et al. 2005). Some species, including rhesus, pigtail, and Tonkean macaques, exhibit variation in the number of repeats present, with various alleles differing in frequencies across species (Wendland et al. 2006; Livak et al. 1995). In humans, the DRD4 7-repeat allele (7R) reduces efficacy of the receptor and is linked with traits such as sensation seeking (Ptacek et al. 2011; Benjamin et al. 1996; Ebstein et al. 1996). A DRD4 length variant has also been associated with novelty seeking in vervet monkeys (Bailey 2007). In the dog, which has been subject to intense artificial selection, a repeat polymorphism predicts social impulsivity and activity-impulsivity endophenotypes (Hejjas et al. 2007a, b). We recently identified 4R, 5R, 6R, and 7R alleles in a colony of rhesus macaques. Compared to the ancestral allele (4R), the loss-of-function 7R allele predicted increased physical aggression toward an unknown age- and sex-matched conspecific (Fig. 2).

2.4 *Anxiety, Reactivity, and the Serotonin Transporter Gene*

Perhaps the most extensively studied genetic variant in primate behavioral genetics and human psychiatric genetics is a polymorphism located upstream from the serotonin transporter gene (*SLC6A4*). In humans, this common polymorphism

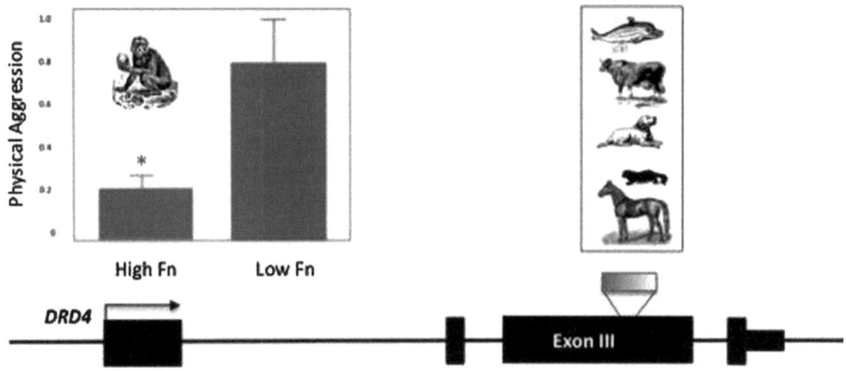

Fig. 2 DRD4 gene is present across species and predicts aggression in rhesus macaques. Tandem repeats present in the third exon of the DRD4 gene are present in a variety of species, including domestic animals. In macaques, variation at this locus predicts individual differences in aggressive responses to an unfamiliar conspecific

reduces expression of the serotonin transporter gene, and variation of this serotonin-transporter-linked polymorphic region predicts certain temperament traits related to anxiety, depression, and aggression, such as neuroticism, harm avoidance, and disagreeableness (Mazzanti et al. 1998). There is variation in the serotonin transporter gene regulatory region in many nonhuman primate species as well (Wendland et al. 2005; Lesch et al. 1997). In rhesus, a similar polymorphism has been shown to alter transcriptional efficiency (Bennett et al. 2002), resulting in decreased serotonin transporter mRNA levels in the brains of animals carrying the variant allele (Lopez and Higley 2002), which may be further regulated by epigenetic mechanisms (Kinnally et al. 2010). Because it might predict alternative strategies, particularly in certain environmental contexts, the serotonin-transporter-linked polymorphic region (5-HTTLPR) polymorphism has been studied extensively in both rhesus and human gene by environment interaction studies (Barr et al. 2004a, b, c). Studies in multiple, independent laboratories demonstrate that the 5-HTTLPR s allele predicts anxiety and responses to stress in rhesus macaques (Bethea et al. 2004; Champoux et al. 2002) and that the 5 HTTLPR genotype can interact with controlled prenatal exposure to stress or alcohol or early life adversity, resulting in long-lasting differences in stress reactivity, sensation seeking, and aggression (Barr et al. 2004a, b, c, d; Kraemer et al. 2008; Schneider et al. 2010, 2011; Schwandt et al. 2010; Spinelli et al. 2007).

3 Environmental Effects on Aggression

Field and laboratory studies of nonhuman primates have demonstrated effects of environmental variables on aggressive behavior. Many of these effects are attributable to the drive to reproduce (Knapp and Innocent 2012). The Challenge

Hypothesis has been put forth as an argument for intra-individual variability in aggression among chimpanzees (Sobolewski et al. 2013). This hypothesis was originally developed in order to understand the biological underpinnings of the transient increases in aggression observed among seasonally breeding male birds (Wingfield et al. 1990). One factor that helps to explain such variability is variation in circulating levels of the reproductive steroid, testosterone. A role for testosterone in the facilitation of aggressive behavior is well established. In fact, in domestic animals, castration is routinely performed in dogs, goats, and horses, to name a few, in order to reduce aggression, particularly toward male conspecifics. There are likely a number of independent mechanisms by which testosterone exerts such effects. First, testosterone binds to nuclear receptors, and the testosterone-receptor complex can then act to modulate expression of testosterone-sensitive genes in both peripheral tissues and in brain. Whether through effects on gene expression or via membrane receptor-mediated alterations in neurotransmitter release, it has been shown that testosterone facilitates attention given to threatening social stimuli. In rhesus macaques studied in the lab, testosterone replacement in hypothalamic-pituitary-gonadal axis-suppressed males increased selective attention to negative social stimuli and interactions without affecting environmental exploration or attention toward novel stimuli (King et al. 2012). Testosterone decreases fear and anxiety and has been shown to act through the dopamine system to increase social reward. As such, the behavioral effects of testosterone, regardless of molecular mechanism, could be attributable to both decreased aversion for negative emotional stimuli and facilitation of approach to social stimuli.

In a number of species, serum levels of testosterone in males increase during the mating season and as a function of female conception cycles (Fichtel et al. 2007; Bales et al. 2006; Ostner et al. 2011). In some seasonally breeding species, testosterone levels correlate with the frequencies of aggressive displays (Higley et al. 1996a), while in chimpanzees, it is proposed that testosterone only facilitates aggressive behavior in certain environmental contexts (Sobolewski et al. 2013). Chimpanzees are nonseasonal breeders, and females only give birth approximately once in a 5-year period. Because of the vigorous competition that exists for females among breeding-aged males and because the external environment does not control the female reproductive cycle, there is variability in the intensity of competition among males over time, and such variability is also dependent on the female over which males are competing. Studies in free-ranging chimpanzees have demonstrated that individual testosterone levels increase with the number of males in a group, particularly so when competing for females who have previously given birth (Sobolewski et al. 2013). Increased circulating testosterone in male chimpanzees exposed to these variables may promote aggressive behavior toward other males and, therefore, make them compete more successfully for breeding females.

Another important factor driving aggressive behavior is environmental stress (discussed in Korte et al. 2005). In humans, there are known links between stress and a variety of psychiatric disorders, including depression, posttraumatic stress disorder, anxiety disorders, and substance abuse disorders (McEwen 2006;

Sapolsky 2001), all of which may be linked to aggressive behavior. Like testosterone, the stress hormone cortisol can bind to nuclear receptors and alter gene expression. It has been demonstrated that expression of many genes that influence behaviors involved in stress response are under the control of glucocorticoids through this mechanism. In some instances, it has been shown that common genetic variants that moderate responses to stressful stimuli alter the corticosteroid response of these genes by disrupting or adding these types of elements to regulatory regions. Genetic variants that influence stress reactivity could predict whether an individual is at risk for psychopathology (Caspi et al. 2002, 2003). This might be assessed by evaluating an individual's sensitivity to environmental stressors, the magnitude and/or duration of the stress response, or the types of responses (including aggression) exhibited following stress exposure. Genetic variations that determine whether individuals habituate or sensitize to repeated or chronic stress would also be expected to play a major role (Spinelli et al. 2012).

4 Gene x Environment Interactions

From an evolutionary perspective, the roots of psychopathology may lie in the different strategies that have evolved for coping with environmental challenges (Korte et al. 2005). Rhesus macaques provide an opportunity to examine genetic-environmental ($G \times E$) interactions in a controlled, prospective manner (Barr et al. 2003b, 2004d; Barr and Goldman 2006). Not only can variables like diet, light–dark cycle, and other factors that are contributors to "noise" in human genetic studies be controlled, but environmental factors that might contribute to $G \times E$ interactions can also be applied in a controlled manner. Such factors commonly studied in behavioral research include alcohol exposure (prenatal, in adolescence, or in adulthood), acute challenges (for example, the presentation of an intruder, or separation from attachment sources), response to certain neuropsychopharmacological agents, or early life stress (variable foraging demand or peer/nursery rearing) (reviewed in Barr and Goldman 2006).

Peer rearing is probably the most extensively studied environmental factor in $G \times E$ experiments in rhesus macaques. As with other primate species, rhesus macaque mothers invest much of their energy into defending, comforting, and caring for their infants, and this maternal "buffering" appears to be critical to normal infant development (Suomi 1982). In the so-called peer rearing condition, subjects are removed from their parents at birth and reared with other age-matched infants, so that they develop in the absence of adult influence (Harlow and Suomi 1974; Chamove et al. 1973). When compared to their mother-reared counterparts, peer-reared subjects exhibit evidence of insecure attachment, higher levels of anxiety, and lower levels of exploration in novel settings (Suomi 1982). Because their peers do not necessarily punish inappropriate behavior, they also can have impaired development of the behavioral inhibition system, and, as in humans, macaques that have been exposed to early adversity (in the form of peer-rearing) show long-lasting differences in brain function and aggressive behavior (Higley et al. 1991; Spinelli et al. 2012). Interactive effects between functional genetic loci and early rearing environment have been demonstrated for the serotonin transporter, Neuropeptide Y (*NPY*), *MAOA*, *CRH*, and *DRD1* genes in rhesus macaques (Barr et al. 2008; Newman et al. 2005, 2009; Lindell et al. 2010).

Because selective pressures differ between the sexes, males and females differ in their responses to environmental challenges (Eme et al. 2007; Wood and Eagly 2002). Variation in the serotonin transporter-linked polymorphism results in sexually dichotomous qualitative and quantitative G $\times$ E interactions, first demonstrated in rhesus studies. Even prior to pubertal development, the 5-HTTLPR s allele predicted increased stress response, but only among females with histories of early adversity (Barr et al. 2004b). Responding to social threat is one domain in which males and females are likely to adopt different adaptive solutions. In terms of allelic effects on aggressive behavior, it has been shown that male adolescents carrying the s allele who were exposed to early life stress are more likely to respond aggressively toward an unfamiliar conspecific (Schwandt et al. 2010). Females, in contrast, are more likely to exhibit redirected aggression toward other members of their group. Both types of aggression are "affective" in nature and likely due to the increased reactivity known to exist among carriers of the 5-HTTLPRs allele.

5 Adaptive Value of Aggressive Behavior and the Role of Selection

In evolutionary biology, a behavior is *adaptive* if it makes the organism more fit to survive and reproduce in comparison to other members of the same species. Aggressive behavior in animals is thought to be adaptive since it increases success in competition for resources and mates and ensures survival during agonistic encounters with both conspecifics and predators. In his writings, Charles Darwin noted that there often appeared to be a conflict of interest between traits that increased survivability and traits that could potentially increase reproductive output (Darwin 1896). In other words, many traits that give advantage in reproductive success have negative consequences for survival. He referred to this as sexual selection, of which there are two types. The first is intersexual selection, a case in which an individual of one sex can choose a mate on the basis of a given trait (e.g., plumage in peacocks). The second is referred to as intrasexual selection, and most frequently refers to male subjects competing over mating opportunities. For the latter, genetic variants that promoted aggression would be most likely to be under selection.

While Brent et al. (2013) found that aggressive behavior predicted reproductive success in rhesus macaques, they also showed that passive, affiliative, nonaggressive subjects were equally successful. It has been shown in other primate species (e.g., baboons and assamese macaques) that individuals with higher levels of sociality have higher infant survival rates and produce more offspring. The fact that high and low aggression phenotypes in rhesus macaques predict increased reproductive output suggests a role for disruptive selection at genes involved in moderating aggressive behavior (Brent et al. 2013; Lande and Arnold 1983).

While the field of behavioral genetics is growing rapidly, most of its research is concerned with the identification of "disease alleles," or gene variation underlying what is considered pathological behavior. Its methods and findings, however, can be applied to a long-standing goal of evolutionary anthropology, namely to understand how changes in allele frequency can affect divergences in primate behavior. Several studies have identified associations between specific alleles and natural features of behavior and life history strategies. For example, the loss-of-function short(s) allele of the serotonin transporter gene promoter length polymorphism, which increases risk for developing depression in the face of adversity, has a functional equivalent in the rhesus macaque (see above). In macaques, this allele is associated with increased endocrine and behavioral stress reactivity as a function of stress exposure, often in a sexually dichotomous manner (Barr et al. 2004; Spinelli et al. 2007; Schwandt et al. 2010). Therefore, this variant appears to increase the risk of developing psychopathology, particularly in the context of stress. Despite this, these variants have been maintained in both humans and in rhesus, in addition to some other nonhuman primate species. Moreover, in human populations in which the s allele is rare, another loss-of-function variant on the L allele background (LA > LG) is present at a higher frequency (Hu et al. 2006). In humans, there is also a VNTR in the second intron, which appears to be functional (Fiskerstrand et al. 1999). This VNTR is present in a number of primate and nonprimate species and is polymorphic in a number of hominoid species (Soeby et al. 2005).

Although SNPs are not necessarily conserved across species, there are instances in which functionally similar SNPs occur in humans and rhesus macaques (Barr et al. 2008a, b; Vallender et al. 2008; Miller et al. 2004). The serotonin transporter gene regulatory region is tremendously variable within species, and it has been demonstrated that gain-of-function *SLC6A4* SNPs have arisen and been maintained in both rhesus and in humans, suggesting that both gain- and loss-of-function variants may be under selection in primates (Vallender et al. 2008). It is interesting that *SLC6A4* variation not only predicts individual differences in impulse control and stress reactivity (Barr et al. 2004; Bennett et al. 2002; Champoux et al. 2002; Schwandt et al. 2010), but that it is also associated with adaptive traits in free-ranging macaques, such as earlier male dispersal (Trefilov et al. 2000) and male reproductive timing (Krawczak et al. 2005). Whether allelic variation at *SLC6A4* predicts "adaptive" traits in humans has not been elucidated (Homberg and Lesch 2011).

An impulsive, aggressive individual that readily approaches novel objects or conspecifics may do well in certain social situations, but may face higher risk of predation or attack than a more cautious, harm-avoidant individual. Such behaviors might, therefore, be predicted to confer selective advantage at particular developmental or life history stages and in certain environmental contexts.

Moreover, because of differences in their behavioral and physiologic responses to stress, the types of stress-related pathology to which bold, proactive individuals and harm-avoidant, reactive individuals are vulnerable are quite different. Whereas the latter are at risk for internalizing disorders, such as depression and anxiety, the former are more likely to develop externalizing conditions, primarily characterized by impaired impulse control and aggression (Korte et al. 2005).

Variation at the *CRH* locus would be expected to increase stress adaptation or modify behavior in a manner that is adaptive, but which may also moderate the risk of stress-related disorders in modern humans. In other words, though genetic variation that increases stress reactivity could assist in the adaptation to stress in the short term, with chronic, repeated or severe stressors it could also make individuals more prone to developing psychopathology. In macaques, the two most common haplotypes are yin-yang, or alternative, haplotypes (Barr et al. 2008b). The persistence of these divergent haplotypes over time suggests that at least one of the alleles on each background is being selected—possibly in a particular environmental context—while the rest are hitchhiking. Several studies in humans (Baerwald et al. 2000; Shimmin et al. 2007) have found evidence for selection at the *CRH* locus, in which alternative, yin-yang haplotype clades are observed. As in the rhesus macaque, the major human *CRH* haplotypes have been shown to vary in terms of their in vitro promoter activity, including differences in glucocorticoid sensitivity (Wagner et al. 2006). In rhesus macaques, carriers of a *CRH*−2232 G allele engage in risky behaviors (Barr et al. 2008b) and exhibit lower levels of the serotonin metabolite 5-HIAA, a neurochemical endophenotype observed both in macaques exposed to early life stress and in humans with antisocial personality disorder (Higley et al. 1991, 1996a). It may be that in humans genetic variations that alter CRH system function could influence multiple behavioral dimensions (i.e., both neuroticism and extraversion), and variants that place an individual at the extremes of these spectra (i.e., inhibited and anxious/stress reactive vs. bold/impulsive and novelty seeking) might increase the risk for psychopathology. Of note, based on its extended haplotype structure, studies that examine effects of the CRH receptor gene (*CRHR1*) haplotype demonstrate molecular evidence for selection (Nelson et al. 2010).

As another example, in both rhesus and in humans, there are nonsynonymous SNPs in the portion of the *OPRM1* gene that encodes the N-terminal domain of the receptor (C77G in rhesus macaque and A118G in human, as discussed above), and these SNPs perform similar functions in vivo (Barr et al. 2007; Chong et al. 2006). In humans, the 118G allele is suspected to increase the likelihood of alcohol abuse because it increases alcohol-induced dopamine release and subjective euphoria (Ramchandani et al. 2011; Ray and Hutchison 2004). We have shown that rhesus carrying the 77G allele exhibit increased alcohol-induced stimulation (a marker for the euphorogenic effects of alcohol) and that G allele carriers also consume more alcohol in the laboratory (Barr et al. 2007). It would stand to reason that *OPRM1* variation might also predict sensitivity to natural rewards, including aggressive or agonistic behavior, which, as stated earlier, can be rewarding. Since these two variants confer similar functional effects and are both observed at relatively high frequencies, and since there is an extended region of LD with the A118G allele in humans (Luo et al. 2008; Zhang et al. 2006; Pang et al. 2009) it might be hypothesized that they have evolved as result of similar selective pressures in the two species, but data to directly address this hypothesis are not yet available. However, studies performed in the macaque demonstrate that this variant predicts behaviors that could theoretically be under selection, including aggressive behavior (Miller et al. 2004), and G allele carriers form stronger attachment bonds with their mothers during infancy (Barr et al. 2008; Higham et al. 2011), especially as a function of repeated maternal separation. It is interesting that the effects of repeated episodes of maternal separation and reunion are similar to those that might be observed during periods of alcohol intake and withdrawal. Similar effects of *OPRM1* genotype on social attachment have recently been demonstrated in human children, showing increased quality of parent–child relations as a function of parental unavailability or inconsistency. These types of studies highlight how traits that could have conferred selective advantage at some point in the evolutionary history of humans can increase the risk of psychopathology in modern society.

6 New Molecular Approaches: The Serotonin Transporter Gene and Genetic Selection

A high-profile meta-analysis performed in 2009 called into question the validity of all of the 5-HTTLPR x stress findings that had been reported in the human literature (Risch et al. 2009). Risch et al. and others who subsequently came out against candidate gene-based and G × E studies claimed that the field needed to look toward whole-genome linkage as a tool. Those who defended the validity of G × E interaction studies argued not only that whole-genome studies are substantially less powerful, but also cited many of the animal studies, including those performed in nonhuman primates (Caspi et al. 2010). This debate is ongoing and calls out for studies to verify the validity of G × E interaction results by elucidating mechanisms through which they might occur. New technologies and approaches in genomics and cell biology may make it possible to identify the mechanisms that underlie G × E interactions. Such studies may also inform us of genetic variation that promotes stress sensitivity or resilience, permitting the identification of candidate loci for the performance of future G × E interaction studies.

Epigenetic mechanisms are one mechanism by which stress could interact with genotype. The emergence of next generation DNA sequencing technologies has increased the potential for discovery of epigenetic effects. Several companies offer platforms that permit high-throughput sequencing, and by combining immunoprecipitation with these approaches, the effects of stress on levels of histone binding or sites of DNA modification (two forms of epigenetic regulation) can be elucidated, and genetic factors that directly promote or inhibit these processes might be identified. Variation within regulatory or coding regions of genes encoding stress-responsive signaling molecules, which may contribute to environmental sensitivity and stress vulnerability, is of particular interest for the study of G $\times$ E interactions.

We recently performed a serotonin-transporter focused study using chromatin immunoprecipitation sequencing (ChIP-SEQ) in order to determine epigenetic stress effects at *SLC6A4* in archived brain tissue from male macaques (Lindell et al. 2012). We found that binding of H3K4me3, a histone protein that marks active promoters, at the serotonin transporter gene was affected by early life stress as a function of 5-HTTLPR genotype and that it varied across adolescent development. These findings support the idea that genetic effects may vary as a function of various environmental factors and that these may act, in part, through epigenetic mechanisms. The ChIP-SEQ methodology employed in this study involved antibody-based isolation of H3K4me3, after which levels of H3K4me3 binding were assessed by sequencing the histone-associated DNA, such that the number of reads for a region are the index for the relative degree of H3K4me3 binding and, therefore, epigenetic regulation via this mechanism. One by-product of approach is that genetic variation can also be identified. We identified 11 SNPs within the serotonin transporter gene and its regulatory region, 3 of which were present in the region of H3K4me3 binding, overlapping with a CpG island—a region that can act as a substrate for a more commonly studied epigenetic regulation, DNA methylation (Kinnally et al. 2010)—and within 100 nucleotides of the transcription start site. Interestingly, there are 3 SNPs present in the corresponding region in humans as well (rs55753714, rs25533, and rs61274396).

As stated earlier, genetic variation in the serotonin-transporter-linked polymorphic region may confer selective advantage in a variety of species. Our identification of abundant, putatively functional polymorphisms in rhesus in this study and the SNP density observed in the corresponding region for humans (http://genome.ucsc.edu/) is relevant to studies examining how variation at or around the *SLC6A4* gene relates to individual differences in environmental reactivity and behavior. Because the region 5′ of the gene includes the promoter and important transcription factor binding sites, these regions are typically under selection. The result is that, relative to intergenic regions, there is both increased conservation of sequence across species and lower levels of intraspecific polymorphism within these regions. We have previously shown that, in both human and macaque, the average SNP density for the region 5 kB upstream of transcription start sites is low relative to intergenic regions, and that it is more similar to that observed in other important regulatory sequences, such as intronic or untranslated regions (which, like the promoter regions, are also under selection) (Yuan et al. 2012). In our 5-HTTLPR epigenetic study, not only did we identify 3 SNPs within the core promoter for the rhesus *SLC6A4* gene (despite sampling a limited number of subjects), but there also appears to be increased diversity in this region in humans, with 25 SNPs within the 1 KB 5′ of the *SLC6A4* transcription start site. This was approximately 10× higher than we had previously reported for variation in human intergenic regions (Yuan et al. 2012).

These findings demonstrate how genetic and environmental effects may interact and how genetic variability could be driven by disruptive and/or balancing selection at genes known to moderate aggressive behavior. It may be that certain genetic variants are only "functional" under certain environmental conditions or during certain developmental phases. It is possible that within-species variation at genes that influence behavior and stress response is favored because these genes promote alternative strategies, particularly at critical developmental time points or under specific environmental conditions.

7 Genetic Selection in Primates: Understanding the Origins of Human Aggression and Psychopathology

Several decades ago, Maynard Smith applied "Game Theory" to animal behavior and found that aggression and fearfulness are traits that tend to balance each other in any social population (Smith and Price 1973). He described the "Hawk-Dove game" in which "Hawks" (aggressive individuals, who were proactive and exhibited a fight-or-flight response to stress) and "Doves" (who were fearful, cooperative and adopted a freeze-and-hide stress response) were likely to co-occur in the same species. He determined that both "Hawk" and "Dove" strategies could potentially be adaptive, perhaps especially so in certain environmental contexts. While "Hawks" would do better when food was abundant and population density high, since they are better at fighting for access to mates than foraging), "Doves" would likely outcompete "Hawks" when the opposite were true since they would excel at getting food during periods of scarcity and avoiding danger during times of increased conflict. The two temperaments must be balanced in any given population; in a population comprised completely of "Hawks," excessive aggression would reduce the chances of an individual surviving to successfully reproduce, while an all "Dove" population would quickly be killed off by some other, more aggressive species. In other words, the presence of both behavioral types may be critical to the survival of the species, while at the same time, there will be selective pressure at the individual level, depending on life history, sex, and environmental context. It then stands to reason that the genetic factors that underlie these two alternative strategies would be likely to be subject to balancing selection and that, as a result, both types of traits or strategies will be observed among individuals of any social species.

There are a number of research groups that have been investigating genetic variations in the rhesus macaque that contribute to the expression of traits that have been linked with human personality and psychiatric disorders, all of which may map well onto the Hawk versus Dove model (i.e., stress reactivity, behavioral dyscontrol, aggression, and reward seeking/sensitivity) (reviewed in Barr 2013).

Many of the variants that have been identified and studied in macaques are functionally similar to those present in humans, and some findings suggest convergent evolution, with the variants maintained by selection in both species (Barr et al. 2008a, b; Vallender et al. 2008). The macaque model has proven useful for learning how relatively common genetic variants, which are associated with traits that may be adaptive in certain environmental contexts, can also increase vulnerability to behavioral pathology.

As stated earlier in this chapter, the more complex the brain of a species, the more severe and organized the aggressive behaviors exhibited among its members becomes. Humans are a highly aggressive species in comparison to other animals, probably as a result of an unusually high benefit-to-cost ratio for intraspecific aggression (Georgiev et al. 2013). Humans are also unique in their capacity for planned murders, and they are the only primate species among which mothers exhibit infanticidal behavior toward their own young (Raine 2013; Hrdy 1999). Several of the genetic variants that have been found across primate species have been linked to human aggression. One of the first genetic variants to be linked with severely aggressive behavior was a coding SNP in the *MAOA* gene, discovered by Han Brunner when he was approached by a woman who wanted genetic counseling because her 10-year-old son was showing signs of aggression (Brunner et al. 1993, discussed in Raine 2013). She reported that many of her male relatives had significant behavior problems, and she described them as being "frightening and aggressive." Brunner et al. tracked down members of the extended family, found 14 male relatives who had a history of violence and impulsive aggression, and were able to identify the variant that had arisen de novo in this familial line. Since the MAOA gene is on the X chromosome, males who were hemizygous for this functional coding variant were unprotected by a "normal" copy of the gene and, thus, all male individuals were affected.

Less than a decade later, a common functional VNTR polymorphism was discovered in the transcriptional control region for the human *MAOA* gene (Sabol et al. 1998). The low activity allele for this variant was found to interact with early adversity to predict later antisocial and violent behavior in a large population of boys (Caspi et al. 2002). Since then, some of the neurocircuitry differences in individuals carrying this allele have been determined. The low activity MAOA-LPR allele predicts decreased prefrontocortical and increased amygdalar responses to emotional stimuli, suggesting impaired ability to control emotional responses during arousal (Meyer-Lindenberg et al. 2006). As stated above, this genetic variant interacts with environmental factors. Not only does early stress exposure moderate the effects of MAOA-LPR genotype, but there is interaction with testosterone levels as well (Sjoberg et al. 2008). This is relevant to the discussion of testosterone and cortisol effects on gene expression and behavior. It is possible that the *MAOA*/testosterone interaction may reflect the direct action of testosterone on the *MAOA* promoter. Testosterone and glucocorticoid hormone/receptor complexes bind with different effects or affinities to enhancer elements in the *MAOA* promoter and, as such, testosterone and cortisol may both alter *MAOA* transcription.

As discussed at the beginning of this chapter, Temple Grandin writes in her book (2005) that animals exhibit different types of aggression, predatory, and affective/emotional. The type of impulsive aggression described by Brunner (1993) and in other studies examining effects of MAOA-LPR variation is most likely affective aggression. In his book *The Anatomy of Violence*, Raine (2013) describes aggression predicted by the *MAOA* genotype as "hot-headed" aggression. What has been found with the 5-HTTLPR genetic variant is quite different. Earlier, we discussed the Hawk versus Dove model and the rhesus macaque 5-HTTLPR variant with regard for the potential for balancing or disruptive selection. What has now been demonstrated is that both the l and the s 5-HTTLPR alleles can predict aggressive behavior in humans, but that the types of aggression are quite different. Consistent with what was found with MAOA-LPR, the low activity 5-HTTLPR s allele was present in individuals who exhibited affective aggression. This is not surprising given that the s allele predicts increased anxiety, environmental sensitivity, and reactivity (Lesch et al. 1997; Caspi et al. 2003, 2010). However, it has also been shown that individuals homozygous for the l allele can also exhibit inappropriate aggression. These subjects are psychopathic individuals, who exhibit lower stress responses than carriers of the s allele and who are more likely to engage in cold-blooded, planned "predatory" aggression (Glenn 2011) (Fig. 3).

Variation in other genes discussed in this chapter has been shown to be under selection in humans and may predict aggressive behavior in this species of primate. Polymorphisms in the *DRD4* and *OPRM1* genes both appear to have been under recent positive selection, and both of these variants have been linked to reward-based behaviors (D'Souza and Craig 2008; Ding et al. 2002). The DRD4 7R allele predicts novelty seeking, impulsivity, thrill seeking, anger, and short temper, traits that may have been under selection. It is likely that various combinations of genetic variants contribute to human violence. Since polymorphisms that affect aggression appear to exist in parallel in humans and some nonhuman primates, these animals will be valuable for determining genetic effects on aggression, for elucidating how environmental factors interact with genetic variation and as potential models for treatment of pathologically aggressive behavior in humans.

8 Future Agendas

8.1 Looking to Other Model Organisms

We have just discussed how violent crimes in humans can be characterized as being hot headed or cold blooded (Raine et al. 2013). The hot-headed crimes are likely similar in etiology to the affective aggression observed in nonhuman primates, and, as such, studies examining impulsive/reactive aggression have been of

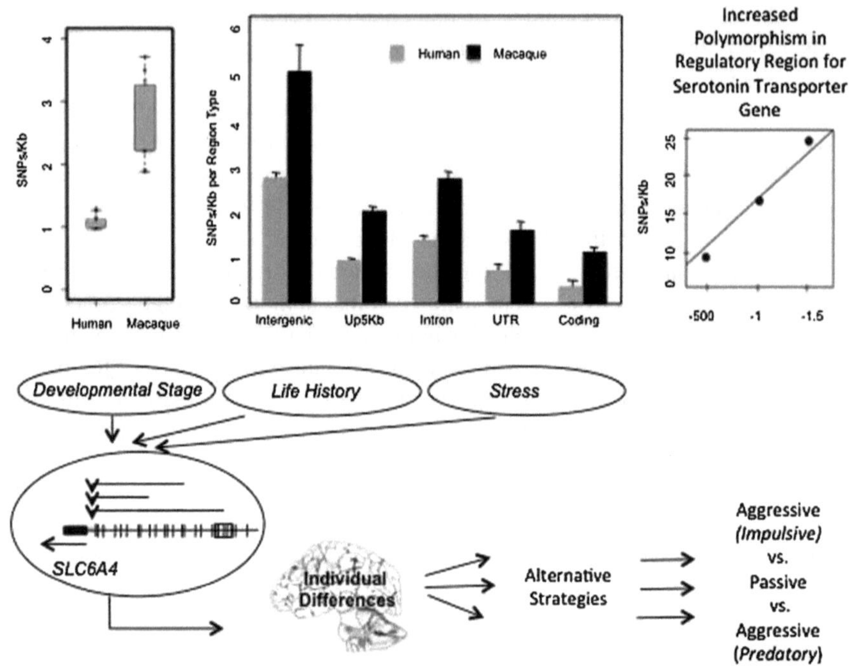

Fig. 3 Density of SNPs within the serotonin transporter regulatory region is higher than that observed for other regulatory regions genome-wide. Shown above are the average number of SNPs in each type of region across the genome as determined by ChIP and RNA-SEQ in comparison to the number of SNPs observed in the 5′ Flanking region for the serotonin transporter gene. Shown is an example of how factors such as developmental stage, life history variables, and stress exposure can differentially regulate genes expressed in brain in a genotype-dependent manner to promote diversity in behavioral strategies within a given population

interest to psychiatrists, evolutionary biologists, psychologists, and primatologists alike. However, other types of aggressive acts committed by humans are less easily modeled in a nonhuman primate. Above, we introduced the idea that, in humans, serotonin transporter genotype can be predictive of impulsive/affective aggression while the alternative genotype increases the risk of psychopathy (Glenn 2011). It could be that a predator species may be a better model for determining the genetic underpinnings of dysregulated predatory behavior and modeling psychopathic, cold-blooded, predatory aggression. This type of killing may have its basis in reward and, though most of us find the idea of harming or killing another aversive, those who do not and who cannot inhibit their prey drives may be at risk (see discussion above relating to prey drive inhibition in domestic animals).

Another area of interest is the aggressive behavior noted among individuals diagnosed with a developmental disorder. In recent years, the incidences of autism and autism spectrum disorders (ASD) have increased rather markedly. Aggression among ASD subjects is, in part, attributable to increased reactivity and impulsivity, both of which can be studied in nonhuman primates. However, other features of the autism spectrum disorders are not easily modeled using nonhuman primates. As with psychopathic, cold-blooded killers, ASD subjects exhibit deficits in empathy and social cognition and some of these have been shown to be dependent on an individual's genotype (for example, 5-HTTLPR; Brune et al. 2006). Unlike the callousness observed among psychopathic individuals, however, ASD subjects have affective empathy that is intact (they can resonate with others' feelings), but they lack the ability to read others' emotions (Jones et al. 2010). Therefore, aggressive behavior observed with ASD is proposed to be aimed at eliciting a clear emotional reaction and, therefore, thought to be attributable to this deficit in social cognition and not aggressiveness per se. In other words, risk for aggression relates to deficits in social cognition among ASD subjects.

One approach that could be used to determine the genetic factors that underlie deficiencies in social cognition is the study of the genetics of domestication (for example, dogs, cats, and horses). At its most basic, domestication is a suite of heritable traits affecting behavior (Belyaev et al. 1981). There are intriguing phenotypic commonalities among domesticates. Most important among these traits, and the only one common to all domesticates, is the ability to coexist with humans. The systems that likely permitted early domestication range from those involving fear and impulse control to those involving reward and sociality. In general, domestic animals have lower levels of aggression than do their nondomesticated ancestors. This is shown not only with aggression toward humans, but with predatory aggression as well. When compared to coyotes, beagle-coyote hybrids exhibit less effective predatory behavior, partially attributable to decreased arousal and increased response threshold. There is also partial inhibition of killing bite in hybrids, and this is completely inhibited in the purebred beagle dog (Fox 1976). At some point in the domestication of some species of animals, there may have also been selection for social cognition, as many domestic animal species are good at looking to humans and for reading human emotions. For example, it has been shown that in studies of wolves, dogs, and hybrid animals, dogs are more likely to look to humans for cues in order to retrieve a reward. It has further been shown that genetic variation at the DRD4 gene contributes to phenotypic variation in this trait (Hori et al. 2013). Whether the latter relates to differences in reward sensitivity or to social cognition are not clear.

In some domestic species, there has been selection among breeds or lines for other desirable traits that have inadvertently resulted in some increased frequency of aggression related to fear, prey drive, territoriality, deference refusal, or impulse control (Overall 1997). These may serve as very powerful models for looking at effects of domestication and reversal of some of those effects through more recent artificial selection. There have been several instances, however, in which wild animals were experimentally domesticated. In Siberia in the late 1950s, Belayev and Trut began an experimental domestication of the silver fox a coat color variant of the red fox (*Vulpes vulpes*) raised industrially on commercial fur farms. Selective breeding began with 130 individuals, 100 females, and 30 males. Using a rather coarse test of mansuetude (see Kukekova 2008) they selected the top 5 % of males and 20 % of females for each successive generation while keeping the population roughly the same size. In order to avoid conditioning the animals to human contact the foxes were never handled prior to phenotyping for the selection of suitable breeders for the next generation. Within a few generations demonstrable changes in behavior toward people were evident in the foxes' vocalizations, position in their cages on approach, the ear and tail positions, gaze toward and willingness to be touched by humans; within 10 generations they had succeeded in breeding a reasonably 'domestic' fox. It is significant that 'domestication' occurred in so few generations, as this is an indication that the number of genes involved in domestication cannot be exceedingly large, perhaps in the range of 10–20 genes of major effect, a conclusion supported by subsequent research which found that large behavioral differences result from limited changes in the transcriptome of genes expressed in the brain (Lindberg et al. 2007). The study of the genetics of domestication across species may, therefore, access a number of traits and underlying genetic variation that are highly relevant to the types of aggression observed in the human condition; not only those relating to fear/anxiety, impaired impulse control, and reward, but those involving individual differences in social cognition as well.

9 Molecular Mechanisms of G × E Interactions

While much has been learned with the use of molecular genetics to dissect behavior in primates, the molecular effects of environment factors are not as widely known. Increasing this knowledge base might aid in prevention and/or development of new pharmacotherapies for the treatment of severe aggression in both domestic animals and humans. One limitation to performing such studies in a nonhuman primate is that it requires the use of a brain sample, which presents ethical concerns to many primatologists and neuroscientists. We discussed in this manuscript the use of massively paralleled sequencing technologies in order to examine epigenetic effects and to discover genetic variation in primates. These techniques can be used to study environmental effects on levels of histone binding or sites of DNA modification (two forms of epigenetic regulation), the advantage being that the entire genome is queried, increasing the yield from a given brain sample. We are currently performing whole exome sequencing in macaques that exhibit extreme variation in aggressivity and have identified a number of candidate alleles using this method.

In this chapter, we also discussed the importance of stress and gonadal steroids as they relate to aggressive behavior in primates. Besides epigenetic modifications, one mechanism by which environmental factors might interact with genetic variation is via hormone receptor-induced regulation of gene expression. Cortisol and testosterone exert pleiotropic effects by altering gene expression in brain regions containing high concentrations of glucocorticoid or testosterone receptors, which can alone or in combination with other factors to influence expression of genes containing glucocorticoid, or androgen response elements (GRE/ARE). Of interest, all of the gene variants presented in this chapter (5-HTTLPR, *CRH*, MAOA-LPR, and *OPRM1* C77G) have AREs or GREs in their regulatory regions, making them putatively responsive to these hormones, and in many of the cases in which we have identified gene by stress interactions in rhesus, glucocorticoid or androgen response is either enhanced or disrupted via polymorphism within AREs/ GREs. Unlike histone proteins, transcription factors are not tightly associated with DNA. However, one might still be able to use next generation sequencing technologies to identify corticosteroid-sensitive genes. By performing X-ChIP-SEQ (with anti-glucocorticoid receptor antibodies) combined with mRNA-SEQ, there may be potential to identify genes that are sensitive to glucocorticoid receptor-mediated activation or repression and to determine if this regulation occurs in a genotype-dependent manner.

Various other nongenomic approaches can be used to assist in elucidating mechanisms through which G × E interactions might occur. The development of induced pluripotent stem cells (IPSCs) could be critically important in determining how G × E interactions occur in brain. These cells, if induced to develop into neurons, could be tremendously informative for examining hormone effects (both sex and stress steroids). The fact that cells can be derived from peripheral tissues affords the opportunity to examine effects of any or all of these variables that are specific to the individuals' genetic makeup and, therefore, can be used to examine individual differences.

References

Bailey JN, Breidenthal SE, Jorgensen MJ, McCracken JT, Fairbanks LA (2007) The association of DRD4 and novelty seeking is found in a nonhuman primate model. Psychiatr Genet 17(1):23–27

Baerwald CG, Mok CC, Tickly M, Lau CS, Wordsworth BP, Ollier B, Panayi GS, Lanchbury JS (2000) Corticotropin releasing hormone (CRH) promoter polymorphisms in various ethnic groups of patients with rheumatoid arthritis. Z Rheumatol 59:29–34

Bales KL, French JA, McWilliams J, Lake RA, Dietz JM (2006) Effects of social status, age, and season on androgen and cortisol levels in wild male golden lion tamarins (*Leontopithecus rosalia*). Horm Behav 49:88–95

Barr CS, Newman TK, Becker ML, Parker CC, Champoux M, Lesch KP, Goldman D, Suomi SJ, Higley JD (2003) The utility of the non-human primate; model for studying gene by environment interactions in behavioral research. Genes Brain Behav 2:336–340

Barr CS, Newman TK, Lindell S, Shannon C, Champoux M, Lesch KP, Suomi SJ, Goldman D, Higley JD (2004a) Interaction between serotonin transporter gene variation and rearing condition in alcohol preference and consumption in female primates. Arch Gen Psychiatry 61:1146–1152

Barr CS, Newman TK, Schwandt M, Shannon C, Dvoskin RL, Lindell SG, Taubman H, Thompson B, Champoux M, Lesch KP, Goldman D, Suomi SJ, Higley JD (2004b) Sexual dichotomy of an interaction between early adversity and the serotonin transporter gene promoter variant in rhesus macaques. Proc Natl Acad Sci USA 101:12358–12363

Barr CS, Newman TK, Shannon C, Parker C, Dvoskin RL, Becker ML, Schwandt M, Champoux M, Lesch KP, Goldman D, Suomi SJ, Higley JD (2004c) Rearing condition and rh5-HTTLPR interact to influence limbic-hypothalamic-pituitary-adrenal axis response to stress in infant macaques. Biol Psychiatry 55:733–738

Barr CS, Schwandt ML, Newman TK, Higley JD (2004d) The use of adolescent nonhuman primates to model human alcohol intake: neurobiological, genetic and environmental variables. Ann NY Acad Sci 1021:221–233

Barr CS, Goldman D (2006) Nonhuman primate models of inheritance of vulnerability to alcohol abuse and addiction. Addiction Biol 11:374–385

Barr CS, Schwandt ML, Lindell SG, Chen SA, Suomi SJ, Goldman D, Heilig M, Higley JD (2007) Mu opioid receptor gene variation is associated with alcohol response and consumption in rhesus monkeys. Arch Gen Psychiatry 64:369–376

Barr CS, Schwandt ML, Lindell SG, Higley JD, Maestripieri D, Goldman D, Suomi SJ, Heilig M (2008a) A functional *OPRM1* variant is associated with attachment behavior in infant rhesus macaques. Proc Natl Acad Sci 105:5277–5281

Barr CS, Dvoskin RL, Yuan Q, Lipsky RH, Gupte M, Hu X, Zhou Z, Schwandt ML, Lindell SG, McKee M, Becker ML, Kling MA, Gold PW, Higley JD, Heilig M, Suomi SJ, Goldman D (2008b) *CRH* haplotype as a factor influencing cerebrospinal fluid levels of corticotropin-releasing hormone, hypothalamic-pituitary-adrenal axis activity, temperament, and alcohol consumption in rhesus macaques. Arch Gen Psychiatry 65(8):934–944

Barr CS (2013) Non-human primate models of alcohol-related phenotypes: the influences of genetic and environmental factors. Curr Top Behav Neurosci 13:223–249

Belyaev DK, Ruvinsky AO, Trut LN (1981) Inherited activation-inactivation of the star gene in foxes; its bearning on the problem of domestication. J Hered 72:267–274

Benjamin J, Li L, Patterson C, Greenberg BD, Murphy DL, Hamer DH (1996) Population and familial association between the D4 dopamine receptor gene and measures of Novelty Seeking. Nat Genet 12:81–84

Bennett AJ, Lesch K-P, Heils A, Long JC, Lorenz JG, Shoaf SE, Champoux M, Suomi SJ, Linnoila MV, Higley JD (2002) Early experience and serotonin transporter gene variation interact to influence primate CNS function. Mol Psychiatry 7:118–122

Bethea CL, Streicher JM, Coleman K, Pau FKY, Moessner R, Cameron JL (2004) Anxious behavior and fenfluramine-induced prolactin secretion in young rhesus macaques with different alleles of the serotonin reuptake transporter polymorphism (5HTTLPR). Behav Genet 34:295–307

Bond C, LaForge KS, Tian MT, Melia D, Zhang SW, Borg L, Gong JH, Schluger J, Strong JA, Leal SM, Tischfield JA, Kreek MJ, Yu L (1998) Single-nucleotide polymorphism in the human mu opioid receptor gene alters beta-endorphin binding activity: possible implications for opiate addiction. Proc Natl Acad Sci USA 95:9608–9613

Brent LJN, Heilbronner SR, Horvath JE, Gonzalez-Martinez J, Ruiz-Lambides A, Robinson AG, Skene JHP, Platt ML (2013) Genetic origins of social networks in rhesus macaques. Scientific Reports 3(1042):1–8

Brune CW, Kim SJ, Salt J, Leventhal BL, Lord C, Cook EH (2006) 5-HTTLPR genotype-specific phenotype in children and adolescents with autism. Am J Psychiatry 163:2148–2156

Brunner HG, Nelen M, Breakefield XO, Ropers HH, van Oost BA (1993) Abnormal behavior associated with a point mutation in the structural gene for monoamine oxidase A. Science 262:578–580

Caspi A, McClay J, Moffitt TE, Mill J, Martin J, Craig IW, Taylor A, Poulton R (2002) Role of genotype in the cycle of violence in maltreated children. Science 297:851–854

Caspi A, Sugden K, Moffitt TE, Taylor A, Craig IW, Harrington H, McClay J, Mill J, Martin J, Braithwaite A, Poulton R (2003) Influence of life stress on depression: moderation by a polymorphism in the 5-HTT gene. Science 301:386–389

Caspi A, Hariri AR, Holmes A, Uher R, Mofitt TE (2010) Genetic sensitivity to the environment; the case of the serotonin transporter gene and its implications for studying complex diseases and traits. Am J Psychiatry 167(5):509–527

Chamove AS, Rosenblum LA, Harlow HF (1973) Monkeys (*Macaca mulatta*) raised only with peers. Anim Behav 21:316–325

Champoux M, Bennett A, Shannon C, Higley JD, Lesch KP, Suomi SJ (2002) Serotonin transporter gene polymorphism, differential early rearing, and behavior in rhesus monkey neonates. Mol Psychiatry 7:1058–1063

Chong RY, Oswald L, Yang X, Uhart M, Lin PI, Wand GS (2006) The mu-opioid receptor polymorphism A118G prttedicts cortisol responses to naloxone and stress. Neuropsychopharmacology 31:204–211

Darwin C (1896) The expression of the emotions in man and animals. Appleton & Co., New York

Ding YC, Chi HC, Grady DL, Morishima A, Kidd JR, Kidd KK, Flodman P, Spence MA, Schuck S, Swanson JM, Zhang YP, Moyzis RK (2002) Proc Nat Acad Sci USA 99:309–314

D'Souza UM, Craig IW (2008) Functional genetic polymorphisms in serotonin and dopamine gene systems and their significance in behavioral disorders. Prog Brain Res 172:73–98

Ebstein RP, Novick O, Umansky R, Priel B, Osher Y, Blaine D, Bennett ER, Nemanov L, Katz M, Belmaker RH (1996) Dopamine D4 receptor (D4DR) exon III polymorphism associated with the human personality trait of Novelty Seeking. Nat Genet 12:78–80

Eme RF (2007) Sex differences in child-onset, life-course-persistent conduct disorder: a review of biological influences. Clin Psychol Rev 27:607–627

Fairbanks LA, Newman TK, Bailey JN, Jorgensen MJ, Breidenthal SE, Ophoff RA, Comuzzie AG, Marin LJ, Rogers J (2004) Genetic contributions to social impulsivity and aggressiveness in vervet monkeys. Biol Psychiatry 55:642–647

Fichtel C, Kraus C, Ganwindt A, Mestermann M (2007) Influence of reproductive season and rank on fecal glucocorticoid levels in free ranging male Verraux's sifakas (*Propithecus verrauzi*). Horm Behav 51:640–648

Fiskerstrand CE, Lovejoy EA, Quinn JP (1999) An intronic polymorphic domain often associated with susceptibility to affective disorders has allele dependent differential enhancer activity in embryonic stem cells. FEBS Lett 458:171–174

Fox MW (1976) Effects of domestication on prey catching and killing in beagles, coyotes and F2 hybrids. Appl Anim Ethol 2:123–140

Georgiev AV, Klimczuk AC, Traficonte DM, Maestripieri D (2013) When violence pays: a cost-benefit analysis of aggressive behavior in animals and humans. Evol Psychol 11:678–699

Gibbons A (2004) TAAPA meeting. Tracking the evolutionary history of a "warrior" gene. Science 304:818

Glenn AL (2011) The other allele: exploring the long allele of the serotonin transporter gene as a potential risk factor for psychopathy: a review of the parallels in findings. Neurosci Biobehav Rev 35:612–620

Grandin T (2005) Animals in translation: using the mysteries of autism to decode animal behavior. Scribner, New York

Harlow HF, Suomi SJ (1974) Induced depression in monkeys. Behav Biol 12:273–296

Hejjas K, Vas J, Kubinyi E, Sasvari-Szekely M, Miklosi A, Ronai Z (2007a) Novel repeat polymorphisms of the dopaminergic neurotransmitter genes among dogs and wolves. Mamm Genome 18(12):871–879

Hejjas K, Vas J, Topal J, Szantai E, Ronai Z, Szekely A, Kubinyi E, Horvath Z, Sasvari-Szekely M, Miklosi A (2007b) Association of polymorphisms in the dopamine D4 receptor gene and the activity-impulsivity endophenotype in dogs. Anim Genet 38(6):629–633

Higham JP, Barr CS, Hoffman CL, Mandalaywala TM, Parker KJ, Maestripieri D (2011) Interaction between *OPRM1* genotype and reproductive status on CSF levels of oxytocin in free-ranging rhesus macaques. Behavioral Neurosci 125(2):131–136

Higley JD, Hasert MF, Suomi SJ, Linnoila M (1991) Nonhuman primate model of alcohol abuse: effects of early experience, personality, and stress on alcohol consumption. Proc Nat Acad Sci USA 88:7261–7265

Higley JD, King ST, Hasert MF, Campoux M, Suomi SJ, Linnoila M (1996a) Stability of interindividual differences in serotonin function and its relationship to severe aggression and competent social behavior in rhesus macaque females. Neuropsychopharmacology 14:67–76

Higley JD, Suomi SJ (1989) Temperamental reactivity in non-human primates. In: Kohnstamm D, Bates HE, Rothbart MK (eds) Temperament in childhood. Wiley, Chichester

Higley JD, Suomi SJ, Linnoila M (1996b) A nonhuman primate model of type II excessive alcohol consumption? Part 1. Low cerebrospinal fluid 5-hydroxyindoleacetic acid concentrations and diminished social competence correlate with excessive alcohol consumption. Alcoholism Clin Exp Res 20:629–642

Homberg JR, Lesch KP (2011) Looking on the bright side of serotonin transporter gene variation. Biol Psychiatry 69(16):513–519

Hori Y, Kishi H, Inoue-Murayama M, Fujita K (2013) Dopamine receptor D4 gene (DRD4) is associated with gazing toward humans in domestic dogs (*Canis familiaris*). Open J Anim Sci 3:54–58. doi:10.4236/ojas.2013.31008

Hrdy SB (1999) Mother nature: a history of mothers, infants and natural selection. Pantheon, New York

Hu XZ, Lipsky RH, Zhu G, Akhtar L, Taubman J, Greenberg BD, Xu K, Arnold PD, Richter MA, Kennedy JL, Murphy DL, Goldman D (2006) Additional functional variation at the SLC6A4 gene. Am J Hum Genet 78:815–826

Jones AP, Happe FGE, Gilbert G, Burnett S, Viding E (2010) Feeling, caring, knowing: different types of empathy deficit in boys with psychopathic tendencies and autism spectrum disorder. J Child Psychol Psychiatry 51:1188–1197

Kalin NH, Shelton SE, Davidson RJ (2000) Cerebrospinal fluid corticotropin releasing hormone levels are stable in monkeys with patterns of brain activity associated with fearful temperament. Biol Psychiatry 47:579–585

King HM, Kurdziel LB, Meyer JS, Lacreuse A (2012) Effects of testosterone on attention and memory for emotional stimuli in male rhesus monkeys. Psychoneuroendocrinol 37:396–409

Kinnally EL, Capitanio JP, Leibel R, Deng L, Leduc C, Haghighi F, Mann JJ (2010) Epigenetic regulation of serotonin transporter expression and behavior in infant rhesus macaques. Genes Brain Behav 9:575–582

Korte SM, Koolhaas JM, Wingfield JC, McEwen BS (2005) The Darwinian concept of stress: benefits of allostasis and costs of allostatic load and the trade offs in health and disease. Neurosci Biobehav Rev 29:3–38

Knapp LA, Innocent SHS (2012) Molecules and mating: positive selection and reproductive behavior in primates. In: Lopez-Larrea C (ed) Sensing in nature. Springer, New York, pp 218–236

Kraemer GW, Moore CF, Newman TK, Barr CS, Schneider ML (2008) Moderate levels of fetal alcohol exposure and serotonin transporter gene promoter polymorphism affect neonatal temperament and LHPA axis regulation in monkeys. Biol Psychiatry 63(3):317–324

Krawczak M, Trefilov A, Berard J, Bercovitch F, Kessler M, Sauermann U, Croucher P, Nürnberg P, Widdig A, Schmidtke J (2005) Male reproductive timing in rhesus macaques is influenced by the HTTLPR promoter polymorphism of the serotonin transporter gene. Biol Reprod 72:1109–1113

Kukekova AV, Trut LN, Chase K, Shepeleva DV, Vladimirova AV, Kharlamova AV, Oskina IN, Stepika A, Klavanov S, Erb HN, Acland GM (2008) Behav Genet 38:185–194

Lande R, Arnold SJ (1983) The measurement of selection on correlated characters. Evolution 37:1210–1226

Larsen SA, Mogensen L, Dietz R, Baagoe HJ, Andersen M, Werge T, Rasmussen HB (2005) Identification and characterization of tandem repeats in exon III of dopamine receptor D4 (DRD4) genes from different mammalian species. DNA Cell Biol 24(12):795–804

Lesch KP, Meyer J, Glatz K, Flügge G, Hinney A, Hebebrand J, Klauck S, Poustka A, Poustka F, Bengel D, Mössner R, Riederer P, Heils A (1997) The 5-HT transporter gene-linked polymorphic region (5-HTTLPR) in evolutionary perspective: alternative biallelic variation in rhesus monkeys. J Neural Transm 104:1259–1266

Lindberg J, Björnerfeldt S, Bakken M, Vilà C, Jazin E, Saetre P (2007) Selection for tameness modulates the expression of heme related genes in silver foxes. Behav Brain Funct 3:18

Lindell SG, Schwandt ML, Sun H, Sparenborg JD, Bjoerk K, Kasckow JW, Sommer WH, Goldman D, Higley JD, Suomi SJ, Heilig M, Barr CS (2010) Functional *NPY* variation as a factor in stress resilience in rhesus macaques. Arch Gen Psychiatry 67(4):423–431

Lindell SG, Yuan Q, Zhou Z, Goldman D, Thompson RC, Lopez JF, Suomi SJ, Higley JD, Barr CS (2012) The serotonin transporter gene is a substrate for age and stress dependent epigenetic regulation in rhesus macaque brain: potential roles in genetic selection and gene x environment interactions. Dev Psychopathol 24:1391–1400

Livak JL, Rogers J, Lichter JB (1995) Variability of dopamine D4 receptor (DRD4) gene sequence within and among nonhuman primate species. Proc Natl Acad Sci USA 92:427–431

Lopez JF, Higley JD (2002) The effect of early experience on brain corticosteroid and serotonin receptors in rhesus monkeys. Biol Psychiatry 51:294

Luo X, Zuo L, Kranzler H, Zhang H, Wang S, Gelernter J (2008) Multiple OPR genes influence personality traits in substance dependent and healthy subjects in two American populations. Am J Med Genet B Neuropsychiatr Genet 147B:1028–1039

Mazzanti CM, Lappalainen J, Long JC, Bengel D, Naukkarinen H, Eggert M, Virkkunen M, Linnoila M, Goldman D (1998) Role of the serotonin transporter promoter polymorphism in anxiety-related traits. Arch Gen Psychiatry 55:936–940

McEwen BS (2006) Protective and damaging effects of stress mediators: central role of the brain. Dialogues Clin Neurosci 8:367–381

Meyer-Lindenberg A, Buckholtz JW, Koachana B, Hariri A, Pezawas L, Blasi G, Wabnitz A, Honea R, Verchinski B, Callicott JH, Egan M, Mattay V, Weinberger DR (2006) Neural mechanisms of genetic risk for impulsivity and violence in humans. Proc Natl Acad Sci USA 103(16):6269–6274

Miller GM, Bendor J, Tiefenbacher S, Yang H, Novak MA, Madras BK (2004) A mu-opioid receptor single nucleotide polymorphism in rhesus monkey: association with stress response and aggression. Mol Psychiatry 9(1):99–108

Nelson EC, Agrawal A, Pergadia ML, Wang JC, Whitfield HB, Saccone FS, Kern J, Grant JD, Schrage AJ, Rice JP, Montgomery GW, Heath AC, Goate AM, Martin NG, Madden PA (2010) H2 haplotype at chromosome 17q21.31 protects against childhood sexual abuse-associated risk for alcohol consumption and dependence. Addict Biol 15:1–11

Newman TK, Parker CC, Suomi SJ, Goldman D, Barr CS, Higley JD (2009) DRD1 5'UTR variation, sex and early infant stress influence ethanol consumption in rhesus macaques. Genes Brain Behav 8:626–630

Newman TK, Syagailo YV, Barr CS, Wendland JR, Champoux M, Graessle M, Suomi SJ, Higley JD, Lesch KP (2005) Monoamine oxidase a gene promoter variation and rearing experience influences aggressive behavior in rhesus monkeys. Biol Psychiatry 57:167–172

Ostner J, Neistermann M, Schulke O (2011) Male competition and its hormonal correlates in Assamese macaques (*Macaca assamensis*). Horm Behav 59:105–113

Overall KL (1997) Clinical behavioral medicine for small animals. Mosby, St Louis

Pang G, Wang J, Wang Z, Goh C, Lee C (2009) The G allele of SNP E1/A118G at the μ-opioid receptor gene locus shows genomic evidence of recent positive selection. Pharmacogenomics 10:1101–1109

Ptacek et al (2011) Dopamine D4 receptor gene DRD4 and its association with psychiatric disorders. Med Sci Monit 17(9):RA215–RA220

Raine A (2013) The anatomy of violence: the biological roots of crime. Pantheon Books, New York

Ramchandani VA, Umhau J, Pavon FJ, Ruiz-Velasco V, Margas W, Sun H, Damadzic R, Eskay R, Schoor M, Thorsell A, Schwandt ML, Sommer WH, George DT, Parsons LH, Herscovitch P, Hommer D, Heilig M (2011) A genetic determinant of striatal dopamine response to alcohol in men. Molec Psychiatry 16:809–817

Ray LA, Hutchison KE (2004) A polymorphism of the mu-opioid receptor gene (*OPRM1*) and sensitivity to the effects of alcohol in humans. Alcohol Clin Exp Res 28:1789–1795

Risch N, Herrell R, Lehner T, Liang KY, Eaves L, Hoh J, Griem A, Kovacs M, Ott J, Merikangas KR (2009) Interaction between the serotonin transporter gene, stressful life events and risk of depression: a meta-analysis. J Am Med Assoc 301:2462–2471

Sabol SZ, Hu S, Hamer D (1998) A functional polymorphism in the monoamine oxidase a gene promoter. Hum Genet 103:273–279

Sapolsky RM (2001) Physiological and pathophysiological implications of social stress in mammals. In: McEwen BS, Goodman HM (eds) Handbook of physiology; Section 7: the endocrine system, vol IV. Coping with the environment: neural and endocrine mechanisms. Oxford University Press, New York

Schneider ML, Moore CF, Larson JA, Barr CS, Dejesus OT, Roberts AD (2010) Timing of moderate level prenatal alcohol exposure influences gene expression of sensory processing behavior in rhesus monkeys. Front Integr Neurosci 3:30–38

Schneider ML, Moore CF, Barr CS, Larson JA, Kraemer GW (2011) Moderate prenatal alcohol exposure and serotonin genotype interact to alter CNS serotonin function in rhesus monkey offspring. Alcohol Clin Exp Res 35(5):912–920

Schwandt ML, Lindell SG, Sjoberg RL, Chisholm KL, Higley JD, Suomi SJ, Heilig M, Barr CS (2010) Gene-environment interactions and response to social intrusion in male and female rhesus macaques. Biol Psychiatry 67(4):323–330

Schwandt ML, Lindell SG, Higley JD, Suomi SJ, Heilig M, Barr CS (2011) Barr OPRM1 gene variation influences HPA axis function in response to a variety of stressors in rhesus macaques. Psychoneuroendocrinology (Epub ahead of print)

Shimmin LC, Natarajan S, Ibarguen H, Montasser M, Kim DK, Hanis CL, Boerwinkle E, Wadhwa PD, Hixson JE (2007) Corticotropin releasing hormone (CRH) gene variation: comprehensive resequencing for variant and molecular haplotype discovery in monosomic hybrid cell lines. DNA Seq 18(6):432–442

Sjoberg RL, Ducci F, Barr CS, Newman TK, Dell'Osso L, Virkkunen M, Goldman D (2008) A non-additive interaction of a functional MAOA VNTR and testosterone predicts antisocial behavior. Neuropsychopharmacol 33(2):425–430

Smith MJ, Price GR (1973) The logic of animal conflict. Nature 246:15–18

Smoller JW, Rosenbaum JF, Biederman J, Kennedy J, Dai D, Racette SR, Laird NM, Kagan J, Snidman N, Hirshfeld-Becker D, Tsuang MT, Sklar PB, Slaugenhaupt SA (2003) Association of a genetic marker at the corticotropin-releasing hormone locus with behavioral inhibition. Biol Psychiatry 54(12):1376–1381

Sobolewski JE, Brown JL, Mitani JC (2013) Female parity, male aggression and the challenge hypothesis in wild chimpanzees. Primates 54:81–88

Soeby K, Larsen SA, Olsen L, Rasmussen HB, Werge T (2005) Serotonin transporter: evolution and impact of polymorphic transcriptional regulation. Am J Hum Genet 136:53–57

Spinelli S, Schwandt ML, Heilig M, Suomi SJ, Higley JD, Barr CS (2012) *HTTLPR* genotype is associated with behavioral response to repeated stress in infant rhesus macaques. Dev Psychopathol 24:157–165

Spinelli S, Schwandt ML, Lindell SG, Newman TK, Heilig M, Higley JD, Suomi SJ, Goldman D, Barr CS (2007) Association between the rh-*5HTTLPR* polymorphism and behavior in rhesus macaques during social separation stress. Dev Psychopathol 19(4):977–987

Suomi SJ (1982) Abnormal behavior in nonhuman primates. In: Fobes JD, King JE (eds) Primate behavior. Academinc Press, New York

Takahashi A, Quadros IM, de Almeida RMM, Miczek KA (2012) Behavioral and pharmacogenetics of aggressive behavior. Curr Top Behav Neurosci 12:73–138

Trefilov A, Berard J, Krawczak M, Schmidtke J (2000) Natal dispersal in rhesus macaques is related to serotonin transporter gene promoter variation. Behav Genet 30:295–301

Vallender EJ, Priddy CM, Hakim S, Yang H, Chen GL, Miller GM (2008) Functional variation in the 3' UTR of the serotonin transporter in human and rhesus macaque. Genes Brain Behav 7:690–697

van Lawick-Goodall J (1968) The behavior of free-living chimpanzees in the gombe stream reserve. Anim Behav Monogr I:161–311

Wendland JR, Hampe M, Newman TK, Syagailo Y, Meyer J, Schempp W, Timme A, Suomi SJ, Lesch KP (2006) Structural variation of the monoamine oxidase A gene promoter repeat polymorphism in nonhuman primates. Genes Brain Behav 5:40–45

Wagner U, Wahle M, Moritz F, Hantzschel H, Baerwald CG (2006) Promoter polymorphisms regulating corticotrophin-releasing hormone transcription in vitro. Horm Metab Res 38(2):69–75

Wendland JR, Lesch KP, Newman TK, Timme A, Gachot-Neveu H, Thierry B, Suomi SJ (2005) Differential functional variability of serotonin transporter and monoamine oxidase A genes in macaque species displaying contrasting levels of aggression-related behavior. Behav Genet 30:1–10

Williamson DE, Coleman K, Bacanu S, Devlin BJ, Rogers J, Ryan ND, Cameron JL (2003) Heritability of fearful-anxious endophenotypes in infant rhesus macaques: a preliminary report. Biol Psychiatry 53:284–291

Wingfield JC, Hegner RE, Dufty AM, Ball GF (1990) The challenge hypothesis- theoretical implications for patterns of testosterone secretion, mating systems and breeding strategies. Am Nat 136:829–846

Wood W, Eagly AH (2002) A cross-cultural analysis of the behavior of women and men: implications for the origins of sex differences. Psychol Bull 128:699–727

Yuan Q, Zhou Z, Lindell SG, Higley JD, Ferguson E, Thompson RC, Lopez JF, Suomi SJ, Baghal B, Baker M, Mash DC, Barr CS, Goldman D (2012) The rhesus macaque is three times as diverse but more closely equivalent in "damaging" coding variation as compared to the human. BMC Genet 13:52

Zhang H, Luo X, Kranzler HR, Lappalainen J, Yang BZ, Krupitsky E, Zvartau E, Gelernter J (2006) Association between two μ-opioid receptor gene (OPRM1) haplotype blocks and drug or alcohol dependence. Hum Mol Genet 15:807–819

The Glucocorticoid/Aggression Relationship in Animals and Humans: An Analysis Sensitive to Behavioral Characteristics, Glucocorticoid Secretion Patterns, and Neural Mechanisms

József Haller

Abstract Glucocorticoids control a wide array of biological processes from glucose homeostasis to neuronal function. The mechanisms mediating their effects are similarly varied and include rapid and transient nongenomic effects on calcium trafficking, various neurotransmitter receptors, and other membrane/cytoplasmic proteins, as well as slowly developing but durable genomic effects that are mediated by a large number of glucocorticoid-sensitive genes that are affected after variable lag-times. Given this complexity, we suggest that the aggression/glucocorticoid relationship cannot be reduced to the simple "stimulation/inhibition" question. Here, we review the effects of glucocorticoids on aggression by taking into account the complexities of glucocorticoid actions. Acute and chronic effects were differentiated because these are mediated by different mechanisms. The effects of chronic increases and decreases in glucocorticoid production were discussed separately, because the activation of mechanisms that are not normally activated and the loss of normal functions should not be confounded. Findings in healthy/normal subjects and those obtained in subjects that show abnormal forms of behavior or psychopathologies were also differentiated, because the effects of glucocorticoids are indirect, and largely depend on the properties of neurons they act upon, which are altered in subjects with psychopathologies. In addition, the conditions of glucocorticoid measurements were also thoroughly evaluated. Although the role of glucocorticoids in aggression is perceived as controversial by many investigators, a detailed analysis that is sensitive to glucocorticoid and behavioral measure as well as to the mediating mechanism suggests that this role is rather clear-cut; moreover, there is a marked similarity between animal and human findings.

Keywords Aggression · Psychopathology · Glucocorticoids · Animals · Humans

---

J. Haller (✉)
Institute of Experimental Medicine, 67, Budapest 1450, Hungary
e-mail: haller@koki.hu

Curr Topics Behav Neurosci (2014) 17: 73–109
DOI: 10.1007/7854_2014_284

Published Online: 11 February 2014

Contents

1 Introduction: A Brief Overview of Glucocorticoid Action

Glucocorticoid hormones, as the name suggests, serve to control blood glucose, have a role in a series of bodily processes like inflammation and immunity, and also affect neurons by a variety of mechanisms. The diversity of glucocorticoid actions shows by itself that the glucocorticoid/aggression relationship cannot be simple. Three aspects of glucocorticoid action seem relevant for aggression research: secretion profiles, mechanisms of action, and the link between neural effects and behavioral control.

1.1 Secretion Profiles

Glucocorticoid secretion shows a variety of autonomous rhythms, is acutely responsive to challenging situations, and shows long-term alterations in response to adverse events. Autonomous rhythms include the awakening response, a surge in glucocorticoid secretion at the beginning of the active phase of the day (light phase in people, dark phase in laboratory rodents). In a recent review, Kudielka and Wüst (2010) argued that the awakening response is the most appropriate measure of HPA-axis basal activity. This glucocorticoid measure is less "contaminated" by accidental stressors, and is a rather uniform response to the sleep–wake transition, which means that its individual variability is rather small under normal conditions. Overall, glucocorticoid levels remain high after awakening, but show rhythmic variations because glucocorticoid secretion is pulsatile (Follenius et al. 1987). Several glucocorticoid secretion rhythms were identified from ultradian ones that have a period of several hours to circahoral rhythms that have a period of about 1 h (Stupfel and Pavely 1990). The amplitude of rhythmic changes is considerable, and due to environmental influences, such rhythms are not synchronized and are not truly periodic. Importantly, ultradian/circahoral fluctuations are clearly detectable when glucocorticoids are measured in the saliva (Trifonova et al. 2012). The existence of secretion rhythms increases individual variability when point measurements are performed. In addition, rapid glucocorticoid responses to even minor stressors and the role of glucocorticoids in plasma glucose control (e.g., in relation with eating) further increase variability. The relative importance of such confounding factors likely increases toward the end of the day when subjects have experienced multiple stressful events and meals, the timing of which is difficult to control. Finally, plasma glucocorticoid levels markedly decrease in the inactive period. The behavioral relevance of glucocorticoids secreted in this period is poorly understood; yet, one cannot exclude the possibility that these have an impact on behavioral traits via genomic mechanisms. Overnight plasma glucocorticoid levels have rarely if ever been studied with respect to behavioral characteristics, except for 24-h urine samples that include glucocorticoids secreted during the night. Taken together, glucocorticoid secretion shows multiple autonomous rhythms that complicate the obtaining of valid glucocorticoid measures.

Two aspects of acute HPA-axis responses to environmental challenges will be discussed here. First, aggression is a powerful stressor that easily overwhelms autonomous rhythms (Koolhaas et al. 1997), so studying acute interactions between aggressive acts and glucocorticoid stress responses is considerably easier than establishing the interaction between oscillating basal levels and aggression-related traits. On the other hand, the HPA-axis responds promptly to even weak stressors; moreover, the mere anticipation of a stressor is sufficient to induce a stress response (Gaab et al. 2005; Tornatzky and Miczek 1994). This aspect of the stress response should be considered when basal levels are being established. Acute stress responses are usually transient and plasma cortisol rapidly returns to basal levels after the termination of the stressor. However, the HPA-axis remains chronically activated when the stressor is recurrent or persistent. The chronic stress response is well known and will not be discussed here. The possibility that the activity of the HPA-axis can be chronically *lowered*—together with the assumption that this condition leads to psychopathology—was recognized by a study on violent antisocially disordered people (Virkkunen 1985), and received much attention in subsequent studies on abnormal forms of aggression (see Sect. 3.3). While the conditions that chronically activate the HPA-axis are well known, this is considerably less true of the factors that have the opposite effect. This endocrine condition is generally attributed to early adversities (Caldji et al. 2000; Gunnar and Vazquez 2001; Saxbe et al. 2012; van Goozen et al. 2007). The same influences usually elicit chronic stress responses in most people, and only a minority of them develops the aforementioned reduction in HPA-axis function (Gunnar and Vazquez 2001). The distinctive features of this subgroup are unknown. Genetic predispositions, epigenetic phenomena triggered by gene–environment interactions, and the functional impairment of the adrenal cortex after-birth complications or other dramatic stressors are suspected, but unproven factors of developmental trajectories associated with chronic HPA-axis hypofunction (Rutter et al. 2006).

1.2 Mechanisms of Action

The mechanisms activated by glucocorticoids are so numerous and varied that they cannot be discussed here in detail. Overall, two basic types of mechanisms have been identified: rapidly developing and transient effects mediated by a variety of cellular mechanisms, and more slowly developing but persistent effects mediated by the genome. Both categories of mechanisms (i.e., nongenomic and genomic) are highly varied. The nongenomic mechanisms activated by glucocorticoids endow them with the ability to induce rapid and transient changes in neuronal function. Such rapid effects are achieved by stimulating the voltage-dependent Ca^{2+} channel and calcium trafficking in general, by binding to and affecting the function of various neurotransmitter receptors, by binding to various membrane and cytoplasmic proteins, and by affecting protein–protein interactions within the neuron (for recent reviews, see de Kloet et al. 2008a; Groeneweg et al. 2011; Haller et al. 2008). In regard to genomic effects, it was found that the number of glucocorticoid responsive elements in the genome is very high (Datson et al. 2008), glucocorticoids also regulate gene expression post-transcriptionally (Ing 2005), and genomic effects are expressed after a variable lag-time, some genomic effects becoming detectable within an hour, while others need days, weeks, and even months of elevated glucocorticoid secretion to emerge (de Kloet et al. 2008a).

Beyond the large differences in the timing of effects and the mechanisms activated, the two types of effects also differ in their ultimate outcome. Rapid nongenomic effects are behavior-activating in nature, and eventually aim to control the critical situation that triggered the stress response (de Kloet et al. 2008b; Koolhaas et al. 2011; Haller et al. 1998a, b). As an example, glucocorticoids promote aggression under conditions of social challenge (see Sect. 2.1.1). Typically, genomic effects are activated by persistent stressors; in such cases, the mere fact that the stress response did not vanish implicitly demonstrates that the subject was unsuccessful in controlling the stressful situation. Overall, such effects are behavior-depressing in nature, and promote survival by limiting risks and costs associated with behavioral activity (Haller et al. 1998a). As an example, chronic social stress (e.g., defeat and low social status) decrease aggressiveness and promote submission (see Sect. 2.1.2), by which the animal avoids unnecessary losses of energy and the danger of harm until conditions become more favorable. Given the time frame of the consequences, the mechanisms mediating the effects of chronic glucocorticoid *deficits* are likely genomic in nature. For example, abnormal features of aggression developed after chronic, but not acute, inhibition of glucocorticoid secretion in animals, while in humans, chronically low basal cortisol levels predict the worsening of aggression-related problems years in advance (see Sects. 2.2 and 3.3.1). Besides likely being genomic in nature, very little is known about the particular mechanisms mediating the effects of glucocorticoid hypofunction. In general, one can hypothesize that this condition involves a deficit in an important genomic controlling factor, which alters a variety of brain mechanisms including serotonergic neurotransmission (Haller et al. 2005b).

1.3 The Link Between Neural Effects and Behavioral Control

None of the vast variety of glucocorticoid mechanisms involves the direct activation of neurons or neuronal circuits, which raises questions regarding the link between nonspecific neural effects and specific behavioral alterations. The most likely explanation is spatiotemporal: glucocorticoids shape the properties of neurons, the consequences of which are felt when and where particular neurons are activated (Joels and de Kloet 1992). For example, glucocorticoids rapidly and nongenomically facilitate the function of voltage-dependent calcium channels (Sze and Iqbal 1994); as a consequence, neuronal circuits activated in a particular situation are facilitated, while nonactivated circuits remain unaltered. Similar interactions between neuronal activation and glucocorticoid action explain why glucocorticoids promote aggressive behavior when animals are socially challenged, but have no effect on aggression under nonchallenging conditions (Mikics et al. 2007; see Sect. 2.1.1). It also explains why the awakening response is not associated with an unconditional increase in aggressiveness. In the absence of a challenge, glucocorticoids do not affect circuits that control challenge responding. Similarly, chronic glucocorticoid deficits down-regulate serotonergic neurotransmission (Haller et al. 2005a), the consequences of which are manifested during conflicts, when the lack of serotonergic control results in a disinhibited aggressive behavior. A weakened controlling mechanism has no effect on aggression when there is no aggression to be controlled.

1.4 Brief Overview and the Structure of the Review

The findings and considerations briefly reviewed above demonstrate that the secretion patterns of, and the mechanisms activated by glucocorticoids preclude simple answers to the question of how the propensity to behave aggressively depends on glucocorticoid secretion. For instance, one cannot expect similar effects from acute surges in glucocorticoid production and chronic HPA-axis hyperactivations, because the mechanisms activated by the two types of glucocorticoid secretion patterns are largely different. In the same vein, the effects of chronic glucocorticoid excesses and deficits, although likely different, are not necessarily opposite; chronic glucocorticoid excesses may activate neural mechanisms that are not normally activated, while glucocorticoid deficits may result in the loss of functions that are normally present. Given, the dependence of glucocorticoid effects on concurrent neuronal activation patterns, one can also expect that these effects are altered by psychopathological states which involve marked changes in brain function. In individuals with psychopathologies, effects emerge from an interaction between glucocorticoids and altered brain functions.

In Sect. 2, we will review the available findings by taking into account the issues discussed above as follows:

(1) Acute and chronic effects of glucocorticoids will be differentiated because these are mediated by different mechanisms.
(2) The effects of chronic increases and decreases in glucocorticoid production will be discussed separately, because the activation of mechanisms that are not normally active and the loss of normal functions should not be confounded.
(3) We will also differentiate findings in healthy/normal subjects from those obtained in subjects that show abnormal forms of behavior or psychopathologies, since the effects of glucocorticoids are indirect and are therefore dependent on neuronal background that may be altered in such individuals.

In addition, the methods and the timing of blood sampling for glucocorticoid measurements will also be considered.

2 Findings in Animals

2.1 *Species-Typical Aggression*

2.1.1 Acute Interactions Between Glucocorticoids and Aggressive Behavior

It was recognized long ago that aggression is stressful and leads to glucocorticoid production in a variety of species (fish: Earley et al. 2006; lizards: Woodley et al. 2000; mice: Bronson and Eleftheriou 1965; quail: Ramenofsky 1985; rats: Schuurman 1980). In the resident-intruder test, the increase in glucocorticoids was similar in the two contestants at the beginning of the encounter. However, once dominance relationships were established plasma corticosterone rapidly returned to basal levels in the winner, while remaining high in the loser.

Disparate early findings suggested that corticosterone secreted in response to the encounter, might have an impact on ongoing behavior, since ACTH and corticosterone treatments increased aggressiveness while inhibiting corticosterone synthesis decreased it (Brain et al. 1971; Bronson and Eleftheriou 1965; Munro and Pitcher 1985; Mainardi et al. 1987). Similar findings were obtained later in fish, where aggressiveness tightly correlated with pre-encounter corticosterone levels, and acute treatment with glucocorticoid receptor antagonists decreased aggressiveness (Chang et al. 2012; Schjolden et al. 2009). In two parallel studies, Hayden-Hixon and Ferris (1991a, b) identified certain subregions of the hypothalamus as the main sites controlling corticosterone effects on attack behavior. Since the hypothalamus is tightly involved in the control of biting attacks in all species studied so far (see Haller 2012 for a review), these findings suggested that corticosterone directly influences the brain mechanisms that control attacks.

The existence of a "positive feedback loop" between aggression and HPA-axis activation was challenged in the early 1980s when the genomic mechanisms of glucocorticoid action were discovered (see above). The general assumption at that time was that, glucocorticoids exert their effects only through genomic mechanisms. As these mechanisms need about 30 min to take effect, and the actual increase in glucocorticoids takes about 5 min more, it was assumed that glucocorticoids cannot influence the aggressive acts that triggered their release, because by the time plasma glucocorticoids increased and the genomic mechanisms were activated, most of the aggressive acts of dyadic encounters had already terminated. As a result, there remained nothing to be influenced by glucocorticoids other than submission, i.e., the consequence of losing the encounter. This belief was reflected in a series of studies, even in those where glucocorticoid synthesis blockade clearly decreased aggressiveness, and glucocorticoids increased aggressiveness (Munro and Pitcher 1985). Later on, a series of studies demonstrated that the rapid effects of glucocorticoids are real and several nongenomic mechanisms were also identified (de Kloet et al. 2008a; Groeneweg et al. 2011; Haller et al. 2008).

In a series of studies on the rapid interactions between glucocorticoids and aggression, we showed first that psychosocial encounters alone are sufficient to increase plasma glucocorticoids, that is, mere sensory contact with opponents can activate the HPA-axis (Haller et al. 1995). Similar findings were obtained in feral animals as well (Landys et al. 2010; Silverin 1993). Since most animal-aggressive encounters are preceded by a series of nonsocial and social behaviors that can be perceived as preparations for fights but are not aggressive themselves, this finding demonstrated that glucocorticoids increase before the actual fights begin. We then proceeded to show that natural variations in glucocorticoid synthesis, e.g., those related to ultradian and diurnal variations, are paralleled by changes in aggressiveness when rats are challenged in the resident/intruder test (Haller et al. 1988b, 2000a, b). Particularly, rats were significantly more aggressive in the increasing phase of their corticosterone fluctuation than counterparts in the decreasing phase. Moreover, glucocorticoid enhancement-induced aggressiveness was normalized when the natural increase in glucocorticoid secretion was counteracted by glucocorticoid synthesis blockade or by glucocorticoid receptor antagonists. As the time frame of these hormonal and behavioral changes was less than 20 min, these findings indirectly supported the notion that glucocorticoids rapidly promote aggressiveness via nongenomic mechanisms. In the next step, we studied the interaction between glucocorticoid secretion and biting attacks elicited by the electrical stimulation of the hypothalamic attack area (Kruk et al. 2004). We found that the stimulation of the attack area dramatically increased corticosterone secretion, while corticosterone injection decreased the electric threshold of attacks, i.e., facilitated the behavioral response. This suggested that there was a fast positive feedback loop between the adrenocortical stress response and brain mechanisms involved in aggressive behavior, the time frame of this interaction being approximately 10 min. Finally, we demonstrated that aggressive behaviors that had been suppressed by corticosterone synthesis inhibition returned to normal within 2 min of corticosterone administration (Mikics et al. 2004). This fast effect of corticosterone was resistant to protein synthesis inhibition, clearly demonstrating both the rapid aggression-promoting effect of glucocorticoids and the nongenomic nature of the mediating mechanism.

Taken together, the findings reviewed above show that (i) the HPA-axis is activated by encounters with unfamiliar individuals even before the actual fights start and (ii) glucocorticoids secreted in response to the encounter rapidly promote aggressiveness. Actually, aggressiveness is expressed concomitantly and with the support of glucocorticoids. In a subsequent study we showed that corticosterone does not affect aggressiveness in established colonies of rats, where the general level of aggressive behavior is low (Mikics et al. 2007). Thus, glucocorticoids rapidly and nongenomically promote aggressiveness under conditions of a social challenge.

2.1.2 Chronic Effects of Glucocorticoids on Aggressiveness

While acute surges in glucocorticoids promote aggressiveness in socially challenged animals, chronic increases in glucocorticoids produce the opposite effect.

As shown above, an aggressive encounter is stressful for both contestants; however, after the encounter plasma glucocorticoids rapidly decrease in the winning animals, but remain elevated in losers (fish: Höglund et al. 2002; lizards: Summers et al. 2003; mice: Bronson and Eleftheriou 1965; monkeys: Yodyingyuad et al. 1985; rat colonies: Blanchard et al. 1995; rat dyadic encounters: Schuurman 1980; snakes: Schuett and Grober 2000). In an experiment involving opponents with contrasting levels of aggressiveness, we showed that the effects of losing do not depend on the prefight status of individuals; plasma corticosterone levels were rapidly reduced in intruders that won the encounter, while the opposite happened in residents that were defeated in their home-cage (Haller et al. 1996).

Elevated levels of glucocorticoids decrease aggression and promote submissiveness; this has been demonstrated in a variety of species treated with glucocorticoids for prolonged periods (from 3 days to 3 weeks) (reduced aggression in birds: Meddle et al. 2002; decreased aggression in fish: Overli et al. 2002; increased submissiveness in fish: DiBattista et al. 2005; competitive disadvantage/reduction of home-range in lizards: DeNardo and Sinervo 1994; decreased aggressiveness in lizards: Tokarz 1987; submission in mice: Leshner et al. 1980; reduced aggression and increased submissiveness in rats: Hayden-Hixson and

Ferris 1991). Similar findings were obtained in experiments where subjects were exposed to stressors. Prolonged and strong single stressors (e.g., restraint for 2–6 h) or chronic social and nonsocial stressors decreased aggressiveness (Blanchard et al. 1995; Corum et al. 1977; Wood et al. 2003; Yohe et al. 2012). The effects of postencounter increases in glucocorticoids appear to be very strong and not necessarily related to the subjective experience of losing; rats showed increased submissiveness if injected with corticosterone one day after an aggressive encounter (Timmer and Sandi 2010).

Taken together, these findings show that the long-lasting increases in circulating glucocorticoids that may result from defeat experience decrease aggressiveness and promote submissive behaviors.

2.1.3 Overview and Interpretation

A mechanistic interpretation of the findings summarized in the two preceding sections might suggest that glucocorticoids both promote and hinder aggression. In fact, however, they do neither. As shown above, glucocorticoids do not affect brain circuits directly, and are not able to elicit motor or social behaviors; rather, they affect the properties of neurons. It is unclear which of the rapid neuronal actions mediate the aggression-promoting effects of glucocorticoids. Nevertheless, the rapidity of such effects and their insensitivity to protein synthesis blockade shows that glucocorticoids induce rapid neural changes nongenomically, and this supports the expression of aggressive behavior (Mikics et al. 2004). The finding that this effect is bound to social challenge (Mikics et al. 2007), supports the hypothesis that glucocorticoids do not unconditionally activate brain mechanisms involved in aggression. One can hypothesize that the neuronal mechanisms activated by social challenges are rapidly affected by glucocorticoids such that the expression of aggression becomes more likely. The contrary seems valid for enduring or chronic glucocorticoid effects. Given the complexity of genomic mechanisms, it is practically impossible at present to speculate on the particular genomic mechanisms that contribute to the aggression-decreasing effects of prolonged HPA-axis hyperfunction. Nevertheless, the lag-time of these effects clearly demonstrates that they are mediated by genomic mechanisms, as nongenomic ones are transient and short-lived.

In functional terms, the aggression-promoting effects of glucocorticoids develop in the early phase of the social challenge, when the animal mobilizes its resources to control the situation. The aggression-decreasing effects of chronic HPA-axis activation develop when the chances of controlling the situation are lost; consequently, the decrease in aggression averts the dangers that aggression involves (e.g., loss of energies and physical harm). In this context, both effects of glucocorticoids are functional, and serve the "interests" of the individual. The transition between the two phases of glucocorticoid action has received little attention so far. Disparate findings suggest that the very early genomic effects of glucocorticoids still promote aggression; glucocorticoids administered 20 min before a 20-min long aggressive encounter promoted aggressiveness, but at this time point (in contrast to earlier ones, see above) the effect was sensitive to protein synthesis inhibition (Mikics et al. 2004). On the other hand, strong stressors inhibit aggression within several hours (Wood et al. 2003; Yohe et al. 2012), suggesting that the aggression-promoting genomic mechanisms relatively rapidly give way to mechanisms that inhibit aggression.

The conclusions formulated above derive from experiments performed under highly controlled conditions. By and large, field experiments involving free-ranging animals or animals living in semi-natural conditions support the conclusions deriving from these controlled experiments. Contradictory findings, however, do exist, as the effects of glucocorticoids on aggression may depend on environmental conditions, reproductive status, social rank, etc., (Summers et al. 2005). We stress here again, that glucocorticoids have no effects on their own; they rather influence processes that are activated by other mechanisms, i.e., particular effects result from the coincidence of several processes. However, the findings demonstrate that under controlled conditions, when interactions from other factors are minimized, glucocorticoids are rapidly secreted in response to social challenges and feed back on their cause by promoting aggressive interactions. When the aggressive encounter is lost or the animal is under the extended influence of another stressor, glucocorticoids decrease aggression. Both responses are functional, as glucocorticoids acutely promote behavior that can be considered an appropriate response to the social challenge, but when the fight is lost (or the subject is under the effect of chronic stress), glucocorticoids prevent useless energy loss and harm by decreasing aggressiveness.

2.2 Abnormal Aggression in Animals

From a human perspective, the control of abnormal, pathological aggression (see below) is critically important. This process may not be adequately modeled by experiments performed under routine conditions, since aggression is integral part of the species-typical, adaptive behavioral repertoire of animals; it ensures access to essential resources (territory, food, etc.), and confers considerable reproductive advantage. So, in animals, the lack of aggression is more of a problem than its existence; in other words, behaviors studied in regularly performed tests of aggression are normal and functional.

Recently, some methods for studying abnormal forms of aggression in animals have been developed. Animal aggression may be considered abnormal if (i) it is elicited by stimuli or treatments that resemble the etiological factors of aggression-related psychopathologies and (ii) the aggressive attacks show one or more of the following features: (a) mismatch between provocation and response (short latency to attack, excessively high attack counts), (b) disregard of species-specific rules (decreased signaling of attacks, attacks on females or on vulnerable targets, attacks in inappropriate situations), (c) insensitivity toward the social signals of the opponent (sustaining attacks despite submissiveness on the other side), and (d) aggression is intermingled with behavioral agitation and/or defensiveness (offensive ambiguity) (Haller and Kruk 2006; Haller 2012). Thus, subjects in models of abnormal aggression are exposed to conditions or treatments that mimic etiological factors of aggression-related psychopathologies, and their behavior breaks species-specific "rules."

The first publication deliberately aimed at identifying abnormal forms of aggression (Haller et al. 2001) showed that low-level glucocorticoid replacement after adrenalectomy (ADXr) does not affect overall attack counts, but dramatically increases attacks on vulnerable body parts of opponents (head throat and belly), and decreases intention signaling in the form of aggressive threats. This model was based on findings in humans (detailed in Sect. 3) indicating that violent antisocially disordered people show chronically low levels of plasma cortisol. Subsequently, we showed that (1) hypoglucocorticoid status (i.e., ADXr) results in behavioral dysfunctions that are similar to those seen in antisocially disordered people in multiple ways (Haller et al. 2004; Haller and Kruk 2006), and (2) the control of this type of aggression is qualitatively different from that of normal rivalry aggression. Briefly, abnormal aggression was associated with complex disturbances at the level of the prefrontal cortex, dramatically altered patterns of amygdala activation and lead to the loss of serotonergic control of aggression (Halasz et al. 2002, 2006; Haller et al. 2005b, 2007). More recently, we showed that, in contrast to species-typical intermale aggression, aggression in glucocorticoid-deficient rats activates brain regions known to control predatory behavior, further outlining the qualitative differences between species-typical aggression and in rats with reduced glucocorticoid function (Tulogdi et al. 2010). Importantly, the effects of acute decreases in plasma glucocorticoids are qualitatively different from those induced by a chronic glucocorticoid deficiency, and the effects of the latter can be abolished by repeated but not single glucocorticoid treatments (Haller et al. 2001, 2004). This suggests that abnormal aggressiveness in the hypoglucocorticoid model is mediated genomically. Particularly, the decreased control of gene expression profiles by glucocorticoids likely induced neural changes that altered species-typical patterns of aggression.

As a counterpart to the hypoglucocorticoid model, we also studied the effects of postweaning social isolation, a model of early social neglect,in which produces exacerbated stress responses and aggression-related psychopathologies in humans (Chapple et al. 2005) and increased aggressiveness in rats (Kikusui et al. 2006). We demonstrated that postweaning social isolation (1) does not alter basal glucocorticoid levels, but dramatically increases the HPA-axis response to social challenge; (2) increases attack counts (not seen in the hypoglucocorticoid model); (3) induces abnormal forms of aggression (attack on vulnerable targets, deficits in social communication, behavioral agitation, and increased defensiveness on the background of increased attacks; the latter two behavioral alterations were not seen in the hypoglucocorticoid model), and (4) overactivates the brain areas that are usually activated by rivalry aggression, but does not induce qualitative changes similar to those seen in the hypoglucocorticoid model (Toth et al. 2008, 2011,

2012). Although the specific role of glucocorticoids was not elucidated in this model, given the important role of acute glucocorticoid stress responses in aggressiveness (see Sect. 2.1.3), one can hypothesize that the exacerbated stress responses seen in these rats contributed to their escalated aggressiveness. Thus, glucocorticoid stress responses enhanced by early social deprivation may enhance aggressiveness through the same mechanisms that promote normal aggressiveness.

These findings suggest that both chronically decreased glucocorticoid levels (hypoglucocorticoid model) and chronically exacerbated stress responses (postweaning isolation model) lead to abnormal aggression in rats. The aggressive behavior elicited by the two models is qualitatively different and has a markedly different neural background. Unfortunately, basal glucocorticoid levels and stress responses were not studied using other models of abnormal aggression. In a recent review, however, we showed that models associated with hypoarousal (e.g., HPA-axis deficits) are similar in terms of neural control to our hypoglucocorticoid model, while models associated with hyperarousal (e.g., increased stress responses) are similar to our postweaning isolation model (Haller 2012). Intraspecies aggression can be elicited by electrical stimulation of the mediobasal hypothalamus. Our recent review of hypothalamic mechanisms of abnormal aggression found that the mediobasal area was over-activated in models that were associated with glucocorticoid hyperfunction. In addition, the serotonergic and vasopressinergic control of the hypothalamus showed the expected changes (decrease and increase, respectively). Such models included rats selected for low anxiety, those submitted to early maternal separation, and hamsters submitted to cocaine administration during adolescence (Neumann et al. 2010; Ricci et al. 2005; Veenema et al. 2006; see Haller 2012 for a detailed analysis). Electrical stimulation of more lateral parts of the hypothalamus can elicit predator aggression; this region was activated in models associated with glucocorticoid hypofunction, and serotonergic and vasopressinergic neurotransmission showed paradoxical changes. Such models included mice selected for aggressiveness, rats selected for high anxiety, and hamsters subjugated or treated with anabolic steroids during adolescence (Benus et al. 1991; Caramaschi et al. 2008; Delville et al. 1998; Melloni et al. 1996; Neumann et al. 2010; Veenema et al. 2004; see Haller 2012 for a detailed analysis). Importantly, glucocorticoids are essential for aggression in mice conditioned to respond for the opportunity to fight another mouse, another important model of abnormal aggression (Fish et al. 2005).

We propose that exacerbated glucocorticoid responses to social challenge predispose rodents to exacerbated levels of aggressiveness, while chronically low plasma glucocorticoids distort behavior, and lead to unpredictable and dangerous forms of aggression. The former results from the pro-aggressive effects of acute glucocorticoid stress responses, while the latter results from plastic neural changes induced by the diminished expression of an important genomic regulator mechanism.

3 Findings in Humans

Most studies evaluating the role of glucocorticoids in humans start with a sentence stating that findings on the glucocorticoid–aggression relationship are confusing. A cursory analysis of the available literature suggests that this statement is true: various studies have associated aggression with low, high and normal glucocorticoid levels. However, the studies differ widely in terms of how "aggression" was conceptualized and how glucocorticoids were measured; therefore, conflicting findings are not surprising. The following constitutes an attempt to make sense of the conflicting findings by focusing on details. Based on this analysis, we conclude that the roles of glucocorticoids identified in animal studies are not much different from those seen in people. The analysis provided below was based on 86 human studies found in PubMed by using multiple search terms. No study was excluded.

3.1 Conceptual and Methodological Issues

3.1.1 Aggressiveness

A typically phenomenological definition states that aggression is a behavior aimed at harming or injuring another living being (Berkowitz 1993, 1994). In these terms, the impact of glucocorticoids on human aggression was rarely if ever studied in its natural form, e.g., in conjunction with real fights. There are, however, a series of studies where surrogates of fights are studied, e.g., the delivery of punishments in competitive situations (money subtraction from, or electric shocks delivered to, real or faked opponents, respectively). Such tests are the closest human analogs of the resident-intruder test and will be called "manifest studies" because aggression is manifested while subjects are studied. In most studies, however, patterns of past behavior (e.g., delinquency records) were investigated. Such protocols in fact study behavioral or personality traits, not specific behaviors per se. These two approaches address different aspects of the glucocorticoid/aggression relationship. "Manifest studies" (e.g., the delivery of punishment) can reveal the effects of aggressive acts on glucocorticoid secretion and the impact of the latter on the former. Analogous animal studies were summarized in Sect. 2.1.1. The interaction between aggressive traits and glucocorticoid secretion patterns was not directly studied in animals. Assuming, however, that the resident/intruder test ultimately reveals traits (the propensity to behave aggressively), human and animal studies can still be compared. Analogous animal studies were summarized in the Sects. 2.1.2 and 2.2.

In addition to basic differences between "manifest" and "trait" studies, there are differences in the traits that were considered in the latter. Surprisingly, studies seldom use past aggressive behavior as a criterion for defining aggressive traits. In most studies, broad behavioral categories (e.g., externalizing behavior, delinquency) or personality or psychiatric disorders are considered (see below). Inferences on aggression are based on the association between broad behavioral categories or diagnoses and glucocorticoid secretion. However, neither externalizing behavior, delinquency nor other diagnoses refer exclusively to aggressive acts. In the studies reviewed below, "externalizing behaviors" included defiance, impulsivity, inattention, hyperactivity, and theft as well as aggression. Similarly, "delinquency" involved nonaggressive acts like theft, violating firework regulations, substance abuse, etc. Psychopathologies do not necessarily involve aggressiveness either. For example, according to DSM-IV antisocial personality disorder can be diagnosed when three out of the seven symptoms of Cluster A are met. However, four of the symptoms are not related to aggression (deceitfulness, failure to plan ahead, irresponsibility, lack of remorse). The same holds true for other aggression-related psychopathologies. While people diagnosed with such disorders are *likely* to be aggressive, that is not necessarily true in all cases. Even when aggression is present, the level and type of aggressive behavior varies widely, which seriously limits the validity of conclusions derived from such studies. Fortunately, aggressive and nonaggressive subgroups were differentiated in some studies.

3.1.2 Glucocorticoids

As suggested already in Sect. 1.1, obtaining valid measures of HPA function is difficult. The reliability of point measurements is decreased by ultradian variations, which result in different values being obtained from the same individual over time (see Sect. 1). As these rhythms are not synchronized, they may also wash out group differences that are within the range of the oscillations. Diurnal variations cause similar problems if the time window of blood sampling is not standardized. Another problem results from stressors that subjects may be exposed to before blood sampling for glucocorticoid measurements. In many studies, blood samples were taken immediately upon arrival. As traveling is often stressful, stress induced rather than basal levels are measured in such studies. In addition, both the duration and and stressfulness of travel may vary by subject, which results in more variable levels of plasma glucocorticoids upon arrival. Such environmental influences can considerably blur the picture.

There are various ways to overcome these difficulties—blood samples may be taken after a short relaxation period, during which glucocorticoid levels return to baseline. This procedure, however, is not employed consistently by researchers. As a blatant example, patients filled out psychometric inventories during the "relaxation period" in one of the studies, while filling out inventories was considered a stressor in another study. And even if performed correctly, the relaxation procedure does not eliminate the effects of natural rhythms.

One useful procedure involves repeated blood sampling, either by establishing glucocorticoid secretion curves over a limited period (e.g., hours) or by averaging levels measured on different days or even on different years. The former approach does not eliminate uncertainties deriving from rhythms, but appropriately handles the problem of travel- and procedure-related stress. The latter approach identifies long-term differences in glucocorticoid levels that are more likely to be correlated with behavioral traits.

Total corticosterone production may also be measured from urine collected over 24 h. While this approach overcomes many difficulties that were discussed above, it ignores particular patterns of diurnal oscillations (e.g., flattened diurnal rhythms) that may be indicative of hormonal disturbances even in the absence of changed total production.

In a review analyzing the problems of glucocorticoid measurements, Kudielka and Wüst (2010) concluded that the best overall measure of HPA-axis function is the awakening response. In addition to this measure being relevant for a variety of psychiatric disorders, it also eliminates the problem of ultradian and diurnal oscillations, as awakening increases glucocorticoid levels invariably. Again, however, this approach fails to consider the diurnal secretion patterns of glucocorticoids, which are relevant to glucocorticoid effects over the day.

We do not want to create the impression that glucocorticoid status cannot be established; most procedures are meaningful, but one must be aware of, and consider the shortcomings of, each procedure. Overall, awakening responses, glucocorticoid secretion "traits," secretion curves and total glucocorticoid production scores seem to be the best approaches.

3.2 Healthy Controls

3.2.1 Acute Interactions Between Glucocorticoids and Aggressive Behavior

The delivery of punishment to imaginary opponents in a competitive situation increases plasma cortisol in subjects. Plasma cortisol levels were increased by both the Taylor Aggression Paradigm, where subjects delivered punishment to opponents in the form of noises, the intensity and duration of which were set by the subject (Böhnke et al. 2010a), and the Point Subtraction Aggression Paradigm, where subjects can subtract points from opponents as a form of punishment in reaction to a loss in gains (Gerra et al. 1997, 2001a, b, 2004, 2007; Geniole et al. 2011). Interestingly, the cortisol response did not depend on the simulated social inclusion or exclusion of subjects (Geniole et al. 2011). These findings suggest that in these paradigms—the closest available human experimental analogs of the rodent resident-intruder test—the responses of animals and humans are similar, and consist of a rapid increase in HPA-axis function. In contrast to animals, the impact of winning and losing were not studied in humans, as the chances of winning were standardized in these studies. Thus, this aspect of the glucocorticoid/aggression relationship is unknown in humans at present.

Laboratory models of human aggression were preceded by glucocorticoid administration or stress exposure in two studies. Böhnke et al. (2010b) treated subjects with cortisol before the Taylor Aggression Paradigm and found that cortisol increased aggressive responses, but, interestingly, only in girls. In the study by Verona and Kilmer (2007), subjects were stressed by an air blast before they were allowed to deliver shocks to fictitious employees. The pretest stressor increased the aggressive response in males, but not in females. These two studies may be perceived as contradictory because of the gender differences observed. Yet, one has to consider that the cortisol dose was low and the stressor was mild, and endogenous cortisol production was not suppressed. In addition, the subjects were inexperienced in the particular aggression-related task. In analogous animal studies (i.e., when endogenous glucocorticoid secretion was not suppressed and the subjects lacked fighting experience) glucocorticoids promoted locomotion without affecting aggression in rats and fish, despite the fact that the inhibition of glucocorticoid secretion or the blockade of glucocorticoid receptors decreased aggression in both species (Haller et al. 1997; Schjolden et al. 2009). When compared to analogous animal studies, the aggression-promoting effects of cortisol and stress appear rather strong in the two human studies. We suggest that the contrasting gender differences were due to small sample sizes, the lack of control over endogenous cortisol secretion, and the subjects' inexperience.

There are several studies where the time interval between the aggressive act and glucocorticoid measurement was not as close as in the procedures discussed above, but where a reciprocal facilitating role of aggression and glucocorticoid production seems likely. Amphetamine challenges increased plasma glucocorticoids more strongly in aggressive than in nonaggressive healthy subjects (White et al. 2006). Similarly, aggressive children showed increased cortisol responses to film viewing (Ivanov et al. 2011). These studies indicate that healthy subjects with a more reactive HPA-axis are also more aggressive, suggesting that aggressive children show increased HPA-axis responses in aggressive situations as well. In other studies, cortisol levels were positively correlated with the number of aggressive acts performed in the following hours (Blair et al. 2005; Gunnar et al. 2010; Murray-Close et al. 2008). These correlations seem to be related to physical aggression, as relational aggression showed the opposite correlation (Murray-Close et al. 2008). Also, the relationship may be valid for aggression between peers, as acute changes in cortisol were not associated with aggressive parent–child interactions (O'Neal et al. 2010).

3.2.2 Chronic Interactions Between Glucocorticoids and Aggressive Behavior

In animals, defeat is followed by a long-lasting increase in plasma glucocorticoids, as well as decreased aggression and increased submissive behavior. Defeat-independent durable stressors and chronic glucocorticoid administrations had a similar effect.

The impact of defeat on plasma cortisol has been poorly studied in humans, and the long-term consequences are even less known. Nevertheless, a few studies performed in athletes suggest that losers have higher postcompetition cortisol levels than winners (Parmigiani et al. 2009), and a similar long-term difference was seen between successful and less successful competitors (Booth et al. 1989). Defeat-induced increases in cortisol interact with testosterone and are associated with decreased dominance and reluctance to compete again (Mehta and Josephs 2010). These disparate findings suggest a similarity between animal and human findings as it regards the effects of losing.

Studies on the role of chronically elevated cortisol levels on human aggression are virtually nonexistent, probably because this condition has such a strong impact on mood and affect. Therefore, "pure" studies on aggression are difficult to run in such subjects because they likely show anxiety and/or depression as well. Although they cannot be considered "healthy" controls, they may still be considered in this section because their psychopathology is not directly related to aggression; in addition, animal subjects discussed in the analogous animal section likely showed similar changes in tests of anxiety and depression. Interestingly, a significant share of depressed patients show outbursts of aggression called anger attacks (Fava 1998; Painuly et al. 2011). Social phobics are also prone to excessive violence under social pressure, including threatening with weapons, assault, and murder (Alden and Taylor 2004; Casiano et al. 2008). As both disorders are associated with chronically increased plasma cortisol, these findings suggest that chronically elevated cortisol levels—in marked conflict with the animal findings—promote aggressiveness. This assumption, however, needs direct experimental support as no studies are available on the involvement of plasma cortisol in depression- or anxiety-related aggressiveness. There are a series of considerations that suggest that the role of cortisol is different in these subjects than in nonanxious and nondepressed people. First, depression and social phobia are associated with increased cortisol levels in general, but increased cortisol is not a "mandatory" concomitant of these disorders. One can hypothesize that aggression occurs in those patients who do not have chronically elevated cortisol levels. Second, depressed or anxious patients may become aggressive in response to provocation, which may induce exacerbated HPA-axis responses in these patients. As such, aggressiveness in depression and social phobia may be a particular case of the reactive aggression–stress response relationship (see below). Third, depression and anxiety bring about important changes in brain functions relevant to emotional control. As the effects of glucocorticoids on aggression are indirect (see above) one can hypothesize that the behavioral consequences of chronic cortisol levels result from an interaction with disorder-specific alterations in brain function.

Taken together, these considerations suggest that aggressiveness seen in depression and social phobia needs further study with respect to the role played by glucocorticoids.

3.2.3 Overview and Interpretation: Healthy Subjects

Acute interactions between aggressiveness and cortisol production are clearly understudied in people, especially the impact of cortisol on aggressiveness. Overall, however, the available information suggests that, as in animals, human aggressiveness and acute increases in plasma glucocorticoids reciprocally facilitate each other. The execution of aggressive acts increases plasma glucocorticoids, which in turn promotes aggressive behaviors. Chronic interactions between cortisol production and aggressiveness are similarly poorly studied in healthy subjects. The available information suggests, however, that defeat increases cortisol production. The impact of chronically elevated glucocorticoid levels on aggression is unknown in people.

3.3 Abnormal Aggression in People

Studies in abnormally aggressive people focus primarily on the "low glucocorticoid hypothesis of aggression," which states that abnormal aggression is associated with decreased cortisol production. The first study on the subject (Virkkunen 1985) showed that 24 h urinary free cortisol secretion was dramatically reduced in habitually violent offenders with antisocial personality disorder as compared to either violent offenders without antisocial personality disorder, patients with antisocial personality disorder who were not habitually violent, or male clinic personnel. This pioneering study prompted a series of investigations that partly confirmed and partly invalidated the early findings. The study by (Virkkunen 1985) shows three major characteristics: (1) the study group was very precisely separated from comparison groups in terms of both psychiatric diagnosis (antisocial personality disorder) and behavior (habitual violence); (2) the glucocorticoid measure was highly reliable, as the author measured 24 h glucocorticoid secretion patterns in the urine, which circumvents problems related to natural oscillations although it is not informative regarding particular periods of the day; (3) the differences between groups in regard to glucocorticoids, diagnosis and behavior were all large. Subsequent studies were often less rigorous in the selection of subjects, the type and magnitude of diagnostic/behavioral differences, and the measurement of HPA-axis function. Nevertheless, these studies help to answer the question of whether the endocrine abnormality specifically seen in violent antisocially disordered people is present in other aggression-related psychopathologies or not.

Responses to acute stress have also been investigated in subjects with psychopathologies shown above, the acute and chronic effects of glucocorticoids are the result of different mechanisms of action. Therefore, studies on acute stress responses will be discussed in a separate section.

3.3.1 Chronic Glucocorticoid Deficits and Aggression-Related Behavioral Problems

In principle, chronic secretion patterns can be identified only by studying subjects over relatively long periods. Nevertheless, baseline levels, if interpreted carefully, may be considered indicative of general secretion patterns. In this section we will compare studies that assessed baseline levels of glucocorticoids in conjunction with a wide spectrum of behavioral problems, from broadband measures like externalizing behavior to studies that focused on more specific aggressive acts.

Broadband Studies: Externalizing Behavior
Externalizing behavior does include aggressiveness, but also other behaviors that are indirectly related to aggression at best. In addition, measures of this behavioral category include symptoms of conduct, oppositional-defiant, and attention deficit-hyperactivity disorders (see above). Thus, the behavioral components of this category are rather mixed. A mechanical analysis suggests that findings are contradictory; while some studies did find a negative association between plasma glucocorticoids and externalizing behaviors (Cicchetti and Rogosch 2001; Ruttle et al. 2011; Shirtcliff and Essex 2008; Shirtcliff et al. 2005), others found no association (Klimes-Dougan et al. 2001), and others found that aggressiveness was associated with *increased* glucocorticoid secretion (Marsman et al. 2008).

Three of the studies in fact extend the original observations by Virkkunen (1985) to externalizing behaviors. These differ from the others because of the thoroughness of the glucocorticoid measurements, the level of externalizing behaviors seen in the sample, and the longitudinal nature of the assessments (Ruttle et al. 2011; Shirtcliff et al. 2005; Shirtcliff and Essex 2008). Glucocorticoid levels were measured repeatedly and averaged, which overcomes interference from ultradian oscillations; moreover, "state"—like (momentary) and "trait"—like (temporally stable) differences in secretion patterns were differentiated (Shirtcliff et al. 2005). The samples showed discrepant levels of externalizing behaviors including highly externalizing subjects, and the studies covered several years. The overall conclusion of these studies is that *concurrently* measured externalizing problems and glucocorticoid measures do not necessarily correlate, but a correlation occurs over the long run. Ruttle et al. (2011) found that externalizing behavior predicted the development of flattened diurnal secretion rhythms, while the other two studies reported that low glucocorticoid levels predicted increased symptom severity longitudinally (Shirtcliff et al. 2005; Shirtcliff and Essex 2008). Interestingly, these long-term developmental trajectories did not differentiate internalizing and externalizing problems, as the increase in symptom severity over time was associated with HPA-axis hypofunction in both.

There are two studies that do not support the negative association between glucocorticoid production and externalizing behaviors. Klimes-Dougan et al. (2001) found no association, while Marsman et al. (2008) found opposite positive association. The glucocorticoid measurements in these studies are similar to those of Cicchetti and Rogosch (2001), who found that glucocorticoids were negatively associated with externalizing behaviors. All three studies investigated morning/awakening glucocorticoid levels, yet their conclusions were sharply different. There was a difference, however, in the populations studied. While Cicchetti and Rogosch (2001) studied subjects with clinical-level externalizing problems, the severity of these problems was milder in the other two studies. Moreover, no purely externalizing subjects were studied by Klimes-Dougan et al. (2001), who compared primarily internalizing and mixed (internalizing/externalizing) children. These findings may suggest that the association between glucocorticoids and externalizing behavior is restricted to subjects with severe behavioral problems. However, the three studies reviewed in the previous paragraph did include highly externalizing subjects but did not find an interaction between *concurrent* glucocorticoid and psychometric measures (Ruttle et al. 2011; Shirtcliff et al. 2005; Shirtcliff and Essex 2008).

These rather contradictory findings do not suggest a clear association between externalizing behavior and concurrent glucocorticoid secretion patterns. However, studies authored by an overlapping set of authors suggest that glucocorticoid levels and the severity of externalizing behaviors correlate longitudinally: low glucocorticoids predict an unfavorable psychological developmental trajectory, while externalizing behaviors predict a long-term glucocorticoid deficit. Tentatively, one can hypothesize that the contradictions in concurrent measurements may be explained by the timing of measurements relative to the developmental trajectories: studies accidentally performed in early phases of the developmental trajectory did not find a negative correlation between glucocorticoid production and externalizing behaviors, while studies performed at later phases did. So the apparently contradictory findings may in fact be due to timing differences, and may be explained by a real interaction between "trait-like" glucocorticoid secretion patterns and behavioral developmental trajectories in terms of internalizing and externalizing behaviors.

Broadband Studies: *Delinquency*

Delinquency studies vary with respect to the level and the behavioral correlates of subjects. In some studies, subjects had committed only petty offenses (Popma et al. 2006, 2007a, b), while aggressive delinquents or violent criminals were studied in others (Brewer-Smyth et al. 2004; Dabbs et al. 1991; Moss et al. 1995). Cortisol measures were taken under resting conditions over the day, or acutely after stressful procedures. Studies addressing the latter relationship will be reviewed in Sect. 3.3.2. In the studies reviewed here, the correlation between a history of aggressive delinquency and point measures of basal cortisol secretion was assessed.

The available data suggest that low awakening cortisol is associated with a higher level of aggressive delinquency. In the study by Brewer-Smyth et al. (2004), the correlation was seen in highly violent subjects, but a similar interaction was seen by Popma et al. (2007a) in subjects who committed petty offenses including but not restricted to aggressive acts. A blunted diurnal decrease in cortisol was noticed in both studies, which supports the negative association between aggressive delinquency and lowered glucocorticoid production, and extends the observations of Virkkunen (1985) to awakening cortisol levels in moderately aggressive delinquent populations.

In studies where cortisol levels were measured throughout the day, low cortisol was not associated with aggressive delinquency but moderated the relationship between testosterone and aggressiveness. Particularly, low cortisol made the proaggressive effects of testosterone conspicuous (Dabbs et al. 1991; Popma et al. 2007b). In a study of male offenders with a history of violent acts, plasma cortisol was negatively associated with the affective-interpersonal features of psychopathy but not with socially deviant lifestyle and behaviors (Holi et al. 2006). There is one study whose findings directly conflict with those reviewed above (Soderstrom et al. 2004), but the circumstances of blood sampling in that study suggest that stress responses rather than basal levels were measured; therefore these findings will be discussed in Sect. 3.3.2.

Overall, the findings reviewed above support a negative association between plasma cortisol and aggressive delinquency, with awakening cortisol predicting aggressive delinquency better than cortisol levels measured later in the day.

Broadband Studies: *Mixed Diagnoses*

The phrase mixed diagnoses stands here for "disruptive behavior" attributed to children that expressed a study-specific mixture of conduct, oppositional-defiant, and attention deficit-hyperactivity disorder symptoms, and "conduct problems," a mixture of conduct and oppositional-defiant disorder symptoms. The level of problems varied. Although this aspect was rarely discussed, in our judgment the subjects of these studies rarely met the conditions of a full diagnosis for any of the diagnoses that made up their "disruptive" or "conduct" symptomatology.

"Disruptive behavior" or "conduct problems" per se did not consistently correlate with particular cortisol secretion profiles. Awakening cortisol and 24 h urinary cortisol secretion were similar in disruptive and control children (Kruesi et al. 1989; Fairchild et al. 2008). When cortisol was measured later in the day, some studies showed higher, others lower, yet other similar levels of basal cortisol in disruptive and control children (Scerbo and Kolko 1994; Kariyawasam et al. 2002; Fairchild et al. 2008).

More consistent findings were obtained when measures of aggressiveness were considered in addition to the "mixed" diagnosis. Low basal cortisol predicted increased overt aggression in disruptive children (van de Wiel et al. 2004; Poustka et al. 2010). In addition, low cortisol was specifically associated with the aggression-related symptoms of disruptive children (Oosterlaan et al. 2005), while the presence of callous unemotional traits accentuated the negative association between disruptive symptoms and morning cortisol (Loney et al. 2006). Long-term associations were also revealed; reliably low basal cortisol levels (e.g., the averages of multiple point measurements) predicted overt aggressiveness and disruptive symptomatology (McBurnett et al. 2000; Sondeijker et al. 2008).

In summary, "disruptive behavior" per se does not correlate with particular cortisol secretion patterns, but aggressiveness seen in disruptive children was consistently associated with low cortisol. In addition, low cortisol levels may predict aggression-inclined developmental trajectories.

Diagnosis-Centered Studies
In this section we review studies in subjects who were diagnosed with particular aggression-related diagnoses.

Although the number of pertinent studies is small, both conduct and oppositional-defiant disorders are associated with low mid-morning cortisol levels (Pajer et al. 2001, 2006; van Goozen et al. 1998; Vanyukov et al. 1993). The same negative association holds true when aggression-related measures were considered (Golubchik et al. 2009; Pajer et al. 2006; van Goozen et al. 1998; Vanyukov et al. 1993), with the following specifications. In the study by Golubchik et al. (2009) the difference between aggressive conduct disordered subjects and healthy controls was not significant, but sample size was small; in the study by Pajer et al. (2006) the dehydroepiandrosterone/cortisol ratio rather than cortisol levels per se showed the expected correlation; in the study by Vanyukov et al. (1993) the negative correlation referred to symptom counts without reference to the particular symptoms, but it is highly likely that higher symptom counts involved a higher number of aggression-related symptoms. Interestingly, Azar et al. (2004) found no correlation between mid-morning cortisol and conduct disorder-associated aggressiveness in pregnant adolescent girls, but since pregnancy brings about large changes in cortisol production, this finding does not seem relevant to this discussion. Unfortunately, no data exist for awakening responses, diurnal rhythms, or 24 h secretion patterns in subjects with these disorders. The study by van Bokhoven et al. (2005) is at variance with the findings presented above; however, these authors studied stress responses rather than basal levels, and as such their findings will be discussed in Sect. 3.3.2.

Psychopathic symptoms were also associated with low awakening or mid-morning basal corticosterone levels (Cima et al. 2008; Dolan et al. 2001; Vaillancourt 2011), although this negative correlation was found in females but not in males in one study (Vaillancourt 2011). Low corticosterone was also seen in aggressive alcoholics (Bergman and Brismar 1994).

In conclusion, studies on "pure" aggression-related diagnoses are relatively scarce but suggest that these patients are characterized by low basal cortisol. The scarcity of studies on awakening cortisol responses, diurnal rhythms, or 24 h secretion patterns is blatant. Similarly, developmental trajectories were not studied in this category of subjects.

Behavior-Centered Studies
There are a few papers where basal cortisol levels were associated with aggressive behavior but no diagnoses were mentioned. In theory, these studies should have been discussed in Sect. 3.2. However, the populations investigated included individuals who showed very high levels of aggressiveness and, in addition underwent highly stressful experiences (lived in refugee camps, or suffered early maltreatment). As such, these subjects were not likely to be healthy in psychological terms, but can be considered "abnormal" as it regards their aggressiveness.

In another study included here, subjects were university students whose aggressiveness was likely within the normal range (Böhnke et al. 2010a). However, aggressiveness was studied after provocation, which increased their aggressiveness to levels that were incongruent with their "trait" aggressiveness.

Low cortisol was associated with higher aggressiveness in three studies involving highly aggressive albeit nonreferred subjects (Gordis et al. 2006; Victoroff et al. 2011; Yu and Shi 2009). In a study by Böhnke et al. (2010a), awakening glucocorticoid levels also correlated negatively with aggressive responses after provocation. Given that the aggressiveness of these subjects was higher than that of nonprovoked controls, this finding also supports the notion that unusually high aggressiveness correlates with low cortisol production patterns even in the absence of an underlying psychopathology.

No association was found in the study by van den Bergh et al. (2008). According to the authors, however, no norms were available for the aggressive behavior measure that they used. Therefore, it is unclear whether this study included subjects with unusually high levels of aggressiveness. In an interesting developmental study (Shoal et al. 2003), low cortisol levels (averaged point measures) at the age of 10–12 years predicted aggressiveness measured 5 years later.

Taken together, these findings show that unusually high levels of aggressiveness are associated with low basal cortisol levels; moreover, low cortisol predicts the development of aggressiveness on the long-run.

Overview and Interpretation: Abnormal Aggression and Low Basal Levels of Cortisol

The findings reviewed above strongly support the notion that plasma cortisol correlates negatively with abnormal aggression in humans. This association is not convincingly seen in subjects belonging to broad categories like "externalizing behavior" and "disruptive behavior." Even in the case of the latter, however, the association becomes rather clear when aggressive subgroups are considered. Overall, low basal glucocorticoid levels were seen in aggressive delinquents, aggressive disruptive children, children diagnosed with conduct or oppositional-defiant disorder, adults with psychopathic traits, and children with an unknown psychopathologic background but who showed unusually high levels of aggressiveness.

In addition to the correlation mentioned above, low glucocorticoid levels seem to predict unfavorable developmental trajectories on the long run. This was shown for externalizing behaviors, disruptive behaviors, and unusually high aggressiveness in nonreferred subjects.

3.3.2 Acute Glucocorticoid Stress Responses and Aggression-Related Behavioral Problems

As shown in the section dealing with animal aggression, glucocorticoids promote aggression in regularly performed resident-intruder tests, and aggression-induced glucocorticoid stress responses are exacerbated in "hyper-arousal" models of abnormal aggression. Taken together, these findings suggest that high glucocorticoid stress responses contribute to the expression of escalated aggression as a response to emotional arousal. Now we will review data on cortisol stress responses in people showing aggression-related psychopathologies.

Differentiating between reactive and proactive aggression would be useful in this respect due to their different emotional backgrounds. Proactive aggression is nonimpulsive, controlled, and instrumental, while reactive aggression is impulsive, emotional (characterized by high levels of psychological and bodily arousal), and occurs in response to threat or provocation (Kempes et al. 2008; Lopez-Duran et al. 2009; van Bokhoven et al. 2005). Based on these definitions and laboratory findings, one would expect to see increased glucocorticoid stress responses after reactive but not proactive aggression. The very few studies that have differentiated between reactive and proactive aggression support the notion that only reactive aggression is associated with increased cortisol stress responses. Lopez-Duran et al. (2009) studied highly externalizing children who were assigned to fear—and frustration-inducing tasks. In this experimental setting, children high in reactive aggression showed increased cortisol responses compared to both children high in proactive aggression and nonaggressive children. Based on a study involving psychopathic individuals, McBurnett et al. (2003, 2005) suggested that aggression in hostile/reactive individuals is associated with high cortisol reactivity, whereas aggression in primary psychopaths is associated with lack of cortisol reactivity. A similar conclusion was reached by O'Neal et al. (2010) in a study involving children at risk for antisocial behavior. The involvement of stress responses in reactive aggression may be also deduced from studies that ostensibly studied baseline levels. van Bokhoven et al. (2005) measured plasma cortisol in subjects immediately after their arrival into the laboratory, despite the fact that multiple investigations have shown that "arrival levels" are higher than those seen after a resting period (van Goozen et al. 1998). As such, this study investigated stress responses rather than true baseline levels. van Bokhoven et al. (2005) found that reactive but not proactive aggression was associated with elevated levels of cortisol, which—given the circumstances of cortisol measurements—supports the notion that reactive but not proactive aggression correlates with increased cortisol stress responses.

While none of the psychiatric constructs discussed here are particularly associated with reactive or proactive aggression, the findings obtained with disordered subjects may still be interpreted in these terms. Conduct disorder, antisocial personality disorder, violent delinquency, and psychopathy usually do imply planned, instrumental aggressive acts that need to be performed in "cold blood," even if people suffering from these disorders are prone to perform reactive aggression as well. Therefore, such people are likely to perform proactive acts of aggression on a regular basis and this habit may be reflected by their stress responses. Based on this assumption and on studies where reactive and proactive aggressions were differentiated, one can hypothesize that these disorders are associated with a less reactive HPA-axis. The available data support this assumption.

When subjects diagnosed with aggressive personality disorder, psychopathy, or oppositional-defiant disorder were challenged with ipsapirone, fenfluramine, or sumatriptan (all of which were used to characterize HPA-axis responsiveness), their cortisol responses were considerably lower than controls, indicating that their HPA-axis was hyporesponsive (Almeida et al. 2010; Dolan et al. 2001; Snoek et al. 2002). The cortisol response to stressful situations also differentiated people with or without aggression-related problems. Aggressive delinquents had lower cortisol responses in anticipation of stressful procedures (auditory event-related potential recording) or to stressful procedures themselves (public speaking) than controls (Moss et al. 1995; Popma et al. 2006). Disruptive children showed lower cortisol than controls in response to frustrating competitive situations (Fairchild et al. 2008; van Goozen et al. 2000), and aggressive children diagnosed with attention deficit-hyperactivity disorder showed diminished cortisol responses after filling out psychological questionnaires (Yang et al. 2007). Although not directly related to these psychopathologies, heroin addicts who responded aggressively in the point subtraction paradigm also showed lower cortisol responses, even when on methadone treatment (Gerra et al. 2001b, 2004).

Contradictory findings also exist, but some of the contradictions are easily explained. For example, van de Wiel et al. (2004) studied disruptive children showing either low or high cortisol responses to competitive frustration and found no significant differences in overt aggression between the two groups (no nondisruptive controls were included in this study). Nevertheless, the two groups did show some differences (the low stress-responsive subgroup tended to be higher in overt aggression) and, in addition, oppositional behavior scores were significantly higher in subjects showing low cortisol stress responses. Since the sample size was relatively low (22 children in total), the marginal differences in overt aggression scores and the significant differences in oppositional scores support rather than deny the negative association between aggression and stress responses in disruptive children.

A stronger case is presented by McBurnett et al. (2005). In this study, conduct problems were associated with increased cortisol stress responses. This finding was inconsistent with the authors' previously published data (McBurnett et al. 2000); they attributed the perceived discrepancies to differences between population and clinically referred samples as well as to differences between reactive-hostile and more instrumental forms of aggression. In our opinion, the two studies are not at variance, because in the first one (McBurnett et al. 2000) the authors investigated the long-term behavioral impact of low *basal* cortisol levels, while in the second (McBurnett et al. 2005) they investigated cortisol *stress responses*. In addition to the reasons outlined by the authors, we see another important reason why conduct problems were associated with increased stress responses in disruptive children. In the studies where decreased stress responses were seen in similar subjects, the stressor was a frustrating competition (Snoek et al. 2002; Fairchild et al. 2008; van Goozen et al. 2000), where as McBurnett et al. (2005) asked their subjects to recall and describe the worst thing that ever happened to them. The two types of stressful situations are obviously different. One can hypothesize that the impact of the worst-event procedure on HPA-axis activation (i.e., the impact of psychological "harm") is different from that of a competitive situation. Another indication that perceived harm may have an impact on stress responses in aggression-related psychopathologies comes from the study by Soderstrom et al. (2004), who suggested that violent offenders have increased basal cortisol. However, their method of corticosterone measurement allows an alternative interpretation. They sampled blood after lumbar puncture, which in addition to being stressful per se, may be perceived by subjects as potentially harmful. The anticipation of harmful experiences can be stressful by itself (see above), so cortisol measured under these conditions is likely to reflect stress responses rather than basal levels. Seen in this light, the authors in fact showed that violent offenders show increased stress responses, contradicting other studies which found the reverse in similar subjects (Moss et al. 1995; Popma et al. 2006). The findings of Soderstrom et al. (2004) raise the interesting possibility that potentially harmful experiences elicit increased stress responses in aggressive delinquents, while the same subjects show decreased stress responses when no harm is involved or when harm is suffered by their opponents.

We propose that two factors should always be considered when interpreting stress responses: the type of aggression (reactive vs. proactive) and the nature of the stressor (competitive vs. potentially harmful). Disparate but consistent findings suggest that reactive and proactive aggression are associated with increased and decreased acute stress responses, respectively. Subjects accustomed to performing proactive acts of aggression are less reactive to pharmacological challenges that affect the HPA-axis, and show decreased stress responses in competitive situations. We also suggest that the impact of perceived harm on stress responses may be altered in subjects with aggression-related psychopathologies.

4 Overall Conclusions and Suggestions for Future Studies

Although the role of glucocorticoids in aggression is perceived as controversial by many investigators, a detailed analysis of the available findings that carefully considers the details of glucocorticoid and behavioral measurements as well as mechanisms of action suggests that this role is rather clear-cut; moreover, there is a marked similarity between animal and human findings. The conclusions of our analysis can be summarized as follows:

(1) Aggressive interactions rapidly increase glucocorticoid production, while glucocorticoids rapidly promote aggression, suggesting the existence of a positive feedback loop between aggressive behavior and acute glucocorticoid responses. This statement appears valid for both untreated animals and healthy humans (i.e., for species-typical aggression).

(2) Defeat results in chronically increased plasma glucocorticoid levels, which suppress animal aggression. Chronic stressors and glucocorticoid administrations have the same effect. Although the issue is clearly understudied, defeat

seems have a similar effect on plasma glucocorticoids and competition-readiness in human subjects over the short run. The long-term consequences of defeat-related or heterotypic stressor-related chronic stress on human aggression are poorly understood.

(3) Chronic low plasma glucocorticoid levels induce abnormal aggression in animals, while aggression-related psychopathologies are associated with chronically decreased basal cortisol in humans. This suggests that a chronic deficit in glucocorticoid production induces brain alterations that promote the development of abnormal forms of aggression.

(4) Decreased stress responses were seen in patients with psychopathologies who regularly engaged in proactive aggression. Comparable data are sparse in animals, but similar phenomena were seen in some models, e.g., in mice selected for high aggressiveness which show abnormally high levels of aggression of the proactive-type (Veenema et al. 2003, 2004).

(5) Exacerbated acute glucocorticoid stress responses were noted in animal models of emotional aggression, while a few data in humans suggest that reactive (emotional) aggression is associated with increased cortisol stress responses. Causal relationships were not established, but the available information suggests that emotionally loaded (reactive) aggression develops on the background of enhanced acute stress responses, while the reverse is true for proactive aggression.

(6) In humans, enduring decreased levels of cortisol production affect developmental trajectories. Particularly, chronically low basal cortisol levels predict the emergence or worsening of aggression-related problems in the long run. It is unclear whether this is valid only for proactive types of aggression, for reactive aggression as well. No analogous animal data are available.

Despite the considerable gaps in current knowledge, the data summarized in this review clearly show that the aggression/glucocorticoid relationship cannot be reduced to the simple "high or low" question. Although much effort has been invested to explain this relationship, there still are many open issues.

There are a few methodological issues that need to be refined in future studies. The first relates to problems associated with glucocorticoid measurements, which were discussed in Sects. 1.1 and 3.1.2. These issues are less troublesome in experiments where acute glucocorticoid/aggression interactions are studied, because the impact of aggressiveness on glucocorticoid production is dramatic, and largely overwhelms the variance resulting from ultradian/diurnal oscillations, or accidental involvement in stressful events or eating, which also increases plasma glucocorticoids. In contrast, accidental variability can easily mask real differences in glucocorticoid levels when the interactions between aggression-related habits ("traits") and glucocorticoid plasma levels are studied. In addition to avoiding trivial errors like measuring plasma cortisol immediately after arrival into the laboratory, highly stressful blood sampling procedures, or measuring glucocorticoids late in the day when they are more likely to be elevated by stressful events, studying the impact of glucocorticoids on aggression-related traits requires that trait-like patterns of glucocorticoid secretion should be studied as well. In more general terms, one should study trait-like patterns of hormone production when explanations for behavioral traits are sought.

The second methodological issue relates to methods for measuring "aggressiveness." Naturally, externalizing behavior and psychopathologies are important study areas, and one may be interested in finding out whether these constructs are associated with particular glucocorticoid secretion patterns. As far as aggression is concerned, however, such broad categories confuse rather than clarify the picture. The symptoms of disorders include but are not restricted to aggressive behavior, and symptom distribution shows large individual variability. As such, one should differentiate aggressive and nonaggressive forms of these constructs if one wants to draw conclusions on the glucocorticoid/aggression relationship. This seems trivial at first sight, but some of the reports reviewed here whose conclusions made reference to aggressive behavior studied psychiatric constructs without investigating the subjects' level of aggressiveness. Differentiating aggressive and nonaggressive forms of a disorder not only helps us understand aggressive behavior but—given their heterogeneity—also helps us understand the disorders themselves.

In addition to differentiating between aggressive and nonaggressive forms of a psychiatric disorder, attention should be paid to the type of aggression shown by subjects. Proactive and reactive aggression differ in terms of etiological origin, behavioral particularities, emotional background, and also in terms of glucocorticoid background and brain mechanisms, according to various human and animal studies. Often psychopathologies involve both types of aggression; nevertheless their relative proportion may show large individual variation.

We believe that in addition to, or in parallel with, refining methodological approaches, there is a need to address several understudied issues. A causal relationship between the execution of aggressive acts and the concomitant increase in glucocorticoid secretion has been established in animals but not in humans. Naturally, glucocorticoids can be administered more freely to animals than to people, and it is similarly easier to explore causal relationships by giving glucocorticoid synthesis blockers or receptor antagonists to animals. Nevertheless, glucocorticoid agonists and glucocorticoid synthesis blockers (e.g., Metopirone) are quite frequently used in the diagnosis and/or treatment of various human diseases; therefore, their use in aggression research does not seem to be overwhelmingly problematic or difficult.

In addition to clarifying the causal relationship between "natural" aggressiveness and the associated HPA-axis response, glucocorticoid agonists and/or glucocorticoid secretion blockers may also be used to investigate causal relationships between altered cortisol stress responses and psychiatrically disordered subjects' propensity to behave aggressively. At present, such relationships are investigated by correlating ongoing behavior and behavior-induced changes in glucocorticoid secretion. Clearer answers would be provided by studies where glucocorticoid secretion was experimentally altered. A differential studying of reactive and proactive aggression appears especially relevant in this respect.

Besides acute interactions, chronic (trait-like) interactions also await further studies. At present, aggression in depressed and social phobic patients is poorly understood in terms of the involvement of glucocorticoids. Human studies on the issue are virtually lacking and appropriate animal models have not yet been developed. The impact of glucocorticoid deficits on aggressive behavior was not entirely clarified either. Refining and diversifying animal models seems advisable, and differences between reactive (emotional, impulsive) and proactive (premeditated, goal-oriented) aggression have not yet been studied in humans. One can imagine several scenarios. Perhaps only proactive aggression is associated with chronically low glucocorticoid levels; if so, then differentiating the two aggression types would result in more clear findings. It is equally possible, however, that both reactively and proactively aggressive people have low basal levels of glucocorticoids but the two groups differ in acute stress responses. Finally, the developmental aspects of altered glucocorticoid secretion patterns (and the etiological factors of such altered patterns) seem understudied although there are several very interesting reports on the subject.

Clarifying the role of glucocorticoids in aggression may help us understand different types of aggression better, may contribute to the elucidation of brain mechanisms, may offer tools for prognosis, and may guide the development of novel treatments. Although glucocorticoids do not elicit aggression themselves, they have a variety of neural effects that affect the probability of being aggressive, the stimuli that will elicit aggression and the normal and abnormal forms of aggressive behavior that are observed. The amount of information obtained so far is impressive, and by and large, the findings are rather consistent when the analysis carefully considers the details of glucocorticoid and behavioral measurements as well as mechanisms of glucocorticoid action. However, the caveats in understanding are rather large.

Acknowledgments This study was supported by the Hungarian National Science and Research Fund (OTKA). Contract grant number: 82069.

References

Alden LE, Taylor CT (2004) Interpersonal processes in social phobia. Clin Psychol Rev 24:857–882

Almeida M, Lee R, Coccaro EF (2010) Cortisol responses to ipsapirone challenge correlate with aggression, while basal cortisol levels correlate with impulsivity, in personality disorder and healthy volunteer subjects. J Psychiatr Res 44(14):874–880

Azar R, Zoccolillo M, Paquette D, Quiros E, Baltzer F, Tremblay RE (2004) Cortisol levels and conduct disorder in adolescent mothers. J Am Acad Child Adolesc Psychiatry 43:461–468

Benus RF, Bohus B, Koolhaas JM, van Oortmerssen GA (1991) Behavioural differences between artificially selected aggressive and non-aggressive mice: response to apomorphine. Behav Brain Res 43:203–208

Bergman B, Brismar B (1994) Hormone levels and personality traits in abusive and suicidal male alcoholics. Alcohol Clin Exp Res 18:311–316

Berkowitz L (1993) The problem of aggression. In: Berkowitz L (ed), McGraw-Hill series in social psychology: aggression: its causes, consequences, and control, McGraw-Hill Book Company, New York, pp 1–23
Berkowitz L (1994) Aggression: definition and perspectives. In: Baron RA, Richardson DR (eds) Perspectives in social psychology: human aggression, 2nd edn. Plenum Press, New York, pp 1–38
Blair C, Granger D, Razza RP (2005) Cortisol reactivity is positively related to executive function in preschool children attending Head Start. Child Dev 76:554–567
Blanchard DC, Spencer RL, Weiss SM, Blanchard RJ, McEwen B, Sakai RR (1995) Visible burrow system as a model of chronic social stress: behavioral and neuroendocrine correlates. Psychoneuroendocrinology 20:117–134
Böhnke R, Bertsch K, Kruk MR, Naumann E (2010a) The relationship between basal and acute HPA axis activity and aggressive behavior in adults. J Neural Transm 117:629–637
Böhnke R, Bertsch K, Kruk MR, Richter S, Naumann E (2010b) Exogenous cortisol enhances aggressive behavior in females, but not in males. Psychoneuroendocrinology 35:1034–1044
Booth A, Shelley G, Mazur A, Tharp G, Kittok R (1989) Testosterone, and winning and losing in human competition. Horm Behav 23(4):556–571
Brain PF, Nowell NW, Wouters A (1971) Some relationships between adrenal function and isolation induced intermale aggression in albino mice. Physiol Behav 6:27–29
Brewer-Smyth K, Burgess AW, Shults J (2004) Physical and sexual abuse, salivary cortisol, and neurologic correlates of violent criminal behavior in female prison inmates. Biol Psychiatry 55:21–31
Bronson FH, Eleftheriou BE (1965) Adrenal response to fighting in mice: separation of physical and psychological causes. Science 147:627–628
Caldji C, Diorio J, Meaney MJ (2000) Variations in maternal care in infancy regulate the development of stress reactivity. Biol Psychiat 48:1164–1174
Caramaschi D, de Boer SF, Koolhaas JM (2008) Is hyper-aggressiveness associated with physiological hypoarousal? A comparative study on mouse lines selected for high and low aggressiveness. Physiol Behav 95:591–598
Casiano H, Belik SL, Cox BJ, Waldman JC, Sareen J (2008) Mental disorder and threats made by noninstitutionalized people with weapons in the national comorbidity survey replication. J Nerv Ment Dis 196:437–445
Chang C, Li CY, Earley RL, Hsu YY (2012) Aggression and related behavioral traits: the impact of winning and losing and the role of hormones. Integrative Comp Biol 52:801–813
Chapple CL, Tyler KA, Bersani BE (2005) Child neglect and adolescent violence: examining the effects of self-control and peer rejection. Violence Victims 20:39–53
Cicchetti D, Rogosch FA (2001) The impact of child maltreatment and psychopathology on neuroendocrine functioning. Dev Psychopathol 13:783–804
Cima M, Smeets T, Jelicic M (2008) Self-reported trauma, cortisol levels, and aggression in psychopathic and non-psychopathic prison inmates. Biol Psychol 78:75–86
Corum CR, Thurmond JB (1997) Effects of acute exposure to stress on subsequent aggression and locomotion performance. Psychosom Med 39:436–443
Dabbs JM Jr, Jurkovic GJ, Frady RL (1991) Salivary testosterone and cortisol among late adolescent male offenders. J Abnorm Child Psychol 19:469–478
Datson NA, Morsink MC, Meijer OC, de Kloet ER (2008) Central corticosteroid actions: search for gene targets. Eur J Pharmacol 583:272–289
de Kloet ER, Han F, Meijer OC (2008a) From the stalk to down under about brain glucocorticoid receptors, stress and development. Neurochem Res 33:637–642
de Kloet ER, Karst H, Joëls M (2008b) Corticosteroid hormones in the central stress response: quick-and-slow. Front Neuroendocrinol 29:268–272
Delville Y, Melloni RH Jr, Ferris CF (1998) Behavioral and neurobiological consequences of social subjugation during puberty in golden hamsters. J Neurosci 18:2667–2672
DeNardo DF, Sinervo B (1994) Effects of corticosterone on activity and home-range size of free-ranging male lizards. Horm Behav 28:53–65

DiBattista JD, Anisman H, Whitehead M, Gilmour KM (2005) The effects of cortisol administration on social status and brain monoaminergic activity in rainbow trout Oncorhynchus mykiss. J Exp Biol 208:2707–2718
Dolan M, Anderson IM, Deakin JF (2001) Relationship between 5-HT function and impulsivity and aggression in male offenders with personality disorders. Br J Psychiatry 178:352–359
Earley RL, Edwards JT, Aseem O, Felton K, Blumer LS, Karom M, Grober MS (2006) Social interactions tune aggression and stress responsiveness in a territorial cichlid fish (Archocentrus nigrofasciatus). Physiol Behav 88:353–363
Fairchild G, van Goozen SH, Stollery SJ, Brown J, Gardiner J, Herbert J, Goodyer IM (2008) Cortisol diurnal rhythm and stress reactivity in male adolescents with early-onset or adolescence-onset conduct disorder. Biol Psychiatry 64:599–606
Fava M (1998) Depression with anger attacks. J Clin Psychiatry 59(Suppl 18):18–22
Fish EW, DeBold JF, Miczek KA (2005) Escalated aggression as a reward: corticosterone and GABA(A) receptor positive modulators in mice. Psychopharmacology 182:116–127
Follenius M, Simon C, Brandenberger G, Lenzi P (1987) Ultradian plasma corticotropin and cortisol rhythms: time-series analyses. J Endocrinol Invest 10:261–266
Gaab J, Rohleder N, Nater UM, Ehlert U (2005) Psychological determinants of the cortisol stress response: the role of anticipatory cognitive appraisal. Psychoneuroendocrinology 30:599–610
Geniole SN, Carré JM, McCormick CM (2011) State, not trait, neuroendocrine function predicts costly reactive aggression in men after social exclusion and inclusion. Biol Psychol 87(1):137–145
Gerra G, Zaimovic A, Ampollini R, Giusti F, Delsignore R, Raggi MA, Laviola G, Macchia T, Brambilla F (2001a) Experimentally induced aggressive behavior in subjects with 3,4-methylenedioxy-methamphetamine ("Ecstasy") use history: psychobiological correlates. J Subst Abuse 13:471–491
Gerra G, Zaimovic A, Avanzini P, Chittolini B, Giucastro G, Caccavari R, Palladino M, Maestri D, Monica C, Delsignore R, Brambilla F (1997) Neurotransmitter-neuroendocrine responses to experimentally induced aggression in humans: influence of personality variable. Psychiatry Res 66:33–43
Gerra G, Zaimovic A, Moi G, Bussandri M, Bubici C, Mossini M, Raggi MA, Brambilla F (2004) Aggressive responding in abstinent heroin addicts: neuroendocrine and personality correlates. Prog Neuropsychopharmacol Biol Psychiatry 28:129–139
Gerra G, Zaimovic A, Raggi MA, Giusti F, Delsignore R, Bertacca S, Brambilla F (2001b) Aggressive responding of male heroin addicts under methadone treatment: psychometric and neuroendocrine correlates. Drug Alcohol Depend 65:85–95
Gerra G, Zaimovic A, Raggi MA, Moi G, Branchi B, Moroni M, Brambilla F (2007) Experimentally induced aggressiveness in heroin-dependent patients treated with buprenorphine: comparison of patients receiving methadone and healthy subjects. Psychiatry Res 149:201–213
Golubchik P, Mozes T, Maayan R, Weizman A (2009) Neurosteroid blood levels in delinquent adolescent boys with conduct disorder. Eur Neuropsychopharmacol 19(1):49–52
Gordis EB, Granger DA, Susman EJ, Trickett PK (2006) Asymmetry between salivary cortisol and alpha-amylase reactivity to stress: relation to aggressive behavior in adolescents. Psychoneuroendocrinology 31:976–987
Groeneweg FL, Karst H, de Kloet ER, Joëls M (2011) Rapid non-genomic effects of corticosteroids and their role in the central stress response. J Endocrinol 209:153–167
Gunnar MR, Kryzer E, van Ryzin MJ, Phillips DA (2010) The rise in cortisol in family day care: associations with aspects of care quality, child behavior, and child sex. Child Dev 81(3):851–869
Gunnar MR, Vazquez DM (2001) Low cortisol and a flattening of expected daytime rhythm: Potential indices of risk in human development. Develop Psychopathol 13:515–538
Halász J, Liposits Z, Kruk MR, Haller J (2002) Neural background of glucocorticoid dysfunction-induced abnormal aggression in rats: involvement of fear- and stress-related structures. Eur J Neurosci 15:561–569

Halász J, Tóth M, Kalló I, Liposits Z, Haller J (2006) The activation of prefrontal cortical neurons in aggression—a double labeling study. Behav Brain Res 175:166–175
Haller J (2012) The neurobiology of abnormal manifestations of aggression—a review of hypothalamic mechanisms in cats, rodents, and humans. Brain Res Bull 2012. doi:S0361-9230(12)00220-1 (Epub ahead of print)
Haller J, Albert I, Makara GB (1997) Acute behavioural effects of corticosterone lack specificity but show marked context-dependency. J Neuroendocrinol 9:515–518
Haller J, Barna I, Baranyi M (1995) Hormonal and metabolic responses during psychosocial stimulation in aggressive and nonaggressive rats. Psychoneuroendocrinology 20:65–74
Haller J, Halasz J, Makara GB, Kruk MR (1998a) Acute effects of glucocorticoids: behavioral and pharmacological perspectives. Neurosci Biobehav Rev 23:337–344
Haller J, Halász J, Mikics E, Kruk MR (2004) Chronic glucocorticoid deficiency-induced abnormal aggression, autonomic hypoarousal, and social deficit in rats. J Neuroendocrinol 16:550–557
Haller J, Halasz J, Mikics E, Kruk MR, Makara GB (2000a) Ultradian corticosterone rhythm and the propensity to behave aggressively in male rats. J Neuroendocrinol 12:937–940
Haller J, Horváth Z, Bakos N (2007) The effect of buspirone on normal and hypoarousal-driven abnormal aggression in rats. Prog Neuropsychopharmacol Biol Psychiatry 31:27–31
Haller J, Kiem DT, Makara GB (1996) The physiology of social conflict in rats: what is particularly stressful? Behav Neurosci 110:353–359
Haller J, Kruk MR (2006) Normal and abnormal aggression: human disorders and novel laboratory models. Neurosci Biobehav Rev 30:292–303
Haller J, Mikics E, Halász J, Tóth M (2005a) Mechanisms differentiating normal from abnormal aggression: glucocorticoids and serotonin. Eur J Pharmacol 526:89–100
Haller J, Mikics E, Makara GB (2008) The effects of non-genomic glucocorticoid mechanisms on bodily functions and the central neural system. A critical evaluation of findings. Front Neuroendocrinol 29:273–291
Haller J, Millar S, Kruk MR (1998b) Mineralocorticoid receptor blockade inhibits aggressive behaviour in male rats. Stress 2:201–207
Haller J, Millar S, van de Schraaf J, de Kloet RE, Kruk MR (2000b) The active phase-related increase in corticosterone and aggression are linked. J Neuroendocrinol 12:431–436
Haller J, Tóth M, Halász J (2005b) The activation of raphe serotonergic neurons in normal and hypoarousal-driven aggression: a double labeling study in rats. Behav Brain Res 161:88–94
Haller J, van de Schraaf J, Kruk MR (2001) Deviant forms of aggression in glucocorticoid hyporeactive rats: a model for 'pathological' aggression? J Neuroendocrinol 13:102–107
Hayden-Hixson DM, Ferris CF (1991a) Steroid specific regulation of agonistic responding in the anterior hypothalamus of male hamsters. Physiol Behav 50:793–799
Hayden-Hixson DM, Ferris CF (1991b) Cortisol exerts site-, context- and dose-dependent effects on agonistic responding in hamsters. J Neuroendocrinol 3:613–622
Höglund E, Balm PH, Winberg S (2002) Behavioural and neuroendocrine effects of environmental background colour and social interaction in Arctic charr (Salvelinus alpinus). J Exp Biol 205:2535–2543
Holi M, Auvinen-Lintunen L, Lindberg N, Tani P, Virkkunen M (2006) Inverse correlation between severity of psychopathic traits and serum cortisol levels in young adult violent male offenders. Psychopathology 39:102–104
Ing NH (2005) Steroid hormones regulate gene expression posttranscriptionally by altering the stabilities of messenger RNAs. Biol Reprod 72:1290–1296
Ivanov I, Yehuda R, Greenblatt E, Davidow J, Makotkine I, Alfi L, Newcorn JH (2011) The effect of trauma on stress reactivity in aggressive youth. Psychiatry Res 189(3):396–402
Joels M, de Kloet ER (1992) Control of neuronal excitability by corticosteroid hormones. Trends Neurosci 15:25–30
Kariyawasam SH, Zaw F, Handley SL (2002) Reduced salivary cortisol in children with comorbid attention deficit hyperactivity disorder and oppositional defiant disorder. Neuro Endocrinol Lett 23:45–48

Kempes M, de Vries H, Matthys W, van Engeland H, van Hooff J (2008) Differences in cortisol response affect the distinction of observed reactive and proactive aggression in children with aggressive behaviour disorders. J Neural Transm 115:139–147

Kikusui T, Nakamura K, Kakuma Y, Mori Y (2006) Early weaning augments neuroendocrine stress responses in mice. Behav Brain Res 175:96–103

Klimes-Dougan B, Hastings PD, Granger DA, Usher BA, Zahn-Waxler C (2001) Adrenocortical activity in at-risk and normally developing adolescents: individual differences in salivary cortisol basal levels, diurnal variation, and responses to social challenges. Dev Psychopathol 13:695–719

Koolhaas JM, Bartolomucci A, Buwalda B, de Boer SF, Flügge G, Korte SM, Meerlo P, Murison R, Olivier B, Palanza P, Richter-Levin G, Sgoifo A, Steimer T, Stiedl O, van Dijk G, Wöhr M, Fuchs E (2011) Stress revisited: a critical evaluation of the stress concept. Neurosci Biobehav Rev 35:1291–1301

Koolhaas JM, De Boer SF, De Rutter AJ, Meerlo P, Sgoifo A (1997) Social stress in rats and mice. Acta Physiol Scand Suppl 640:69–72

Kruesi MJ, Schmidt ME, Donnelly M, Hibbs ED, Hamburger SD (1989) Urinary free cortisol output and disruptive behavior in children. J Am Acad Child Adolesc Psychiatry 28:441–443

Kruk MR, Halász J, Meelis W, Haller J (2004) Fast positive feedback between the adrenocortical stress response and a brain mechanism involved in aggressive behavior. Behav Neurosci 118:1062–1070

Kudielka BM, Wüst S (2010) Human models in acute and chronic stress: assessing determinants of individual hypothalamus-pituitary-adrenal axis activity and reactivity. Stress 13:1–14

Landys MM, Goymann W, Schwabl I, Trapschuh M, Slagsvold T (2010) Impact of season and social challenge on testosterone and corticosterone levels in a year-round territorial bird. Horm Behav 58:317–325

Leshner AI, Korn SJ, Mixon JF, Rosenthal C, Besser AK (1980) Effects of corticosterone on submissiveness in mice: some temporal and theoretical considerations. Physiol Behav 24:282–288

Loney BR, Butler MA, Lima EN, Counts CA, Eckel LA (2006) The relation between salivary cortisol, callous-unemotional traits, and conduct problems in an adolescent non-referred sample. J Child Psychol Psychiatry 47:30–36

Lopez-Duran NL, Olson SL, Hajal NJ, Felt BT, Vazquez DM (2009) Hypothalamic pituitary adrenal axis functioning in reactive and proactive aggression in children. J Abnormal Child Psychol 37:169–182

Mainardi M, Valenti G, Valsecchi P, Parmigiani S, Brain PF, Vescovi P (1987) Acute effects of low doses of two ACTH preparations on fighting behavior in male mice. Med Sci Res 15:649–650

Marsman R, Swinkels SH, Rosmalen JG, Oldehinkel AJ, Ormel J, Buitelaar JK (2008) HPA-axis activity and externalizing behavior problems in early adolescents from the general population: the role of comorbidity and gender The TRAILS study. Psychoneuroendocrinology 33:789–798

McBurnett K, Lahey BB, Rathouz PJ, Loeber R (2000) Low salivary cortisol and persistent aggression in boys referred for disruptive behavior. Arch Gen Psychiatry 57:38–43

McBurnett K, Raine A, Stouthamer-Loeber M, Loeber R, Kumar AM, Kumar M, Lahey BB (2005) Mood and hormone responses to psychological challenge in adolescent males with conduct problems. Biol Psychiatry 57:1109–1116

McBurnett K, King J, Scarpa A (2003) The HPA and the development of aggressive, antisocial, and substance abuse disorders. In: Cicchetti D, Walker E (eds) Neurodevelopmental mechanisms in psychopathology. Cambridge University Press, New York, pp 324–344

Meddle SL, Romero LM, Astheimer LB, Buttemer WA, Moore IT, Wingfield JC (2002) Steroid hormone interrelationships with territorial aggression in an Arctic-breeding songbird, Gambel's whitecrowned sparrow, Zonotrichia leucophrys gambelii. Horm Behav 42:212–221

Mehta PH, Josephs RA (2010) Testosterone and cortisol jointly regulate dominance: evidence for a dual-hormone hypothesis. Horm Behav 58(5):898–906

Melloni RH Jr, Ferris CF (1996) Adolescent anabolic steroid use and aggressive behavior in golden hamsters. Ann N Y Acad Sci 794:372–375

Mikics E, Barsy B, Haller J (2007) The effect glucocorticoids on aggressiveness in established colonies of rats. Psychoneuroendocrinology 32:160–170

Mikics E, Kruk MR, Haller J (2004) Genomic and non-genomic effects of glucocorticoids on aggressive behavior in male rats. Psychoneuroendocrinology 29:618–635

Moss HB, Vanyukov MM, Martin CS (1995) Salivary cortisol responses and the risk for substance abuse in prepubertal boys. Biol Psychiatry 38:547–555

Munro AD, Pitcher TJ (1985) Steroid hormones and agonistic behavior in a cichlid teleost, Aequidens pulcher. Horm Behav 19:353–371

Murray-Close D, Han G, Cicchetti D, Crick NR, Rogosch FA (2008) Neuroendocrine regulation and physical and relational aggression: the moderating roles of child maltreatment and gender. Dev Psychol 44:1160–1176

Neumann ID, Veenema AH, Beiderbeck DI (2010) Aggression and anxiety: social context and neurobiological links. Front Behav Neurosci 4:12

O'Neal CR, Brotman LM, Huang KY, Gouley KK, Kamboukos D, Calzada EJ, Pine DS (2010) Understanding relations among early family environment, cortisol response, and child aggression via a prevention experiment. Child Dev 81(1):290–305

Oosterlaan J, Geurts HM, Knol DL, Sergeant JA (2005) Low basal salivary cortisol is associated with teacher-reported symptoms of conduct disorder. Psychiatry Res 134:1–10

Overli O, Kotzian S, Winberg S (2002) Effects of cortisol on aggression and locomotor activity in rainbow trout. Horm Behav 42:53–61

Painuly NP, Grover S, Gupta N, Mattoo SK (2011) Prevalence of anger attacks in depressive and anxiety disorders: implications for their construct? Psychiatry Clin Neurosci 65(2):165–174

Pajer K, Gardner W, Rubin RT, Perel J, Neal S (2001) Decreased cortisol levels in adolescent girls with conduct disorder. Arch Gen Psychiatry 58:297–302

Pajer K, Tabbah R, Gardner W, Rubin RT, Czambel RK, Wang Y (2006) Adrenal androgen and gonadal hormone levels in adolescent girls with conduct disorder. Psychoneuroendocrinology 31:1245–1256

Parmigiani S, Dadomo H, Bartolomucci A, Brain PF, Carbucicchio A, Costantino C, Ferrari PF, Palanza P, Volpi R (2009) Personality traits and endocrine response as possible asymmetry factors of agonistic outcome in karate athletes. Aggress Behav 35:324–333

Popma A, Doreleijers TA, Jansen LM, van Goozen SH, van Engeland H, Vermeiren R (2007a) The diurnal cortisol cycle in delinquent male adolescents and normal controls. Neuropsychopharmacology 32:1622–1628

Popma A, Jansen LM, Vermeiren R, Steiner H, Raine A, van Goozen SH, van Engeland H, Doreleijers TA (2006) Hypothalamus pituitary adrenal axis and autonomic activity during stress in delinquent male adolescents and controls. Psychoneuroendocrinology 31:948–957

Popma A, Vermeiren R, Geluk CA, Rinne T, van den Brink W, Knol DL, Jansen LM, van Engeland H, Doreleijers TA (2007b) Cortisol moderates the relationship between testosterone and aggression in delinquent male adolescents. Biol Psychiatry 61:405–411

Poustka L, Maras A, Hohm E, Fellinger J, Holtmann M, Banaschewski T, Lewicka S, Schmidt MH, Esser G, Laucht M (2010) Negative association between plasma cortisol levels and aggression in a high-risk community sample of adolescents. J Neural Transm 117(5):621–627

Ramenofsky M (1985) Acute changes in plasma steroids and agonistic behavior in male Japanese quail. Gen Comp Endocrinol 60:116–128

Ricci LA, Grimes JM, Knyshevski I, Melloni RH Jr (2005) Repeated cocaine exposure during adolescence alters glutamic acid decarboxylase-65 (GAD65) immunoreactivity in hamster brain: correlation with offensive aggression. Brain Res 1035:131–138

Rutter M, Moffitt TE, Caspi A (2006) Gene-environment interplay and psychopathology: Multiple varieties but real effects. J Child Psychol Psychiat 47:226–261

Ruttle PL, Shirtcliff EA, Serbin LA, Fisher DB, Stack DM, Schwartzman AE (2011) Disentangling psychobiological mechanisms underlying internalizing and externalizing behaviors in youth: longitudinal and concurrent associations with cortisol. Horm Behav 59(1):123–132

Saxbe DE, Margolin G, Spies Shapiro LA, Baucom BR (2012) Does dampened physiological reactivity protect youth in aggressive family environments? Child Dev 83:821–830

Scerbo AS, Kolko DJ (1994) Salivary testosterone and cortisol in disruptive children: relationship to aggressive, hyperactive, and internalizing behaviors. J Am Acad Child Adolesc Psychiatry 33:1174–1184

Schjolden J, Basic D, Winberg S (2009) Aggression in rainbow trout is inhibited by both MR and GR antagonists. Physiol Behav 98:625–630

Schuett GW, Grober MS (2000) Post-fight levels of plasma lactate and corticosterone in male copperheads, *Agkistrodon contortrix* (Serpentes, Viperidae): differences between winners and losers. Physiol Behav 71:335–341

Schuurman T (1980) Hormonal correlates of agonistic behaviour in adult male rats. Prog Brain Res 53:415–420

Shirtcliff EA, Essex MJ (2008) Concurrent and longitudinal associations of basal and diurnal cortisol with mental health symptoms in early adolescence. Dev Psychobiol 50:690–703

Shirtcliff EA, Granger DA, Booth A, Johnson D (2005) Low salivary cortisol levels and externalizing behavior problems in youth. Dev Psychopathol 17:167–184

Shoal GD, Giancola PR, Kirillova GP (2003) Salivary cortisol, personality, and aggressive behavior in adolescent boys: a 5-year longitudinal study. J Am Acad Child Adolesc Psychiatry 42:1101–1107

Silverin B (1993) Territorial aggressiveness and its relation to the endocrine system in the pied flycatcher. Gen Comp Endocrinol 89:206–213

Snoek H, van Goozen SH, Matthys W, Sigling HO, Koppeschaar HP, Westenberg HG, van Engeland H (2002) Serotonergic functioning in children with oppositional defiant disorder: a sumatriptan challenge study. Biol Psychiatry 51:319–325

Soderstrom H, Blennow K, Forsman A, Liesivuori J, Pennanen S, Tiihonen J (2004) A controlled study of tryptophan and cortisol in violent offenders. J Neural Transm 111:1605–1610

Sondeijker FE, Ferdinand RF, Oldehinkel AJ, Tiemeier H, Ormel J, Verhulst FC (2008) HPA-axis activity as a predictor of future disruptive behaviors in young adolescents. Psychophysiology 45:398–404

Stupfel M, Pavely A (1990) Ultradian, circahoral and circadian structures in endothermic vertebrates and humans. Comp Biochem Physiol A Comp Physiol 96:1–11

Summers CH, Summers TR, Moore MC, Korzan WJ, Woodley SK, Ronan PJ, Höglund E, Watt MJ, Greenberg N (2003) Temporal patterns of limbic monoamine and plasma corticosterone response during social stress. Neuroscience 116:553–563

Summers CH, Watt MJ, Ling TL, Forster GL, Carpenter RE, Korzan WJ, Lukkes JL, Overli O (2005) Glucocorticoid interaction with aggression in non-mammalian vertebrates: reciprocal action. Eur J Pharmacol 526:21–35

Sze PY, Iqbal Z (1994) Glucocorticoid action on depolarization-dependent calcium influx in brain synaptosomes. Neuroendocrinology 59:457–465

Timmer M, Sandi C (2010) A role for glucocorticoids in the long-term establishment of a social hierarchy. Psychoneuroendocrinology 35:1543–1552

Tokarz RR (1987) Effects of corticosterone treatment on male aggressive behavior in a lizard (*Anolis sagrei*). Horm Behav 21:358–370

Tornatzky W, Miczek KA (1994) Behavioral and autonomic responses to intermittent social stress: differential protection by clonidine and metoprolol. Psychopharmacology 116:346–356

Tóth M, Halász J, Mikics E, Barsy B, Haller J (2008) Early social deprivation induces disturbed social communication and violent aggression in adulthood. Behav Neurosci 122:849–854

Toth M, Mikics E, Tulogdi A, Aliczki M, Haller J (2011) Post-weaning social isolation induces abnormal forms of aggression in conjunction with increased glucocorticoid and autonomic stress responses. Horm Behav 60:28–36

Toth M, Tulogdi A, Biro L, Soros P, Mikics E, Haller J (2012) The neural background of hyper-emotional aggression induced by post-weaning social isolation. Behav Brain Res 233: 120–129

Trifonova ST, Gantenbein M, Turner JD, Muller CP (2012) The use of saliva for assessment of cortisol pulsatile secretion by deconvolution analysis. Psychoneuroendocrinology. doi:10.1016/j.psyneuen.2012.10.016 (Epub ahead of print)

Tulogdi A, Toth M, Halasz J, Mikics E, Fuzesi T, Haller J (2010) Brain mechanisms involved in predatory aggression are activated in a laboratory model of violent intra-specific aggression. Eur J Neurosci 32:1744–1753

Vaillancourt T, Sunderani S (2011) Psychopathy and indirect aggression: the roles of cortisol, sex, and type of psychopathy. Brain Cogn 77:170–175

van Bokhoven I, van Goozen SH, van Engeland H, Schaal B, Arseneault L, Séguin JR, Nagin DS, Vitaro F, Tremblay RE (2005) Salivary cortisol and aggression in a population-based longitudinal study of adolescent males. J Neural Transm 112:1083–1096

van de Wiel N, van Goozen S, Matthys W, Snoek H, van Engeland H (2004) Cortisol and treatment effect in children with disruptive behavior disorders: a preliminary study. J Am Acad Child Adoloscent Psychiatry 43:1011–1018

van den Bergh BR, van Calster B, Pinna Puissant S, van Huffel S (2008) Self-reported symptoms of depressed mood, trait anxiety and aggressive behavior in post-pubertal adolescents: associations with diurnal cortisol profiles. Horm Behav 54:253–257

van Goozen SH, Fairchild G, Snoek H, Harold GT (2007) The evidence for a neurobiological model of childhood antisocial behavior. Psychol Bull 133:149–182

van Goozen SH, Matthys W, Cohen-Kettenis PT, Buitelaar JK, van Engeland H (2000) Hypothalamic-pituitary-adrenal axis and autonomic nervous system activity in disruptive children and matched controls. J Am Acad Child Adolesc Psychiatry 39:1438–1445

van Goozen SH, Matthys W, Cohen-Kettenis PT, Gispen-de Wied C, Wiegant VM, van Engeland H (1998) Salivary cortisol and cardiovascular activity during stress in oppositional-defiant disorder boys and normal controls. Biol Psychiatry 43:531–539

Vanyukov MM, Moss HB, Plail JA, Blackson T, Mezzich AC, Tarter RE (1993) Antisocial symptoms in preadolescent boys and in their parents: associations with cortisol. Psychiatry Res 46:9–17

Veenema AH, Blume A, Niederle D, Buwalda B, Neumann ID (2006) Effects of early life stress on adult male aggression and hypothalamic vasopressin and serotonin. Eur J Neurosci 24:1711–1720

Veenema AH, Koolhaas JM, de Kloet ER (2004) Basal and stress-induced differences in HPA axis, 5-HT responsiveness, and hippocampal cell proliferation in two mouse lines. Ann N Y Acad Sci 1018:255–265

Veenema AH, Meijer OC, de Kloet ER, Koolhaas JM, Bohus BG (2003) Differences in basal and stress-induced HPA regulation of wild house mice selected for high and low aggression. Horm Behav 43:197–204

Verona E, Kilmer A (2007) Stress exposure and affective modulation of aggressive behavior in men and women. J Abnormal Psychol 116:410–421

Victoroff J, Quota S, Adelman JR, Celinska B, Stern N, Wilcox R, Sapolsky RM (2011) Support for religio-political aggression among teenaged boys in Gaza: part II: neuroendocrinological findings. Aggress Behav 37(2):121–132

Virkkunen M (1985) Urinary free cortisol secretion in habitually violent offenders. Acta Psychiatr Scand 72:40–44

White TL, Grover VK, de Wit H (2006) Cortisol effects of D-amphetamine relate to traits of fearlessness and aggression but not anxiety in healthy humans. Pharmacol Biochem Behav 85:123–131

Wood GE, Young LT, Reagan LP, McEwen BS (2003) Acute and chronic restraint stress alter the incidence of social conflict in male rats. Horm Behav 43:205–213

Woodley SK, Matt KS, Moore MC (2000) Neuroendocrine responses in free-living female and male lizards after aggressive interactions. Physiol Behav 71:373–381

Yang SJ, Shin DW, Noh KS, Stein MA (2007) Cortisol is inversely correlated with aggression for those boys with attention deficit hyperactivity disorder who retain their reactivity to stress. Psychiatry Res 153:55–60

Yodyingyuad U, de la Riva C, Abbott DH, Herbert J, Keverne EB (1985) Relationship between dominance hierarchy, cerebrospinal fluid levels of amine transmitter metabolites (5-hydroxyindole acetic acid and homovanillic acid) and plasma cortisol in monkeys. Neuroscience 16:851–858

Yohe LR, Suzuki H, Lucas LR (2012) Aggression is suppressed by acute stress but induced by chronic stress: immobilization effects on aggression, hormones, and cortical 5-HT(1B)/striatal dopamine D(2) receptor density. Cogn Affect Behav Neurosci 12:446–459

Yu YZ, Shi JX (2009) Relationship between levels of testosterone and cortisol in saliva and aggressive behaviors of adolescents. Biomed Environ Sci 22:44–49

Sensory, Hormonal, and Neural Basis of Maternal Aggression in Rodents

Rosa Maria Martins de Almeida, Annabel Ferreira and Daniella Agrati

Abstract We review existing knowledge of the neural, hormonal, and sensory basis of maternal aggression in the female rat. Although females may express different kinds of aggression, such as defense or dominance, the most frequent and conspicuous form of aggressive behavior among females is the one associated with motherhood. Maternal aggression occurs in various vertebrate and invertebrate species; however, our emphasis will be on maternal aggression in rats because most of the physiological investigations have been performed in this species. Firstly, we address those factors that predispose the female to attack, such as the endocrine profile, the maternal state, and the stimulation provided by the pups, as well as those that trigger the aggressive response, as the intruder's characteristics and the context. As the postpartum aggression is a fundamental component of the maternal repertoire, we emphasize its association with maternal motivation and the reduction of fear and anxiety in dams. Finally, we outline the neurocircuitry involved in the control of maternal aggression, stressing the role of the ventro-orbital region of prefrontal cortex and the serotoninergic system.

Keywords Maternal aggression · Reproductive hormones · Sensory cues · Brain · Neurotransmitters

---

R. M. M. de Almeida (✉)
Laboratório de Psicologia Experimental, Neurociências e Comportamento (LPNeC), Instituto de Psicologia do Desenvolvimento e da Personalidade da Universidade Federal do Rio Grande do Sul (UFRGS), Porto Alegre, RS, Brazil
e-mail: rosa_almeida@yahoo.com; rosa.almeida@ufrgs.br

A. Ferreira · D. Agrati
Faculdad de Ciencias, Universidad de la Republica, Montevideo, Uruguay
e-mail: ferreira.annabel@gmail.com

D. Agrati
e-mail: dagrati@fcien.edu.uy

Curr Topics Behav Neurosci (2014) 17: 111–130
DOI: 10.1007/7854_2014_312

Published Online: 20 May 2014

Contents

1 Maternal Aggression: Definition and General Features

Maternal aggression can be defined as the pattern of aggressive responses displayed by mothers in confrontation toward intruders, usually unfamiliar adult conspecifics.[1] Typically the female rat approaches the intruder, lunges at it (frontal attack) or approaches sideways pushing it with the hip (sideways threat, lateral attack), bites the intruder's the head, body, and neck, and adopts an aggressive posture (Erskine et al. 1978b, 1980a, b; Haney et al. 1989; Stern and Koluine 1991; see also Svare 1977; Svare and Gandelman 1976a, b for studies in mice) (Fig. 1). These aggressive behaviors elicit in the intruder submissive and escape responses (Ferreira and Hansen 1986; Koluine and Stern 1990; Mos et al. 1987a, b).

The aggressive behavior of the female toward an intruder in the nest cage is similar to the dominant males' aggression in rat colonies, although some important differences exist (Blanchard and Blanchard 1984). Like dominant males, lactating females display lateral attacks (sideways threats) and attack bites toward the neck and back and they exhibit pilo-erection, but unlike males, females also engage in frontal attacks, which are considered a defensive behavior (Blanchard and

[1] Maternal aggression in the postparturient rat is routinely assessed in the female's home cage as its expression requires a recent contact with the pups (Ferreira and Hansen 1986; Haney et al. 1989). Usually, an adult male from the same or different strain is employed as an intruder, but also cycling and lactating females have been used to test aggression (Haney et al. 1989; Neumann et al. 2001). Different components of the females' agonistic behavior as well as of the submission responses of the intruder are used to assess maternal aggression (Haney et al. 1989; Stern and Koluine 1991; Mos et al. 1987a, b); therefore, it is important to attend to the specific behavioral measures when comparing results from different studies.

Fig. 1 Components of maternal aggression repertoire. When an adult co-specific is introduced in a lactating dam's home cage, the female investigates the intruder (**a**), approaches it sideways, usually pushing it with the hip (sideways threat, lateral attack) (**b**), and bites directed toward the head, neck, or body. **c** These responses evoke submission postures from the intruder

Blanchard 1984). Nevertheless, the frontal attack of lactating dams toward intruders is not preceded by an escape response as is the case in defensive aggression (Albert et al. 1992a; Olivier et al. 1985). Although maternal aggression comprises components that have been defined as defensive and offensive (Blanchard and Blanchard 1984; Lucion and De Almeida 1996), it is a proactive behavior driven by a strong tendency to approach and initiate with very short latency even before the intruder approaches the nest (Olivier et al. 1985), indicating its offensive nature.

Dams' aggressive behavior follows the time course of maternal responsiveness in rats (Mayer and Rosenblatt 1987; Mayer et al. 1987b; Haney et al. 1989) and mice (Mann and Svare 1982; Svare and Gandelman 1976b). Thus, it sharply increases during the last days of gestation, peaks around parturition, and remains high throughout the first 2 weeks, to finally decline gradually until weaning. Similar to maternal behavior, maternal aggression is a complex motivated behavior whose expression is modulated by multiple physiological factors, such as the hormonal profile of gestation and parturition (Ferreira et al. 2002; Mayer and Rosenblatt 1987), the sensory stimulation provided by the pups and the intruder (Hansen and Ferreira 1986a; Kolunie and Stern 1990; Stern and Kolunie 1993; Svare 1977; Svare et al. 1980), and the maternal motivation (Agrati et al. 2011). Moreover, as discussed below, the strength of each factor in determining females' aggressive behavior is dynamic, changing from pregnancy and throughout the postpartum period.

2 Factors Determining the Expression of Maternal Aggression

Several physiological and contextual factors modulate the expression of maternal aggression (Fig. 2); however, it is important to distinguish between those that predispose females to be aggressive (distal) from those triggering the aggressive responses (proximal). For instance, while the ovarian steroid changes of late

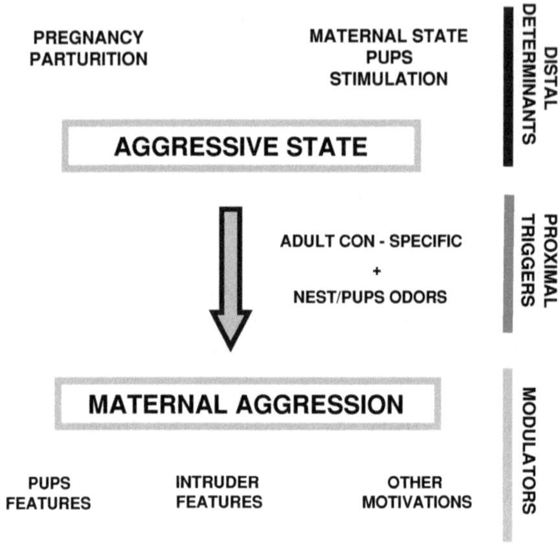

Fig. 2 Summary of the factors that render females aggressive during the postpartum period and those that trigger and modulate the expression of maternal aggression

gestation and parturition phases as well as the ventral stimulation delivered by the pups during lactation promote a female's aggressive behavior (Factor et al. 1993; Mayer et al. 1987a; Stern and Kolunie 1993), cues from the nest, the pups, and the intruder trigger the aggressive responses (Erskine et al. 1978a, b; Ferreira and Hansen 1986).

2.1 Hormonal Changes of Late Gestation Control the Onset of a Dam's Aggressive Behavior

Mayer and co-workers (1987b) reported that a dam's aggressive behavior toward an adult intruder in her home cages increases from day 16 of gestation until parturition, pointing toward hormonal factors in determining the onset of this behavior. The aggressive behavior observed before parturition has been called *home cage aggression* to differentiate it from the *maternal aggression* observed during the postpartum period when pups are usually present (Mayer et al. 1987b).

The importance of ovarian steroids for the onset of aggression has been proven in a series of experiments showing that females that have been hysterectomized and ovariectomized (HO) on day 16 of gestation and received an injection of estradiol (*E*) (Mayer and Rosenblatt 1987), as well as virgin ovariectomized rats that received a prolonged regimen of *E* and progesterone (*P*) and a final injection of *E* (Mayer et al. 1990a, b), increase their aggressive behavior compared to females treated with oil.

Interestingly, the mechanism through which ovarian steroids promotes aggressive behavior seems to be related to those eliciting maternal behavior, as both procedures stimulate a short-latency onset of maternal responsiveness (Bridges 1984; Siegel and Rosenblatt 1975) and aggression (Mayer and Rosenblatt 1987; Mayer et al. 1990a) in nonparturient animals.[2] Also, even in the absence of hormonal treatment, sensitized females (nongestating rats that develop maternal behavior through being constantly exposed to pups, Footnote 2) display aggressive responses, although with markedly lower intensity than those of postparturient rats (Erskine et al. 1980b; Ferreira et al. 2002; Mayer and Rosenblatt 1993a). Together, these results highlight the close relationship between maternal care and aggressive behavior.

Although the onset of maternal aggression is strongly associated with the ovarian steroid changes during late pregnancy, ovariectomy does not eliminate a dam's aggressive behavior once established after parturition (Albert et al. 1992b), indicating that its maintenance does not depend on ovarian influences. However, a recent study found that ovariectomy on day 2 of lactation reduces both maternal behavior and aggressive responses of postparturient females (de Sousa et al. 2010), suggesting a direct or indirect (i.e., modifying the maternal state) influence of ovarian secretions on the expression of maternal aggression.

As discussed later, once established, maternal aggression mainly depends on the interaction between the mother and the pups. However, it is worth to note that a dam's separation from the pups for 1 h reduces maternal aggression on day 7 of lactation, but not on days 2 or 4 (Mayer et al. 1987b), indicating that the postpartum aggression is still strongly influenced by reproductive hormones in this early phase and gradually loses its hormonal control and becomes more dependent on the maternal state and the pups' stimulation (Fig. 2). Thus, it seems that the optimal expression of maternal aggression during the postpartum period depends on the priming action of ovarian steroids and its subsequent influence on the effect of the interaction with the pups.

2.2 *Stimuli from the Pups Maintain Aggression During the Postpartum*

The importance of the mother–pups interactions on the maintenance of maternal aggression is supported by several studies showing that after removing the litter from the home cage, progressively declines until complete cessation after 24 h of separation (Erskine et al. 1978a; Ferreira and Hansen 1986; Mayer et al. 1987b;

---

[2] Not only pregnant and peri-parturient rats display maternal responsiveness to pups, but also nongestating rats can develop maternal behavior by being constantly exposed to newborn pups, a procedure called *sensitization* (Cosnier and Courturier 1966; Rosenblatt 1967). The sensitized animal is a useful model to examine the role of the maternal state in different aspects of maternity independently of the endocrine changes that characterize gestation, parturition and lactation. The sensitization latency, the period of time required to develop maternal behavior after exposure to pups begins, is employed to evaluate the role of different factors on maternal behavior expression.

Stern and Koulnie 1993). Returning the pups to the postparturient female restores the high levels of aggressive behavior that characterize the lactation period (Gandelman and Simon 1980).

Which sensory factors, related to the mother–litter interaction, maintain a dam's aggressive behavior? A main source of sensory information that the mother receives during the interaction with the pups is the somatosensory stimulation of the ventral trunk provided by pups when searching for the nipples and during suckling. Accordingly, the lesions of the peripeduncular nucleus, that forms part of the ascending milk-ejection pathway during the maintaining phase (days 4–8), completely abolish maternal aggression (Hansen and Ferreira 1986a; Factor et al. 1993). Similarly, the cutaneous anesthesia of the dam's nipples and surrounding skin, but not thelectomy (removing the nipples), impairs maternal aggression, demonstrating that ventral stimulation, but not the suckling stimulus per se, is essential for the maintenance of maternal aggression in the rat (Factor et al. 1993; Mayer et al. 1987a; Stern and Koluine 1993). In contrast to the rat, thelectomy completely abolishes this behavior in mice (Svare 1977; Svare and Gandelman 1976a; Svare and Mann 1983), indicating that suckling stimulation of the nipples is essential for the induction of aggressive behavior during the first postpartum days in this species.

2.3 *Stimuli from the Pups and the Intruder Trigger and Modulate Maternal Aggression*

The pups' somatosensory stimulation of the ventral trunk is important for maintaining dams' aggressive behavior; however, females attack an intruder before suckling stimulation begins. Besides, maternal aggression is only displayed when dams are in the home cage with the pups or immediately after removing them, indicating that other cues coming from the pups or the nest and the intruder trigger this behavior. Ferreira and Hansen (1986) found a progressive reduction of maternal aggression, similar to that observed when pups are removed, when placing the pups in glass bottles that allow the female to see but not to smell them. Conversely, placing the pups inside of mesh bags that allow the dam to smell the pups prevents the decrease of attacks toward the intruder, stressing the importance of olfactory cues from the pups on a female's aggression. In support of the importance of the olfactory modality, it was shown that different procedures that completely abolish olfaction (i.e., removing the olfactory epithelium and infusions of $ZnSO_4$) significantly decrease maternal aggression during the maintenance phase (Ferreira et al. 1987; Koulnie and Stern 1995; Mayer and Rosenblatt 1993b).

Lactating females spent one-third of the time smelling the intruder before initiating an attack (Ferreira and Hansen 1986; Kolunie and Stern 1990), suggesting that the intruder's pheromones are also important for triggering a female' attack. In support of this mechanism, maternal aggression is severely impaired by lesions of the mediodorsal thalamic nucleus and prefrontal insular cortex (Fig. 3,

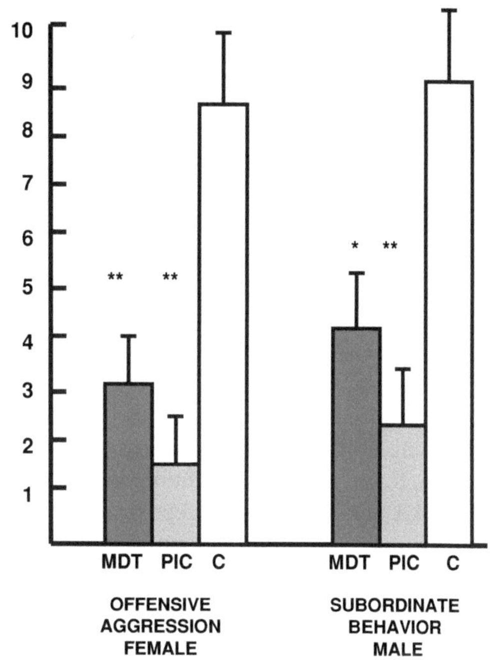

Fig. 3 Aggressive behavior of female rats bearing electrolytic or sham lesions (C) in the mediodorsal thalamic nucleus (MDT) or prefrontal insular cortex (PIC). (Means ± SEM). Offensive aggression includes number of attacks, bites, and lateral postures. Overall group differences by Kruskal–Wallis analysis of variance: aggression. $H = 11.6$, $p < 0.003$; submission $H = 8.9$, $p < 0.012$; $^{*}p < 0.05$; $^{**}p < 0.02$ by Mann–Whitney U test. Adapted and reproduced with permission from Ferreira et al. 1987

Ferreira et al. 1987) that eliminate the ability to discriminate between odors without making rats' anosmic (Eichenbaum et al. 1980; Slotnick and Kaneko 1981). This evidence has prompted the proposal that it is the contrast between the "attractive" odors of the pups and the nest and the "aversive" odor of an intruder that triggers maternal aggression (Ferreira et al. 1987; Lynds 1976).

It is noteworthy that the acute perioral anesthesia of mystacial pads or the trigeminal denervation efficiently disrupts attacks directed toward intruders by lactating females (Kolunie and Stern 1990; Stern and Kolulne 1991), indicating that the somatosensory stimulation of the mouth and snout's regions, while the mother explores and bites the intruder, plays a major role in promoting maternal aggression. The same perioral somatosensory cues are necessary to elicit normal nursing, retrieving, and licking of the pups (Stern and Johnson 1989; Stern and Kolunie 1989, 1991), once again indicating the close relationship between processes underling maternal care and aggression. On the other hand, distal stimuli from the pups and the intruder—visual and auditory—may play a modulating role in the aggressive encounter, but are not essential for the expression of maternal aggression, as blind and deaf dams attack male intruder in their home cage with the same intensity as control animals (Kolunie et al. 1994).

In addition to the role of hormonal and sensory factors in determining and triggering dam's aggressive response, other cues from the pups, such as their age (Giovenardi et al. 2000), and from the intruder, as its sex and size (Hardely et al. 1989; Mos et al. 1987b), modulate the intensity as well as the topology of the aggressive response.

3 Maternal Aggression and Reduction of Fear and Anxiety

Concomitant with the increase in aggressive behavior, lactating rats exhibit less fear (Hård and Hansen 1985) and anxiety-related behaviors (Ferreira et al. 1989; Fleming and Luebke 1981; Lonstein 2005; Pereira et al. 2005) in several experimental models. This reduction in fear and anxiety can be associated with the offensive nature of maternal aggression (Hansen et al. 1985; Lonstein and Gammie 2002; Olivier et al. 1985), because dams attack "without attempts to escape from the object being attacked" (Moyer 1976); in contrast to defensive aggression typically associated with fear, in which the individual attacks only if escape is precluded (Adams 2006; Blanchard and Blanchard 1984, Blanchard et al. 2003). Accordingly, similar hormonal treatments with E and P that reduce the anxiety-like behavior of virgin female rats in an open field and promote maternal behavior (Fleming et al. 1989), also increase a female's aggressive behavior (Mayer et al. 1990a, b, see above). In addition, low doses of the positive allosteric modulator of GABAA receptors chlordiazepoxide (CDP) known to reduce fear and anxiety exert pro-aggressive effects on postparturient females confronted with an intruder (Mos et al. 1987a, b; Olivier et al. 1985). Complimentary to this evidence, the treatment of lactating females with a partial inverse benzodiazepine agonists FG 7142 or with caffeine significantly decreases a female's aggressive responses toward intruders and increases her fear reaction (Hansen et al. 1985). Moreover, recent studies in mice suggest that down-regulation of corticotrophin releasing factor neurotransmission, implicated in anxiety behavior, during lactation is necessary for the optimal display of maternal aggression (Gammie et al. 2004).

Maternal aggression and reduction in fear or anxiety are usually concurrently expressed and share partly overlapping control mechanisms even though both are two independent processes controlled by different mechanisms. For instance, dams bred for high anxiety (HAB)-related behavior exhibit greater levels of both anxiety-like behavior and maternal aggression compared to those LAB-related behaviors (Bosch et al. 2005; Bosch and Neumann 2012). Lesions of the peripeduncular nucleus or manipulations of GABAergic neurotransmission in the ventromedial nucleus of the hypothalamus and the amygdala reduce maternal aggression without modifying the immobility fear reaction (Hansen and Ferreira 1986a, b). Moreover, the pro-aggressive and immobility effects of benzodiazepines seem to require the activity of different subunits of the $GABA_A$ receptor complex (de Almeida et al. 2004).

Even if some controversies continue to exist, most studies point toward a close relationship between maternal aggression and reduction in fear which is consistent with the hypothesis that maternal aggression is linked to an approach motivation.

4 Maternal Aggression as an Affective State and a Motivated Behavior

Our focus in this chapter is on maternal aggression as a behavior rather than as an affective experience.[3] However, the presence of an intruder in the home cage of a lactating dam likely elicits a negative affective state in the female as it interferes with her territory and goal directed care-giving activities toward the pups, and affects her well-being. Under these conditions, the intruder may acquire a negative affective valence for the dam and promote an approach motivation to attack. The direction of this motivation can be better defined as "against" rather than "toward" the intruder, as the mother makes an effort to inflict pain or harm on the intruder to remove it from the nest. As a result, the intruder escapes from the nest or performs submissive postures. Therefore, maternal aggression or its consequences—defeating the opponent—may generate a rewarding state in the individual. Interestingly, male mice can be trained to perform operant behaviors that are reinforced by the opportunity to attack (Fish et al. 2002), and female hamsters can be conditioned to prefer a place where they had fought (Meisel and Joppa 1994). We are currently considering this possibility in the case of the rat's maternal aggression.

In summary, it could be proposed that maternal aggression occurs as a result of a negative affective state or emotion in the mother elicited by the presence of a negative-valenced intruder in her nest, which promotes an approach motivation "against" the intruder and may have a "reinforcing" effect on the female (Fig. 4). The hypothesis that maternal aggression involves an approach motivation associated with mothering is supported by studies showing that mothers with low maternal motivation, as sensitized virgin females and lactating dams treated with a low dose of the D1/D2 receptor antagonist haloperidol (Pereira and Ferreira 2006), exhibit considerably lower levels of aggression than highly motivated lactating dams. Interestingly, when assessing maternal aggression in the presence of demanding pups, which have an increased incentive value for mothers (Pereira and Ferreira 2006), the deficits in maternal aggression of sensitized virgin females and haloperidol-treated lactating rats are overridden (Ferreira et al. 2012).

The conclusion that aggression represents an appetitive motivation generated from a negative affective state casts doubt on the often reiterated proposal that positive affects derive from the appetitive system and negative affects originate from the aversive system (Davidson 1998, 2000) and highlights the ambiguity of many studies that confound motivational direction with affective valence. Aggression, like anger (Carver and Harmon-Jones 2009), may provide a case in which the affective valence does not coincide with the direction of motivation and may help to reduce this ambiguity as well as to question the theories that propose

[3] Affect is defined to imply a hedonic experience, a sense of valence, and a subjective sense of positivity or negativity arising from an event (Cardinal et al. 2004; Carver and Harmon-Jones 2009; Louilot and Besson 2000).

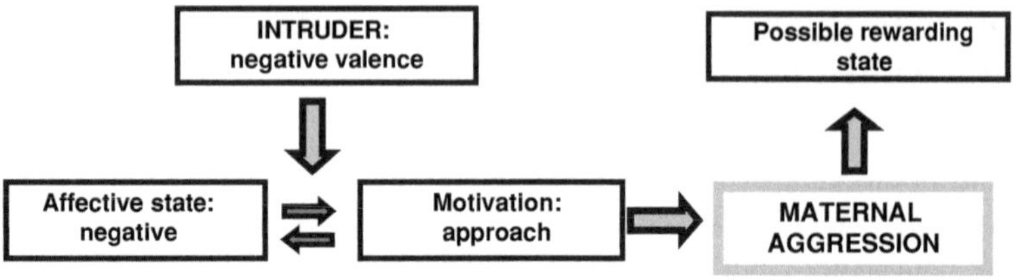

Fig. 4 Summary of affective and motivation processes that may be implicated in the expression of maternal aggression in the rat

that affects are positive and negative dimensions arising from appetitive or aversive motivational systems (Lang et al. 1990).

In contrast to the negative affective value that a male attaches to a lactating female (Erskine et al. 1978b; Ferreira and Hansen 1986), it acquires an attractive value and elicits a positive affective state and an approach motivation in a sexually active female when she can pace the sexual interactions (Bermant and Westbrook 1988; Paredes and Alonso 1997). Therefore, maternal aggression and sexual solicitation may be considered as mutually exclusive behaviors (Lin et al. 2011), supporting the theories that propose that the expression of a motivated behavior implies the inhibition of competitive motivations and that negative and positive affective states (i.e., aggressive and sexual behaviors) are reciprocal opposites. Interestingly, in the physiological context of the postpartum estrus (PPE), females tested in their home cage co-express sexual and maternal aggressive responses toward a male without inhibition (Agrati et al. 2011). The same stimulus, a male, can acquire a double affective valence (positive and negative), indicating that positive and negative valences are not reciprocally opposites in this situation. Moreover, the increase in the negative value of the male, by increasing maternal motivation (i.e., separating the mother from the pups), enhances maternal aggression without reducing sexual behavior (Agrati et al. 2011), as would be expected if they were reciprocal opposites. These observations question the usefulness of the notion of valence to describe the affective system (Colombetti 2005).

5 Neural Control of Maternal Aggression

Few studies have explored the neural basis of maternal aggression and most of them show that brain areas implicated in the control of other forms of aggressive behavior also modulate maternal aggression. Thus, the medial preoptic area of the hypothalamus (MPOA), the anterior hypothalamus (AHA), the ventromedial hypothalamic nucleus (VMH), the lateral septum (LS), several nuclei in the amygdaloid complex, the periaqueductal gray (PAG), and the prefrontal cortex (PFC), which are related to the control of male offensive aggression (De Boer and Koolhaas 2005; Kruk, this volume), also regulate maternal aggression.

5.1 Subcortical Areas

Several hypothalamic regions have been implicated in the control of maternal aggression. Particularly lesions of the *ventromedial hypothalamus (VMH)* considered as a part of the neuronal defensive system (Lin et al. 2011), during the postpartum period reduce maternal aggression both in rats (Hansen 1989) and in mice (Gammie 2005; Lin et al. 2011). Genetic methods for neural circuit manipulation in mice provided a potentially powerful approach in addition to electrical brain-stimulation-evoked aggression. It was shown that optogenetic, but not electrical, stimulation of neurons in the VMH, ventrolateral subdivision (VMHvl) caused male mice to attack both females and inanimate objects, as well as males. Pharmacogenetic silencing of VMHvl reversibly inhibited inter-male aggression. Immediate early gene analysis and single unit recordings from VMHvl during social interactions revealed overlapping but distinct neuronal subpopulations involved in fighting and mating. Neurons activated during attack were inhibited during mating, suggesting a potential neural substrate for competition between these opponent social behaviors (Lin et al. 2011).

On the other hand, the *paraventricular nucleus of the hypothalamus (PVN)* seems to modulate the expression of maternal aggression in the rat, although contradictory data exist proposing either a facilitatory (Consiglio and Lucion 1996) or an inhibitory (Giovenardi et al. 1998) role, or questioning its participation (Olazabal and Ferreira 1997). Nevertheless, several studies show that the PVN is activated during the display of maternal aggression (Gammie and Nelson 2001; Hasen and Gammie 2005) and that the neuropeptide oxytocin (OT) is released in this nucleus during aggressive encounters (Bosch et al. 2005). Consistent with this evidence, the intracerebroventricular administration of antagonists of OT and vasopressin (AVP) to females HAB reduces their levels of aggressive behavior, whereas the chronic central administration of OT or AVP to females LAB enhances their maternal aggression (Bosch and Neumann 2012), suggesting that these neuropeptides promote a dam's aggressive behavior. However, in other studies, an increase in maternal aggression was found after antagonizing OT transmission in the central amygdala or in the paraventricular nucleus (Giovenardi et al. 1998), as well as after the central administration of an AVP antagonist (Nephew and Bridges 2008; Nephew et al. 2010). These studies point toward an inhibitory role of both neuropeptides on a dam's aggressive behavior. It appears that the predisposition to engage in aggressive behavior maybe a critical determinant for the role of OT and AVP in discrete brain areas that extends also to the control of maternal aggression.

Curiously, the role of the *medial preoptic area (MPOA)*, a key area in the control of maternal motivation and behavior (Numan 1974, 2006; Pereira and Morrell 2011), has not been systematically explored regarding maternal aggression. Nevertheless, some studies report a significant increase in the neural activity c-Fos of the MPOA when postpartum mice exhibit maternal aggression (Gammie and Nelson 2001; Gammie et al. 2004; Hasen and Gammie 2005), suggesting a link between the activation of this region and the agonistic behavior of dams.

Table 1 Effects of stimulating 5-HT_{1A} and 5-HT_{1B} receptor activity in different brain regions on the expression of maternal aggression

Brain areas	Receptor	Maternal aggression	Species	References
MRN	5-HT1A	↓ Offensive and defensive	Rats	de Almeida and Lucion 1997
SL	5-HT1A	↑ Offensive	Rats	de Almeida and Lucion 1997
CoMeA	5-HT1A	↓ Offensive and defensive	Rats	de Almeida and Lucion 1997
Dorsal PAG	5-HT1A	↓ Offensive and defensive	Rats	de Almeida and Lucion 1997
VO PFC	5-HT1B	↓ Offensive	Rats	Veiga et al. 2011
DRN	5-HT1A	↑ Offensive	Rats	Veiga et al. 2011

The normal expression of maternal aggression depends critically on intact activity in the *amygdaloid complex*, specially the medial amygdala (MeA), which receives information from pheromones, via the accessory olfactory bulb (Gammie 2005). Accordingly, lactating mice (Gammie and Nelson 2001) and rats (Popeski and Woodside 2004) with high aggressive behavior relative to animals with low levels of aggression show increased Fos expression in MeA following the exposure to a male intruder. GABAergic and serotonergic systems appear to be critical in the amygdaloid complex as evidenced by the reduction in maternal aggression after microinfusion of bicuculline and 8-OH-DPAT into this region (5-HT1A agonist, Table 1) (Hansen and Ferreira 1986b; De Almeida and Lucion 1997).

The *septum* and the *periaqueductal gray (PAG),* neural structures directly involved in the control of defensive and aggressive behaviors in males (Miczek and Grossman 1972; Blanchard et al. 1977; Lonstein and Gammie 2002), also regulate maternal aggression. Thus, lesions in subregions of both areas impair maternal aggression in the rat (Flannelly et al. 1986; Lonstein and Stern 1997; Lonstein and Stern 1998). Besides, the microinfusion of 5-HT_{1A} receptor agonists in the medial septum increases the frequency of attacks in lactating rats, while the microinfusion of these agonists in the caudal PAG reduces aggression (Table 1) (de Almeida and Lucion 1997). Studies in mice show that the administration of GABAergic antagonists in the lateral septum and in the caudal PAG (Lee and Gammie 2010) decreases a dam's aggressive behavior.

5.2 Prefrontal Cortex

The PFC is a crucial area in the modulation of aggressive behavior in several mammalian species including also humans (Blair 2004). While questions have been raised about the functional role of the PFC in rodents versus primates (Ghashghaei and Barbas 2002), compelling arguments have been presented to support these structures to subserve several key functions such as behavioral inhibition and cognitive deficits (Boulougouris et al. 2007). With regard to maternal aggression, Veiga et al. (2007) demonstrated that the ventro-orbital prefrontal cortex (VO PFC) has an important role in maternal aggressive behavior

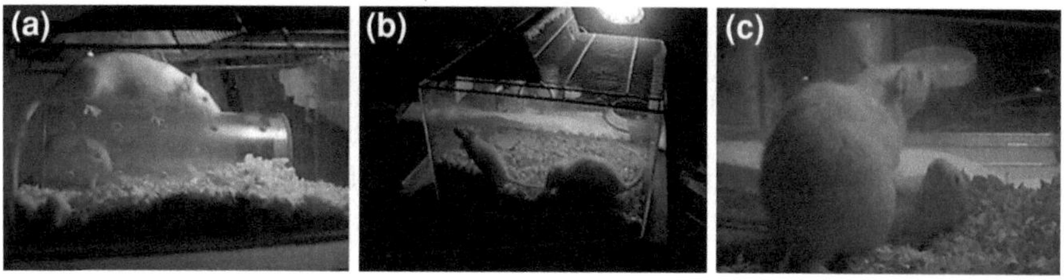

Fig. 5 Resident lactating female investigating the instigator that is protected in acrylic tube for social provocation, on the fifth day postpartum, in the presence of pups (**a**). Female sniffing the intruder during the aggression test (**b**). Female is dominating the male intruder during the aggression test (**c**)

of rats submitted to social provocation[4] (Fig. 5). Importantly, after social instigation lactating rats exhibit an increase in Fos expression in the VO PFC compared to the control group (i.e., no social instigation and no aggressive behavior) without significant changes in Fos immunoreactivity in the DRN. These results point to the VO PFC as an important region in the modulation of maternal aggressive behavior (Veiga et al. 2013).

A main role of the PFC in the control of aggressive behavior is sustained by its key position in neuronal processing circuits, as the PFC projects to several subcortical structures, such as the amygdala (Barbas and Olmos 1990), the ventral striatum (Kunishio and Haber 1994), and the thalamus (Barbas et al. 1991) and receives afferent projections from these regions. Also the different subdivisions of the PFC are strongly connected between them, as well as with other associative cortices (Cavada and Goldman-Rakic 1989). Finally, the PFC is innervated, and also innervates, several neuromodulatory systems including the serotonergic system (Smiley and Goldman-Rakic 1996; Sesack et al. 1989).

5.3 The Serotoninergic System: Dorsal Raphe-PFC

The serotonergic system (5-HT) has been linked more often to aggressive behavior than any other. The simple generalization, that this evolutionary old neurotransmitter is necessary for the inhibition of aggressive behavior, is being replaced with a more differentiating view of serotonergic subsystems (for a review see De Boer and Koolhaas 2005; Takahashi et al. 2011). Only few studies have analyzed the role of different 5-HT receptor subtypes and transporters in maternal aggression.

[4] This procedure consists of placing a clear perforated glass cylinder containing an opponent male ("instigator"), for 5 min in the center of the resident female's home cage. The residents typically threaten the protected instigator and attack the perforated glass cylinder. In this social instigation procedure, rodents initiate attacks with very short latency and high frequency when tested with an intruder in their home cage with the pups after having been provoked previously by an opponent (Potegal 1991).

Veiga et al. (2011) showed that microinjection of the 5-HT_{1A} agonist 8-OH-DPAT into the Dorsal Raphé Nucleus (DRN) increases maternal aggression. This result was attributed to the stimulation of the somatodendritic autoreceptors by 8-OH-DPAT that reduced 5-HT impulse flow and decreased the release of 5-HT in terminal areas including the PFC (Sprouse and Aghajanian 1986; Sinton and Fallon 1988; VanderMaelen et al. 1986; Adell et al. 1993; Bosker et al. 1994; Casanovas et al. 1997; Kreiss and Lucki 1994).

Interestingly, when 8-OH-DPAT was microinjected into DRN and the selective 5-HT_{1B} agonist CP-93,129 was microinjected into VO PFC in female rats concurrently, the increased aggression due to the activation of 5-HT_{1A} autoreceptors in DRN was prevented (Veiga et al. 2011). The participation of 5-HT_{1B} receptors was verified by the reversal of the anti-aggressive effects using the selective antagonist SB-224,289.

These studies confirm the hypothesis that the stimulation of the somatodendritic 5-HT_{1A} autoreceptors in the DRN, that reduces the 5-HT neurotransmission, increases maternal aggression, while the activation of 5-HT_{1B} receptors in the VO PFC exerts anti-aggressive effects probably by increasing 5-HT activity (Table 1).

Table 1 summarizes the effects of stimulating 5-HT_{1A} and 5-HT_{1B} receptor activity in different brain regions such as lateral septum (LS), corticomedial amygdala (CoMeA), dorsal PAG, ventro-orbital prefrontal cortex (VO PFC), median raphé nucleus (MRN), and DRN on the expression of maternal aggression; only microinjections into the LS and DRN result in increased maternal aggressive behavior.

6 Final Remarks

While the endocrine profile of late gestation and parturition, the maternal condition, and stimuli originating from the pups and intruders promote and regulate the display of maternal aggression, other factors such as maternal motivation or other competing motivations also modulate its expression. Several cortical and subcortical areas have been identified in the control of maternal aggression in rodents, and of particular interest, recent studies highlight a role of the PFC, which has been extensively related to aggressive behaviors in humans. We consider maternal aggression as a complex affective behavior, which relies on conserved neural circuits and is dynamically modulated by multiple factors, making it an excellent model for studying the neural basis of aggressive behavior in general.

References

Adams DB (2006) Brain mechanisms of aggressive behavior: an updated review. Neurosci Biobehav Rev 30:304–318

Adell A, Carceller A, Artigas F (1993) In vivo brain dialysis study of the somatodendritic release of serotonin in the Raphe nuclei of the rat: effects of 8-hydroxy-2-(di-n-propylamino)tetralin. J Neurochem 60:1673–1681

Agrati D, Fernández-Guasti A, Ferreño M, Ferreira A (2011) Coexpression of sexual behavior and maternal aggression: the ambivalence of sexually active mother rats toward male intruders. Behav Neurosci 125(3):446–451
Albert DJ, Jonik RH, Walsh ML (1992a) Hormone-dependent aggression in male and female rats: experiential, hormonal, and neural foundations. Neurosci Biobehav Rev 16:177–192
Albert DJ, Jonik RH, Walsh ML (1992b) Ovariectomy does not attenuate aggression by primiparous lactating female rats. Physiol Behav 52:1043–1046
Barbas H, De Olmos J (1990) Projections from the amygdala to basoventral and mediodorsal prefrontal regions in the rhesus monkey. J Comp Neurol 300(4):549–71
Barbas H, Henion TH, Dermon CR (1991) Diverse thalamic projections to the prefrontal cortex in the rhesus monkey. J Comp Neurol 313(1):65–94
Bermant G, Westbrook WH (1988) Peripheral factors in the regulation of sexual contact by female rats. J Comp Physiol Psychol 61:244–250
Blair RJ (2004) The roles of orbital frontal cortex in the modulation of antisocial behavior. Brain Cogn 55:198–208
Blanchard DC, Blanchard RJ, Takahashi LK, Takahashi T (1977) Septal lesions and aggressive behavior. Behav Biol 21(1):157–161
Blanchard DC, Blanchard RJ (1984) Affect and aggression: an animal model applied to human behavior. Adv Study Aggress 1:1–62
Blanchard RJ, Wall PM, Blanchard DC (2003) Problems in the study of rodent aggression. Horm Behav 44(3):161–170
Boulougouris V, Dalley JW, Robbins TW (2007) Effects of orbitofrontal, infralimbic and prelimbic cortical lesions on serial spatial reversal learning in the rat. Behav Brain Res 179(2):219–228
Bosch OJ, Meddle SL, Beiderbeck DI, Douglas AJ, Neumann ID (2005) Brain oxytocin correlates with maternal aggression: link to anxiety. J Neurosci 25:6807–6815
Bosch OJ, Neumann ID (2012) Both oxytocin and vasopressin are mediators of maternal care and aggression in rodents: from central release to sites of action. Horm Behav 61:293–303
Bridges RS (1984) A quantitative analysis of the roles of dosage, sequence, and duration of estradiol and progesterone exposure in the regulation of maternal behavior in the rat. Endocrinology 114:930–940
Bosker F, Klompmakers A, Westenberg H (1994) Extracellular 5-hydroxytryptamine in median raphe nucleus of the conscious rat is decreased by nanomolar concentrations of 8-hydroxy-2-(di-n-propylamino)tetralin and is sensitive to tetrodotoxin. J Neurochem 63:2165–2171
Cardinal RM, Winstanley CA, Robbins TW, Everitt BJ (2004) Limbic corticostriatal systems and delayed reinforcement. Ann N Y Acad Sci 1021:33–50
Casanovas JM, Lésourd M, Artigas F (1997) The effect of the selective 5-HT1A agonists alnespirone (S-20499) and 8-OH-DPAT on extracellular 5-hydroxytryptamine in different regions of rat brain. Br J Pharmacol 122:733–741
Carver CS, Harmon-Jones E (2009) Anger is an approach-related affect: evidence and implications. Psychol Bull 135:183–204
Cavada C, Goldman-Rakic PS (1989) Posterior parietal cortex in rhesus monkey: II. Evidence for segregated corticocortical networks linking sensory and limbic areas with the frontal lobe. J Comp Neurol 287(4):422–45
Colombetti G (2005) Appraising valence. J Conscious Stud 12:103–126
Consiglio AR, Lucion AB (1996) Lesion of hypothalamic paraventricular nucleus and maternal aggressive behavior in female rats. Physiol Behav 59(4–5):591–596
Cosnier J, Couturier C (1966) Comportement maternal provoqué chez les rattes adultes castrees. C R Seances Soc Biol Ses Fil 160:789–791
Davidson RJ (1998) Anterior electrophysiological asymmetries, emotion, and depression: conceptual and methodological conundrums. Psychophysiology 35:607–614
Davidson RJ (2000) Affective style, psychopathology, and resilience: brain mechanisms and plasticity. Am Psychol 55:1196–1214

de Almeida RMM, Lucion AB (1997) 8-OH-DPAT in the median raphe, dorsal periaqueductal gray and corticomedial amygdala nucleus decreases, but in the medial septal area it can increase maternal aggressive behavior in rats. Psychopharmacology 134:392–400

de Almeida RM, Rowlett JK, Cook JM, Yin W, Miczek KA (2004) GABAA/alpha1 receptor agonists and antagonists: effects on species-typical and heightened aggressive behavior after alcohol self-administration in mice. Psychopharmacology 172(3):255–263

De Boer SF, Koolhaas JM (2005) 5-HT_{1A} and 5-HT_{1B} receptor agonists and aggression: a pharmacological challenge of the serotonin deficiency hypothesis. Eur J Pharmacol 526: 125–139

de Sousa FL, Lazzari V, de Azevedo MS, de Almeida S, Sanvitto GL, Lucion AB, Giovenardi M (2010) Progesterone and maternal aggressive behavior in rats. Behav Brain Res 212(1):84–89

Eichenbaum H, Shedlack KJ, Eckrnan KW (1980) Thalamocortical mechanisms in odor guided behavior. I. Effects of lesions of the medial thalamus and frontal cortex on olfactory discrimination in the rat. Brain Behav Evol 17:255–275

Erskine MS, Barfield RJ, Goldman BD (1978a) Intraspecific fighting during late pregnancy and lactation in rats and effects of litter removal. Behav Biol 23:206–218

Erskine MS, Denenberg ZV, Goldman BD (1978b) Aggression in the lactating rat: effects of intruder age and test arena. Behav Biol 23:52–66

Erskine MS, Barfield RJ, Goldman BD (1980a) Postpartum aggression in rats. I. Effects of hypophysectomy. J Comp Physiol Psychol 94:484–494

Erskine MS, Barfield RJ, Goldman BD (1980b) Postpartum aggression in rats. II. Dependence on maternal sensitivity to young and effects of experience with pregnancy and parturition. J Comp Physiol Psychol 94:495–505

Factor EM, Mayer AD, Rosenblatt JS (1993) Peripeduncular nucleus lesions in the rat: I. Effects on maternal aggression, lactation, and maternal behavior during pre- and postpartum periods. Behav Neurosci 107:166–185

Ferreira A, Hansen S (1986) Sensory control of maternal aggression en *Rattus norvegicus*. J Comp Psychol 2:173–177

Ferreira A, Dahlöf LS, Hansen S (1987) Olfactory mechanisms in the control of maternal aggression, appetite, and fearfulness: effects of lesions to olfactory receptors, mediodorsal thalamic nucleus, and insular prefrontal cortex. Behav Neurosci 101:709–717

Ferreira A, Hansen S, Nielsen M, Archer T, Minor BG (1989) Behavior of mother rats in conflict tests sensitive to antianxiety agents. Behav Neurosci 103:193–201

Ferreira A, Pereira M, Agrati D, Uriarte N, Fernández-Guasti A (2002) Role of maternal behavior on aggression, fear and anxiety. Physiol Behav 77:197–204

Ferreira A, Uriarte N, Pereira M (2012) Maternal aggression is a maternally motivated behavior modulated by the pups (Abstract). 8th The Federation of European Neuroscience Societies (FENS) Forum of Neuroscience, Spain

Flannelly KJ, Kemble ED, Blanchard DC, Blanchard RJ (1986) Effects of septal-forebrain lesions on maternal aggression and maternal care. Behav Neural Biol 45(1):17–30

Fleming AS, Luebke C (1981) Timidity prevents the virgin female rat from being a good mother: emotionality differences between nulliparous and parturient females. Physiol Behav 27:863–868

Fleming AS, Cheung U, Myhal N, Kessler Z (1989) Effects of maternal hormones on 'timidity' and attraction to pup-related odors in female rats. Physiol Behav 46:449–453

Fish EW, De Bold JF, Miczek KA (2002) Aggressive behavior as a reinforcer in mice: activation by allopregnanolone. Psychopharmacology 163(3–4):459–466

Gammie SC, Nelson RJ (2001) cFOS and pCREB activation and maternal aggression in mice. Brain Res 898:232–241

Gammie SC, Negron A, Newman SM, Rhodes JS (2004) Corticotropin-releasing factor inhibits maternal aggression in mice. Behav Neurosci 118(4):805–814

Gammie SC (2005) Current models and future directions for understanding the neural circuitries of maternal behaviors in rodents. Behav Cogn Neurosci Rev 4:119–135
Gandelman R, Simon NG (1980) Postpartum fighting in the rat: nipple development and the presence of young. Behav Neural Biol 28:350–360
Giovenardi M, Padoin MJ, Cadore LP, Lucion AB (1998) Hypothalamic paraventricular nucleus modulates maternal aggression in rats: effects of ibotenic acid lesion and oxytocin antisense. Physiol Behav 63:351–359
Giovenardi M, Consiglio AR, Barros HM, Lucion AB (2000) Pup age and aggressive behavior in lactating rats. Braz J Med Biol Res 33:1083–1088
Ghashghaei HT, Barbas H (2002) Pathways for emotion: interactions of prefrontal and anterior temporal pathways in the amygdala of the rhesus monkey. Neuroscience 115(4):1261–1279
Haney M, DeBold JF, Miczek KA (1989) Maternal aggression in mice and rats towards male and female conspecifics. Aggress Behav 15:443–453
Hansen S, Ferreira A, Selart ME (1985) Behavioural similarities between mother rats and benzodiazepine-treated non-maternal animals. Psychopharmacology (Berlin) 86:344–347
Hansen S, Ferreira A (1986a) Food intake, aggression, and fear behavior in the mother rat: control by neural systems concerned with milk ejection and maternal behavior. Behav Neurosci 100:64–70
Hansen S, Ferreira A (1986b) Effects of bicuculline infusions in the ventromedial hypothalamus and amygdaloid complex on food intake and affective behavior in mother rats. Behav Neurosci 100:410–415
Hasen NS, Gammie SC (2005) Differential fos activation in virgin and lactating mice in response to an intruder. Physiol Behav 84(5):681–695
Hård E, Hansen S (1985) Reduced fear behavior in the lactating rat. Physiol Behav 33:641–643
Kolunie JM, Stern JM (1990) Maternal aggression: disruption by perioral anesthesia in lactating Long-Evans rats (*Rattus norvegicus*). J Comp Psychol 104:352–360
Kolunie JM, Stern JM, Barfield RJ (1994) Maternal aggression in rats: effects of visual or auditory deprivation of the mother and dyadic pattern of ultrasonic vocalizations. Behav Neural Biol 62:41–49
Kolunie JM, Stern JM (1995) Maternal aggression in rats: effects of olfactory bulbectomy, ZnSo4-induced anosmia, and vomeronasal organ removal. Horm Behav 29:492–518
Kreiss DS, Lucki I (1994) Differential regulation of serotonin (5-HT) release in the striatum and hippocampus by 5-HT1A autoreceptors of the dorsal and median raphe nuclei. J Pharmacol Exp Ther 269:1268–1279
Kunishio K, Haber SN (1994) Primate cingulostriatal projection: limbic striatal versus sensorimotor striatal input. J Comp Neurol 350(3):337–356
Lang PJ, Bradley MM, Cuthbert BN (1990) Emotion, attention, and the startle reflex. Psychol Rev 97:377–395
Lee G, Gammie SC (2010) GABAA receptor signaling in caudal periaqueductal gray regulates maternal aggression and maternal care in mice. Behav Brain Res 213(2):230–237
Lin D, Boyle MP, Dollar P, Lee H, Lein ES, Perona P, Anderson DJ (2011) Functional identification of an aggression locus in the mouse hypothalamus. Nature 470:221–226
Lonstein JS, Stern JM (1997) Role of the midbrain periaqueductal gray in maternal nurturance and aggression: c-fos and electrolytic lesion studies in lactating rats. J Neurosci 17(9):3364–3378
Lonstein JS, Stern JM (1998) Site and behavioral specificity of periaqueductal gray lesions on postpartum sexual, maternal, and aggressive behaviors in rats. Brain Res 804(1):21–35
Lonstein JS, Gammie SC (2002) Sensory, hormonal, and neural control of maternal aggression in laboratory rodents. Neurosci Biobehav Rev 26:869–888
Lonstein JS (2005) Reduced anxiety in postpartum rats requires recent physical interactions with pups, but is independent of suckling and peripheral sources of hormones. Horm Behav 47(3):241–255

Louilot A, Besson C (2000) Specificity of amygdalostriatal interactions in the involvement of mesencephalic dopaminergic neurons in affective perception. Neuroscience 96:73–82

Lucion AB, De Almeida RMM (1996) On the dual nature of maternal aggression in rats. Aggress Behav 22(5):365–373

Lynds PG (1976) Olfactory control of aggression in lactating female house mice. Physiol Behav 17:157–159

Mann MA, Svare B (1982) Factors influencing pregnancy-induced aggression in mice. Behav Neural Biol 36(3):242–258

Mayer AD, Rosenblatt JS (1987) Hormonal factors influence the onset of maternal aggression in laboratory rats. Horm Behav 21:253–267

Mayer AD, Carter L, Jorge WA, Mota MJ, Tannu S, Rosenblatt JS (1987a) Mammary stimulation and maternal aggression in rodents: thelectomy fails to reduce pre- or postpartum aggression in rats. Horm Behav 21:501–510

Mayer AD, Reisbick S, Siegel HI, Rosenblatt JS (1987b) Maternal aggression in rats: changes over pregnancy and lactation in a Sprague–Dawley strain. Aggress Behav 13:29–43

Mayer AD, Ahdieh HB, Rosenblatt JS (1990a) Effects of prolonged estrogen-progesterone treatment and hypophysectomy on the stimulation of short-latency maternal behavior and aggression in female rats. Horm Behav 24:152–173

Mayer AD, Monroy MA, Rosenblatt JS (1990b) Prolonged estrogen-progesterone treatment of nonpregnant ovariectomized rats: factors stimulating home-cage and maternal aggression and short-latency maternal behavior. Horm Behav 24:342–364

Mayer AD, Rosenblatt JS (1993a) Persistent effects on maternal aggression of pregnancy but not of estrogen/progesterone treatment of nonpregnant ovariectomized rats revealed when initiation of maternal behavior is delayed. Horm Behav 27:132–155

Mayer AD, Rosenblatt JS (1993b) Contributions of olfaction to maternal aggression in laboratory rats (*Rattus norvegicus*): effects of peripheral deafferentation of the primary olfactory system. J Comp Psychol 107:12–24

Meisel RL, Joppa MA (1994) Conditioned place preference in female hamsters following aggressive or sexual encounters. Physiol Behav 56(5):1115–1118

Miczek KA, Grossman SP (1972) Effects of septal lesions on inter- and intraspecies aggression in rats. J Comp Physiol Psychol 79(1):37–45

Mos J, Olivier B, van der Poel AM (1987a) Modulatory actions of benzodiazepine receptor ligands on agonistic behaviour. Physiol Behav 41:265–278

Mos J, Olivier B, van Oorschot R (1987b) Maternal aggression towards different sized male opponents: effect of chlordiazepoxide treatment of the mothers and d-amphetamine treatment of the intruders. Pharmacol Biochem Behav 26:577–584

Moyer KE (1976) The psychobiology of aggression. Harper & Row, New York

Nephew BC, Bridges RS (2008) Central actions of arginine vasopressin and a V1a receptor antagonist on maternal aggression, maternal behavior, and grooming in lactating rats. Pharmacol Biochem Behav 91(1):77–83

Nephew BC, Byrnes EM, Bridges RS (2010) Vasopressin mediates enhanced offspring protection in multiparous rats. Neuropharmacology 58(1):102–106

Neumann ID, Toschi N, Ohl F, Torner L, Krömer SA (2001) Maternal defence as an emotional stressor in female rats: correlation of neuroendocrine and behavioural parameters and involvement of brain oxytocin. Eur J Neurosci 13:1016–1024

Numan M (1974) Medial preoptic area and maternal behavior in the female rat. J Comp Physiol Psychol 87(4):746–759

Numan M (2006) Hypothalamic neural circuits regulating maternal responsiveness toward infants. Behav Cogn Neurosci Rev 5(4):163–190

Olazabal DE, Ferreira A (1997) Maternal behavior in rats with kainic acid-induced lesions of the hypothalamic paraventricular nucleus. Physiol Behav 61(5):779–784

Olivier B, Mos J, van Oorschot R (1985) Maternal aggression in rats: effects of chlordiazepoxide and fluprazine. Psychopharmacology (Berlin) 86:68–76
Paredes RG, Alonso A (1997) Sexual behavior regulated (paced) by the female induces conditioned place preference. Behav Neurosci 111:123–128
Pereira M, Uriarte N, Agrati D, Zuluaga MJ, Ferreira A (2005) Motivational aspects of maternal anxiolysis in lactating rats. Psychopharmacology (Berlin) 180:241–248
Pereira M, Ferreira A (2006) Demanding pups improve maternal behavioral impairments in sensitized and haloperidol-treated lactating female rats. Behav Brain Res 175:139–148
Pereira M, Morrell JI (2011) Functional mapping of the neural circuitry of rat maternal motivation: effects of site-specific transient neural inactivation. J Neuroendocrinol 23(11):1020–1035
Popeski N, Woodside B (2004) Central nitric oxide synthase inhibition disrupts maternal behavior in the rat. Behav Neurosci 118(6):1305–1316
Potegal M (1991) Attack priming and satiation in female golden hamsters: tests of some alternatives to the aggression arousal interpretation. Aggress Behav 17:327–335
Rosenblatt JS (1967) Nonhormonal basis of maternal behavior in the rat. Science 156:1512–1514
Sesack SR, Deutch AY, Roth RH, Bunney BS (1989) Topographical organization of the efferent projections of the medial prefrontal cortex in the rat: an anterograde tract-tracing study with Phaseolus vulgaris leucoagglutinin. J Comp Neurol 290(2):213–242
Siegel HI, Rosenblatt JS (1975) Hormonal basis of hysterectomy-induced maternal behavior during pregnancy in the rat. Horm Behav 6:211–222
Sinton CM, Fallon SL (1988) Electrophysiological evidence for a functional differentiation between subtypes of the 5-HT1 receptor. Eur J Pharmacol 157:173–81
Slotnick BM, Kaneko N (1981) Role of mediodorsal thalamic nucleus in olfactory discrimination learning in rats. Science 214:91–92
Smiley JF, Goldman-Rakic PS (1996) Serotonergic axons in monkey prefrontal cerebral cortex synapse predominantly on interneurons as demonstrated by serial section electron microscopy. J Comp Neurol 367(3):431–443
Sprouse JS, Aghajanian GK (1986) Propranolol blocks the inhibition of serotonergic dorsal raphe cell firing by 5-HT1A selective agonists. Eur J Pharmacol 128:295–298
Stern JM, Johnson SK (1989) Perioral somatosensory determinants of nursing behavior in Norway rats (*Rattus norvegicus*). J Comp Psychol 103(3):269–280
Stern JM, Kolunie JM (1989) Perioral anesthesia disrupts maternal behavior during early lactation in Long-Evans rats. Behav Neural Biol 52(1):20 38
Stern JM, Kolunie JM (1991) Trigeminal lesions and maternal behavior in Norway rats: I. Effect of cutaneous rostral snout denervation on maintenance of nurturance and maternal aggression. Behav Neurosci 105:984–997
Stern JM, Kolunie JM (1993) Maternal aggression of rats is impaired by cutaneous anesthesia of ventral trunk but not by nipple removal. Physiol Behav 54:861–868
Svare B, Gandelman R (1976a) Postpartum aggression in mice: the influence of suckling stimulation. Horm Behav 7(4):407–416
Svare B, Gandelman R (1976b) A longitudinal analysis of maternal aggression in Rockland-Swiss albino mice. Dev Psychobiol 9(5):437–446
Svare B (1977) Maternal aggression in mice: influence of the young. Biobehav Rev 1(3):151–164
Svare B, Mann M, Samuels O (1980) Mice: suckling stimulation but not lactation important for maternal aggression. Behav Neural Biol 29(4):453–462
Svare B, Mann MA (1983) Hormonal influences on maternal aggression in mammals. In: Svare B (ed) Hormones and aggressive behavior. Plenum Press, New York
Vandermaelen CP, Matheson GK, Wilderman RC, Patterson LA (1986) Inhibition of serotoninergic dorsal raphe neurons by systemic and iontophoretic administration of buspirone, a non-benzodiazepine anxiolytic drug. Eur J Pharmacol 129:123–130
Veiga CP, Miczek KA, Lucion AB, de Almeida RMM (2007) Effect of 5-HT$_{1B}$ receptor agonists injected into the prefrontal cortex on maternal aggression in rats. Braz J Med Biol Res 40:825–830

Veiga CP, Miczek KA, Lucion AB, de Almeida RMM (2011) Social instigation and aggression in postpartum female rats: role of 5-HT_{1A} and 5-HT_{1B} receptors in the dorsal raphé nucleus and prefrontal cortex. Psychopharmacology 213:475–487

Veiga CP, Winkelmann-Duarte E, Pavesi E, Louzada S, Miczek KA, Lucion AB, de Almeida RMM (2013) *Fos* expression in the prefrontal cortex and dorsal raphe nucleus mesencephalic in lactating rats after social instigation. Psychol Neurosci 6(1):115–121

Takahashi A, Quadros IM, de Almeida RMM, Miczek KA (2011) Brain serotonin receptors and transporters: initiation versus termination of escalated aggression. Psychopharmacology 213:183–212

Nitric Oxide and Serotonin Interactions in Aggression

Tracy A. Bedrosian and Randy J. Nelson

Abstract Violence is a significant public health problem worldwide. Neurobiological research on violence and aggression attempts to elucidate the cellular and molecular pathways that increase the propensity toward this behavior. Research over the past 40 years has implicated several brain regions and neurotransmitters in aggression, mainly using rodent models. Perhaps the strongest association is the link between serotonin and aggression, which has compelling interactions with the nitric oxide system. Recently, new insights into these relationships have been added as modern techniques allow more sophisticated analyses. This chapter will discuss current developments implicating serotonin and nitric oxide in aggressive behavior. Recently developed high-resolution methods for examining the neurobiological basis of aggression will be considered, with emphasis on future directions for the field.

Keywords 5-HT · nNOS · NO · Impulsivity · Resident-intruder · Violence

Contents

T. A. Bedrosian (✉)
Department of Neuroscience, The Ohio State University Wexner Medical Center, 636 Biomedical Research Tower, 460 W. 12th Avenue, Columbus, OH 43210, USA
e-mail: Bedrosian.2@osu.edu

R. J. Nelson
Department of Neuroscience, The Ohio State University Wexner Medical Center, 4084 Graves Hall, 333 W. 10th Avenue, Columbus, OH 43210, USA
e-mail: Randy.Nelson@osumc.edu

Curr Topics Behav Neurosci (2014) 17: 131–142
DOI: 10.1007/7854_2013_273

Published Online: 25 December 2013

1 Introduction

Violence represents a significant public health problem worldwide. In 2010, the Federal Bureau of Investigation (FBI) reported over 1 million violent crimes (i.e., involving force or the threat thereof) in the U.S. alone, with murders comprising almost 15,000 of these cases. Neurobiological research on violence and aggression attempts to elucidate the cellular and molecular pathways that increase the propensity for aggressive behavior. Research over the past 40 years has implicated a number of brain regions and neurotransmitters in aggression, mainly using rodent models. Perhaps the strongest association is the link between serotonin (5-HT) and aggression, which has compelling interactions with the nitrergic (nitric oxide [NO]) system.

Historically, evidence implicating serotonin in aggression came from studies investigating the effects of isolation on mice. The first proposed link was reported over 40 years ago (Garattini et al. 1967; Giacalone et al. 1968; Maas 1962). Since that time, serotonin has been implicated in aggressive behavior more than any other neurotransmitter. In general, 5-HT is negatively correlated with aggression in animals and violent behavior in humans. These relationships have been demonstrated in diverse species ranging from lobsters to rodents to humans (Simon and Lu 2005). Pharmacological evidence suggests that increasing 5-HT tone with serotonin precursors, reuptake inhibitors, or receptor agonists reduces aggression in rodents (Chiavegatto et al. 2001; Fish et al. 1999; Lyons et al. 1999; Miczek et al. 1998; Olivier et al. 1995). Genetic evidence confirms that lack of the gene for 5-HTR1B increases rodent aggression (Saudou et al. 1994). In humans, excessive aggression and the related trait of impulsivity are correlated with low CSF concentrations of 5-HT metabolites (reviewed in (Lesch and Merschdorf 2000)).

In contrast, the role of nitric oxide in aggressive behavior was discovered serendipitously. NO is a gaseous neurotransmitter capable of diffusing directly into cells and acting as a messenger. It is extremely labile, with a half-life of <5 s, and is therefore synthesized as needed. In the brain, this is accomplished by the enzyme neuronal nitric oxide synthase (*nNOS* or NOS1), using the precursor L-arginine (Fig. 1). Given its lability, NO is more easily manipulated indirectly by targeting *nNOS*. The role of NO in aggression in mammals was discovered this way. While establishing a colony of mice lacking the gene for *nNOS*, it was observed by laboratory staff that mice were suffering apparently violent attacks by their cage mates. When the mice were tested in a resident-intruder model, it became clear that *nNOS* knockout mice were violently and inappropriately

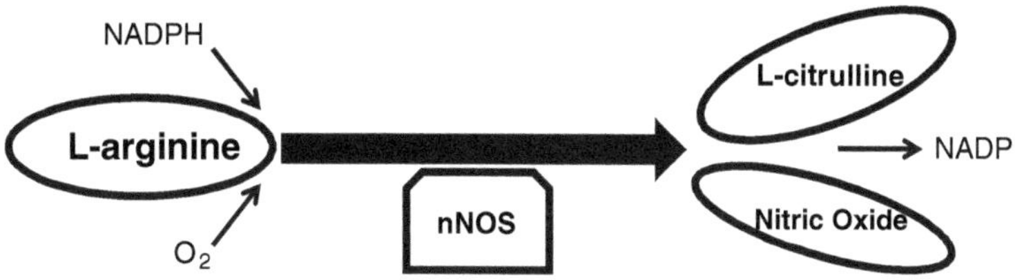

Fig. 1 Synthesis of nitric oxide from L-arginine in the brain by neuronal nitric oxide synthase (*nNOS*).

aggressive, displaying 3–4 times more aggression than wild-type residents and initiating attacks six times more frequently. *nNOS* knockout mice continued to attack even after the intruder had shown signs of submission (Nelson et al. 1995).

Recently, interactions between the 5-HT and NO systems have been identified and new insights on these relationships have been added as modern techniques allow more sophisticated analyses. This chapter will discuss current evidence implicating serotonin and NO in aggressive behavior. Attention will be given to recent highlights and serotonin-NO interactions, as the role of serotonin in aggression is being reviewed comprehensively elsewhere in this edition. Recently developed high-resolution methods for examining the neurobiological basis of aggression will also be considered, with emphasis on future directions for the field.

2 Serotonin and Aggression: Recent Highlights

2.1 Animal Models

Impulsive aggression represents a behavior to which serotonin and nitric oxide signaling both contribute. Recent evidence has delineated some of the serotonergic contributions in more detail using animal models. In addition to other forms of aggression, serotonin transmission has been associated with aggression related to trait impulsivity. In humans, behavioral disinhibition tends to be related to conduct or personality disorders, which may include aggression or violence. Two recently developed rodent models may be useful for studying this relationship. For one, a genetic knock-out mouse for the *Tph2* gene was developed. These mice lack the neuronal form of tryptophan hydroxylase, the rate-limiting enzyme in the synthesis of serotonin, and thus lack brain serotonin. The 5-HT neuronal system remains intact, however, as administration of the downstream intermediate 5-hydroxytryptophan (5-HTP) leads to the synthesis of serotonin. *Tph2* knockout mice demonstrate strong compulsive and impulsive responses in well-accepted behavioral tests, along with heightened resident-intruder aggression. Even when socialized, only 30 min of isolation is sufficient to produce intense intruder-directed aggression (Angoa-Perez et al. 2012). This study employs a “gene to behavior” approach that is useful because of the specificity of the genetic manipulation; the disadvantages of this approach include the potential developmental effects of the deletion, as well as low resolution in terms of which brain regions contribute to the behavioral phenotype. Another approach would be to examine *Tph2* expression in particular brain regions using animals that naturally exhibit high impulsivity and aggression, as well as examining mice in which *Tph2* gene expression could be deleted in adulthood.

An example of the former study was recently performed using rats selectively bred for innate response to a novel environment. Selectively bred High (bHR) and Low (bLR) Responder rats differ in emotionality and behavioral control. bHRs exhibit high levels of exploration when placed in a novel environment, along with impulsivity in a variety of behavioral tasks. bLRs, in contrast, tend to be inhibited in a novel environment and exhibit depressive- and anxiety-like behaviors. In this experiment, bHR rats display high levels of aggression toward a male intruder, whereas bLRs show low levels of aggression. Expression of baseline *Tph2* and *Sert* increased among bHRs in select raphe nuclei (Kerman et al. 2011). Furthermore, bHRs exhibit distinct patterns of brain *c-fos* expression and reduced expression of *5HT-1B* receptor in emotion-regulating brain regions (Clinton et al. 2011). These studies, taken together, demonstrate from both "gene to behavior" and "behavior to gene" approaches that altered serotonergic transmission is a common molecular pathway mediating impulsivity and aggression.

Besides actions of serotonin itself, modulatory neurotransmitters have gained more attention recently for their ability to indirectly regulate aggressive behavior by fine-tuning serotonergic tone. GABA(B) receptor was recently implicated in the control of serotonergic dorsal raphe neurons, leading to heightened aggression in mice. Administration of the GABA(B) agonist baclofen into the dorsal raphe escalates aggressive behavior, whereas administering baclofen into the median raphe, or a GABA(A) agonist into either region, does not affect aggression. It is likely that this behavioral effect is due to modulation of serotonin neurons because co-administering baclofen with a 5-HT(1A) agonist that inhibits 5-HT activity blocks the increase in aggression. Further, measuring 5-HT levels using in vivo microdialysis confirms that GABA(B) activation increases extracellular serotonin in medial prefrontal cortex (Takahashi et al. 2010).

Other examples of the importance of modulatory neurotransmission are becoming apparent. For example, the role of monoamine oxidase (MAOA), the key enzyme involved in degrading serotonin and norepinephrine, in aggression was recently shown to be regulated by NMDARs. Changes in NMDAR subunit composition in the prefrontal cortex mediate the role of MAOA in aggressive behaviors (Bortolato et al. 2012). Mice lacking an important CREB coactivator (CRTC1) exhibit impulsive aggression and other altered emotional behaviors. These mice display reduced serotonin turnover in the prefrontal cortex, implicating CRTC1 in mood and aggression regulation, possibly through altered serotonin transmission (Breuillaud et al. 2012).

2.2 Humans

Recent studies in humans have also identified candidates putatively involved in impulsive aggression. Neuropeptide Y (NPY) was previously implicated in aggression solely on the basis of animal studies. A recent study comparing patients with personality disorders to healthy individuals provided the first evidence that NPY has a role in human aggression too. NPY-like immunoreactivity in CSF samples is elevated in personality disorders compared to healthy individuals. Furthermore, NPY-like immunoreactivity is also increased in subjects with higher impulsivity and aggression scores, as well as in those with intermittent explosive disorder (Coccaro et al. 2012). NPY acts through Y1 receptors to regulate the serotonin system (Karl et al. 2004), though it is not quite clear yet whether this finding is directly related to the elevated NPY in aggressive subjects or a modulatory effect on the 5-HT system, as *Y1* knockout mice also have reduced *Tph2*, the rate-limiting synthetic enzyme for serotonin.

Developmental interactions between serotonin and aggression have been documented, but how altered levels of 5-HT during development might contribute to later life brain function and behavior remains a central question. In rodents, epigenetic effects of early life have been shown to have a major influence on adult phenotype. In humans, there is currently no way to assay epigenetic modifications in living brain tissue, so other proxy measures must be found. One such recent study reported levels of DNA methylation on certain genes in the 5-HT pathway in human peripheral white blood cells. Using a sample of adult male subjects with a history of childhood physical aggression, brain serotonin synthesis was evaluated using positron emission tomography (PET). Males with childhood aggression have reduced 5-HT synthesis in the orbitofrontal cortex, along with higher methylation of a serotonin transporter gene. In vitro methylation of this gene suppresses its transcriptional activity. This is the first study to identify a potential proxy for brain methylation states in humans that is linked to a history of physical aggression and possible vulnerability to impulsive behavior (Wang et al. 2012). Each of these recent findings, both in human studies and rodent models, highlights an important contribution to our understanding of the role of 5-HT in impulsive aggression.

3 Nitric Oxide and Aggression: Recent Findings

3.1 Animal Models

Following the initial observation of aggression among *nNOS* knockout mice, other studies have confirmed the behavioral role for *nNOS* in aggression using pharmacological methods. Treatment of wild-type mice with 7-nitroindazole, a pharmacological *nNOS* inhibitor, increases aggression and dramatically reduces brain citrulline, a byproduct of nitric oxide synthesis that is produced at a 1:1 ratio and typically used as a marker for NO in vivo (Demas et al. 1997). Additionally, as some aggressive personality disorders coincide with social withdrawal, it was investigated whether such a phenotype was observed in *nNOS* deficient mice. Mice with either a genetic deletion of *nNOS* or a pharmacological knockdown using 3-bromo-7-nitroindazole spend less time investigating a novel intruder mouse behind a barrier, but are more aggressive (Trainor et al. 2007). This evidence strengthens the case of *nNOS* involvement in aggression, but why male *nNOS* knockout mice exhibit violent aggression, whereas females do not, remains a central question. Even lactating females, who typically exhibit high levels of maternal aggression to protect their young, display significant deficits in aggression when they lack the *nNOS* gene (Gammie and Nelson 1999). Male *nNOS* knockout mice have higher levels of testosterone than their wild-type counterparts, so an experiment was performed to determine whether testosterone was necessary and sufficient to produce an aggressive phenotype. Indeed, castration reduces aggression in wild-type and *nNOS* knockout mice, whereas testosterone replacement restores aggression to pre-castration levels in both groups (Kriegsfeld et al. 1997).

Further studies have extended the role for nitric oxide signaling in rodent aggression. It was known that housing in short day lengths exaggerated aggressive behavior in Siberian hamsters (*Phodopus sungorus*), a seasonally breeding species, and that this behavioral change was associated with increased brain expression of neuronal nitric oxide synthase (*nNOS*), the synthetic enzyme for NO in the brain (Wen et al. 2004). It was unknown, however, whether directly manipulating levels of *nNOS* might in turn have a behavioral effect. Treatment with 3-bromo-7-nitroindazole, a selective inhibitor of *nNOS*, increases intruder-directed aggression in male hamsters housed in long-days, but not in short-days, when *nNOS* levels are already reduced (Bedrosian et al. 2012).

3.2 Humans

Until recently, NO signaling was implicated in aggression on the basis of rodent studies alone. Work by Reif and colleagues has now provided the link in human subjects as well. First, a NOS1 promoter repeat length variation (NOS1 Ex1f variable number tandem repeat [VNTR]) was identified and its association with impulsive/aggressive characteristics was assessed. The short repeat variant is associated more frequently with adult ADHD, personality disorder, and aggression, which is interesting because short variants lead to reduced transcriptional activity of the NOS1 exon 1f promoter (Reif et al. 2009). Another question was to determine whether adverse environmental events interact with the short variant to increase risk for impulsive and aggressive behavior. In a sample of Estonian pupils, followed from age 9 to 18, the short variant allele was again associated with more impulsivity. Furthermore, stressful life events and adverse family environment interact with the short variant to increase the risk of maladaptive impulsivity (Reif et al. 2010). These studies are the first to provide evidence in humans to support a link between NOS1 and impulsive and aggressive behaviors.

4 Serotonin–NO Interactions

The mechanism whereby NO influences aggressive behavior was unclear, but recent evidence demonstrates that NO modulates serotonergic function to affect behavior in rodents. Male mice with a genetic deletion of *nNOS* have altered levels of 5-HT and its metabolite, 5-HIAA, as well as reduced serotonin turnover in distinct brain regions involved in aggression, such as the amygdala and hypothalamus (Table 1), whereas treatment with the serotonin precursor, 5-hydroxytryptamine, to increase serotonin transmission reduces aggression in these mice. The serotonin receptors $5HT_{1A}$ and $5HT_{1B}$ also show hypofunction in *nNOS* knockout mice (Chiavegatto et al. 2001). These data clearly demonstrate the important interaction between NO and 5-HT in regulating rodent aggression. It remains to be determined whether or not the decreased $5HT_{1A}$ and $5HT_{1B}$ receptor function reflects only the mechanisms of elevated aggression in the *nNOS* knockout mice, as these receptors have not been assessed in mice with pharmacological inhibition of NO.

Whether such an interaction is at play in humans remains unclear, but a recent study provided some evidence in support of this notion. Platelet activity of monoamine oxidase (MAO), an enzyme responsible for degrading serotonin and other transmitters, has been associated with impulsivity in clinical samples. Whether the activity of MAO interacts with NOS1 genetic variants remained undetermined. Laas and colleagues compared genotype with platelet MAO activity and self-reported impulsivity scores, and reported that short variant NOS1 alleles are related to elevated impulsivity. Furthermore, if short variant individuals have mid range MAO activity, then they have the highest impulsivity scores (Laas et al. 2010). Certainly using self-reports of impulsivity is one limitation of this study, but the interaction bears similarity to results in rodents previously described, making the notion of an NO-serotonin interaction in impulsive or aggressive traits among humans plausible.

5 Advanced Methods and Future Agenda

Violence and aggression present a universal challenge for which there is no satisfactory solution. Modern technology is allowing for higher resolution methods, which will pave the way to a better understanding of the neural mechanisms of aggression. Traditionally, studies using electrical stimulation or brain lesioning have been used to identify brain regions implicated in aggression or attack behavior. But these methods lack specificity in terms of targeting very specific cell

Table 1 Serotonin transmission in *nNOS* knockout mice

Change relative to WT			
Brain region	5-HT	5-HIAA	5-HIAA/5-HT
Cerebral cortex	↑	↓	↓
Hypothalamus	↑	↓	↓
Hippocampus	↑	↓	↓
Amygdala	↓	↓	↓
Midbrain	↑	↓	↓
Cerebellum	↑	↑	↓

groups. Using these approaches, it is very difficult to restrict the stimulation or lesion to one group of cell bodies, without also affecting adjoining areas or fibers of passage. Similarly, it is problematic making temporal connections between the lesioned region and a particular behavior, as one cannot turn on and off the lesion. Recently, however, new genetic approaches for targeting very specific cell groups have made it possible to obtain the type of spatial and temporal resolution necessary for probing the neural basis of aggression.

Optogenetics is among the most innovative new techniques of the past decade. This approach combines optical stimulation or silencing with precise genetic targeting of cells, which has allowed unprecedented control over neuronal function in awake and behaving animals. Briefly, light-activated channels are introduced and targeted to specific cell groups using genetic manipulations. Certain channels can be used for neuronal stimulation (e.g., channelrhodopsins), whereas others may be used for neuronal inhibition (e.g., halorhodopsins). By implanting a light source, such as fiber optics, directly into the brain of an animal, it is possible to stimulate or silence very precise cell groups in order to observe their direct behavioral output. The technology has the potential to significantly advance our understanding of neural circuits governing behavior, and will undoubtedly contribute to elucidating pathways underlying aggression.

Lin and colleagues (Laas et al. 2010) applied an optogenetic approach combined with in vivo electrophysiological recording to identify an aggression locus in the mouse hypothalamus. Stimulation with a fiber optic cable of neurons in the VMHvl expressing channelrhodopsin rapidly elicited attack behavior when the male mouse was presented with an intruder, regardless of whether it was a male, female, or castrated male intruder. Even when presented with an anesthetized intruder or inanimate object, light stimulation provoked attack behavior, especially if the intruder or object was moved by the experimenter. In the absence of an intruder, optic stimulation did not provoke any significant behavioral changes. And when other cell groups were targeted, specifically the VMHdl and VMHc, flight and freezing behaviors were initiated. Electrophysiological recording from the mice during the encounter with an intruder demonstrated that spiking activity of neurons in the VMHvl increased after introduction of the intruder and as the encounter progresses (Lin et al. 2011). This study was the first of its kind to harness the temporal and spatial precision of optogenetics to probe the neural circuitry of attack behavior.

Certainly we have only begun to broach the capacity of this technology for studying aggression. For example, genetic targeting of specific serotonergic cell groups combined with optogenetic stimulation or silencing in behaving animals may be one avenue for research aimed at unraveling the complex role of serotonin in aggressive behavior. Further, defining such circuits may lead us to identify specific pathologies, or "circuitopathies," that contribute to specific forms of aggression (Anderson 2012). Identifying such circuitry may open the possibility of human treatments using modern neuromodulation techniques. Currently, deep brain stimulation is successfully used to treat a variety of disorders, including Parkinson's disease, tremor, chronic pain, and major depression. In this surgical procedure, an implantable pacemaker device is introduced into a specific brain region to modulate electrical activity. The possibility of extending the current use of this procedure to treating aggressive or violent behavioral disorders is an intriguing notion, but may be met with negative perceptions. In 1964, Dr. José Delgado conducted an experiment in which he implanted a Spanish fighting bull with an electrical stimulator in the caudate nucleus of the brain. Delgado engaged in a traditional bullfight, swinging his red cloth, but just as the bull charged he activated the brain stimulator using a remote control. The electrical impulse stopped the bull in its tracks and entirely eliminated the aggressive instinct. This was a landmark experiment, but it was met with contention. The media portrayed Delgado's work as 'mind control' and suggested such treatment might separate humans from their individuality. Delgado contended, however, that it would be more ethical to treat an individual with violent tendencies than to keep them locked in prisons for the criminally insane. In any case, there is certainly much work to be done on identifying the individual brain circuitry underlying aggression before such technology could be explored.

RNA interference (RNAi) is another genetic technique with tremendous potential for the field of aggression research. RNAi refers to the silencing of genes by RNA-dependent mechanisms. One of these mechanisms is the cleavage of double-stranded RNA to form short-interfering RNA (siRNA). Single-stranded siRNA then integrates into an RNA-induced silencing complex (RISC), which cleaves the target mRNA, preventing it from being translated. This system can be targeted to certain cell groups by microinjection into the brain using viral-vector mediated delivery. Recently, this technique was used to silence estrogen receptor-alpha (ERα) expression in the preoptic area of female mice, given the role of this brain region in maternal care and aggression. Suppression of ERα strongly inhibited maternal care, but had no effect on maternal aggression toward a male intruder (Ribeiro et al. 2012). This study is a good example of how RNAi can be a higher resolution technique for studying aggressive behavior. siRNA makes it possible to manipulate specific genes in precise cell populations in order to observe very specific behavioral output. RNAi has proven useful for studying the aggression of insects as well (Hoyer et al. 2008; Kravitz and Huber 2003).

MicroRNA (miRNA) is another form of RNAi for regulating gene expression. miRNAs are genomically encoded RNAs that can be integrated into the RISC complex to silence gene expression. The particular mRNA that is targeted by miRNA depends partly on the degree of complementarities between the two. In that regard, gene polymorphisms may influence the extent of miRNA action on a specific mRNA. Under that premise, a recent study examined an A to G polymorphism in the human serotonin receptor 1B (HTR1B) gene that strongly influenced the interaction with a miRNA called miR-96. Mice with a genetic knock-out of this receptor show an aggressive phenotype (Saudou et al. 1994); and consistent with those data, an A to G polymorphism at the gene was related to more aggression related conduct-disorder behaviors in a sample of college students (Jensen et al. 2009). These results suggest that an inhibitor of miR-96 might modulate aggressive behavior. In rodent studies, modifying miRNA expression in the brain may be useful for delineating some of the specific gene effects on aggressive behavior.

Aggression is used by individuals to defend or obtain resources. Inappropriate aggression causes problems whether evoked by other people or even by their pet dogs. The goal of this line of research is to understand and manage inappropriate aggression. Considered together, the field of aggression is on the threshold of new techniques to advance what has been established in the cause and persistence of inappropriate aggression and violence.

References

Anderson DJ (2012) Optogenetics, sex, and violence in the brain: implications for psychiatry. Biol Psychiatry 71:1081–1089

Angoa-Perez M, Kane MJ, Briggs DI, Sykes CE, Shah MM, Francescutti DM et al (2012) Genetic depletion of brain 5HT reveals a common molecular pathway mediating compulsivity and impulsivity. J Neurochem 121:974–984

Bedrosian TA, Fonken LK, Demas GE, Nelson RJ (2012) Photoperiod-dependent effects of neuronal nitric oxide synthase inhibition on aggression in Siberian hamsters. Horm Behav 61:176–180

Bortolato M, Godar SC, Melis M, Soggiu A, Roncada P, Casu A et al (2012) NMDARs mediate the role of monoamine oxidase A in pathological aggression. J Neurosci 32:8574–8582

Breuillaud L, Rossetti C, Meylan EM, Merinat C, Halfon O, Magistretti PJ et al (2012) Deletion of CREB-regulated transcription coactivator 1 induces pathological aggression, depression-related behaviors, and neuroplasticity genes dysregulation in mice. Biol Psychiatry 72:528–536

Chiavegatto S, Dawson VL, Mamounas LA, Koliatsos VE, Dawson TM, Nelson RJ (2001) Brain serotonin dysfunction accounts for aggression in male mice lacking neuronal nitric oxide synthase. Proc Natl Acad Sci U S A 98:1277–1281

Clinton SM, Kerman IA, Orr HR, Bedrosian TA, Abraham AD, Simpson DN et al (2011) Pattern of forebrain activation in high novelty-seeking rats following aggressive encounter. Brain Res 1422:20–31

Coccaro EF, Lee R, Liu T, Mathe AA (2012) Cerebrospinal fluid neuropeptide Y-like immunoreactivity correlates with impulsive aggression in human subjects. Biol Psychiatry 72(12):997–1003

Demas GE, Eliasson MJ, Dawson TM, Dawson VL, Kriegsfeld LJ, Nelson RJ et al (1997) Inhibition of neuronal nitric oxide synthase increases aggressive behavior in mice. Mol Med 3:610–616
Fish EW, Faccidomo S, Miczek KA (1999) Aggression heightened by alcohol or social instigation in mice: reduction by the 5-HT(1B) receptor agonist CP-94,253. Psychopharmacology 146:391–399
Gammie SC, Nelson RJ (1999) Maternal aggression is reduced in neuronal nitric oxide synthase-deficient mice. J Neurosci 19:8027–8035
Garattini S, Giacalone E, Valzelli L (1967) Isolation, aggressiveness and brain 5-hydroxytryptamine turnover. J Pharm Pharmacol 19:338–339
Giacalone E, Tansella M, Valzelli L, Garattini S (1968) Brain serotonin metabolism in isolated aggressive mice. Biochem Pharmacol 17:1315–1327
Hoyer SC, Eckart A, Herrel A, Zars T, Fischer SA, Hardie SL et al (2008) Octopamine in male aggression of Drosophila. Curr Biol 18:159–167
Jensen KP, Covault J, Conner TS, Tennen H, Kranzler HR, Furneaux HM (2009) A common polymorphism in serotonin receptor 1B mRNA moderates regulation by miR-96 and associates with aggressive human behaviors. Mol Psychiatry 14:381–389
Karl T, Lin S, Schwarzer C, Sainsbury A, Couzens M, Wittmann W et al (2004) Y1 receptors regulate aggressive behavior by modulating serotonin pathways. Proc Natl Acad Sci U S A 101:12742–12747
Kerman IA, Clinton SM, Bedrosian TA, Abraham AD, Rosenthal DT, Akil H et al (2011) High novelty-seeking predicts aggression and gene expression differences within defined serotonergic cell groups. Brain Res 1419:34–45
Kravitz EA, Huber R (2003) Aggression in invertebrates. Curr Opin Neurobiol 13:736–743
Kriegsfeld LJ, Dawson TM, Dawson VL, Nelson RJ, Snyder SH (1997) Aggressive behavior in male mice lacking the gene for neuronal nitric oxide synthase requires testosterone. Brain Res 769:66–70
Laas K, Reif A, Herterich S, Eensoo D, Lesch KP, Harro J (2010) The effect of a functional NOS1 promoter polymorphism on impulsivity is moderated by platelet MAO activity. Psychopharmacology 209:255–261
Lesch KP, Merschdorf U (2000) Impulsivity, aggression, and serotonin: a molecular psychobiological perspective. Behav Sci Law 18:581–604
Lin D, Boyle MP, Dollar P, Lee H, Lein ES, Perona P et al (2011) Functional identification of an aggression locus in the mouse hypothalamus. Nature 470:221–226
Lyons WE, Mamounas LA, Ricaurte GA, Coppola V, Reid SW, Bora SH et al (1999) Brain-derived neurotrophic factor deficient mice develop aggressiveness and hyperphagia in conjunction with brain serotonergic abnormalities. Proc Natl Acad Sci U S A 96:15239–15244
Maas JW (1962) Neurochemical differences between two strains of mice. Science 137:621–622
Miczek KA, Hussain S, Faccidomo S (1998) Alcohol-heightened aggression in mice: attenuation by 5-HT1A receptor agonists. Psychopharmacology 139:160–168
Nelson RJ, Demas GE, Huang PL, Fishman MC, Dawson VL, Dawson TM et al (1995) Behavioural abnormalities in male mice lacking neuronal nitric oxide synthase. Nature 378:383–386
Olivier B, Mos J, van Oorschot R, Hen R (1995) Serotonin receptors and animal models of aggressive behavior. Pharmacopsychiatry 28(Suppl 2):80–90
Reif A, Jacob CP, Rujescu D, Herterich S, Lang S, Gutknecht L et al (2009) Influence of functional variant of neuronal nitric oxide synthase on impulsive behaviors in humans. Arch Gen Psychiatry 66:41–50
Reif A, Kiive E, Kurrikoff T, Paaver M, Herterich S, Konstabel K, et al (2010) A functional NOS1 promoter polymorphism interacts with adverse environment on functional and dysfunctional impulsivity. Psychopharmacology 44(3):222–238

Ribeiro AC, Musatov S, Shteyler A, Simanduyev S, Arrieta-Cruz I, Ogawa S et al (2012) siRNA silencing of estrogen receptor-alpha expression specifically in medial preoptic area neurons abolishes maternal care in female mice. Proc Natl Acad Sci U S A 109:16324–16329

Saudou F, Amara DA, Dierich A, LeMeur M, Ramboz S, Segu L et al (1994) Enhanced aggressive behavior in mice lacking 5-HT1B receptor. Science 265:1875–1878

Simon NG, Lu SF (2005) Androgens and aggression. In: Nelson RJ (ed) Biology of aggression. Oxford University Press, New York

Takahashi A, Shimamoto A, Boyson CO, DeBold JF, Miczek KA (2010) GABA(B) receptor modulation of serotonin neurons in the dorsal raphe nucleus and escalation of aggression in mice. J Neurosci 30:11771–11780

Trainor BC, Workman JL, Jessen R, Nelson RJ (2007) Impaired nitric oxide synthase signaling dissociates social investigation and aggression. Behav Neurosci 121:362–369

Wang D, Szyf M, Benkelfat C, Provencal N, Turecki G, Caramaschi D et al (2012) Peripheral SLC6A4 DNA methylation is associated with in vivo measures of human brain serotonin synthesis and childhood physical aggression. PLoS ONE 7:e39501

Wen JC, Hotchkiss AK, Demas GE, Nelson RJ (2004) Photoperiod affects neuronal nitric oxide synthase and aggressive behaviour in male Siberian hamsters (Phodopus sungorus). J Neuroendocrinol 16:916–921

Hypothalamic Attack: A Wonderful Artifact or a Useful Perspective on Escalation and Pathology in Aggression? A Viewpoint

Menno R. Kruk

Abstract W.R. Hess' early demonstration of aggressive responses evoked by electrical stimulation in the cat's hypothalamus had a significant impact on the development of psychological and behavioral concepts. Many ideas on behavioral routines, allegedly organized in the brainstem, derive from his observation. Similar responses have since been evoked from the hypothalamus of many different species, suggesting that the mechanism mediating these responses is evolutionarily well preserved. However, these effects have also been portrayed as artificial responses to an artificial stimulus in an artificial environment. True enough; after many years of research, crucial questions on the underlying mechanism remain unanswered. Questions such as: How do they emerge in the first place? What neuronal elements mediate these responses? What is their role in "spontaneous" aggression? In the first part of this chapter we show methodology to study such questions in a consistent way using behavioral, physiological, anatomical, and pharmacological findings on hypothalamic attack in rats. In the second part we suggest that one important function of the underlying mechanism is to match the dynamics of the endocrine stress response with the dynamics of the behavioral and physiological requirements of coping with conflicts. This neuroendocrine-behavioral matching seems crucial right from the first emergence of the aggressive response in inexperienced animals, up to the full-blown violent responding in fully experienced animals. Impeding these essential functions results in inadequate coping with conflicts. The stress response during a first conflict in an inexperienced individual in an unfamiliar environment seems to rapidly initialize a crucial change in a mechanism involved in the appraisal of social signals during conflict. That change has enduring consequences for future conflict strategies. This concept opens another perspective on "escalated" or "pathological" aggression, especially so in individuals with a dysfunctional stress response.

M. R. Kruk (✉)
Gorleus Lab, Department of Medical Pharmacology,
Leiden Academic Center for Drug Research (LACDR), Leiden University,
Einsteinweg 55, 2333 CC Leiden, Netherlands
e-mail: Info@Mennokruk.nl

Curr Topics Behav Neurosci (2014) 17: 143–188
DOI: 10.1007/7854_2014_313

Published Online: 23 May 2014

Keywords Hypothalamus · Violence · Conflict appraisal · Stress dynamics · Rat

Contents

1 Introduction

The question of whether aggressive behavior reflects a "prewired" or "innate" property of the brain or whether it is a consequence of learning during development has intrigued scientists as well as laypersons for a long time. It still is an important and often controversial issue within the perpetual nurture–nature debate. Walter Rudolf Hess' (1928) Nobel Prize-winning demonstration of aggressive responses evoked by electrical stimulation in the cat's hypothalamus only seemed to settle the issue. If a natural aggressive pattern could be evoked by simple electrical stimulus of a specific brain area, then it surely must derive from some kind of pre-existing physiological organization. That consideration had a seminal impact on the development of later psychological and behavioral concepts. Many ideas on "innate", "hardwired", "fixed", "instinctive", "impulsive" or "hierarchical" behavioral routines allegedly organized at the level of the brainstem derive from Hess' early findings (Anderson 2012; Tinbergen 1951). Aggressive responses have since been evoked from the hypothalamus of many different species, testifying to their evolutionarily well-preserved origins (Bejjani et al. 2002; Demski 1973; Ettrup et al. 2012; Herndon et al. 1979; Hess 1928; Lin et al. 2011; Lipp and Hunsperger 1978; Martin 1976; Panksepp 1971; Putkonen 1966; Roberts et al. 1967; Vergnes and Karli 1970). A recent cross-species comparison is given in Goodson and Kingsbury (2013). The close resemblance of the evoked responses to elements of the behavioral repertoire under natural conditions, suggests that they derive from the same neural mechanism. But how electrical stimulation of an area known to be essential to autonomous and endocrine regulation translates into the complex sensory and motor components of aggressive behavior is still far from clear.

Others have argued that these electrically evoked responses were behavioral artifacts, produced by an artificial agent in an artificial environment. In order to demonstrate that these responses are caused by the stimulation, most studies on hypothalamic aggressive responses have been evoked in artificial environments where "spontaneous" aggressive responses are unlikely. Also, the many different electrical stimulation parameters used to induce hypothalamic aggression hardly mimic natural firing patterns of hypothalamic neurons. Moreover, stimulation only evokes a limited subset of the behavioral elements observed under natural conflict conditions. Often one obtains only the most violent part, while other, more moderate elements used to handle conflicts, are absent or even suppressed. In addition, it is not uncommon that a mixture of aggressive, sexual, and flight responses is evoked, or that unusual targets are attacked (Kruk et al. 1979; Lin et al. 2011). Such criticism requires serious discussion. In the first part of this chapter we will discuss the validity of these criticisms.

Resolving the controversy requires unraveling the underlying neural and behavioral mechanism. A direct approach would be to identify the neural elements involved, and record their activity in natural experimental settings conducive to social conflict. However, that goal remains elusive, due to the technical difficulties of recording in such settings. Also, the many circuits converging in the hypothalamus complicate deciding which elements are directly involved in the evoked aggressive response. Hence, it is not precisely known what function this artificially activated "aggressive" neural circuit has in aggression under natural conditions. The same uncertainty surrounds its role in aggression associated with dysfunctional behavioral, autonomic or endocrine regulation.

Here we discuss how the arguments in the controversy on the existence of behaviorally specific circuits do apply in hypothalamic aggression in the Norway rat. We will discuss the typology, the neuroanatomical distribution, the stability, and endocrine concomitants and the first emergence of these responses while trying to clarify its function. We will show that hypothalamic attack is induced in a specific part of the hypothalamus. However, its induction apparently depends on the experimental setting, the behavioral consequences and endocrine conditions under which it is first evoked. These conditions suggest how violent responses could contribute to escalation and pathology in aggression and could become an overriding element within the normal aggressive repertoire.

2 Caused by Association: The Prepotency Hypothesis

The idea that hypothalamic behavioral responses derive from behaviorally specific "hardwired" neural systems has been seriously questioned (Valenstein et al. 1968, 1969, 1970). It was derided as a new kind of "phrenology" in an otherwise justified effort to criticize the medical practice of "psychosurgery" in psychological disordered humans fashionable at that time (Valenstein 1973). It was shown that a specific hypothalamically induced behavioral response to a particular object could be changed into another response by removing that first object, and subsequently stimulating the animal repeatedly—literally overnight—in the exclusive presence of another object while the first object had been removed. Hence, it was argued that hypothalamic responses derive from the association between brain stimulation and an object that happened to be present or "prepotent" in the environment during stimulation. The phenomenon was labeled "stimulation experience" and the concept took hold as the "prepotency hypothesis". Also, probably due to the relatively crude technology used at the time, it also appeared impossible to identify distinct response-specific hypothalamic areas in the rat. Therefore, the idea of behaviorally specific neural systems in the hypothalamus was rejected (Valenstein et al. 1968, 1969, 1970). Whether these arguments also applied to hypothalamic aggression was not clear (Kruk and van der Poel 1980).

3 Threshold Changes

The "stimulation experience" idea or "prepotency" hypothesis was, in its turn, criticized. It was asserted that the stimulation procedures used in those days lacked sufficient resolution to allow such extreme conclusions, given the anatomical complexity of the hypothalamus. Later neuroanatomical studies have strengthened this criticism by revealing the perplexing complexity of hypothalamic circuits. It was also postulated that the reported changes in response could be due to threshold decreases upon repeated stimulation of spatially overlapping but behaviorally specific neuronal systems (Roberts et al. 1967). Later it was shown that forced changes in response caused by repeated hypothalamic stimulation in the presence of a particular object could indeed be due to differential changes in the thresholds of behaviorally distinct responses (Wise 1968, 1969).

Both sides in this early controversy raised valid points that are still relevant to our understanding of the origins of hypothalamic, "natural" and "escalated" aggression. This controversy addresses the classical scientific problem of reproducibility and specificity. In this respect aggression is an especially difficult case, since it is intrinsically unstable. Serious conflict has lasting consequences for future conflicts: there usually is a winner and a loser, and after a conflict their behavior will be changed accordingly. After developing a valid method to determine thresholds for hypothalamic aggression repeatedly, it became clear that the facilitation of hypothalamic attacks upon repeated winning are a robust, persistent characteristic of hypothalamic aggression (Kruk et al. 1979, 1984b, 2004; Lammers et al. 1988a). However, the mechanism mediating that facilitation was not seriously studied. Hence, the early "stimulation experience" of a naive animal during the first stimulations in the presence of an opponent could still be a factor affecting its subsequent aggressive behavior. A possible neuroendocrine mechanism underlying such attack facilitation upon repeated winning was only recently reported (Kruk et al. 2013 and this chapter).

4 Delimiting the Hypothalamic Attack Area

Early studies in the rat already suggested that aggression could be elicited from a specific area in the hypothalamus (Kruk et al. 1979; Panksepp 1971; Panksepp and Trowill 1969; Vergnes and Karli 1970; Woodworth 1971). However, the precise delimitation of the area remained arbitrary. Aggression eliciting sites seemed to be lie between sites that did not evoke aggression but other responses such as flight. Hence a statistical procedure was needed to falsify or confirm the hypothesis of a specific aggression-related area in the hypothalamus. Discriminant analysis comparing the 3D histological coordinates (frontal, dorsal and lateral) of a large number of histologically verified hypothalamic electrode positions where aggression could be elicited with the coordinates of sites that failed to elicit attacks, clearly falsified the hypothesis that hypothalamic localization was immaterial to the behavioral effect (Kruk et al. 1983). A continuous area where the probability to evoke aggression was higher than 80 % could be delimited. The most violent attacks at the lowest stimulation intensities were most likely to be evoked near the core of that area. This area has since been referred to as the "hypothalamic attack area" or HAA. It is surrounded by areas where the chance to induce aggression diminishes progressively from the core of the HAA outwards. Intriguingly the behaviorally defined HAA did not coincide with a classical neuroanatomical distinction of the hypothalamus current at that time. But it did encompass parts of virtually all areas that were mentioned in previous studies of hypothalamic aggression, such as the subfornical hypothalamic area, the mediobasal hypothalamic area, the ventrolateral pole of the ventromedial nucleus, and the area just in front of the ventromedial nucleus of the hypothalamus (Goodson and Kingsbury 2013). Since electrical stimulation affects cell bodies, axons as well as dendrites, in this anatomical extremely complex area this finding may not come as a surprise. More selective methods such as optogenetic stimulation and gene expression patterns such as c-Fos (Halasz et al. 2002a, b; Haller 2013; Lin et al. 2011; Tulogdi et al. 2010) will be required to determine the specific contribution of the different sub-areas within the HAA in aggression.

5 Mapping the "Behavioral" Hypothalamus

A subsequent systematic distribution study (Lammers et al. 1987, 1988a, b), using "movable" electrodes (van der Poel et al. 1983) and an enriched test environment, containing many different objects in addition to another live rat, confirmed and specified the distribution of the HAA in more detail. In this study the moveable electrodes descended along tracks in 75 μ steps, and three thresholds were determined of every clear behavioral response elicited in every site tested. The trade-off between threshold current intensity and electrode position along individual electrode tracks, suggests that the effective field of these electrodes at intensities between 50 and 150 μA was in the order of 150–200 μ. Figure 1 shows a summary 3D-representation of the distinct areas where the different responses (attack, flight, or self-grooming) were evoked. The figure clearly demonstrates that self-grooming, attacks, and flight, were evoked from distinct, though slightly overlapping, areas in the hypothalamus summarized and adapted from (Lammers et al. 1987, 1988a, b). Apparently the site of stimulation is an important determinant of the type of behavioral response evoked.

6 Anatomical Characteristics of HAA

The delimitation of the HAA to a circumscribed area enabled the comparison of its anatomical connections (Roeling et al. 1993, 1994; Toth et al. 2010) and it facilitated cross-species comparisons (Anderson et al. 2012; Goodson and Kingsbury 2013; Haller 2013; Siegel et al. 1999). It also allowed the comparison of c-Fos activation patterns following hypothalamic attack and "spontaneous" aggression in the hypothalamus and the rest of the brain (Halasz et al. 2002b; Haller 2013; Tulogdi et al. 2010; Veening et al. 2005). Surprisingly, these anatomical studies do

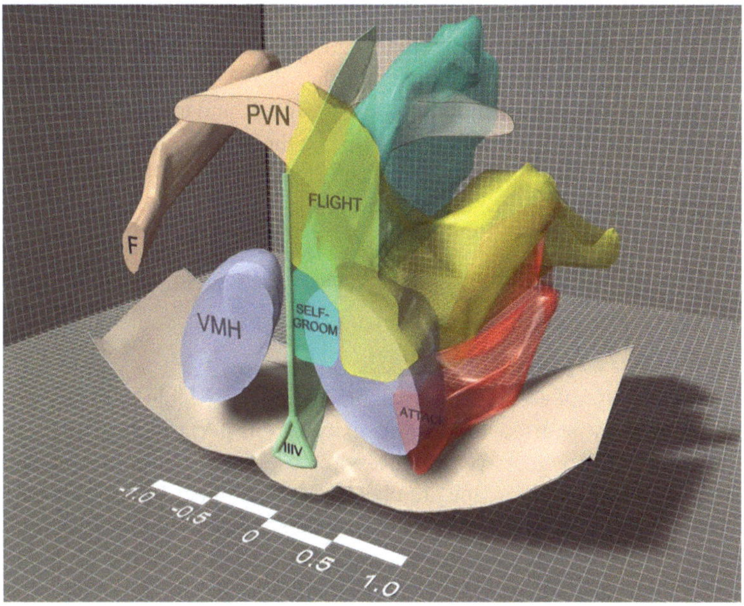

Fig. 1 Pseudo 3D-mapping of behavioral responses to hypothalamic stimulation in an "enriched" environment with several different objects in addition to another rat. A hypothalamic section is depicted from a caudal viewpoint. The cloud-like structures depict areas where the chance to induce a particular response is higher than 80 %. The *green*, frontally ascending cloud, close to the 3rd ventricle and traversing the paraventricular nucleus (PVN) depicts the self-grooming area. The *yellow*, frontally descending cloud encompassing the medial part of the ventromedial nucleus depicts flight or escape. The *red*, basal cloud encompassing the lateral part of the ventromedial nucleus (VMH) and extending frontally around it shows the hypothalamic attack area or HAA. A 150 μ shell around each area has been "peeled of" to account for current spread and to eliminate some of the overlap for the sake of clarity. (Reconstruction based on Lammers et al. 1987, 1988a, b)

not reveal direct connections between the HAA and areas directly involved in motor control. Rather they reveal the importance of areas involved in attention, assessment and appraisal of the relevance of external stimuli. These connections often seem to be organized as reciprocal loops that convey environmental and social features that may be important in conflict and other behavioral responses. It is beyond the scope of this paper to discuss the possible involvement of the more than 35 brain areas presently known to be connected to the HAA. For an inspiring blueprint on the possible flow of information appraising the social environment in escalated conflict between these frontal areas and the brainstem see (Potegal 2012). This interesting blueprint does not specify the specific role of the HAA. To assess the function of the HAA itself it seems interesting to consider its "strategic" position between paraventricular nucleus (PVH), a crucial node in the regulation of the stress response, and the pituitary, the brain's main endocrine gateway to the rest of the body. Intriguingly many fibers in the HAA are loaded with synapse-like structures ("boutons en passant" or varicosities), while many fibers emerging from the PVN in all directions are similarly loaded with such varicosities (Roeling et al. 1993, 1994). These characteristics suggest that the HAA functions as cross-road between the brain circuitry that assesses social signals and a mechanism controlling the stress response. Interestingly, stimulation of the HAA elicits a strong and rapid rise in circulating ACTH and corticosterone (Kruk et al. 1998, 2004) that rapidly facilitates aggression.

7 The First Induction of Attacks

Testing whether stimulation via an implanted electrode is able to induce attacks in a naïve rat usually starts at a low intensity of 50 μA which is then stepwise increased either until aggressive responses are observed, or responses incompatible with attacks such as flight/escape are observed, or an upper limit of about 300 μA is reached. It is quite conceivable that during those initially ineffective stimulation efforts preceding the very first attacks, the animal has some kind of "stimulation experience" that helps it to "learn" how or what to attack. An early experiment failed to confirm this hypothesis. Groups of rats in which stimulation was started at 50 μA and rats in which stimulation was started at 100 or 200 μA finally yielded about 50 % attack-eliciting electrodes, although the animals started at the lower intensities did receive considerably more of such hypothetical sub-liminal "stimulation experience" (Fig. 2). Moreover, attack topology does not develop gradually during the first confrontations—the first attacks are just as goal-directed as later attacks.

8 Hypothalamic Attack Topology

Figure 3a–c shows some "typical" examples of the various forms of hypothalamic attack. It should be emphasized that these postures do not reflect fixed stereotypes in any respect. These attack patterns are examples taken from of a continuum of forms, and many intermediate forms can be observed. The actual form depends on the strain of rats used (Kruk et al. 1979, 1990) and the strength of the stimulation. It also depends on the rapidly changing defensive actions of the opponent. Attacks are characterized by a bite directed at the head, or the neck, or the frontal part of the back. A back bite is most often accompanied by a forceful blow with the hind paws against the flank of the opponent and followed by a clinch fight, in which both animals roll over while trying to get on top of each other. If an opponent takes a defensive upright position in a corner (Fig. 3f) an "attack jump" can be observed (Fig. 4). While jumping the attacker delivers a forceful kick against the chest of the opponent. Most often both rats fall and a clinch fight with additional kicks and bites follows. Thus, the constant elements of an attack are a bite and a kick.

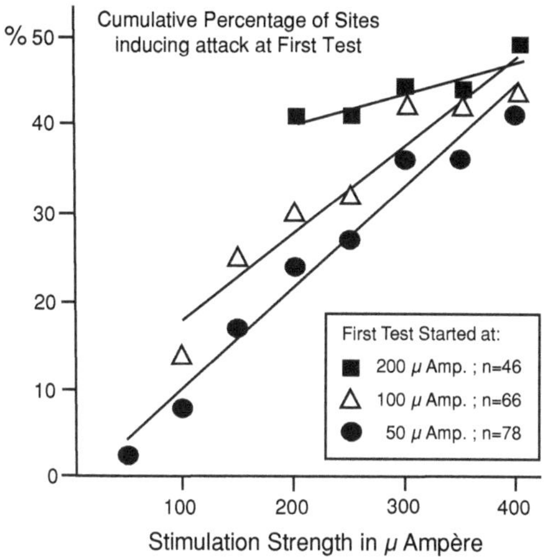

Fig. 2 The cumulative percentage of electrodes yielding an attack at or below a given stimulation intensity in three groups of 'inexperienced' rats. Stimulation was started at the different intensities of 50, 100 and 200 μA. All groups yield about 50 % successfully attack evoking sites. Most sites not evoking attacks are outside the HAA. The first attacks are similar to subsequent attacks (Adapted from Kruk, Behaviour 1979 with permission)

The precise form—attack jump, bite and kick attack or clinch-fight—depends on stimulation intensity and the precise posture and movements of the opponent. The accompanying kick may be lacking at low stimulation intensities. Attackers seem to lose interest in their opponents immediately (Fig. 3e) when stimulation ends. These violent attacks are the only aggressive responses observed. Other elements of the agonistic repertoire, like aggressive grooming and sideways threat, are absent. They are even suppressed when hypothalamic attack is evoked in a more natural territorial setting (Koolhaas 1978). The limited subset of aggressive responses evoked by HAA stimulation suggests that an important part of the mechanism controlling aggression in natural settings is overruled or bypassed. Though hypothalamic aggression is deficient in these respects when compared with territorial aggression, the attack patterns themselves are remarkably similar to the most violent part of the behavioral repertoire in inter-male fighting or maternal aggression in defense of pups.

9 Attack Directing Stimuli

In the absence of an opponent or an "attackable" object no aggression is induced, but rather locomotion, sometimes teeth-chattering, sometimes flight and transient pilo-erection (Kruk 1991; Kruk et al. 1979). Apparently some kind of opponent-related

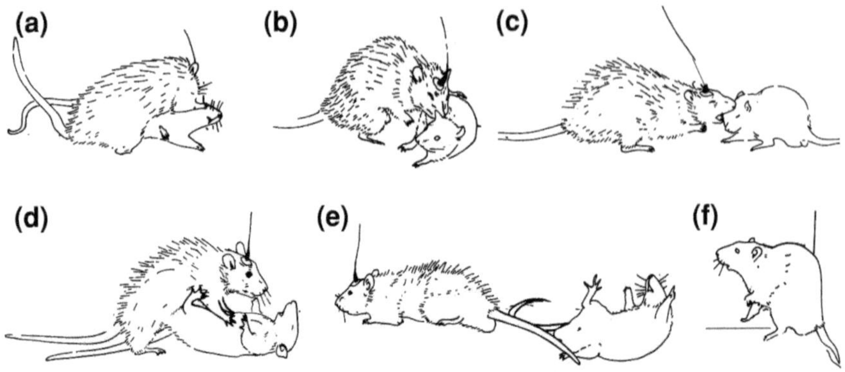

Fig. 3 "Typical" examples of hypothalamic attack: **a** start of a clinch fight; **b** bite attack on nape of the neck; **c** bite attack on snout, common in WE-zobs, rare in Wistars; **d** sub-dom position, an efficient way to avoid back biting; **e** at the cessation of stimulation; **f** defensive position, opponent in corner. The differences between attacks arise from differences in stimulation efficiency and the defensive actions of the opponent. To bite the back seems the general aim of all attacks, with its effect at high intensities enhanced by a hind paw-kick against the opponent's flank. The various forms are not evidence for different "behavioral systems", and derive from the same general area. (Adapted from Kruk, Behaviour 1979 with permission)

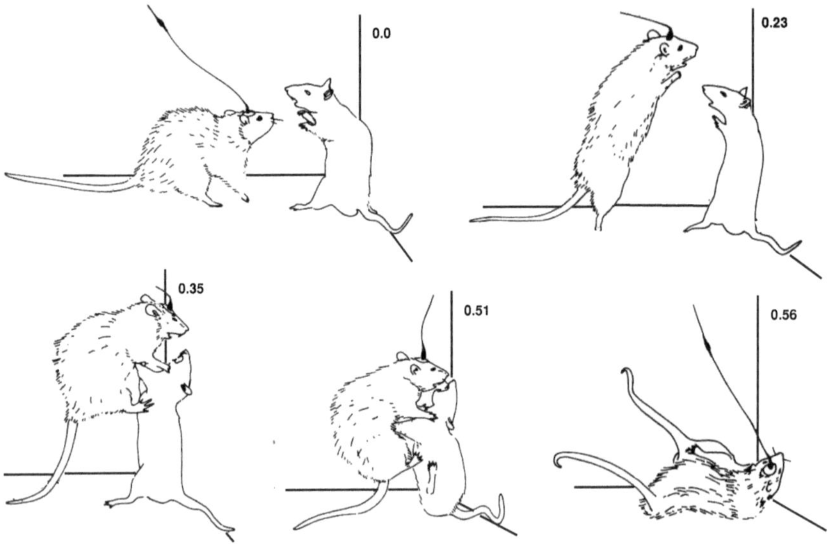

Fig. 4 The spectacular and rapid attack jump is seen only when the opponent assumes an upright defensive posture, often in a corner. In subsequent experiments opponents routinely received a strong immobilizing analgesic; no attack jumps are observed toward such opponents. *Numbers* refer to seconds after start of attack (Adapted from Kruk, Behaviour 1979 with permission)

sensory information is required to enable a response. One describes these attacks as motor patterns, following the classical ethological approach, but the HAA is not known as motor pattern generator, nor does it communicate directly to such generators as far as we know now. So what might be guiding and structuring these attacks? The only position that can last longer than a few seconds provides an indication. (Figure 3d). When the opponent is lying in a supine position with the attacker on top, the opponent can effectively prevent the attacker reaching the dorsal side of the head, the nape of the neck, or its back by placing its hind paws firmly into the groin of the attacker. Those parts receive most of the bites in both territorial fighting and in hypothalamic attack. Interestingly, the most vulnerable parts of its body, the belly and throat, are clearly exposed in that position, but these parts are hardly ever attacked. It seems as if the attacker still tries to reach the back of the opponent. This position is known as the "sub-dom" element in rats and many other species. In natural settings this supine position is considered a submission signal that effectively prevents further attacks. It seems as if some of these particular "rules of engagement" of territorial aggression are still observed in hypothalamic aggression. However, it could also be that the frontal part of the opponent's back and the nape of its neck serve as directing signals crucial to the release of an attack. One rather gets the impression that the opponent "knows" what the intended "strategic" target of the attack is, and that it adjusts its behavior accordingly by keeping that target away from the attacker, by assuming an upright position in a corner, or by adopting of a supine position with its back on the substrate. Interestingly, different rat strains have distinct hypothalamic attack patterns and targets (Kruk et al. 1990). Rather than viewing these different attacks as distinct motor patterns—an early ethological concept—one can conceive these targets as critical coordinates essential for directing and releasing effective attacks. In this concept the pattern will then be shaped by the moment-to-moment dynamics of directing and releasing stimuli on the back and nape of the neck of both adversaries. "Directing" and "releasing" stimuli are ethological concepts often implied in animal studies of aggression. Viewing hypothalamic aggression in this perspective has implications for the neural mechanism supposedly mediating these attacks. It suggests that stimulation primarily changes a mechanism involved in the animal's appraisal of its adversary rather than activating a set of fixed motor patterns.

10 Changed Sensory Modalities

Supporting this appraisal hypothesis is the observation that during stimulation, attack-evoking hypothalamic sites in the cat produce remarkable, immediate changes in the way the animal reacts to tactile and visual stimuli that are relevant for effective attacking. Stimulation produces changes in the way visual stimuli are followed (Pott and MacDonnell 1986). It also produces precisely delimited areas around vibrissae which upon touch produce a movement toward the touch. Similar fields arise on the lips that release biting upon touch (MacDonnell and Flynn 1966). Stimulation also produces stimulation-dependent fields on the paws of the cat which upon touching evoke striking out (Bandler 1982; Bandler and Flynn 1972). Increasing stimulation expands such fields in an orderly stepwise fashion, suggesting that these changes are mediated by precise neural mechanisms. The important inference from these observations is that stimulation of hypothalamic sites does not directly organize a fixed motor pattern for attack, but rather changes thresholds of sensory modalities that are crucial for attack. The observed motor patterns elicited in hypothalamic attack seem well adapted for fighting. But their expression is conditional on receiving specific aggression-related sensory stimuli. In the absence of such stimuli, and by extension, in the absence of an opponent, there is no pattern. Similar findings have been reported in the rat (Smith 1972).

11 Changed Social Appraisal

Subliminal stimulation does not evoke visible signs of aggression, but it activates the neural mechanism involved in attack, because simultaneous subliminal stimulation of two sites in the same animal at contralateral sites still produces attacks. The subliminal effects of stimulation apparently add up somehow. Subliminal stimulation of one of those sites in the presence of a potential opponent, while not evoking attacks, does change the behavioral interactions between the animals. Both the stimulated animal and its companion are affected by the stimulation. Both opponents withdraw from physical contact, and the behavior of the stimulated rat becomes more sensitive to changes in the behavior of its companion as current intensity increases towards attack threshold. The opponent seems to minimize contact, as if it tries to avoid drawing attention to itself. There is obviously much signaling going on between rats that escape direct human observation. In anthropomorphic terms, stimulation seems to promote a shift from friendly social contact towards a more apprehensive and "touchy" social attitude (Haccou et al. 1988). These observations support the idea that HAA stimulation primarily changes the appraisal of the social environment.

12 Attackable Objects

Many studies have explored the range of objects that will release hypothalamic attack. The first studies on hypothalamic attack used "mouse killing" as paradigm, that is they used mice as opponents. (King and Hoebel 1968; Panksepp 1971; Vergnes and Karli 1970; Woodworth 1971). Mouse killing is a natural predatory response of the rat, first described by Karli (1956). It was assumed that hypothalamic mouse killing was activating circuits involved in the natural predatory response as well as intra-specific attacks of the rat, in analogy of the "affective attack" and "silent" or "predatory" attack responses evoked in the cat's hypothalamus (Siegel et al. 1999). However, the first published picture of hypothalamically evoked mouse killing does show a clear bite-jump-and-kick attack, typical of an attack on another rat and rather unlike the natural mouse killing responses (Woodworth 1971). Comparison of the localization of hypothalamic mouse killing sites of these earlier mouse-killing studies with the distribution of the HAA, suggests that both hypothalamic responses derive from the same neural substrate (Kruk 1991). However, a recent comparison of hypothalamic neural activation after spontaneous mouse killing with neural activation following territorial aggression in the rat, shows that spontaneous territorial aggression and predation activate different, adjacent areas of the hypothalamus (Haller 2013; Tulogdi et al. 2010) These observations suggest that hypothalamic stimulation apparently may force a rat to attack an unusual target—the mouse—using an intra-specific pattern rather than its own natural predatory attack pattern. Interestingly, and in contrast to territorial settings, both estrous female opponents and dominant opponents are attacked during HAA stimulation (Koolhaas 1978). Moreover, attacks can be evoked in female rats from the same hypothalamic area. These females attack both males and females, at the same current intensity (Kruk et al. 1984b). When compared with attack observed under natural conflict conditions, the range of objects attacked under HAA stimulation is apparently wider. These findings suggest that a mechanism constraining attacking in natural settings is overruled by hypothalamic stimulation.

13 Opponent Feedback for Sustained Responding

Intriguingly, the rat's hypothalamic attacks on mice are not sustained upon repeated testing. Panksepp et al. stated early on that "*the experience that causes some rats to stop attacking mice in the present situation is not clear. Perhaps the rat learns that mice are not an adequate threat*" (Panksepp and Trowill 1969). Such observations suggest that some kind of feedback from an opponent is required to sustain hypothalamic attack upon repetition, but what kind of feedback was not known: "*although some type of learning seems to modulate centrally elicited attack in rats, the crucial factors remain to be defined*" (Panksepp and Trowill 1969). Other students of hypothalamic attack have reported similar findings following their efforts to replace life opponents by lifeless objects (Kruk et al. 1979). However, attacks on anaesthetized rats are sustained upon repeated testing. The standard procedure (Kruk 1991).

14 Stimulation: Activation or Disruption?

The aim of brain stimulation in behaving animals is to identify mechanisms controlling behavior by changing the signaling in a specific neural system. Therefore, it is an artificial method by definition. Table 1 presents examples of the

Table 1 Early stimulation technology used

Author	Species	Year	Wave form	Frequency, cps
Hess et al.	Cat	1928	Slow ramp 25 ms	2–12
Hunsperger et al.	Cat	1967	Wave of 4 ms on a 5,000 cps carrier	50
King et al.	Norway rat	1968	Monophasic square 0.1 ms	100
Panksepp et al.	Norway rat	1969	Sine wave	60
Robinson et al.	Rhesus monkey	1969	Monophasic square 1.0 ms	50
Cox et al.	Norway rat	1969	Sine wave	60
Valenstein et al.	Norway rat	1968	Sine wave/0.2 ms biphasic	60/100
Woodworth	Norway rat	1969	Square wave 0.2 ms	100
Smith	Norway rat	1971	Sine wave	60
Flynn et al.	Cat	1977	Biphasic square 1–2 ms	60
Watson	Norway rat	1977	Sine wave	60
Sugerman et al.	Lizard	1978	Biphasic 2.0 ms	50
Koolhaas	Norway rat	1978	Sine wave	50
Lipp et al.	Marmoset	1978	Wave of 4 ms on carrier of 5000 cps	50
Kruk et al.	Norway rat	1979	Biphasic 0.2 and 12.3 ms interval	40
Roberts	Norway rat	1980	Monophasic square 0.1 ms	50
Lammers et al.	Norway rat	1987	Biphasic 0.2 and 12.3 ms interval	40

Examples of the many different stimulation parameters used to induce hypothalamic responses, including aggression, since Hess in 1928. Many of these may have been unsuitable for evoking similar responses repeatedly and reliably

different technologies used to induce aggression in earlier studies. Most early investigators clearly were not trying to mimic the "natural" firing pattern of hypothalamic neurons. It seems rather unlikely that the neurons in the hypothalamus communicate using by long trains of 50 Hz sinus waves or long square wave pulses. Some investigators adjusted the parameters of stimulation to minimize the activation of "passing fibers" (Hess 1928; Lipp 1978; Lipp and Hunsperger 1978). Others were apparently more bent on getting the aggressive response in the first place rather than on selective activation of a particular set of neuronal elements. Whether these responses were obtained by mimicking a natural signaling process or by disrupting an ongoing controlling mechanism was seldom discussed. However, behavioral observations such as poor discrimination of the type of opponent and suppression of less violent aggressive responses seemed to suggest that stimulation probably creates a kind of temporary functional lesion. Studies on the effects of deep brain stimulation in humans suggest that it indeed causes temporal, partial functional lesions (Benazzouz and Hallett 2000; Benazzouz et al. 2000). Interestingly, electrolytic or chemical lesions of hypothalamic areas involved in aggression do not abolish territorial aggression, but rather change its nature (Albert and Walsh 1982; Halasz et al. 2009; Mos et al. 1983; Olivier et al. 1983; van den Berg et al. 1982). Such observations, and the effects of subthreshold stimulation on social interactions (Haccou et al. 1988), suggest that the "aggressive mechanism" in the hypothalamus is involved in shaping the appraisal of the social environment rather than in activating motor commands of "instinctive" or "impulsive" responses.

15 Parameters for Stable Responses

Studying the behavioral stability of hypothalamic responses upon repeated stimulation, or studying the effect of drugs or hormones, is critically dependent on the stability of the electrode and the underlying brain tissue. Some early stimulation methods were clearly ill suited in this respect. Many early stimulation techniques almost certainly promoted the formation of highly reactive and damaging molecules and charged ions at the electrode tip (Ranck 1975) due to excessive charge transfer and polarization. Monophasic impulses will easily polarize at the electrode-tissue interface, causing local lesions or metallic deposits. Sine waves derived from main supply (e.g. 50–60 Hz) rarely cause electrode polarization. But they force about 350 times more charged molecules through the electrode-brain interface than shorter (e.g. 0.2 ms) and equally effective biphasic rectangular pulses of the same frequency and intensity.

16 Quantifying Attack Propensity

Aggressive propensity is usually assessed by measures such as attack latency, total time spent on aggression (either defensive or offensive), or attack frequency. Counting the number of attacks per stimulation is ill suited for the purpose of studying the stability of hypothalamic attack upon repeated testing. The relationship between current intensity and attack frequency is different at each electrode site, even within one individual rat (Fig. 5). Moreover, the nature of the attacks changes from mild to violent as stimulation intensity of the electrode site is increased (Kruk et al. 1981, 1984a). Latency of response onset is also a poorly reproducible measure of the propensity to attack, since hypothalamic attack is apparently episodic in nature (Fig. 6). It appears impossible to evoke attacks within 3 s of stimulation onset, and stimulation periods longer than 10 s are no more effective than shorter periods. Attacks are most likely between 4 and 10 s after stimulation onset (Kruk et al. 1981); such a narrow time window does not allow changes in attack propensity to be detected.

17 Determining Attack Thresholds

Mimicking the "natural" pattern of the underlying mechanism may not be possible yet. However, determining the current intensity required to evoke an attack in 50 % of the stimulation trials, appeared to be the most reliable and sensitive behavioral measure of the attack propensity. It allowed optimizing the parameters of stimulation, by minimizing the total charge transfer per evoked attack, preventing electrode polarization by carefully balancing positive and negative pulses

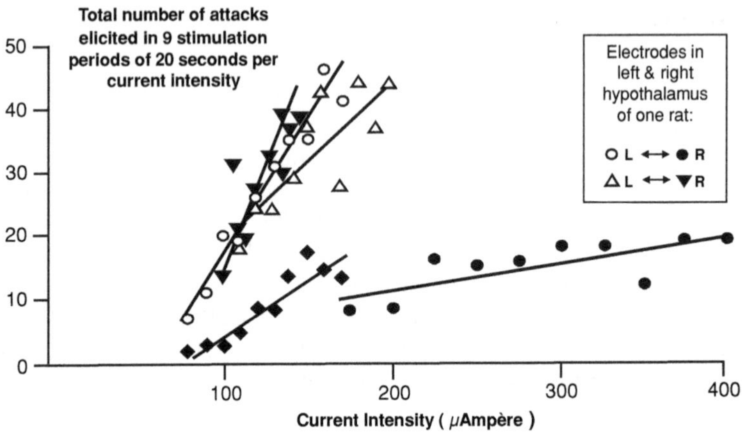

Fig. 5 Number of attacks elicited from individual electrodes as a function of increasing current intensity. Depending on the exact position of the electrode tip, slopes may differ, even within one animal. Electrodes in the core of the HAA tend to have steeper slopes. (Adapted with permission from Kruk et al., The biology of aggression 1981)

Fig. 6 Relation of attack thresholds and the duration of a stimulation train. Attacks are hard to induce with train durations of 3 s or shorter, and prolonging trains beyond 10 s does not yield lower thresholds or more attack, illustrating the episodic, impulsive nature of hypothalamic attack. (Adapted with permission from Kruk et al., The biology of aggression 1981)

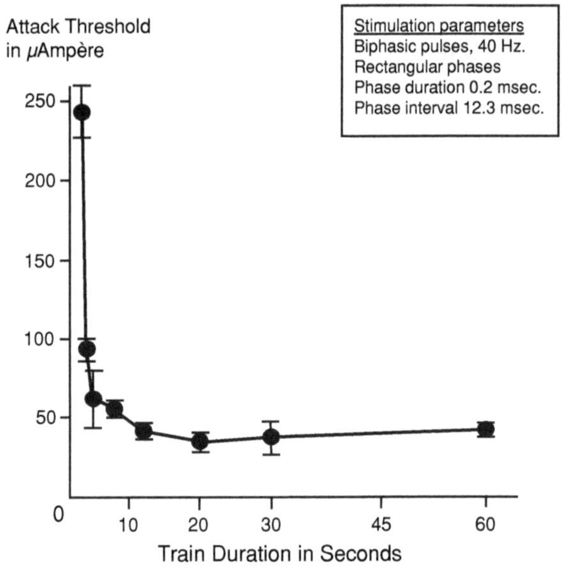

and by taking into account the estimated refractory period of the neural elements involved (Kruk et al. 1981; Kruk and van der Poel 1980). Depending on electrode position, such attack thresholds vary between 30 and 200 μA. The stimulation intensity that will evoke attack in a specific hypothalamic site in a naïve animal is initially unknown. To minimize injury to the animals, the lowest intensity in a site is found by increasing the current intensity from zero, stepwise until a first attack occurs. From that intensity, the attack threshold in a particular site is calculated using a sequential up-and-down procedure (Wetherill 1966). It is a rapid method for operationalizing the propensity for aggression with a minimum of attacks. It is also the least invasive method, causing the least animal distress. Attack thresholds for a wide range of stimulation parameters—pulse duration and frequency, train duration, and phase interval—allowed the selection of stimulation parameters best suited to obtain sustained responses in repeated testing, at minimal charge transfer per evoked attack. Hence, 10 s trains (trails) of 0.2 ms biphasic pulses with an interphase interval of 12.3 ms were used as standard parameters to assess the effects of drugs, hormones and brain lesions on hypothalamic attack. When testing is repeated on subsequent days, attack thresholds initially decrease until reaching about 50 % of their initial value. That initial threshold decrease occurs reliably in both sexes of different strains of rats (Kruk et al. 1984b). Once thresholds have decreased after three threshold determinations on subsequent days, they remain stable for periods up to several weeks and even months (Kruk et al. 1984a). Such stable thresholds in experienced animals have been used extensively in studies on the effects of drugs, hormones and lesions on attack propensity (Kruk 1991; Kruk et al. 2004, 2013; Mos et al. 1983, 1987; van der Poel et al. 1982).

18 Complex Stimulation Patterns

It seems still improbable that hypothalamic neurons would fire in the artificial pattern used in these studies, in natural settings. Vasopressinergic neurons in the hypothalamus fire in discrete short lasting bursts (Yamashita et al. 1979). Accordingly we could evoke hypothalamic aggression using burst-like patterns by breaking up the 10 s trains into shorter volleys. Figure 7 shows that attacks can still be evoked if one splits 10 s trains into short volleys of 25, 100, and 500 ms duration and reduces the number of volleys by up to 50 % compared with an uninterrupted 10 s train (right-most point in the three curves shown), without any change in threshold. Note that the horizontal scale in the figure is logarithmic. These results suggest that a natural mode of communication in the "aggressive" hypothalamus might be a bursting pattern. This complicated "volley" procedure was not further explored, but it provided an early hint at a neuroendocrine involvement in hypothalamic attack. Hypothalamic vasopressin involvement in aggression has since been demonstrated in the rat and the golden hamster (Cervantes and Delville 2007; Compaan et al. 1993; Delville et al. 1996; Ferris 2000, 2005; Ferris et al. 1997, 2006). However, even this more "efficient" but complicated pattern probably still does not mimic the natural pattern of the neurons mediating the aggressive consequences of stimulation at the electrode tip.

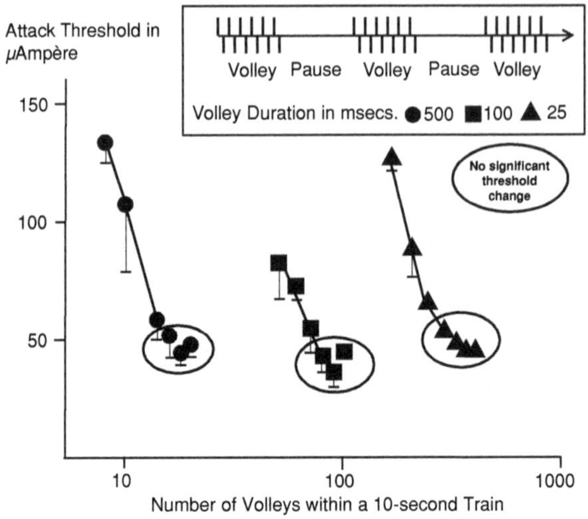

Fig. 7 Effects of a bursting pattern of HAA stimulation. Continuous 10 s stimulation trains of 40 biphasic pulses (i.e. 80 balanced phases, 0.2 ms/s) constitutes an artificial pattern that most certainly does not mimic the natural pattern of the underlying network (right-most points of the *three curves* in the figure). Splitting the train in a series of volleys, cutting the total number of pulses by about half does not change attack thresholds. Note that the *horizontal scale* is logarithmic (Adapted with permission from Kruk et al., The biology of aggression 1981)

19 Repeated Testing Facilitates Attacking

Specific responses, such as aggressive responses, clearly derive from specific areas. However, that finding does not exclude the possibility that "stimulation experience" within the experimental setting, or with the first aggressive responses, may still be a factor in the expression of aggressive responses. There is some evidence supporting that idea. Facilitation of hypothalamic attack after repeated testing on subsequent days has been well documented in different strains of both male and female rats (Kruk 1991; Kruk et al. 1979, 1984b, 2004, 2013; Mos et al. 1983). Such facilitation shows up as an average decrease of up to 50 % of the attack thresholds from the first till the fifth threshold determination, with the largest decrease taking place between the first and second determinations. Most studies on hypothalamic attack were done in animals with such "stabilized" aggressive response baselines (Bermond et al. 1982; Halasz et al. 2002b; Kruk 1991; Kruk et al. 2004, 2013; Lammers et al. 1988a; Mos et al. 1987; van der Poel et al. 1982). Such hypothalamic attack facilitation was often, incorrectly, attributed to a transient, local hypothalamic tissue reaction to implantation, or to a physical change caused by the first passage of current through the electrode tip. However, mimicking the entire first stimulation procedure in rats with no hypothalamic attack experience, i.e. stimulating their HAA electrodes in the absence of an opponent, and hence with no resulting fighting, does not abolish the subsequent facilitation of hypothalamic attack by repeated threshold determination (Kruk et al. 1979). Hypothalamic attack facilitation apparently requires the execution of attacks.

20 Generalized Attack Facilitation

Interestingly, the facilitation seems to generalize to other sites within the same animal. Figure 8 (Kruk et al. 2013) shows the facilitation of attack in animals who have received an electrode in both the right and left HAA. The upper panel gives an outline of the experimental procedure. Panel A shows the well-known attack facilitation over the first three threshold determinations evoked from the first tested side. It also shows that there is hardly any decrease in the three attack thresholds that are subsequently evoked from the other side in the same animals. However, depicting these two series of threshold determinations in these animals as one single series, in the order they have been determined, disregarding the fact that they were obtained from contralateral sides, suggests that the facilitation derives from one single, long-lasting generalized process (Fig. 8, panel B) (Kruk et al. 2013).

Paradoxically, this seemingly slow, progressive facilitation is already initialized immediately after the very first induction of hypothalamic aggression. That could be shown in a similar design, i.e. in animals with an electrode in both the left as well as an electrode in the right HAA. It appeared that the thresholds in the very first tested side were significantly higher than the first threshold in the second tested side, when the thresholds at the contralateral side were determined within one hour after the first. The latter threshold is of course the second attack threshold determination experienced by the animals Fig. 9 (Kruk et al. 2013). Intriguingly, there is no downward trend in the attack-evoking current intensities during the about 20 min it takes to determine a threshold (Kruk et al. 1979). The facilitation of attack apparently arises in the interval between tests in the absence of fighting. The effect only becomes evident as a difference between first and subsequent thresholds.

21 Drugs and Hypothalamic Attack

Stable attack thresholds in experienced rats have successfully been used to determine the effects of drugs on hypothalamic aggression in repeated tests extending over many weeks. Surprisingly, hypothalamic attack thresholds are not affected by many drugs that do alter territorial aggression, such as benzodiazepines, scopolamine, alcohol and amphetamine. However, haloperidol, direct and indirect serotonin agonists, and L-propranolol do inhibit hypothalamic attack thresholds (Kruk 1991; Olivier et al. 1990). It is also surprising that drugs have only inhibitory effects on hypothalamic attack thresholds; no facilitating effects

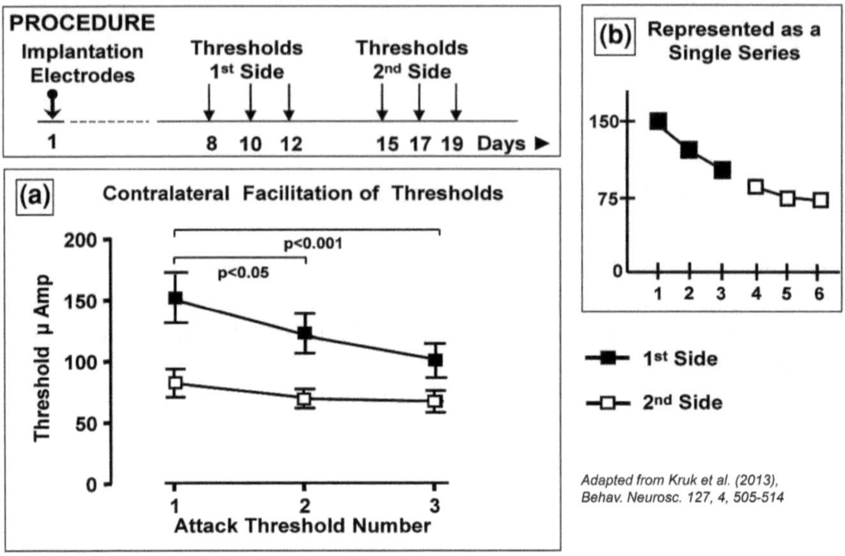

Fig. 8 Progressive facilitation of hypothalamic attack during repeated testing on subsequent days in animals having one single electrode in each contralateral HAA. *Top panel* design of the experiment. *Panel A filled squares* three decreasing attack thresholds on the first tested side; *closed squares* three thresholds at the opposite side in the same animals determined after the first three in the same animal. *Panel B* depicts both series as one series in time, disregarding the fact that these were obtained from opposite sides. Attack facilitation apparently is a general characteristic of the animal and not a consequence of a local change at the electrode tip (Reprinted with permission from Kruk et al. 2013)

Fig. 9 Rapid contralateral facilitation of attack in rats, when the contralateral side is tested within 1 h after the first tested ipsilateral side. *Black square* first tested side; *open square* second tested sides in the same rats. *Top panel* design (Reprinted with permission from Kruk et al. 2013)

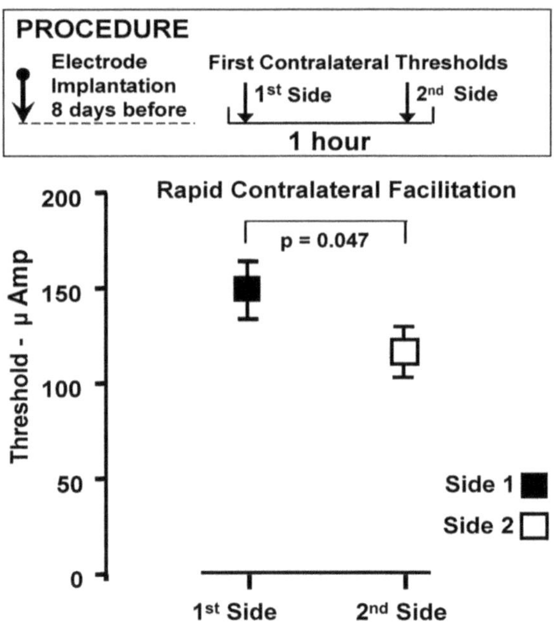

were found. Other hypothalamic responses such as locomotion and stimulation switch-off are not affected by propranolol and the serotonergic agonists used. Increasing the dose of L-propranolol not only increases the attack thresholds, but it also shifts the form of the attack from more violent towards milder forms. It is the only drug to do so. Haloperidol was least selective, since it inhibits all hypothalamic responses. We suggest that this relative pharmacological selectivity is due to the fact that hypothalamic stimulation bypasses several drug sensitive mechanisms involved in the control of aggression in natural settings. Interestingly propranolol, serotonergic agonists, and haloperidol are also used in the symptomatic control of socially unacceptable aggression in clinical contexts.

22 Neuroendocrine Crosstalk Within the HAA?

The hypothalamus is not known as a motor control area, but rather as a structure crucial to autonomic and endocrine regulation. Discussing the function of the many specific anatomical connections of the HAA is beyond the scope of this chapter. However, together these connections suggest that the HAA serves as a node in a network that compares and matches social and environmental information with the physiological state and requirements of the animal. The HAA seems well placed for that purpose. It is situated between the pituitary and the PVN, a nucleus crucial to the control of the stress response (Fig. 1).

23 The Endocrine Hypothalamus and Attack

Its localization, its local anatomical features and its extra-hypothalamic connections suggest that the HAA serves as a kind of bi-directional link between brain mechanisms appraising the challenges and potential of the social environment and the neuroendocrine stress response (de Kloet et al. 1996). In the classical view, these appraisal mechanisms adjust the dynamics of the neuroendocrine stress response to the behavioral requirements of the social environment. In that view it "prepares the body for fight or flight" by activating energy resources and shutting down non-essential metabolic activities. More recent evidence suggests that the dynamics of the neuroendocrine stress response in turn rapidly modulate the appraisal mechanisms to bring the internal "state of readiness" of the animal's physiology to bear on the behavioral response to the environment. Such a feedback could, for instance, signal whether the physiological resources available allow the animal to engage in a protracted conflict, or whether the lack of such resources favors one single all-or-nothing violent attack that could open an escape route for subsequent flight. In the same way, since circulating corticosteroids increase rapidly in all sorts of emergencies, not just in social conflict, the same mechanism could serve as an incentive for aggression should an opponent prevent a behavioral response adaptation required to cope with that emergency. The induction of aggression by direct injection of cortisol into the hypothalamus of golden hamsters was one of the early indications of such an integrative function of the hypothalamic link in the brain's aggressive circuitry (Hayden-Hixson and Ferris 1991a, b).

24 Rapid Bidirectional Feedback

Stimulation of the HAA rapidly provokes a large increase in circulating ACTH and corticosterone in the rat (Kruk et al. 1998, 2004). Similar responses have been found in territorial conflict (Schuurman 1980). In hypothalamic attack such an adrenocortical stress response is not primarily due to the physical stress of the conflict and fighting, since the response is quite similar in either the presence or the absence of an opponent (Kruk et al. 1998, 2004). In experienced rats, interestingly, chronic stabilization of their circulating corticosteroids at low, basal levels by adrenalectomy and implantation of a slow corticosteroid-releasing pellet does not change the attack thresholds (ADX+r). However, in such rats an acute injection of corticosteroids that mimics the corticosteroid response to HAA stimulation, when given shortly before an aggression test, produces an immediate, but short-lasting facilitation of hypothalamic attack thresholds, Fig. 10. The rapidity and short duration of this effect suggests that it is of a non-genomic origin. The rapid adrenocortical response and the rapid but short-lasting facilitation allow a mutual positive facilitating feedback within the time fame of one single conflict (Fig. 11).

Similar effects of a corticosterone injection were observed in territorial aggression when the natural adrenocortical response was blocked by a pharmacological blockade. In those settings it appeared that aggression facilitation by corticosterone was a two-stage process. The first and fastest phase was not inhibited by inhibition of protein synthesis, and hence was probably of a non-genomic origin. The second phase was sensitive to such inhibition, and therefore most probably based on a genomic effect of the corticosteroid response (Mikics et al. 2004). Together these findings point to a rapid mutual feedback between the adrenocortical stress response and the brain's aggression circuitry. The observation that a non-genomic, rapid—within the short time frame of an actual conflict—positive feedback response between an aggression-control mechanism and adrenocortical stress could engender a longer lasting genomic process, suggests a pathway by which a successful violent conflict strategy may an become an entrenched characteristic in similar future conflicts and in escalated aggression.

25 Adrenalectomy in "First Offenders"

Fast, non-genomic facilitation of aggression prompting a genomic process with long lasting consequences for future conflict behavior in experienced fighters, could evidently also support the emergence of aggression in naïve, inexperienced

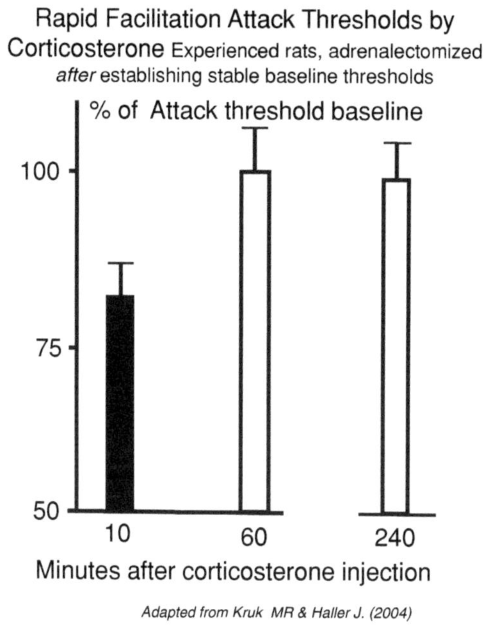

Fig. 10 Rapid, short lasting attack facilitation by an injection of corticosterone. In experienced rats with stable attack thresholds following repeated testing, chronic low basal levels of circulating corticosterone were achieved by adrenalectomy and sub-cutaneous implantation of a slow-release corticosterone pellet (ADX+r). Such adrenalectomy in these experienced rats does not in itself change their attack thresholds. However, an acute injection 10 min before testing rapidly facilitates the attack thresholds (Adapted with permission from Kruk et al. 2004)

fighters. Hypothalamic attack is well suited to test that idea, since it is first evoked in animals unfamiliar with such attacks in a novel setting, outside their home cage. Adrenalectomy and chronically stabilizing circulating corticosteroids at low basal levels one week before such naïve animals were tested for hypothalamic attack for the first time greatly increases the number of electrodes that fail to elicit attack. In many of these "failing electrodes" flight is induced instead. Moreover, Fig. 12 shows that in the few animals that still do attack, thresholds do not show the customary decrease upon repeated testing. Both findings suggest that a dynamic adrenal feedback has an important function in the emergence of novel aggressive responses and their development into a stable behavioral routine in subsequent conflicts. Figures 13 and 14 show the localizations of the tips of the electrodes within the HAA in this experiment. The HAA is depicted as a red transparent shape in a pseudo 3D representation. Electrodes that do evoke attacks are represented by dark red balls, electrodes that fail to induce attacks are represented by white balls. Clearly, there are many more ineffective electrodes in adrenalectomized animals compared with intact animals, even within the very core of the HAA, Fig. 14. These findings suggest that signals from the adrenal may be even more important in the emergence of hypothalamic attack in inexperienced animals

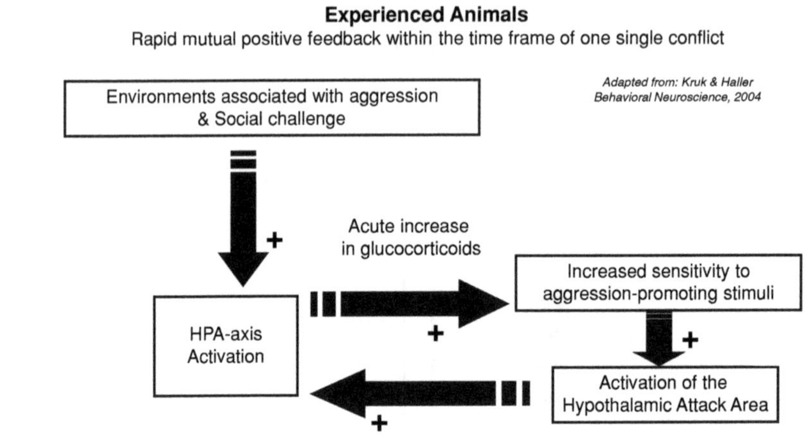

Fig. 11 Hypothetical schedule, depicting the concept that the anticipation of an impending conflict activates the Hypothalamic-Pituitary-Adrenal (HPA) axis, producing an adrenocortical response that promotes an increased sensitivity for aggression releasing and directing stimuli by a rapid feedback to appraisal mechanisms in the brain. Fighting could then activate the HPA axis even more, producing a mutual positive feedback between appraisal mechanisms and the adrenocortical stress response within one single conflict in a way that is conducive to the escalation of aggressive behavior (Adapted with permission from Kruk et al. 2004)

than they are in facilitating aggression in experienced animals. It seems possible that the failure to induce attacks in the HAA of some adrenally intact rats (Fig. 13) is due to a lack of adrenal responsivity at the moment these animals were first tested in. In territorial settings aggression is much more likely in the ascending phase than in the descending phase of the ultradian rhythm of the adrenocortical stress response (Haller et al. 2000a). Large individual differences exist between rats in adrenocortical responsiveness, and these may be related differences in aggressive responsivity or "coping" in territorial settings (de Boer et al. 2003).

26 Tipping the Fight–Flight Balance Towards Attack

The fact that stimulating the HAA in adrenalectomized, inexperienced animals triggers flight rather than attack responses suggests a fundamental change in the operation of the aggression-controlling circuitry in the brain. It seems to bring together in the HAA three old, well established concepts: (1) the hypothalamus as the chief regulator of the neuroendocrine and autonomic stress response; (2) the stress response as a mechanism to prepare the body for fight or flight; (3) the concept based on many behavioral observations in many species that in moments of intense conflict, individuals strike a balance between fight and flight, based on the internal requirements and the resources available to cope with the environmental

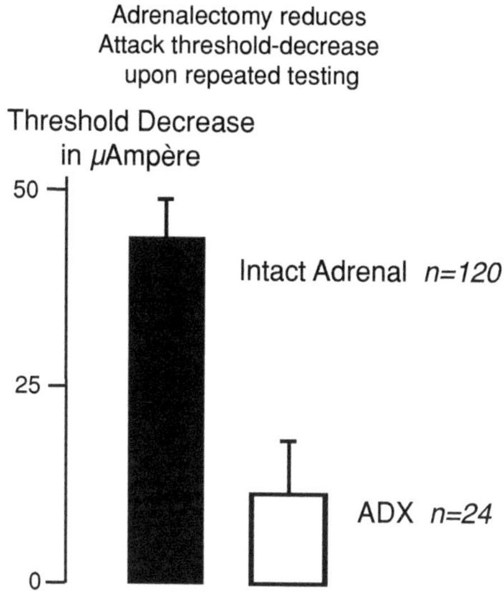

Fig. 12 In the minority of inexperienced rats that do still attack when tested for the first time first after prior adrenalectomy, attack thresholds hardly decrease, suggesting that adrenal signals have a special enabling role during the first induction of attacks in inexperienced animals. Difference between 1st and 3rd attack threshold in inexperienced animals with intact adrenals (*black bar*) and adrenalectomized rats (*white bar*) (ADX+r)

challenges and opportunities they are faced with. This observation suggests that in adrenally intact animals the dynamics of the adrenocortical stress response, in addition to readying the body for fight and flight, also operates as a feedback that switches the balance between fight and flight. It could do so either directly in the HAA or elsewhere in the aggression-controlling circuitry since corticosteroids rapidly enter the entire brain.

This idea of corticosteroid feedback tipping the balance between fight and flight at the level of the brain seems to be supported by the results presented in Fig. 15 (J. Haller and M.R. Kruk, unpublished data used with permission). It represents an experiment with four groups of animals that were all adrenalectomized and supplied with a subcutaneous corticosterone pellet that ensured a low, basal level of circulating corticosterone before the first test on behavioral responses from the HAA. One group was tested in the presence of an opponent, and attack thresholds were determined on four subsequent days. Another group was tested in the absence of an opponent and hence could not attack. In these animals four flight thresholds (operationalized as an effort to escape from the stimulation cage) were determined instead. Half of the animals received an intra-peritoneal injection of 0.25 mg/kg corticosterone in HBC, the other half receiving HBC vehicle, 10 min before the first test and again before every subsequent threshold determination. Attack thresholds in animals confronted with opponents are already low. The group receiving corticosterone shows a gradual but marginal facilitation of hypothalamic attack. The effect of corticosterone on flight thresholds is much more pronounced. In animals receiving vehicle flight thresholds are low and constant, on average even somewhat below the attack thresholds, which may explain why it is so difficult to induce attacks in adrenalectomized animals. By contrast, flight thresholds in animals

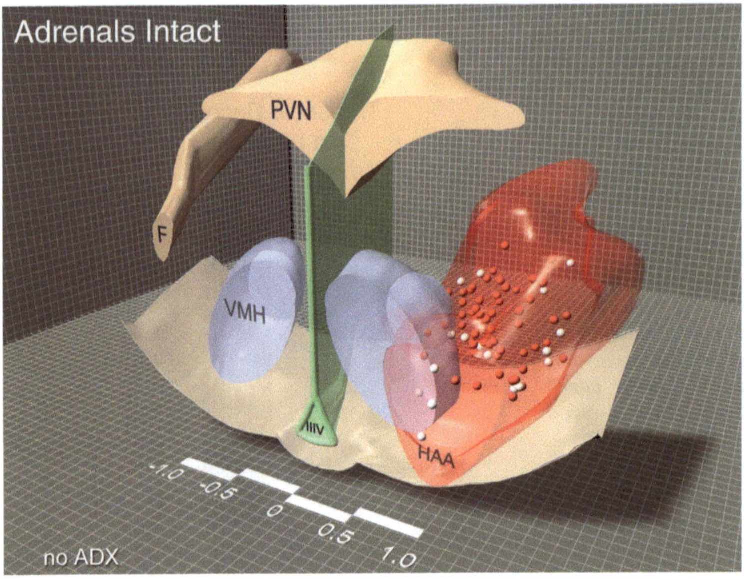

Fig. 13 Electrode positions within the HAA in a series of animals that had intact adrenals during their first aggression test. The HAA is depicted as a *red, semitransparant structure* viewed from a caudal position as in Fig. 1. Individual electrode positions from left or right hypothalamus are projected on the right hypothalamus in a pseudo-3D fashion. Each *red ball* represents a site from where attacks could be elicited. Each *white ball* represents a site from where stimulation failed to elicit attacks despite repeated testing. Most sites (54) elicited attacks, but a minority (17) failed to do so. Most failures lie at the perifery of the HAA

receiving corticosterone are more than twice as high on average compared to the vehicle group. Flight thresholds do not seem to change upon repeated testing.

These findings suggest an interesting scenario for what might happen during the first hypothalamic attacks in initially inexperienced, adrenally intact animals. During the very first test of these animals in the stimulation cage—a novel environment, where flight would be one of the behavioral options—the rapid increase of corticosterone accompanying HAA stimulation could increase thresholds for flight, i.e. suppress flight, and enable the expression of aggressive responses. These enabled aggressive responses could then become an entrenched behavioral routine upon repeated testing in that environment. In this concept the rapid adrenocortical response accompanying a first conflict would operate as an agent that tips the hypothetical fight-flight balance mechanism in the brain towards attack. This concept could be tested in other conflict paradigms in animals and humans.

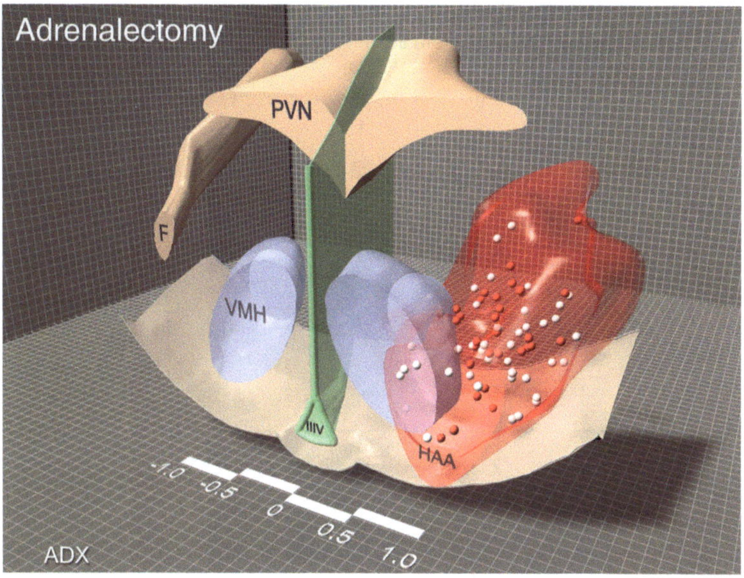

Fig. 14 Electrode positions within the HAA in a series of animals adrenalectomized (ADX+r) one week after electrode implantation and first tested for aggression one week after adrenalectomy. The HAA is depicted as a *red, semitransparant structure* viewed from a caudal position. Attack-eliciting sites are depicited as *red balls*. Sites failing to elicit attacks are shown as *white balls*. Compared with adrenaly intact animals (Fig. 13) far fewer sites (30) elicited attacks. The majority of sites (41) failed to elicit attacks. Moreover, these failures are distributed throughout the entire HAA and clearly present in its core

27 Mechanisms Elsewhere in the Brain?

The mechanism involved in these effects is largely unknown. It could very well be that it is not caused in the HAA itself, but by changes elsewhere in the aggression-controlling circuitry of the brain. Immediate early gene expression in the brain (c-Fos) indicative of regional brain activity after territorial and hypothalamic aggression is similar in most, but not all brain areas. Surprisingly, in adrenally intact animals, there are no large differences between c-Fos patterns in the brains of animals sacrificed directly after the first attack threshold, and animals sacrificed after the third and "stabilized" attack threshold. However, in the central amygdala (CeA), a structure involved in the appraisal of threatening stimuli, there is a large difference in activity. Immediately after the first test the CeA is strongly activated, while after the third test (CeA). c-Fos activity is almost absent (Fig. 16, black bars) (Halasz and Kruk, unpublished data, used with permission). White bars depict the approximately 40 % decrease in c-Fos in the HAA, which is in good agreement with a similar difference in between first and second attack threshold. It appears that the mechanism mediating the facilitation of attack is not necessarily confined

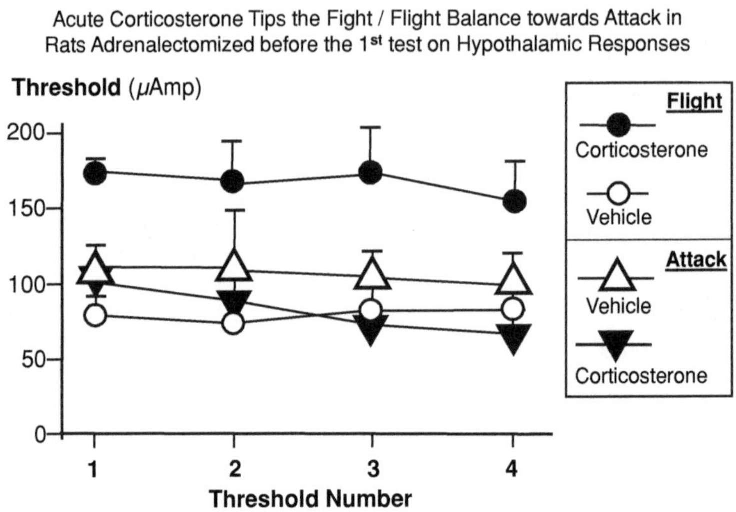

Fig. 15 Effect of repeated acute corticosterone treatment on the hypothalamic "fight/flight-balance" in inexperienced rats that were adrenalectomized (ADX+r) before the first test on hypothalamic responses. Acute corticosterone given just before each subsequent flight threshold determination inhibits flight considerably and immediately (i.e. higher thresholds), suggesting a rapid non-genomic effect on hypothalamic flight. Acute corticosterone given just before each subsequent attack threshold produces a small, gradual and marginal decrease of attack thresholds (J. Haller and M.R. Kruk, unpublished data used with permission)

to the hypothalamus itself. It could possibly be related to a decreased sensitivity for threatening stimuli, a concept to be explored in the context of escalated aggression in territorial settings (Halasz et al. 2002a). It was only recently shown that the adrenocortical stress response is crucially involved in the facilitation of aggression during repeated conflicts (Kruk et al. 2013 and this chapter), offering a new method for its study, and possibly opening a new perspective on this type of escalated aggression.

28 Mineralocorticoid Receptor Blockade in Naive Animals

Chronic stabilization of circulating corticosteroids at basal levels by adrenalectomy and implantation of a pellet slowly releasing corticosterone (ADX+r), is a drastic but utterly unselective method for impairing adrenocortical responses. Among its many other effects, it impairs the working of the autonomic nervous system, and it changes the settings and responsivity of the HPA axis. Moreover, the adrenal is also a source of other signals and hormones such as adrenaline, which conceivably could also affect hypothalamic attack.

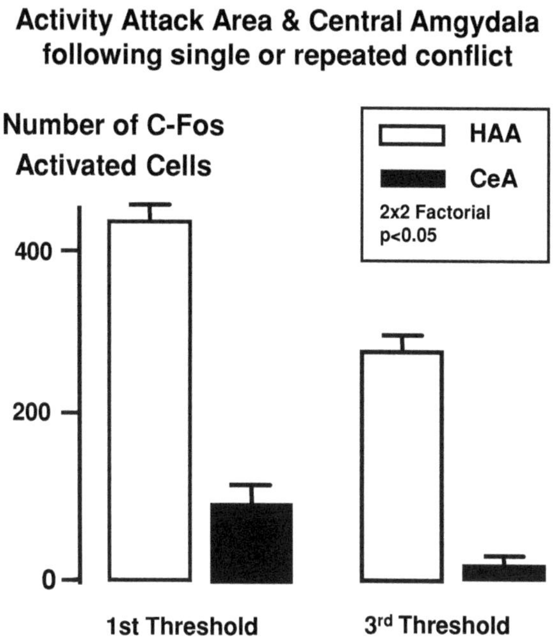

Fig. 16 Pilot experiment suggesting that the effects of attack threshold decreases upon repeated testing are not necessarily caused by a local change in the sensitivity of HAA neural elements. There are no large differences in immediate early gene expression (c-Fos) patterns in the brains of animals sacrificed directly after the first test on aggression, and animals sacrificed after the third test on aggression, with the exception of the central amygdala (CeA) a structure involved in the appraisal of threatening stimuli. Immediately after the first test the central amygdala is strongly activated, while after the third test CeA c-Fos is almost absent (*black bars*) (unpublished data, J. Halasz with permission). *White bars* depict the decrease in c-Fos in the HAA which decreases by about 40 % in good agreement with a similar threshold difference in threshold between first and second attack

Figures 17, 18 and 19 show the effects of more a selective impairment of adrenocortical response. In this experiment, spironolactone, which blocks both the genomic and rapid non-genomic effects of the mineralocorticoid receptor (de Kloet et al. 2008; Karst et al. 2005), was given intraperitoneally one hour before attack thresholds were determined in inexperienced rats. Spironolactone inhibits resident-intruder aggression in rats in a time dependent way (Haller et al. 1998b, 2000b).

Spironolactone rapidly enters the brain but by itself it has little activity. MR-blockade is caused by its only active metabolite, canrenone (Tokumura et al. 2005), which reaches its maximum concentration in about 1 h, when threshold determinations start. Both spironolactone and canrenone are rapidly cleared from the body (Kaukonen et al. 1998). The control group receives vehicle. Subsequent attack thresholds are determined spaced one day apart and a last threshold is determined 3 weeks later. 1 h before second, third and fourth threshold determination all

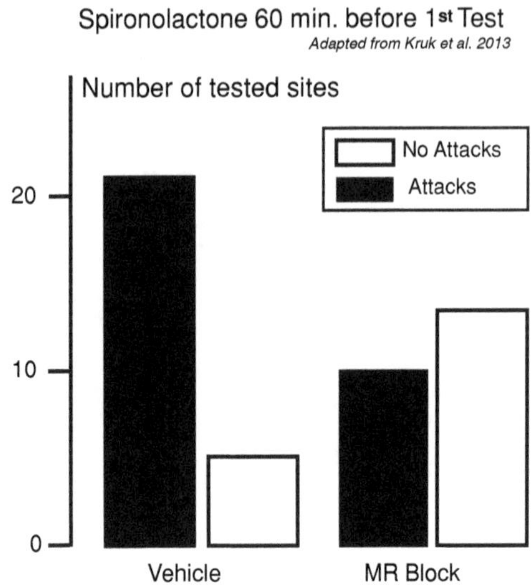

Fig. 17 Blockade of MR signaling by spironolactone increases the number of electrodes that fail to induce attacks (despite repeated re-tests) in inexperienced animals during their first aggression test, compared with control animals (Reprinted with permission from Kruk et al. 2013)

groups receive vehicle, with the exception of the "control" group that initially received vehicle; for the last threshold this group was given spironolactone instead. The procedure of the experiment is depicted in the top panel of Fig. 18.

During the first attack threshold determination in such initially inexperienced rats, MR blockade does not appreciably affect thresholds in rats which attacked. However, Fig. 17 shows that the fraction of rats that do attack under MR blockade is significantly lower than in those receiving vehicle. All rats that failed to attack when stimulated were re-tested twice though unsuccessfully on subsequent days in the absence of spironolactone. The behavioural effect apparently outlasts the blocker's availability in the body. Subsequent attack thresholds in the control group do show the usual decrease with repeated testing. By contrast, thresholds in animals that received spironolactone before their first aggression test do not decrease at all. Since spironolactone and canrenone are completely cleared from the body after 48 h (Kaukonen et al. 1998), therefore the second, third and fourth thresholds in this group are determined under drug-free conditions, but attack thresholds remain at their original high level. Intriguingly, MR blockade apparently does not inhibit thresholds during the fourth threshold in the control group. In the experimental group we are apparently observing a drug effect that interferes with process that is special to the initial phases of fighting in a novel environment. It is likely that the high failure rate in that group is an immediate expression of the same underlying process. Intriguingly, the consequence of temporary MR blockade—lack of attack facilitation—only shows up after repeated testing. The limited time window during which MR blockade is able to suppress the initialization of subsequent attack facilitation became clear in an experiment where spironolactone was given 15 min before the first test on attacks in naïve animals. In this case

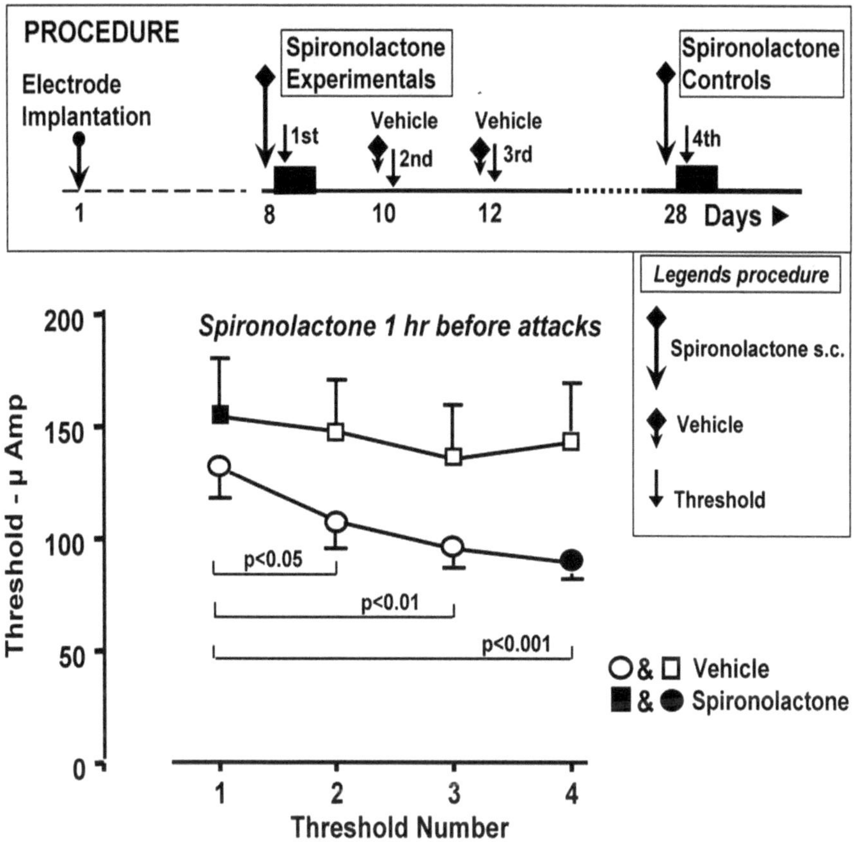

Fig. 18 The figure shows the involvement of MR signaling in the facilitation of attack during repeated testing. *Upper panel* for the design. Blocking MR signaling by spironolactone in inexperienced animals that still do attack abolishes attack facilitation upon repeated testing. Curiously MR is only temporary blocked around the moment of the very first attacks. During the second attack test, spironolactone and its active metabolite are already cleared from the body, and the corticosterone response accompanying hypothalamic attack is free to bind to the MR. Yet during the third and fourth aggression test, attack thresholds remain high. Surprisingly spironolactone does not inhibit attack thresholds during the fourth test in the control group that did receive vehicle before the first, second and third test, suggesting that the effect of MR blockade is confined to a narrow time-window around the first test (Reprinted with permission from Kruk et al. 2013)

canrenone, the active metabolite of spironolactone, only reaches the brain after the first aggressive responses. Figure 19 shows that this treatment does not block subsequent attack facilitation (Kruk et al. 2013).

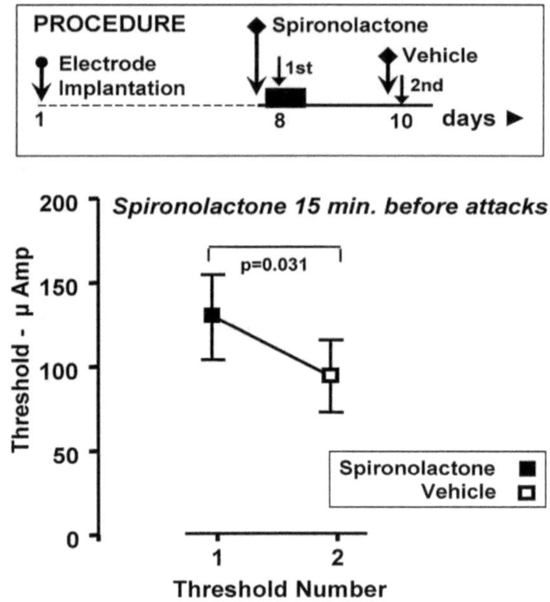

Fig. 19 First and second attack thresholds when spironolactone was given 15 min before the first aggression test. In this case spironolactone and its active metabolite will only reach effective concentrations after the aggression test is over. The figure shows that attack thresholds under such conditions do decrease as usual, confirming the idea that the MR-mediated corticosteroid signal that initializes subsequent attack facilitation acts within a narrow time-window around the first attacks (Reprinted with permission from Kruk et al. 2013)

29 Discussion

29.1 Timing of MR-Dependent Initializing

The effects on hypothalamic attack of both methods of impairing the adrenocortical stress response, adrenalectomy or MR-blockade, share several intriguing characteristics. Both are ineffective in *experienced* attackers. In *naïve inexperienced* attackers both methods do suppress the first induction of attacks and both impair the facilitation of attacks upon repeated testing in animals that still do attack. These findings suggest that the rapid corticosteroid response accompanying a first conflict in a novel environment involves an appraisal process that initializes the subsequent progressive facilitation of the brain's aggression controlling process. The term "initializing" was used to designate this hypothetical process that is associated with the first expression of aggressive response in an unfamiliar environment, and distinct from the "organizing" and "activating" processes that change aggression control mechanisms around birth and puberty, respectively (Kruk et al. 2013). It is as if a routine that has been lying dormant in a computer and has never yet been used needs to be unzipped and its settings initialized during its first use in an unfamiliar environment. Interestingly, short-lasting social defeat causes a long-lasting down regulation of MRs in septal and hippocampal areas (Buwalda et al. 2001) suggesting further studies on MRs in the emergence and aftermath of conflict.

29.2 A Two-pronged View on the HAA in Appraising Conflict

In our view the hypothalamic attack area (HAA) serves as a crucial node in a two-pronged mechanism that enables the emergence of adaptive responses to novel social challenges in an unfamiliar environment and their subsequent development into stable behavioral routines in subsequent similar challenges. It does so via its reciprocal neural connections with brain areas involved in the appraisal of social and environmental challenges and opportunities (Roeling et al. 1994; Toth et al. 2010). HAA stimulation strongly activates c-Fos activity in such areas. Surprisingly, "*attacks induced by the electrical stimulation of the HAA are associated with the activation of brain centers upstream relative to the HAA. This suggests that the upstream connections of the HAA are functionally relevant for the induction of attacks*" (Halasz et al. 2002b). Also, the HAA and the upstream areas involved in the appraisal of environmental stimuli are strongly activated by resident intruder conflict (Halasz et al. 2002a, b, 2006; Haller 2013; Tulogdi et al. 2010; Veening et al. 2005).

29.3 Timing is Crucial in MR Effects on Appraisal

The rapid adrenocortical feedback to the brain via the general circulation caused by HAA activation of the HPA-axis is critically involved in the adaptive response to novel social challenges. This adrenocortical signal—corticosterone in case of the rat, cortisol in humans and other animals—activates the brain's mineralocorticoid receptor. Our findings suggest that MR-activation is an essential step in the emergence and first expression of aggressive responses to a novel social challenge. In aggression it probably involves a two-stage non-genomic, genomic process that supports an immediate adaptation to a social challenge followed by a consolidation of the behavioral response to be used in future conflicts. Similar observations on the involvement of MRs in appraisal of novel situations have been reported in other behavioral paradigms (Joels and Baram 2009; Joels et al. 2008). The timing of the MR mediated signal with respect to the behavioral event seems a critical variable (Wiegert et al. 2006). Both neural and neuroendocrine branches of this "appraisal" or "social assessment" mechanism are necessary, as disturbance of each branch in itself leads to contextually inadequate coping with social challenges and a failure to adapt to similar challenges on future occasions.

29.4 An Alternative View on HAA "Activation" and Its "Natural" Function

This "appraisal supporting" view on the function of the HAA, is a sharp departure from more classical interpretations of hypothalamic aggression, that depicted the HAA as a center coordinating an innate hierarchical "downstream" neural structure that controls impulsive, instinctive responses (Anderson 2012; Kruk et al. 1979; Tinbergen 1951). However, this new, alternative viewpoint has the advantage that it could reconcile seemingly paradoxical findings on hypothalamic and adrenocortical function in aggression. Hypothalamic electrolytic or chemical lesions in do not abolish aggression, but rather change its nature often making animals more reactive (Olivier et al. 1983; Albert and Walsh 1982), Some chemically specific lesions promote offensive behavior (Vergnes et al. 1988) other chemically specific lesions reduce violence (Halasz et al. 2009). Some lesions disturb the normal sequential nature of the repertoire: "*Whereas intact animals usually terminate the agonistic interaction after a few clinches and keeping down the opponent, the [PMV] lesioned animals continue to perform sideways posture throughout the experiment. It seems as if they get locked into some behaviors and have difficulty in terminating an agonistic sequence*" (van den Berg et al. 1982). Human hamartomas, a bewildering disorganization of hypothalamic neuroanatomical structure, is accompanied by aggressive symptoms (de Almeida et al. 2008; Weissenberger et al. 2001), and there are numerous case reports on hypothalamic trauma or tumors being accompanied by aggressive symptoms. For a review of studies on the neurobiology of abnormal manifestations of aggression and hypothalamic mechanisms see (Haller 2013).

29.5 Stimulation as Functional Lesioning

Taken together these findings suggest that such disruptions cause a misreading of social signals, inadequate coping with social challenges which results in aggressive responses. Similarly, the various, arbitrary stimulation patterns used in many studies, suggest that hypothalamic aggression is also caused by a disruption of neural mechanisms involved in appraisal of social stimuli. The sudden, violent form of hypothalamic attack, the absence of introductory or intention-signaling patterns, and the poorer discrimination of objects suitable for attacking all support the view that HAA stimulation produces a severe temporary lesion that disables appraisal mechanisms. Advanced analysis of behavioral interactions shows that the appraisal of the social environment is already changed during HAA stimulation below attack threshold. Animals seem to become more apprehensive of the opponent's actions, and withdraw from social interactions (Haccou et al. 1988). However, the first attacks are executed quite suddenly and these attacks are immediately well targeted and well adapted to the opponent's postures and actions.

Animals apparently do not have to "learn" how to attack. However, while attacking they could "learn" that attacking this particular kind of opponent is successful, or, at least, that they suffer no negative consequences from attacking. During stimulation of the HAA animals seem to revert to an elementary "default" routine to deal with another individual in its surroundings: violent attack, or sometimes flight. The idea that brain stimulation may create functional lesions in addition to activating neurons has been reported earlier (Benazzouz and Hallett 2000; Benazzouz et al. 2000).

29.6 Consolidating an Inappropriate Violent Response

While HAA stimulation probably disturbs neural appraisal mechanisms it still produces an adrenocortical stress response that is very similar to the response observed in territorial aggression (Schuurman 1980). Under those conditions the neuroendocrine branch of the proposed "appraisal mechanism" would still be able to affect the emergence, or initialization of this elementary or "artificial" response and its development into an entrenched behavioral routine in future conflicts. The different underlying mechanisms of neuroendocrine support of the appraisal mechanisms most probably act in rather precise, limited time windows (Joels and Baram 2009).

29.7 Timing is Crucial in Hypothalamic Attack

The precise experimental control of the HAA stimulation paradigm over timing, opponent, and the use of repeated attack threshold determinations as a convenient measure of aggressivity, made it possible to tease apart the timing of the different effects of the adrenocortical stress response in aggression. Its first effects on the emergence in naive animals in an unfamiliar setting, and the influence of that first emergence on its subsequent facilitation into an entrenched violent behavioral routine in subsequent conflicts, and its support of aggressive responding in experienced animals have now been studied in more complex aggression paradigms such as the territorial resident–intruder paradigm. It appears that action of the MR signal that enables the expression and subsequent facilitation into a behavioral routine acts within a narrow time window around the first expression of attacks. Without the MR signal the majority of animals do not attack and the attack thresholds in that minority of animals that still attack do not decrease when they are tested again. So MR-signaling within a short time window is necessary, even crucial for the development of adaptive responses to a social challenge. However, it is not sufficient in itself. Stimulation of the HAA, and the expression of aggressive behavior is equally crucial. Together these events, stimulation, the execution of attacks and the simultaneous adrenocortical signal seem to create a

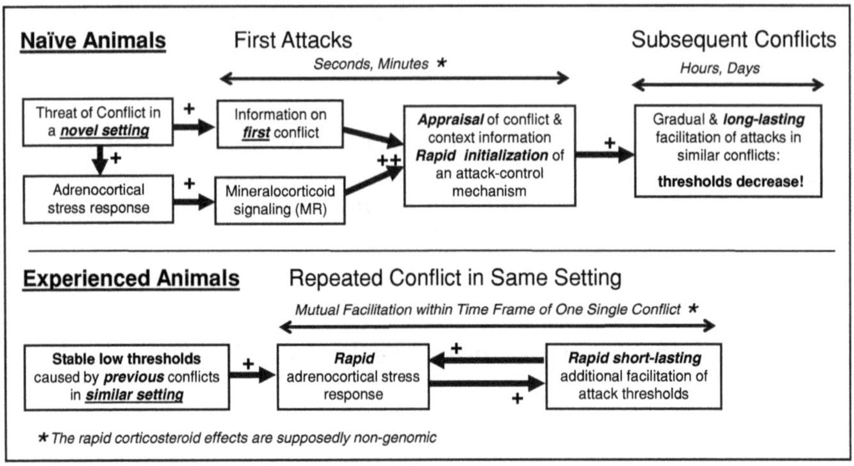

Fig. 20 Simplified summarizing representation of the proposed interactions between the adrenocortical stress response in naive, inexperienced animals facing a new conflict in an unfamiliar setting, the hypothetical appraisal mechanisms involved in initializing aggressive responding (*upper panel*). *Lower panel* Representation of the proposed facilitating interaction between the adrenocortical stress response and aggression control in fully experienced animals in which the appraisal of the well-known social challenge and the appropriate responses has already been initialized

sufficient condition allowing the development of an easily recruitable response in similar conflicts in the future. Both the initializing role of the adrenocortical signal, during the first expression of hypothalamic attack in naive animals, as well as its facilitating role during hypothalamic attack in experienced animals, act within the short time frame of one single conflict. Figure 20 gives an outline of the proposed mechanisms involved in the neuroendocrine and neural factors involved in these behavioral processes. The short time-frame suggests that these some of these effects depend on a non-genomic, possibly epigenetic mechanism (Kruk et al. 2004, 2013).

29.8 Timing is Crucial in Territorial Aggression

How do these ideas on the role of rapid non-genomic effects of the adrenocortical response on appraisal mechanisms apply to "spontaneous" territorial aggression in a "natural" setting such as the resident-intruder paradigm? In such natural settings, aggressive behavior is much more likely when the natural ultradian corticosterone rhythm is in its rising phase than when it is in a decreasing phase (Haller et al. 2000a). These findings confirm the powerful short-term influence of the adreno-cortical stress response over aggression. Temporary blockade of the synthesis of corticosterone, inhibits territorial aggression, in a two-staged process. The first phase is insensitive to protein synthesis inhibition, suggesting a short non-genomic initial phase. The second phase is sensitive to protein synthesis inhibition, suggesting that the non-genomic effect of corticosterone is engaging a genomic process that helps to consolidate the behavioral response for future conflicts (Mikics et al. 2004).

In these studies with experienced animals in natural settings, the neural branch of the appraisal mechanism connected to the HAA is presumably still functioning, and the animals can still act according to their regular aggressive repertoire. However, when initially naive animals which have previously been adrenalectomized are faced with their first territorial conflict in the resident-intruder paradigm, the powerful effect of the adrenocortical stress response on initial appraisal processes becomes clear again. Control animals with intact adrenals develop the normal aggressive repertoire upon repeated testing as usual. That includes threats, intention signaling, and avoidance of attacks to vulnerable body parts on the opponent. However, when animals are adrenalectomized before their first conflict they do not gradually develop the normal aggressive repertoire. They attack immediately with a high frequency during the very first conflict, they fail to signal aggressive intent, and they do not follow the normal "rules of engagement", attacking the most vulnerable parts of their opponent. Significantly, these animals persist in this abnormal pattern in subsequent tests on subsequent days. Clearly, these animals cannot assess the true nature of the conflict they are facing, and adapt to it. However, this abnormal aggressive pattern can be completely reversed by a corticosterone injection 10 minutes before the first and subsequent tests, which again suggests a rapid non-genomic mechanism (Haller et al. 2001).

29.9 Initialization and Experience in Natural Settings

These observations strongly suggest that corticosteroids in naive animals in natural settings also enable a rapid "initialization" process with long-lasting effects on behavior in subsequent conflicts, rather similar to the findings observed in hypothalamic attack. Apparently the adrenocortical influence on the appraisal processes in stressful conflicts is more important than understood until recently. However, the significance of that signal needs to be defined carefully. These effects cannot be understood in simple terms like "high levels of circulating corticosteroids promote aggression, and low levels inhibit aggression". Such simple ideas do not explain the results discussed above. In one example (Mikics et al. 2004) circulating corticosteroids are low and "aggression" is inhibited. In the another case discussed above (Haller et al. 2001) circulating levels of corticosteroids are chronically low and "aggression" is increased. One has to consider the different time domains in the emergence of aggressive responses and the behavioral context, as well as the nature of the aggressive response. In inexperienced animals during a first confrontation in an unfamiliar setting, we suggest that the rising phase of the adrenocortical stress response enables the animal to appraise or asses the impending conflict properly, supported by a rapid non-genomic corticosteroid-dependent mechanism that may "convey a sense of urgency" to the analysis of the environment. It allows the animal to employ an appropriate behavioral response during that first conflict. That short-lasting initialization of the appropriate behavioral response is subsequently developed into a engrained scenario or script during similar conflicts, which is also supported by the adrenocortical stress response within the timeframe of one single conflict, and probably consolidated by a later genomic follow-up that is not fully understood yet.

29.10 Experience in Other Aggressive Paradigms

Interestingly, direct activation of the HAA, an area rich in glutamate receptors (Hrabovszky et al. 2005), by a glutamate-agonist in combination with a GABA antagonist only induced aggression in animals with recent experience with territorial fighting, and not in naive animals (Haller et al. 1998a). This finding gave an early indication that the conflict appraisal mechanism associated with the HAA needs to be initialized. However, the adrenocortical effects are not necessarily due to local effects in the HAA. As far as we know at this moment, mineralocorticoid receptors are absent in the HAA. However, they are expressed in areas involved in the appraisal of social and environmental signals such as the hippocampus, the frontal and orbital cortex, the septum, and the amygdala. Adrenalectomy (ADX+r) in conflict naive animals which promotes "abnormal" aggression, a pattern that can be reversed by corticosterone injections, produces excessive c-Fos activation in the central amygdala (CeA) during conflict. The CeA is an area involved in the appraisal of threatening information. Injections of corticosterone shortly before a the resident-intruder test reverse both the abnormal attack pattern and the excessive activation of the CeA. The CeA is also the structure that is strongly activated in animals after the first hypothalamic attacks, but is hardly activated in experienced attackers (this chapter). In an extensive c-Fos study of the brain it was stated that "*an acute glucocorticoid treatment abolished both behavioral and neural consequences of glucocorticoid hypofunction. These data suggest that glucocorticoid hypofunction-induced abnormal forms of aggressiveness are related to increased sensitivity to stressors and fear-eliciting stimuli*" (Halasz et al. 2002a). Such findings further support the idea that a rapid adrenocortical stress response supports an appraisal mechanism that is crucial for adequate responding to social challenges.

29.11 "Stimulation Experience" Revisited

Adequate coping with novel social challenges requires the simultaneous activation of both neural and endocrine signals. In this context the older ideas on "stimulation experience" and "decreasing thresholds" (Roberts et al. 1967; Valenstein et al. 1968,

1969; Wise 1968, 1969) acquire a new meaning here, not as opposing explanations of hypothalamic responses, but as necessary factors for the emergence of adaptive behavioral responses and their enduring incorporation into the behavioral repertoire. The crucial role of the adrenocortical stress response in aggression appears to be a prime mover in the first emergence of aggressive responding reconciling these early ideas. Without the stress response there can be no adequate behavioral response, and without the behavioral response ("stimulation experience") there can be no adaptation or facilitation ("threshold decreases") during similar conflicts in the future. In the case of hypothalamic attack when both branches of the appraisal mechanism are impaired by ADX+r or MR-blockade, a profound disturbance becomes apparent, and flight rather than aggression results. In such an adrenalectomized condition flight is suppressed by acute injection of corticosterone, suggesting a role of the adrenocortical stress response in rapidly switching the flight/fight balance during conflicts (this chapter).

29.12 A "Somatic Marker" conveying a "Sense of Urgency"

We suggest that the adrenocortical feedback loop may convey a "sense of urgency" or "metabolic readiness" to the brain, facilitating the expression of a coping response that matches the internal physiological state of the individual with the challenges and opportunities of the social environment. This "matching" concept brings to mind the "somatic marker" hypothesis first proposed by Damasio (1999). That hypothesis states that in humans a "somatic marker" from the peripheral body allows people to decide how to react to a novel challenge. In the absence of such a signal, behavioral reactions to a challenge are difficult, even though all the considerations that go in such a decision are fully understood. A more extensive discussion on role of the stress response and the possible rapid molecular mechanisms involved in such appraisal mechanisms can be found in (de Kloet et al. 2008; Joels and Baram 2009; Karst et al. 2005, 2010; Kruk et al. 2013). An engaging blueprint on the flow of information in such appraisal mechanisms in escalated aggression in humans can be found in (Potegal 2012).

The public view on hormonal factors in aggression is often expressed in popular catchy one-liners such as "I felt my adrenaline rising" as an explanation of anger and aggressivity. Such ideas have an intuitive appeal, and may still be correct. However, adrenaline may be just one of the factors, and it does not easily enter into the brain. The other adrenal hormone, cortisol in humans or corticosterone in rodents, does rapidly enter the brain, and in contrast to earlier assumptions, it can exert almost immediate effects on brain neurons via a non-genomic process. Future studies on social appraisal and conflict assessment in humans may therefore prove useful.

29.13 Adrenocortical and Social Dysregulation in Humans

Our findings suggest that the processing of conflict-related stimuli is probably disordered in groups of psychiatric disorders where adrenocortical dysregulation and inappropriate conflict handling are common, such as PTSD (Yehuda et al. 2014) and borderline personality disorder (Rinne et al. 2002). Recent experimental studies in healthy persons on the effects of cortisol indicate that both basal and exogenous cortisol affect aggression in humans (Bohnke et al. 2010a, b). Effects of social stress and stress-induced cortisol on the preconscious selective attention to social threat were reported (Roelofs et al. 2007). Exogenous cortisol facilitates responses to social threat under high provocation (Bertsch et al. 2009b, c, 2011). Aggression produced significant effects on information processing in the emotional Stroop test, which lasted beyond the aggressive encounter even in healthy participants (Bertsch et al. 2009a). In addition, the processing of emotional facial expressions is affected by cortisol in healthy subjects (Bertsch et al. 2009b). Taken together such results suggest an important role of the adrenocortical stress response in the processing and appraisal of social signals in conflicts. If such effects are already apparent in the moderately "aggressive" paradigms used in these experimental studies, one should expect that the adrenocortical stress response has even more influence in real life experiences of an extremely stressful or traumatic nature. As our animal experiments predict such effects could be very rapid in their initial phase during a first conflict, and then become engrained in behavior and in aggression controlling mechanisms in the brain, resulting in long-lasting structural and functional changes. Accordingly, morphometric differences in hypothalamus and hippocampus were reported in women with borderline personality disorder, and these differences were positively related to traumatic experience (Kuhlmann et al. 2013). Further testing of such predictions awaits the advent of ever more powerful imaging and genomic technology, allowing the study of the HAA and its connections in individuals suffering from inadequate conflict handling. Our results suggest that studying the effect of the neuroendocrine stress response could contribute to understanding of the emergence and consolidation of escalated aggression under stressful conditions in humans. This may especially apply to individuals, with damage to frontal areas, a vulnerable adrenocortical stress response and individuals with a history of social trauma, or a combination of these risk factors.

Acknowledgments Many persons contributed with ideas and experiments during the more than 40 years we have tried to understand hypothalamic aggression as a starting point to define and clarify pathology and escalation in social conflict. However, while building on such ideas and experiments, this chapter reflects my own views, and its inevitable errors and shortcomings are my responsibility only.

We are very much indebted to the early workers in the Ethopharmacology Group in Leiden guided by Guus (A. M.) van der Poel who initiated this research line. We are equally indebted to the stimulating colleagues of Medical Pharmacology in Leiden guided by Ron (E. R.) de Kloet.

The long and pleasant cooperation with Jozsef Haller, Jozsef Halasz and Eva Mikics of the Hungarian Intitute of Experimental Medicine in Budapest, allowed us to succesfully explore a fresh neuroendocrine perspective on hypothalamic aggression.

Our very special thanks go to Wout Meelis, who meticulously performed many of our experiments and actively supported all others over a period of almost 40 years.

Without the constant encouragement, support, clear editorial eye and scientific understanding of my wife, Martje Kruk-de Bruin, our modest contribution to the behavioral neuroscience of aggression would never have been possible.

Over these years our work was supported by a grant from the European Training Program for Brain and Behavior Research (ETP), by several grants from Netherlands Organization for Scientific Research (NWO), grants from the Hungarian Scientific Research Fund (OTKA), a grant from the Royal Netherlands Academy of Arts and Sciences (KNAW), and by several grants from the Harry Frank Guggenheim Foundation in New York to Jozsef Haller and me.

References

Albert DJ, Walsh ML (1982) Medial hypothalamic lesions in the rat enhance reactivity and mouse killing but not social aggression. Physiol Behav 28:791–795

Anderson DJ (2012) Optogenetics, sex, and violence in the brain: implications for psychiatry. Biol Psychiatry 71:1081–1089. doi:10.1016/j.biopsych.2011.11.012

Bandler RJ (1982) Neural control of aggressive behavior. Trends Neurosci 5:390–394

Bandler RJ Jr, Flynn JP (1972) Control of somatosensory fields for striking during hypothalamically elicited attack. Brain Res 38:197–201

Bejjani BP, Houeto JL, Hariz M, Yelnik J, Mesnage V, Bonnet AM, Pidoux B, Dormont D, Cornu P, Agid Y (2002) Aggressive behavior induced by intraoperative stimulation in the triangle of Sano. Neurology 59:1425–1427. doi:10.1212/01.WNL.0000031428.31861.23

Benazzouz A, Hallett M (2000) Mechanism of action of deep brain stimulation. Neurology 55:S13–S16

Benazzouz A, Piallat B, Ni ZG, Koudsie A, Pollak P, Benabid AL (2000) Implication of the subthalamic nucleus in the pathophysiology and pathogenesis of Parkinson's disease. Cell Transplant 9:215–221

Bermond B, Mos J, Meelis W, van der Poel AM, Kruk MR (1982) Aggression induced by stimulation of the hypothalamus: effects of androgens. Pharmacol Biochem Behav 16:41–45. doi:10.1016/0091-3057(82)90010-7

Bertsch K, Bohnke R, Kruk MR, Naumann E (2009a) Influence of aggression on information processing in the emotional stroop task—an event-related potential study. Front Behav Neurosci 3:28. doi:10.3389/neuro.08.028.2009

Bertsch K, Bohnke R, Kruk MR, Richter S, Naumann E (2011) Exogenous cortisol facilitates responses to social threat under high provocation. Horm Behav 59:428–434. doi:10.1016/j.yhbeh.2010.12.010

Bertsch K, Khan R, Kruk MR, Richter S, Naumann E (2009b) The Influence of cortisol and aggression on the processing of emotional facial expressions. Psychophysiology 46:S25–S25

Bertsch K, Trenz J, Khan R, Kruk MR, Richter S, Naumann E (2009c) The influence of trait anger, cortisol, and aggression resting cortical activity. Psychophysiology 46:S50–S50

Bohnke R, Bertsch K, Kruk MR, Naumann E (2010a) The relationship between basal and acute HPA axis activity and aggressive behavior in adults. J Neural Transm 117:629–637. doi:10.1007/s00702-010-0391-x

Bohnke R, Bertsch K, Kruk MR, Richter S, Naumann E (2010b) Exogenous cortisol enhances aggressive behavior in females, but not in males. Psychoneuroendocrinology 35:1034–1044. doi:10.1016/j.psyneuen.2010.01.004

Buwalda B, Felszeghy K, Horvath KM, Nyakas C, de Boer SF, Bohus B, Koolhaas JM (2001) Temporal and spatial dynamics of corticosteroid receptor down-regulation in rat brain following social defeat. Physiol Behav 72:349–354

Cervantes MC, Delville Y (2007) Individual differences in offensive aggression in golden hamsters: a model of reactive and impulsive aggression? Neuroscience 150:511–521. doi:10.1016/j.neuroscience.2007.09.034

Compaan JC, Buijs RM, Pool CW, De Ruiter AJ, Koolhaas JM (1993) Differential lateral septal vasopressin innervation in aggressive and nonaggressive male mice. Brain Res Bull 30:1–6

Damasio AR (1999) The feeling of what happens: body and emotion in the making of consciousness. Harcourt Brace and Company, New York

de Almeida AN, Fonoff ET, Ballester G, Teixeira MJ, Marino R Jr (2008) Stereotactic disconnection of hypothalamic hamartoma to control seizure and behavior disturbance: case report and literature review. Neurosurg Rev 31:343–349. doi:10.1007/s10143-008-0142-8

de Boer SF, van der Vegt BJ, Koolhaas JM (2003) Individual variation in aggression of feral rodent strains: a standard for the genetics of aggression and violence? Behav Genet 33:485–501

de Kloet ER, Karst H, Joels M (2008) Corticosteroid hormones in the central stress response: quick-and-slow. Front Neuroendocrinol 29:268–272. doi:10.1016/j.yfrne.2007.10.002

de Kloet ER, Korte SM, Rots NY, Kruk MR (1996) Stress hormones, genotype, and brain organization. Implications for aggression. Ann N Y Acad Sci 794:179–191

Delville Y, Mansour KM, Ferris CF (1996) Testosterone facilitates aggression by modulating vasopressin receptors in the hypothalamus. Physiol Behav 60:25–29

Demski LS (1973) Feeding and aggressive behavior evoked by hypothalamic stimulation in a cichlid fish. Comp Biochem Physiol, A: Comp Physiol 44:685–692. doi:10.1016/0300-9629(73)90134-5

Ettrup KS, Sorensen JC, Rodell A, Alstrup AK, Bjarkam CR (2012) Hypothalamic deep brain stimulation influences autonomic and limbic circuitry involved in the regulation of aggression and cardiocerebrovascular control in the Gottingen minipig. Stereotact Funct Neurosurg 90:281–291. doi:10.1159/000338087

Ferris CF (2000) Adolescent stress and neural plasticity in hamsters: a vasopressin–serotonin model of inappropriate aggressive behaviour. Exp Physiol 85 Spec No: 85s–90s

Ferris CF (2005) Vasopressin/oxytocin and aggression. Novartis Found Symp 268:190–198 (discussion 198–200, 242–253)

Ferris CF, Lu SF, Messenger T, Guillon CD, Heindel N, Miller M, Koppel G, Robert Bruns F, Simon NG (2006) Orally active vasopressin V1a receptor antagonist, SRX251, selectively blocks aggressive behavior. Pharmacol Biochem Behav 83:169–174. doi:10.1016/j.pbb.2006.01.001

Ferris CF, Melloni RH Jr, Koppel G, Perry KW, Fuller RW, Delville Y (1997) Vasopressin/serotonin interactions in the anterior hypothalamus control aggressive behavior in golden hamsters. J Neurosci 17:4331–4340

Goodson JL, Kingsbury MA (2013) What's in a name? Considerations of homologies and nomenclature for vertebrate social behavior networks. Horm Behav 64:103–112. doi:10.1016/j.yhbeh.2013.05.006

Haccou P, Kruk MR, Meelis E, Vanbavel ET, Wouterse KM, Meelis W (1988) Markov-models for social interactions—analysis of electrical-stimulation in the hypothalamic aggression area of rats. Anim Behav 36:1145–1163. doi:10.1016/S0003-3472(88)80074-5

Halasz J, Liposits Z, Kruk MR, Haller J (2002a) Neural background of glucocorticoid dysfunction-induced abnormal aggression in rats: involvement of fear- and stress-related structures. Eur J Neurosci 15:561–569. doi:10.1046/j.0953-816x.2001.01883.x

Halasz J, Liposits Z, Meelis W, Kruk MR, Haller J (2002b) Hypothalamic attack area-mediated activation of the forebrain in aggression. NeuroReport 13:1267–1270

Halasz J, Toth M, Kallo I, Liposits Z, Haller J (2006) The activation of prefrontal cortical neurons in aggression—a double labeling study. Behav Brain Res 175:166–175. doi:10.1016/j.bbr.2006.08.019

Halasz J, Zelena D, Toth M, Tulogdi A, Mikics E, Haller J (2009) Substance P neurotransmission and violent aggression: the role of tachykinin NK(1) receptors in the hypothalamic attack area. Eur J Pharmacol 611:35–43. doi:10.1016/j.ejphar.2009.03.050

Haller J (2013) The neurobiology of abnormal manifestations of aggression—a review of hypothalamic mechanisms in cats, rodents, and humans. Brain Res Bull 93:97–109. doi:10.1016/j.brainresbull.2012.10.003

Haller J, Abraham I, Zelena D, Juhasz G, Makara GB, Kruk MR (1998a) Aggressive experience affects the sensitivity of neurons towards pharmacological treatment in the hypothalamic attack area. Behav Pharmacol 9:469–475

Haller J, Halasz J, Mikics E, Kruk MR, Makara GB (2000a) Ultradian corticosterone rhythm and the propensity to behave aggressively in male rats. J Neuroendocrinol 12:937–940

Haller J, Millar S, Kruk MR (1998b) Mineralocorticoid receptor blockade inhibits aggressive behaviour in male rats. Stress 2:201–207. doi:10.3109/10253899809167283

Haller J, Millar S, van de Schraaf J, de Kloet ER, Kruk MR (2000b) The active phase-related increase in corticosterone and aggression are linked. J Neuroendocrinol 12:431–436. doi:10.1046/j.1365-2826.2000.00470.x

Haller J, van de Schraaf J, Kruk MR (2001) Deviant forms of aggression in glucocorticoid hyporeactive rats: a model for 'pathological' aggression? J Neuroendocrinol 13:102–107. doi:10.1111/j.1365-2826.2001.00600.x

Hayden-Hixson DM, Ferris CF (1991a) Cortisol exerts site-, context- and dose-dependent effects on agonistic responding in hamsters. J Neuroendocrinol 3:613–622. doi:10.1111/j.1365-2826.1991.tb00326.x

Hayden-Hixson DM, Ferris CF (1991b) Steroid-specific regulation of agonistic responding in the anterior hypothalamus of male hamsters. Physiol Behav 50:793–799. doi:10.1016/0031-9384(91)90020-O

Herndon JG, Perachio AA, McCoy M (1979) Orthogonal relationship between electrically elicited social aggression and self-stimulation from the same brain sites. Brain Res 171:374–380. doi:10.1016/0006-8993(79)90346-9

Hess WR (1928) Stammganglien-Reizversuche. Berichte über die gesamte Physiologie und experimentelle Pharmakologie 42:554–555

Hrabovszky E, Halasz J, Meelis W, Kruk MR, Liposits Z, Haller J (2005) Neurochemical characterization of hypothalamic neurons involved in attack behavior: glutamatergic dominance and co-expression of thyrotropin-releasing hormone in a subset of glutamatergic neurons. Neuroscience 133:657–666. doi:10.1016/j.neuroscience.2005.03.042

Joels M, Baram TZ (2009) The neuro-symphony of stress. Nat Rev Neurosci 10:459–466. doi:10.1038/nrn2632

Joels M, Karst H, de Rijk RH, de Kloet ER (2008) The coming out of the brain mineralocorticoid receptor. Trends Neurosci 31:1–7. doi:10.1016/j.tins.2007.10.005

Karli P (1956) The Norway rat's killing response to the White Mouse. Behaviour 10:81–103

Karst H, Berger S, Erdmann G, Schutz G, Joels M (2010) Metaplasticity of amygdalar responses to the stress hormone corticosterone. PNAS 107:14449–14454. doi:10.1073/pnas.0914381107

Karst H, Berger S, Turiault M, Tronche F, Schutz G, Joels M (2005) Mineralocorticoid receptors are indispensable for nongenomic modulation of hippocampal glutamate transmission by corticosterone. PNAS 102:19204–19207. doi:10.1073/pnas.0507572102

Kaukonen AM, Lennernas H, Mannermaa JP (1998) Water-soluble beta-cyclodextrins in paediatric oral solutions of spironolactone: preclinical evaluation of spironolactone bioavailability from solutions of beta-cyclodextrin derivatives in rats. J Pharm Pharmacol 50:611–619. doi:10.1111/j.2042-7158.1998.tb06894.x

King MB, Hoebel BG (1968) Killing elicited by brain stimulation in rats. Communs Behav Biol A2:173–177

Koolhaas JM (1978) Hypothalamically induced intraspecific aggressive behaviour in the rat. Exp Brain Res 32:365–375

Kruk MR (1991) Ethology and pharmacology of hypothalamic aggression in the rat. Neurosci Biobehav Rev 15:527–538

Kruk MR, Halasz J, Meelis W, Haller J (2004) Fast positive feedback between the adrenocortical stress response and a brain mechanism involved in aggressive behavior. Behav Neurosci 118:1062–1070. doi:10.1037/0735-7044.118.5.1062

Kruk MR, Haller J, Meelis W, de Kloet ER (2013) Mineralocorticoid receptor blockade during a rat's first violent encounter inhibits its subsequent propensity for violence. Behav Neurosci 127:505–514. doi:10.1037/a0033553

Kruk MR, Meelis W, van der Poel AM, Mos J (1981) Electrical stimulation as a tool to trace physiological properties of the hypothalamic network in aggression. In: Brain PF, Benton D (eds) The biology of aggression, vol D8. Sijthoff and Noordhoff International Publishers B.V., Alphen aan den Rijn, pp 383–395

Kruk MR, Van der Laan CE, Meelis W, Phillips RE, Mos J, van der Poel AM (1984a) Brain-stimulation induced agonistic behaviour: a novel paradigm in ethopharmacological aggression research. Prog Clin Biol Res 167:157–177

Kruk MR, Van der Laan CE, Mos J, van der Poel AM, Meelis W, Olivier B (1984b) Comparison of aggressive behaviour induced by electrical stimulation in the hypothalamus of male and female rats. Prog Brain Res 61:303–314. doi:10.1016/s0079-6123(08)64443-x

Kruk MR, van der Poel AM (1980) Is there evidence for a neural correlate of an aggressive behavioural system in the hypothalamus of the rat? Prog Brain Res 53:385–390. doi:10.1016/S0079-6123(08)60077-1

Kruk MR, van der Poel AM, de Vos-Frerichs TP (1979) The induction of aggressive behaviour by electrical stimulation in the hypothalamus of male rats. Behaviour 70:292–322

Kruk MR, van der Poel AM, Meelis W, Hermans J, Mostert PG, Mos J, Lohman AH (1983) Discriminant analysis of the localization of aggression-inducing electrode placements in the hypothalamus of male rats. Brain Res 260:61–79. doi:10.1016/0006-8993(83)90764-3

Kruk MR, Van der Laan CE, van der Poel AM, Vanerp AMM, Meelis W (1990) Strain differences in attack patterns elicited by electrical-stimulation in the hypothalamus of male CPBWEzob and CPBWI Rats. Aggress Behav 16:177–190. doi:10.1002/1098-2337(1990)16:3/4<177:Aid-Ab2480160305>3.0.Co;2-V

Kruk MR, Westphal KG, Van Erp AM, van Asperen J, Cave BJ, Slater E, de Koning J, Haller J (1998) The hypothalamus: cross-roads of endocrine and behavioural regulation in grooming and aggression. Neurosci Biobehav Rev 23:163–177

Kuhlmann A, Bertsch K, Schmidinger I, Thomann PA, Herpertz SC (2013) Morphometric differences in central stress-regulating structures between women with and without borderline personality disorder. J Psychiatry Neurosci 38:129–137. doi:10.1503/jpn.120039

Lammers JH, Kruk MR, Meelis W, van der Poel AM (1988a) Hypothalamic substrates for brain stimulation-induced attack, teeth-chattering and social grooming in the rat. Brain Res 449:311–327

Lammers JH, Kruk MR, Meelis W, van der Poel AM (1988b) Hypothalamic substrates for brain stimulation-induced patterns of locomotion and escape jumps in the rat. Brain Res 449:294–310

Lammers JH, Meelis W, Kruk MR, van der Poel AM (1987) Hypothalamic substrates for brain stimulation-induced grooming, digging and circling in the rat. Brain Res 418:1–19

Lin D, Boyle MP, Dollar P, Lee H, Lein ES, Perona P, Anderson DJ (2011) Functional identification of an aggression locus in the mouse hypothalamus. Nature 470:221–226. doi:10.1038/nature09736

Lipp HP (1978) Aggression and flight behaviour of the marmoset monkey *Callithrix jacchus*: an ethogram for brain stimulation studies. Brain Behav Evol 15:241–259. doi:10.1159/000123781

Lipp HP, Hunsperger RW (1978) Threat, attack and flight elicited by electrical stimulation of the ventromedial hypothalamus of the marmoset monkey *Callithrix jacchus*. Brain Behav Evol 15:260–275. doi:10.1159/000123782

MacDonnell MF, Flynn JP (1966) Control of sensory fields by stimulation of hypothalamus. Science 152:1406–1408

Martin JR (1976) Motivated behaviors elicited from hypothalamus, midbrain, and pons of the guinea pig (*Cavia porcellus*). J Comp Physiol Psychol 90:1011–1034. doi:10.1037/h0078658

Mikics E, Kruk MR, Haller J (2004) Genomic and non-genomic effects of glucocorticoids on aggressive behavior in male rats. Psychoneuroendocrinology 29:618–635. doi:10.1016/S0306-4530(03)00090-8

Mos J, Lammers JHCM, van der Poel AM, Bermond B, Meelis W, Kruk MR (1983) Effects of midbrain central gray lesions on spontaneous and electrically induced aggression in the rat. Aggress Behav 9:133–155. doi:10.1002/1098-2337(1983)9:2<133:Aid-Ab2480090205>3.0.Co;2-H

Mos J, Olivier B, Lammers JHCM, van der Poel AM, Kruk MR, Zethof T (1987) Postpartum aggression in rats does not influence threshold currents for EBS-induced aggression. Brain Res 404:263–266. doi:10.1016/0006-8993(87)91377-1

Olivier B, Mos J, Rasmussen D (1990) Behavioural pharmacology of the serenic, eltoprazine. Drug Metabol Drug Interact 8:31–83

Olivier B, Olivier-Aardema R, Wiepkema PR (1983) Effect of anterior hypothalamic and mammillary area lesions on territorial aggressive behaviour in male rats. Behav Brain Res 9:59–81

Panksepp J (1971) Aggression elicited by electrical stimulation of the hypothalamus in albino rats. Physiol Behav 6:321–329. doi:10.1016/0031-9384(71)90163-6

Panksepp J, Trowill J (1969) Electrically induced affective attack from the hypothalamus of the albino rat. Psychon, Sci 16

Potegal M (2012) Temporal and frontal lobe initiation and regulation of the top-down escalation of anger and aggression. Behav Brain Res 231:386–395. doi:10.1016/j.bbr.2011.10.049

Pott CB, MacDonnell MF (1986) Enhancement of optokinetic responses by lateral hypothalamic areas associated with attack in cats. Physiol Behav 38:415–422

Putkonen PT (1966) Attack elicited by forebrain and hypothalamic stimulation in the chicken. Experientia 22:405–407. doi:10.1007/BF01901167

Ranck JB Jr (1975) Which elements are excited in electrical stimulation of mammalian central nervous system: a review. Brain Res 98:417–440

Rinne T, de Kloet ER, Wouters L, Goekoop JG, de Rijk RH, van den Brink W (2002) Hyperresponsiveness of hypothalamic-pituitary-adrenal axis to combined dexamethasone/corticotropin-releasing hormone challenge in female borderline personality disorder subjects with a history of sustained childhood abuse. Biol Psychiatry 52:1102–1112

Roberts WW, Steinberg ML, Means LW (1967) Hypothalamic mechanisms for sexual, aggressive, and other motivational behaviors in the opossum, *Didelphis virginiana*. J Comp Physiol Psychol 64:1–15. doi:10.1037/h0024805

Roeling TA, Veening JG, Peters JP, Vermelis ME, Nieuwenhuys R (1993) Efferent connections of the hypothalamic "grooming area" in the rat. Neuroscience 56:199–225

Roeling TAP, Veening JG, Kruk MR, Peters JP, Vermelis ME, Nieuwenhuys R (1994) Efferent connections of the hypothalamic "aggression area" in the rat. Neuroscience 59:1001–1024. doi:10.1016/0306-4522(94)90302-6

Roelofs K, Bakvis P, Hermans EJ, van Pelt J, van Honk J (2007) The effects of social stress and cortisol responses on the preconscious selective attention to social threat. Biol Psychol 75:1–7. doi:10.1016/j.biopsycho.2006.09.002

Schuurman T (1980) Hormonal correlates of agonistic behavior in adult male rats. Prog Brain Res 53:415–420. doi:10.1016/S0079-6123(08)60079-5

Siegel A, Roeling TA, Gregg TR, Kruk MR (1999) Neuropharmacology of brain-stimulation-evoked aggression. Neurosci Biobehav Rev 23:359–389

Smith DA (1972) Increased perioral responsiveness: a possible explanation for the switching of behavior observed during lateral hypothalamic stimulation. Physiol Behav 8:617–621

Tinbergen N (1951) The study of instinct. Clarendon Press/Oxford University Press, Oxford
Tokumura T, Muraoka A, Masutomi T, Machida Y (2005) Stability of spironolactone in rat plasma: strict temperature control of blood and plasma samples is required in rat pharmacokinetic studies. Biol Pharm Bull 28:1126–1128. doi:10.1248/bpb.28.1126
Toth M, Fuzesi T, Halasz J, Tulogdi A, Haller J (2010) Neural inputs of the hypothalamic "aggression area" in the rat. Behav Brain Res 215:7–20. doi:10.1016/j.bbr.2010.05.050
Tulogdi A, Toth M, Halasz J, Mikics E, Fuzesi T, Haller J (2010) Brain mechanisms involved in predatory aggression are activated in a laboratory model of violent intra-specific aggression. Eur J Neurosci 32:1744–1753. doi:10.1111/j.1460-9568.2010.07429.x
Valenstein ES (1973) Brain control, a critical examination of brain stimulation and psychosurgery. Wiley Interscience, New York
Valenstein ES, Cox VC, Kakolewski JW (1968) Modification of motivated behavior elicited by electrical stimulation of the hypothalamus. Science 159:1119–1121
Valenstein ES, Cox VC, Kakolewski JW (1969) Hypothalamic motivational systems: fixed or plastic neural circuits? Science 163:1084. doi:10.1126/science.163.3871.1084
Valenstein ES, Cox VC, Kakolewski JW (1970) Reexamination of the role of the hypothalamus in motivation. Psychol Rev 77:16–31
van den Berg MJ, Ter Horst GJ, Koolhaas JM (1982) The nucleus premammillaris ventralis (PVM) and aggressive behavior in the rat. Aggress Behav 9:41–47
van der Poel AM, Olivier B, Mos J, Kruk MR, Meelis W, van Aken JH (1982) Anti-aggressive effect of a new phenylpiperazine compound (DU27716) on hypothalamically induced behavioural activities. Pharmacol Biochem Behav 17:147–153
van der Poel AM, van der Hoef H, Meelis W, Vletter G, Mos J, Kruk MR (1983) A locked, non-rotating, completely embedded, moveable electrode for chronic brain stimulation studies in freely moving, fighting rats. Physiol Behav 31:259–263
Veening JG, Coolen LM, de Jong TR, Joosten HW, de Boer SF, Koolhaas JM, Olivier B (2005) Do similar neural systems subserve aggressive and sexual behaviour in male rats? Insights from c-Fos and pharmacological studies. Eur J Pharmacol 526:226–239. doi:10.1016/j.ejphar.2005.09.041
Vergnes M, Karli P (1970) Elicitation of aggressive behavior by electric stimulation of the medial hypothalamus in the rat. Physiol Behav 5:1427–1430. doi:10.1016/0031-9384(70)90131-9
Vergnes M, Depaulis A, Boehrer A, Kempf E (1988) Selective increase of offensive behavior in the rat following intrahypothalamic 5,7-DHT-induced serotonin depletion. Behav Brain Res 29(1–2):85–91
Weissenberger AA, Dell ML, Liow K, Theodore W, Frattali CM, Hernandez D, Zametkin AJ (2001) Aggression and psychiatric comorbidity in children with hypothalamic hamartomas and their unaffected siblings. J Am Acad Child Adolesc Psychiatry 40:696–703. doi:10.1097/00004583-200106000-00015
Wetherill GB (1966) Sequential estimation of points on quantal response curves. In: Wetherill GB (ed) Sequential methods in statistics. Methuen & Co Ltd, London, pp 162–179
Wiegert O, Joels M, Krugers H (2006) Timing is essential for rapid effects of corticosterone on synaptic potentiation in the mouse hippocampus. Learn Mem 13:110–113. doi:10.1101/lm.87706
Wise RA (1968) Hypothalamic motivational systems: fixed or plastic neural circuits? Science 162:377–379
Wise RA (1969) Plasticity of hypothalamic motivational systems. Science 165:929–930
Woodworth CH (1971) Attack elicited in rats by electrical stimulation of the lateral hypothalamus. Physiol Behav 6:345–353
Yamashita H, Koizumi K, Brooks CM (1979) Rhythmic patterns of discharge in hypothalamic neurosecretory neurons of cats and dogs. Proc Natl Acad Sci U S A 76:6684–6688
Yehuda R, Flory JD, Bierer LM, Henn-Haase C, Lehrner A, Desarnaud F, Makotkine I, Daskalakis NP, Marmar CR, Meaney MJ (2014) Lower methylation of glucocorticoid receptor gene promoter 1 in peripheral blood of veterans with posttraumatic stress disorder. Biol Psychiatry. doi:10.1016/j.biopsych.2014.02.006

The Role of Serotonin, Vasopressin, and Serotonin/Vasopressin Interactions in Aggressive Behavior

Thomas R. Morrison and Richard H. Melloni Jr.

Abstract Aggression control has been investigated across species and is centrally mediated within various brain regions by several neural systems that interact at different levels. The debate over the degree to which any one system or region affects aggressive responding, or any behavior for that matter, in some senses is arbitrary considering the plastic and adaptive properties of the central nervous system. Nevertheless, from the reductionist point of view, the compartmentalization of evolutionarily maladaptive behaviors to specific regions and systems of the brain is necessary for the advancement of clinical treatments (e.g., pharmaceutical) and novel therapeutic methods (e.g., deep brain stimulation). The general purpose of this chapter is to examine the confluence of two such systems, and how their functional interaction affects aggressive behavior. Specifically, the influence of the serotonin (5HT) and arginine vasopressin (AVP) neural systems on the control of aggressive behavior will be examined individually and together to provide a context by which the understanding of aggression modulation can be expanded from seemingly parallel neuromodulatory mechanisms, to a single and highly interactive system of aggression control.

Keywords AAS · Steroids · Serotonin · 5HT · Vasopressin · Aggression

T. R. Morrison · R. H. Melloni Jr. (✉)
Program in Behavioral Neuroscience, Department of Psychology, Northeastern University, 125 Nightingale Hall, 360 Huntington Ave, Boston, MA 02155, USA
e-mail: r.melloni@neu.edu

T. R. Morrison
e-mail: morrison.t@husky.neu.edu

Curr Topics Behav Neurosci (2014) 17: 189–228
DOI: 10.1007/7854_2014_283

Published Online: 5 February 2014

Contents

1 The Role of 5HT in Aggressive Behavior

The 5HT system has been found to primarily have an inhibitory influence over aggressive behavior in numerous species (see Olivier and Van Oorschot 2005 and Olivier 2004 for reviews). Selective 5HT reuptake inhibitors (SSRI) elevate 5HT concentrations in the pre- and postsynaptic regions of 5HT neurons by binding to the serotonin transporter and blocking the removal of 5HT from the extracellular space (Blier and De Montigny 1994; Invernizzi et al. 1992, 1994), and this acute elevation in 5HT is believed to account for reduced aggressive behavior observed in adult mice (Maj et al. 1981; Pinna et al. 2006, 2009; Roliński and Herbut 1981; Sánchez and Hyttel 1994; Wagner et al. 1993), rats (Datla and Bhattacharya 1990; Ho et al. 2001; Homberg et al. 2007; Kostowski et al. 1984; Matto et al. 1998; Mitchell et al. 1991), fish (Lynn et al. 2007), lizards (Deckel 1996), hamsters (Deckel 1996; DeLeon et al. 2002a; Ferris 1996, 1997; Grimes and Melloni 2002; Taravosh-Lahn et al. 2006), and sparrows (Sperry et al. 2003, 2005). After long-term exposure to SSRIs, elevations in 5HT desensitize inhibitory presynaptic 5HT receptors, allowing for a significant increase in serotonin neuronal activity and subsequent postsynaptic release of 5HT (De Montigny et al. 1990; El Mansari et al. 2005). Elevated postsynaptic 5HT release has also been found to decrease aggressive responding in numerous species including mice (Ibi et al. 2008; Maj et al. 1981; Pinna et al. 2003b), voles (Villalba et al. 1997), fish (Kania et al. 2012; Perreault et al. 2003), Vervet monkeys (Fairbanks et al. 2001), rats (Barreto-Medeiros et al. 2004), and human adults (Coccaro and Kavoussi 1997). Similarly, acute administration of the SSRI fluoxetine has been shown to suppress the effects of various agents that increase aggression in multiple animal models, including synthetic anabolic–androgenic steroids (AAS; Grimes and Melloni 2002), testosterone (Delville et al. 1996), cocaine (DeLeon et al. 2002a), AVP (Ferris et al. 1997), alcohol (Wagner et al. 1993), and apomorphine (Datla et al. 1991; Matto et al. 1998). This suggests that direct infusion of 5HT into 5HT sensitive brain areas (Clotfelter et al. 2007; Datla et al. 1991; Munro 1986; Puciłowski et al. 1985) or supplementing the diets of mice (Thurmond et al. 1979), dogs (DeNapoli et al. 2000), or normal human males (Cleare and Bond 1995) with chemical precursors to 5HT might reduce aggressive behavior. In contrast to enhancing the activity of 5HT in the CNS, reducing 5HT activity, either by lesioning 5HT neurons or through the systemic administration of agents that inhibit 5HT synthesis, increases aggressive responding in mice, rats, and firemouth cichlids (Adams et al. 1996; Keele 2001; Vergnes et al. 1988; Vergnes and Kempf 1982).

SSRIs elevate postsynaptic levels of 5HT in various brain regions associated with aggression and impulse control. For example, the display of defensive aggressive behaviors has been linked to neuronal activity in the amygdala and ventromedial hypothalamus in cats (Adamec 1990); regions where 5HT afferents are known to exist (Azmitia and Segal 1978; DeLeon et al. 2002a; Duan and Ju 1998; Ferris 1996; Fraga et al. 2005; Grimes and Melloni 2002, 2006; Perry and Fuller 1993; Qu et al. 2003; Taravosh-Lahn et al. 2006; Vertes 1991) and where fluoxetine-induced elevations in 5HT stimulated the expression of the immediate early gene c-Fos, an indicator of neuronal activity (Duan and Ju 1998; Ferris et al. 1997; Fraga et al. 2005; Grimes and Melloni 2002, 2006; Qu et al. 2003; Vertes 1991). In a highly aggressive mouse strain, baseline levels of hippocampal 5HT are higher than in mice genetically altered to exhibit low levels of aggression (Kulikov et al. 2005), suggesting that 5HT signaling in the hippocampus also plays a role in aggression control. In line with these data and similar to the effects of SSRI exposure on 5HT concentrations within the amygdala and hypothalamus, it is not surprising that a moderate density of 5HT afferents exists within various regions of the hippocampus (Vertes 1991) and that the concentration of 5HT is elevated in response to fluoxetine exposure (Sabol et al. 1992), activating postsynaptic neuronal activity (Qu et al. 2003).

In addition to the amygdala and hippocampus, the lateral septum (LS) and bed nucleus of the stria terminalis (BNST), the paraventricular nucleus (PVN) of the hypothalamus, and the prefrontal cortex (PFx) are all believed to contain neurons that control aggressive behaviors (Ricci et al. 2004; Summers et al. 2005; Veenema et al. 2010). More specifically, microdialysis experiments have revealed that 5HT release into the PFx is decreased during and following an aggressive encounter in rats (Van Erp and Miczek 2000), and high-performance liquid chromatography has shown that highly aggressive mice have lower baseline levels of 5HT in the PVN (Summers et al. 2005). Similarly, highly aggressive hamsters treated with cocaine over developmentally critical time periods express more punctate (axon terminal) 5HT3 receptors in the BNST and fewer in the LS, compared to saline-treated controls (Ricci et al. 2004). In untreated rats, hamsters or nonhuman primates, the LS, BNST, PFx, and PVN have all been shown to contain 5HT afferents (Azmitia and Segal 1978; Cunningham et al. 2005; DeLeon et al. 2002a; Duan and Ju 1998; Lewis 1990; Ricci et al. 2004; Sawchenko et al. 1983; Taravosh-Lahn et al. 2006; Vertes 1991) and have altered levels of 5HT in response to systemic SSRI exposure (Fraga et al. 2005; Qu et al. 2003). Neurons within both the nucleus circularis (NC) and medial supraoptic nucleus (mSON) send their projections to hypothalamic regions important for the regulation of aggressive behavior (Ferris et al. 1989, 1990; Mahoney et al. 1990; Melloni and Ricci 2010). Like the structures mentioned above, the NC and mSON have an intermediate density of 5HT afferents within or immediately surrounding their structural demarcations (Duan and Ju 1998; Sawchenko et al. 1983), and are regionally sensitive to enhanced 5HT neurotransmission induced by SSRI exposure (Auerbach et al. 1989; Fraga et al. 2005).

Results from the functional and localization studies above generally indicate that brain areas implicated in aggression control are sensitive to alterations in 5HT neuronal activity. Both pre- and postsynaptic neurons within these regions possess various types of 5HT receptors, and thus influence the activity of both 5HT and non-5HT neurons. For example, in situ hybridization has confirmed the presence of 5H1A, 5HT2A, and 5HT2C receptor mRNA in the LS, MeA, AH, and the LH (Wright et al. 1995; Zhang et al. 1999), and inhibitory postsynaptic 5HT1A receptors exist in the lateral subdivision of the AH (Ricci et al. 2006). Similarly, 5HT1B and 5HT2A receptors are located postsynaptically on somata in the LAH brain region (Ferris et al. 1997; Grimes and Melloni 2005; Melloni and Ricci 2010; Ricci et al. 2006; Schwartzer et al. 2009). Also, 5HT1B receptors function as heteroreceptors in the MeA and VLH (Grimes and Melloni 2005), and also as terminal autoreceptors in the MeA, AH, LAH, and VLH (Grimes and Melloni 2005; Wright et al. 1995). Presynaptically, the excitatory 5HT3 subtype receptors can be found on neuronal puncta located in the LS, MeA, AH, and LH at varying degrees of density (Carrillo et al. 2010).

The inhibition of aggressive responding is generally accepted to be a function of 5HT receptor activation, although the neuromodulatory effects differ depending on neuronal type, receptor location, and receptor type. Specifically, inhibition of aggression primarily occurs through 5HT interactions with the 5HT 1A and 1B receptor subtypes (de Almeida et al. 2001; Olivier et al. 1995; Sijbesma et al. 1991). It has been shown previously that the 5HT1A receptor plays a regulatory role in the aggression-modulating circuitry of mice that display enhanced aggressive behaviors after alcohol exposure. Following ethanol exposure, resident male mice preselected for their high levels of aggressive display were administered a selective 5HT1A receptor agonist (either 8-OH-DPAT or flesinoxan) or saline. Both of the 5HT1A agonists decreased various measures of aggression in alcohol-exposed resident mice compared to controls, with only minor impairments in nonaggressive behaviors (e.g., reduced motor activity) observed during a resident-intruder test (Miczek et al. 1998). In the presence of the selective 5HT1A antagonist WAY 100,635, the dose-response curves for 5HT1A agonists for both bites and sideways threats were shifted to the right, thus suggesting that 5HT1A receptor activation is involved in alcohol heighted aggressive responding (Miczek et al. 1998).

In wild-type rats noted for their easily elicited aggressive response, De Boer et al. (1999) showed that systemic administration of various individual 5HT1A agonists, including 8-OH-DPAT, inhibits offensive aggressive behaviors in residents males, but not defensive behaviors in intruder males. Administration of WAY 100,635 15 min prior to agonist exposure blocked the aggression-inhibiting effects in resident males (De Boer et al. 1999), showing that 5HT1A receptor activation modulates the display of offensive, but not defensive, agonistic behaviors in rats. The role of the 5HT1A receptor in inhibiting aggression is supported further by a study that reported that 8-OH-DPAT dose-dependently increased the latencies of isolation-induced attacks in highly aggressive male mice that had depleted levels of central nervous system (CNS) 5HT (Sánchez and Hyttel 1994), indicating that in the absence of all other activity regulated by the activation of other 5HT receptor subtypes, 5HT1A receptor stimulation is enough to inhibit aggressive behavior.

Interestingly, 5HT1A receptor activation seems to affect aggressive behaviors in a sex-specific manner, as 8-OH-DPAT has been shown to produce different results in females and males. For example, in female hamsters, the dose range of systemic 8-OH-DPAT necessary to reduce attack rate and increase attack latency has also been found to produce 5HT syndrome-like behaviors, e.g., body shakes and flattening (Joppa et al. 1997). Thus, although aggressive behaviors are reduced in reaction to high concentrations of 8-OH-DPAT, normal motor function is reduced as well; suggesting 5HT1A receptor stimulation alone is not sufficient to affect aggressive responding in female hamsters. Further data indicate that the 5HT1A receptor differentially regulates behaviors related to aggression between sexes and may be site-specific; intraventricular (IVC) delivery of 8-OH-DPAT was shown to reduce aggressive responding in males, but not ovariectomized females (Joppa et al. 1997). These sexually dimorphic effects of 8-OH-DPAT indicate that the expression of aggressive-mediating 5HT1A receptors may be different in density and/or location between sexes.

Studies using direct infusion methods (e.g., microinjection, IVC) have helped researchers investigate and identify the functionally important loci for aggression control, and have elucidated the specific regional effects of 5HT using agents that directly target specific 5HT receptors (e.g., agonists and antagonists). For example, the raphe nuclei are known to have the highest density of 5HT-containing neurons in the brain (Azmitia and Whitaker-Azmitia 1995) and in this region 5HT1A receptors can be found on 5HT cell bodies functioning as autoreceptors, providing a homeostatic regulatory mechanism for 5HT neuronal firing and release (Aghajanian and Wang 1978; Sprouse and Aghajanian 1987). Direct application of 5HT1A agonists into this region decreased both maternal-aggression and species-typical aggression measures in male residents (de Almeida and Lucion 1997; Van Der Vegt et al. 2003), presumably through 5HT1A autoreceptor activation.

Postsynaptically, direct delivery of 5HT1 receptor agonists into various brain regions has been shown to reduce aggressive behavior via activation of 5HT1 receptors located on non-5HT neurons (or 5HT1 heteroreceptors). In one study, Ferris et al. (1999) hypothesized that since systemic exposure to either 5HT1A or 5HT1B agonists reduces aggressive responding in a variety of animal models (e.g., De Boer et al. 1999; Joppa et al. 1997; Sánchez and Hyttel 1994), the direct application of these agonists may elucidate which regions are important for aggression control. In their study, 8-OH-DPAT and the 5HT1B agonist CGS-12066A were microinjected into the AH to identify whether they could inhibit the enhanced levels of aggressive behavior shown to follow direct microinfusion of AVP (Ferris et al. 1997). The results revealed that, although 8-OH-DPAT potently inhibited AVP-induced aggressive responding, CGS-12066A had no effect, showing that the 5HT1A heteroreceptor is the 5HT1 receptor primarily responsible for aggression behavioral control in the AH (Ferris et al. 1999). Microinjection of 8-OH-DPAT into various postsynaptic areas, including the MeA, has been shown to reduce attack behavior in female rats. Further, de Almeida and Lucion (1997) demonstrated that individual microinjections of various doses of 8-OH-DPAT into the MeA region and the dorsal periaqueductal gray decreased behavioral measures of maternal aggression, including frontal attacks, lateral attacks, or bites in lactating female rats 7 days postpartum. In these same animals, however, an intermediate dose of 8-OH-DPAT delivered into the medial septal nucleus increased the number of bites compared to saline controls (de Almeida and Lucion 1997). Combined with the results from Ferris et al. (1999) and data showing the sexually dimorphic effects of 5HT1A stimulation, these data suggest that the 5HT1A receptor stimulation can inhibit or facilitate aggression, depending on region and/or sex.

Similar to 5HT1A receptor stimulation, 5HT1B hetero- and auto-receptor stimulation also inhibits aggressive behaviors. Systemically, administered 5HT1B receptor agonists have potent anti-aggressive properties across a variety of aggression paradigms and species, including alcohol-heightened (Fish et al. 1999; Miczek and de Almeida 2001), social-instigation (de Almeida and Miczek 2002; Fish et al. 1999), frustration-induced (de Almeida and Miczek 2002), intermale (Bell et al. 1995), isolation induced (Rilke et al. 2001), and territorial aggression (Grimes and Melloni 2005). The effects of the 5HT1B receptor agonist CGS-12066A are similar to 8-OH-DPAT, in that, when delivered IVC, measures of territorial aggression are reduced in males, but not ovariectomized females (Joppa et al. 1997), suggesting that like activation of the 5HT1A receptor, 5HT1B receptor-induced inhibition of aggression may be regionally specific and sexually dimorphic.

Direct application of partial and full 5HT1B receptor agonists into the dorsal raphe nucleus (DRN) inhibits aggressive responding and decreases motor activity in both resident male rats (Mos et al. 1993) and alcohol-treated mice (Faccidomo et al. 2008). Similarly, direct application of 5HT1B agonists into postsynaptic locations, including the ventral orbitofrontal prefrontal cortex (VO PFx) decreases maternal aggression in rats (Veiga et al. 2007), and also territorial (de Almeida et al. 2006) and socially instigated aggression in male mice (Centenaro et al. 2008). Interestingly, when 5HT1B receptor agonists are delivered into alternate areas outside the VO PFx, such as the infralimbic (IL) region or the region lateral to the VO PFx of the mouse brain, there are no changes in aggression (de Almeida et al. 2006). As in the VO PFx, application of 5HT1B receptor agonists to the medial preoptic area of the hypothalamus has been shown to decrease aggressive displays in castrated androgen- and estrogen-dependent resident male mice (Cologer-Clifford et al. 1997). Although in the presence of AVP microinfusion of CGS-12066A into the AH of male hamsters produced only a trend towards decreased aggressive acts (Ferris et al. 1999). These data indicate that, as with inhibitory 5HT1A receptors, the effects of 5HT1B receptor activation on aggression are sex, location, and/or species-dependent. Combined with immunocytochemical data showing that synthetic steroid exposure throughout adolescence increases 5HT1B receptor expression within the AH region of aggressive hamsters (Grimes and Melloni 2005), these data may also suggest that 5HT1B receptors in the AH are likely involved in normal aggressive responding, though to a lesser extent than 5HT1A receptors.

Compared to the number of studies that have focused on 5HT1 receptors, fewer studies have investigated the link between the activation of other 5HT receptors and aggression. One such study found that extremely high doses of the 5HT2A receptor partial agonist DOI prolonged attack latency in male mice trained for aggressive responding (Sánchez et al. 1993). However, DOI's weak affinity for the 5HT2A receptor (Arnt and Hyttel 1989), and weak inhibition of aggressive behavior compared to more efficacious substances (e.g., 8-OH-DPAT), provide little evidence for the involvement of 5HT2A receptors in controlling aggression (Sánchez et al. 1993). Another 5HT2A receptor partial agonist, DOB, has a marginally higher affinity for the 5HT2A receptor (in its low affinity state) than DOI (Roth et al. 1997), and in the water competition (WC) test it has been shown to block defensive aggression in rats. Interestingly, DOI also reduced the number of offensive aggressive behaviors (i.e., attacks, greater latency to first attack, shorter attack duration) in the same animals that exhibited DOI-induced reductions in defensive behaviors during the WC test (Muehlenkamp et al. 1995).

The results of 5HT receptor agonist studies notwithstanding, blockade of 5HT2 or 5HT3 receptors by systemic administration of either the 5HT2A antagonist ritanserin or the 5HT3 antagonist ondansetron had no effect on isolation-induced aggressive behavior in resident mice during a resident-intruder test (Sánchez et al. 1993). In contrast, heightened aggressive behaviors induced by chronic developmental exposure to low-dose cocaine are reduced after administration of the 5HT3 receptor antagonist, tropisetron, in resident hamsters (Ricci et al. 2004). Also, in saline-treated control resident hamsters, a single acute dose of the 5HT3 receptor agonist mCPBG increases the number of bites, upright offensives, and lateral attacks, while also reducing attack and bite latencies in resident males during aggression tests (Ricci et al. 2005). These findings suggest that stimulation of 5HT2 and 5HT1 receptors, and blockade of 5HT3 receptors, inhibit offensive aggression. Further, activation of 5HT3 receptor enhances and/or facilitates aggressive responding, and modulation of the 5HT2 receptor may be more closely associated with defensive aggressive display than other 5HT receptor types.

Data from laboratory animals are mostly in line with the results of human studies regarding the role of 5HT in aggressive behavior, Bjork et al. (2000) demonstrated that depletion of dietary tryptophan, the chemical precursor to 5HT, enhances aggression-like responding in human males with high trait hostility, which is similar to the results of numerous animal studies (e.g., Chamberlain et al. 1987; Giammanco et al. 1990; Vergnes et al. 1986). Moreover, given that high levels of plasma free tryptophan (pfTryp) reflect greater 5HT synthesis activity in the CNS (Shoaf et al. 1998), males with low trait hostility had significantly higher pfTryp following bolus tryptophan injection than males with high trait hostility, suggesting that aggressive men have lower levels of CNS 5HT than nonaggressive men (Bjork et al. 2000). In line with these results, acute depletion of dietary tryptophan in patients with ADHD has been found to produce increased aggression-like responding after only low social provocation (Zimmermann et al. 2012). In a separate study, elevated 5HT levels following brief exposure to the SSRI paroxetine reduced aggressive responding in both males and females with self-reported histories of aggression (Berman et al. 2009). From these human studies, it can be generally assumed that aggressive tendencies are higher in some, but not others. This difference may be genetically mediated as aggressive human and nonhuman primates have been found to express a specific polymorphism in the promoter region of the 5HT transporter gene (Jarrell et al. 2008; Lesch and Merschdorf 2000) that may produce a hypofunctioning 5HT system (Whale et al. 2000).

The influence of 5HT receptor expression on the modulation of aggressive display is further supported in part by data from genetically altered mutant knockout mice that lack specific 5HT components. For example, in the isolation-induced aggression paradigm, homozygous knockout mice lacking the 5HT1B receptor have shorter attack latencies, higher attack frequencies, and display more intense aggressive acts than both wild-type and heterozygous mice (Ramboz et al. 1996; Saudou et al. 1994). Similar effects were found in matched pairs of wild-type and 5HT1B knockout mice, which also had shorter attack latencies and greater levels of aggressive behavior than wild-types during a resident-intruder encounter (Bouwknecht et al. 2001). Moreover, in the latter study, nonentrained stimuli (i.e., experimenter handling, resident-intruder testing) had a greater effect on physiological parameters (i.e., heart rate, body temperature) in knockouts than in wild-type mice, and it was hypothesized that 5HT1B modulates impulse control processes (Bouwknecht et al. 2001).

In knockout mice that lack a key enzyme necessary for intercellular inactivation of 5HT, high levels of aggression (as measured by the assessment of aggression directed toward a human gloved hand) are associated with the increased expression of 5HT1A receptor mRNA in portions of the amygdala and the frontal cortex when compared to wild-type controls (Popova et al. 2007). Not surprisingly, rats found to be highly aggressive using this method possessed a lower density of 5HT1A receptors in the midbrain compared to their nonaggressive counterparts (Popova et al. 2007). These results suggest that in highly aggressive animals, 5HT-containing midbrain neurons express fewer 5HT1A autoreceptors, while postsynaptic neurons in specific regions of aggression control contain a greater number of inhibitory 5HT1A heteroreceptors. Along similar lines, preliminary data indicate that knockout mice lacking 5HT1A receptors have lower levels of aggression compared to wild-type mice, though it is currently unclear whether these effects are specifically receptor-mediated or due to indirect homeostatic effects that are activated during development in the absence of the 5HT1A receptor.

The translational validity of knockout animal studies was investigated by Lappalainen et al. (1998), who examined aggressive/impulsive behavioral parallels between 5HT1B knockout mice and 5HT1B gene expression in violent alcoholic criminals. Lappalainen et al. found that, in violent alcoholics, specific marker loci in the 5HT1B gene sequence were more common than in nonviolent controls. In a separate study, seemingly contrary results showed that the same gene marker loci were associated with decreased 5HT1B receptor binding in the postmortem brain tissue of humans who had either committed suicide or had died from non-self injurious factors (Huang et al. 1999). Furthermore, decreased frequencies of marker loci and 5HT1B receptor binding have been found in both pathologically aggressive and nonaggressive patients (Huang et al. 1999). When these data are combined, some of the findings from Lappalainen et al.'s human studies fall in line with 5HT1B knockout animal studies in that the 5HT1B knockout mouse is known to possess impulsive tendencies (Bouwknecht et al. 2001) and, relatedly, a heightened preference for alcohol (Crabbe et al. 1996). More data are necessary to clarify the link between genetic markers for 5HT receptors as well as their impact on both receptor function and aggressive behaviors in humans.

Together, data from both human and animal studies suggest that 5HT1B receptors may play a larger role in some, but not all forms of aggression; or that the 5HT1B receptor's role in aggression is species-specific. Nevertheless, pharmacotherapies that target the 5HT system are effective in treating various forms of aggression. For example, chronic exposure to lithium carbonate is known to elevate postsynaptic 5HT (Treiser et al. 1981) and is effective in reducing aggression in violent prisoners (Sheard et al. 1976; Tupin et al. 1973), children (Campbell et al. 1984, 1995; Malone et al. 2000), mentally handicapped patients (Micev and Lynch 1974), and psychiatric patients with violent tendencies (Dorrego et al. 2002; Van Putten and Sanders 1975). Considering the effects of lithium on 5HT levels in postsynaptic regions, it is not surprising that chronic SSRI exposure, which elevates pre- (Invernizzi et al. 1992) and postsynaptic 5HT (Invernizzi et al. 1996; Sprouse et al. 2001) is effective in reducing aggressive behaviors in human psychiatric patients (Blader 2006; Bond 2005; Reist et al. 2003). However, it should be noted that the mechanisms involved in reducing aggressive behaviors go well beyond elevated 5HT levels—complex interactions between neurons that express 5HT receptors, as well as the particular effects of each 5HT receptor on neurotransmission and the regulation of observable animal behavior make up the 5HT-mediated portion of the aggression control system.

Various drugs, including those that interact with 5HT receptors, can alter receptor expression and subsequently change behavior. The aggression-inducing effects of long-term adolescent exposure to both AAS and cocaine are mediated in part by anatomical changes to the 5HT system (DeLeon et al. 2002a; Knyshevski et al. 2005b; Melloni et al. 1997; Ricci et al. 2006). Both these drugs reduce the density of 5HT afferents and thus presumably decrease 5HT tone, while also altering the expression of 5HT receptors in brain areas related to aggression (DeLeon et al. 2002a; Grimes and Melloni 2002, 2005, 2006; Ricci et al. 2004, 2006). For instance, compared to control animals, hamsters administered AAS during adolescence showed heighted levels of aggression that temporally emerged in parallel with decreased 5HT1A receptors within the AH (Ricci et al. 2006). In a separate study using the same conditions, AAS-treated hamsters showed decreased density of 5HT1B receptors on neuronal puncta in the AH, CeA, MeA, and VLH, while also exhibiting an increased number of 5HT1B receptors on cell bodies in the

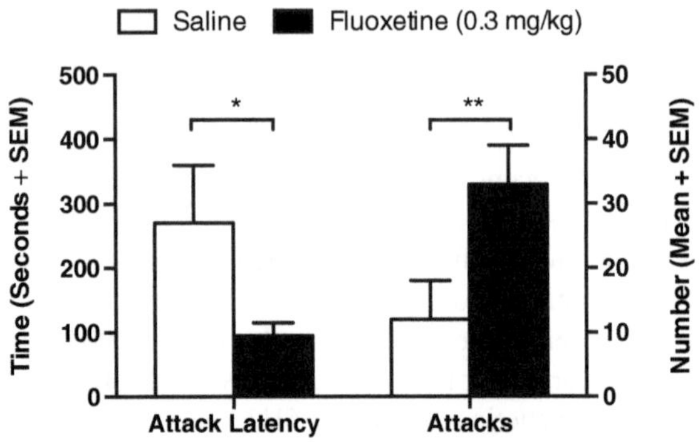

Fig. 1 Effects of fluoxetine administration during adolescence on general measures of offensive aggression. Low-dose fluoxetine (0.3 mg/kg/day) exposure significantly increased the total number of attacks and decreased the latency to first attack of flouxetine-treated residents toward intruders compared to saline-treated control animals. *Bars* denote *SEM;* $*p < 0.05$. $**p < 0.01$

AH, MeA, and VLH (Grimes and Melloni 2005). Similarly, compared to nonaggressive vehicle controls, aggressive hamsters chronically treated with low doses of cocaine during adolescence had more 5HT3 receptors on neuronal afferents in the AH, CeA, and VLH (Ricci et al. 2004). Recent research found that long-term exposure to clinically relevant low doses of fluoxetine during adolescence increased the attack behaviors and decreased attack latencies, while increasing the density of 5HT afferent fibers in various brain regions that modulate aggressive behavior, including the LAH (Ricci and Melloni 2012; see Figs. 1 and 2). Fluoxetine-induced anatomical alterations within the CNS are not yet as well characterized or understood as those produced by illicit drugs of abuse. Fluoxetine's effects are contrary to the reported 5HT afferent-reducing effects of AAS and cocaine that co-occur with the emergence of the aggressive phenotype in hamsters, suggesting that inhibition of aggressive behaviors does not always follow the enhanced presence of 5HT in postsynaptic aggression regions.

The fluoxetine effects reported by Ricci and Melloni (2012) are seemingly contrary to the results of most studies that have investigated the link between 5HT neurotransmission and aggressive responding. For example, both short-term and chronic treatment with fluoxetine and other SSRIs are known to increase 5HT tone in the AH (Pergola et al. 1993) and throughout the CNS (Invernizzi et al. 1992, 1996). When fluoxetine is administered shortly before an aggression test, the purported increases in 5HT inhibit aggressive responding facilitated by long-term exposure during adolescence to both AAS (Grimes and Melloni 2002) and cocaine (Jackson et al. 2005). These effects are most likely due to 5HT's interactions with the 5HT-1, 2, and/or 3 receptor subtypes; like fluoxetine, 5HT1 and 2 agonists as well as 5HT3 antagonists have all been found to inhibit aggressive behavior (Bell et al. 1995; De Boer and Koolhaas 2005; Ricci et al. 2005). These data make it difficult to functionally identify why developmental exposure to fluoxetine, but not treatment in early adulthood, facilitates increased, but not reduced, aggressive responding.

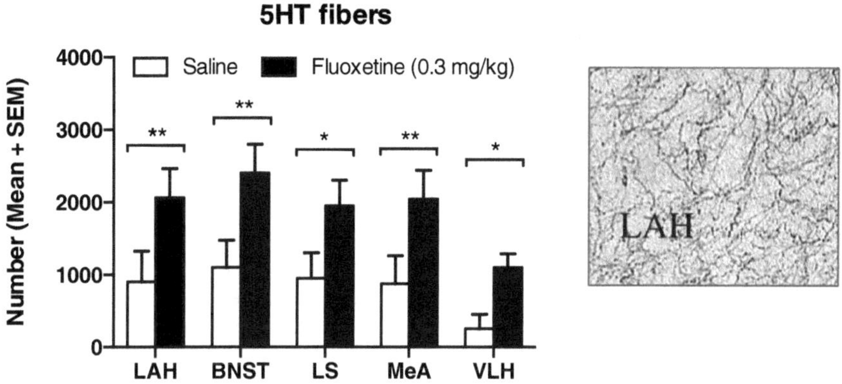

Fig. 2 Effects of low-dose (0.3 mg/kg) fluoxetine administration during adolescence on 5HT and AVP neural development in brain areas implicated in aggression control. Fluoxetine exposure significantly increased the total number of 5HT and AVP afferent fibers compared with saline-treated control animals. Fluoxetine-treated animals possessed significantly more 5HT afferent fibers (as depicted in the inset—40X magnification) in the latero-anterior hypothalamus (*LAH*), bed nucleus of the stria terminals (*BNST*), lateral septum (*LS*), medial amygdala (*MeA*), and ventrolateral hypothalamus (*VLH*) than did their saline-treated counterparts. *Bars* denote *SEM;* * $p < 0.05$. ** $p < 0.01$

In addition to the conflicting studies on aggressive behavior, other experiments investigating the effects of fluoxetine exposure during early critical periods have shown that both depressive and anxiety-related behaviors, which are reduced by late-life exposure to SSRIs in highly anxious mice (Dulawa et al. 2004), may be either increased or decreased by fluoxetine exposure early in life (Ansorge et al. 2008; Ansorge et al. 2004; Maciag et al. 2006; Popa et al. 2008). In neonatal mice, prolonged exposure to SSRIs from postnatal day 4 (P4) to P21 increased behaviors related to depression and anxiety in the suppressed feeding paradigm during early adulthood compared to saline-treated controls (Ansorge et al. 2004). Neonatal rats treated with the SSRI citalopram from P8 to P21 exhibited lower levels of sexual behaviors when tested in adulthood (P90; Maciag et al. 2006). Using the sucrose preference test, forced swim test, and tail suspension test, Popa et al. (2008) found that wild-type adult mice treated chronically with SSRI from P5 to P19 showed increased depressive-like behaviors compared to controls. Although in contrast, both the elevated plus maze and the dark/light box test revealed that these same animals had levels of anxiety-like responding that were similar to controls, and exhibited significantly less anxiety-like responding than genetically modified serotonin transporter knockout mice (Popa et al. 2008). These findings suggest that fluoxetine exposure during early critical periods of development may result in different anatomical alterations than those that occur after later-life exposure, and lend support to the notion that fluoxetine exposure may have differing effects on aggression depending on the developmental time period over which they are administered.

Various behavioral and anatomical studies indicate that repeated exposure to cocaine, AAS, and fluoxetine during critical developmental periods alters brain structures and has long-lasting behavioral effects that are present in early adulthood. Since the behavioral effects of chronic exposure to cocaine and AAS on aggression are similar to the effects of SSRIs while the anatomical alterations are not, perhaps the latter enhance aggressive behaviors through activation of an excitatory receptor. The 5HT3 receptor is the only excitatory 5HT receptor, and its up-regulation has been correlated with increased cocaine-induced aggressive responding (Ricci et al. 2004). A possible 5HT3-mediated aggression mechanism may involve the release of the inhibitory neurotransmitter γ-aminobutyric acid (GABA). 5HT3 receptors have been colocalized to GABA cells in the hypothalamus (Morales and Bloom 1997), and these GABA cells may synapse with 5HT terminals (as proposed previously in Ricci et al. 2004), and actively inhibit 5HT release onto 5HT1A-containing cells in the hypothalamus. The currently accepted hypothesis is that activation of 5HT responsive cells in the AH underlies aggressive behavior (Ferris et al. 1997; Melloni and Ricci 2010) and in the absence of 5HT, these cells are disinhibited, thereby increasing aggressive behavior. Further studies that investigate the interaction between 5HT3 receptors and the GABA system may clarify the physiology and anatomy that underlie fluoxetine's aggression-stimulating effects.

Aside from a primarily 5HT-based mechanism, other neurotransmitter systems may be responsible for facilitating fluoxetine-increased aggression. Indeed, studies investigating alternative mechanisms of SSRIs suggest that, at doses too low to affect reuptake, fluoxetine enhances the presence of the neurosteroid allopregnanolone, which may affect GABA-A receptor-mediated inhibition of aggression (Pinna et al. 2006). This suggests that 5HT alone may have limited effects on aggressive activity; a hypothesis further supported by data showing that tianeptine inhibits 5HT neurotransmission (Kuroda et al. 1994), elevates CNS neurosteriod biosynthesis, and decreases aggressive behaviors in socially isolated mice (Pinna et al. 2003a).

Regardless of their specific mechanisms of action, drugs that primarily target the 5HT system alter aggressive responding through various means. For instance, besides altering the 5HT system, enhanced aggressive responding following early fluoxetine exposure is correlated with elevated AVP afferent development within the LAH (Ricci and Melloni 2012), suggesting increased AVP tone within the LAH after prolonged fluoxetine exposure. Microdialysate data from aggressive hamsters indicate that elevated AVP within the AH correlates with heightened aggression (Melloni and Ricci 2010), and microinjection of AVP directly into the AH significantly increases aggressive behavior (Ferris 1996). These effects are believed to be mediated by AVP V1A receptors bound to neurons within the AH (Ferris et al. 1989, 1997; Mocaer et al. 1988). Considering the effects that chronic fluoxetine exposure has on both AVP and 5HT afferents, as well as AVP's facilitatory role in aggression, and 5HT's generally inhibitory role in aggression, it is important to consider the interacting elements of these two systems. However, before doing so we will review the AVP system's involvement in aggressive behavior.

2 The Role of AVP in Aggressive Behavior

Arginine vasopressin (AVP) is most commonly known as the antidiuretic hormone, for its role in suppressing urine flow (Berlinger et al. 1958), and plays a part in various physiological functions throughout the body including vasoconstriction (Vincent and Su 2008) and water homeostasis (Birnbaumer 2000). Clinically, drugs that interact with the AVP system are effective in reducing symptoms associated with diabetes insipidus (Drincic and Robertson 1999), and may be useful in treating shock (Malay et al. 1999; Patel et al. 2002), cardiac arrest (Holmes et al. 2001), and hypertension (Militante and Lombardini 2002). In the brain, AVP plays an important role in memory formation (Engelmann et al. 1996) and modulating socially relevant behaviors (Bluthé et al. 1993; Engelmann and Landgraf 1994), but is particularly known for facilitating aggressive behavior across various species including rats (Koolhaas et al. 1998), voles (Stribley and Carter 1999; Winslow et al. 1993), mice (Bester-Meredith et al. 1999; Koolhaas et al. 1998; Scordalakes and Rissman 2004; Wersinger et al. 2007a), hamsters (Albers and Bamshad 1998; Ferris and Potegal 1988; Ferris et al. 1997; Melloni and Ricci 2010; Potegal and Ferris 1989), and humans (Coccaro et al. 1998).

AVP cell bodies can be found throughout the CNS in similar regions across many different species. AVP cell bodies have been localized to the BNST, MeA, SCN, SON, and PVN (De Vries et al. 1981, 1985; De Vries and Buijs 1983; Dubois-Dauphin et al. 1989, 1990; Mahoney et al. 1990; Rood and De Vries 2011) in rats, hamsters, and guinea pigs, and are also found in the NC and the ventrolateral and ventromedial preoptic area in the mouse and hamster (Dubois-Dauphin et al. 1990; Ferris et al. 1989; Mahoney et al. 1990; Rood and De Vries 2011). The SON and PVN are densely populated with AVP immunoreactive cells (Dubois-Dauphin et al. 1990; Ferris et al. 1989; Mahoney et al. 1990; Rood and De Vries 2011) and are the two primary sites of AVP synthesis in the CNS (Gainer 1998; Rittig et al. 1996). The AVP nuclei send the majority of their axonal projections and varicosities to the posterior pituitary gland (Ferris et al. 1989; Swanson and Kuypers 1980), while fewer than half innervate brain structures, including various areas within the hippocampus, BNST, amygdala, preoptic nucleus, and the latero-anterior portion of the hypothalamus of the mouse (Rood and De Vries 2011), rat (De Vries et al. 1985; Mason et al. 1984; Zhang and Hernández 2013) and hamster (Dubois-Dauphin et al. 1990; Ferris et al. 1989; Grimes et al. 2006, 2007). Binding sites for AVP have been found in the hippocampus, BNST, the CeA and MeA, the lateral and ventrolateral hypothalamus, and subdivisions of the thalamus within the hamster (DeLeon et al. 2002b; Dubois-Dauphin et al. 1990), rat (Dorsa et al. 1984; Giri et al. 1990; Saito et al. 1995; Tribollet et al. 1988), and mouse (Campbell et al. 2009; Young et al. 1999).

Throughout the body, AVP selectively activates the AVP V1A (V1A), AVP V1B (V1B), and AVP V2 (V2) receptors (Barberis et al. 1998). Peripherally, V1A receptor activation is responsible for the vasoconstriction effects of AVP (Share 1988) and can be found in the kidneys and the liver (Ostrowski et al. 1992), while the V1B receptor is primarily responsible for the antidiuretic effects of AVP (Antoni 1993), and can be found throughout the pituitary gland, spleen, thymus, heart, breast, and kidneys (Antoni 1993; Gerstberger and Fahrenholz 1989; Lolait et al. 1995). V2 receptors are primarily located in the kidneys (Fenton et al. 2007) and thus, play a primary role in renal function, water reabsorption and retention (Nielsen et al. 2002). Centrally, AVP interacts with the V1A and V1B receptors (Barberis et al. 1998; Birnbaumer 2000). Within the CNS, activation of the V1A receptor influences maternal care, mate-bonding, and social recognition (Bielsky et al. 2004; Pitkow et al. 2001; Wersinger et al. 2002) while the V1B receptor is associated with depression, the stress response, and anxiety (Griebel et al. 2012, 2003; Salomé et al. 2006; Stemmelin et al. 2005). As mentioned previously, both of these receptors are important components of the aggression circuit and can be found in various regions that are associated with aggression control. V1A receptors are the most widely distributed AVP receptor in the CNS (Gerstberger and Fahrenholz 1989; Ostrowski et al. 1994) and are found in the hippocampus, LS, CeA, MeA, BNST, PVN, SCN, and various other portions of the hypothalamus across several species (Campbell et al. 2009; DeLeon et al. 2002b; Dubois-Dauphin et al. 1990; Tribollet et al. 1988; Young et al. 1999), while V1B receptors are more sparsely distributed, but can also be found in many aggression areas, including the hippocampus, BNST, amygdala, and hypothalamus (Hernando et al. 2001; Tribollet et al. 1988; Young et al. 2006).

AVP is involved in various CNS circuits that underlie many different processes, including those related to the regulation of aggressive behavior, and various studies have used AVP agonists and antagonists to promote or block aggressive behaviors. For example, in resident male hamsters, oral administration of the V1A antagonist SRX251 reduces bite frequencies, and increases attack latencies during resident-intruder testing (Ferris et al. 2006). Similarly, IP administration of V1A antagonists to bluehead wrasse fish decreases aggressive behaviors in territorial males (Semsar et al. 2001). V1A receptor density in the PVN, AH, and LH positively correlates with aggression intensity in socially isolated hamsters, but not in hamsters with prior social experience (Albers et al. 2006). In sexually inexperienced voles, developmental exposure to systemic AVP increases V1A receptor binding and attacks toward both unfamiliar male and female conspecifics, while developmental exposure to V1A antagonist reduces the likelihood of aggressive behavior (Gobrogge et al. 2009; Stribley and Carter 1999). Conversely, V1A knockout mice exhibit levels of aggression similar to wild-type mice after being socially isolated for 2 weeks prior to resident-intruder testing (Wersinger et al. 2007b); however, these behavioral similarities are likely a function of developmental homeostatic mechanisms that are activated in the absence of the V1A receptor gene, and are not necessarily due to a lack of V1A receptor activation (Wersinger et al. 2007b).

In contrast to V1A knockout studies, V1B knockout mice show reduced levels of aggression in a variety of situations, including maternal aggression (Wersinger et al. 2007a) and the resident-intruder tests, as well as neutral arena and food competition paradigms in males (Caldwell and Young 2009; Wersinger et al. 2002,

2007a). More precisely, V1B knockout mice show reduced defensive aggression in the reversed resident-intruder paradigm, and also show no impairment in predatory behaviors (Wersinger et al. 2007a). The association between V1B receptor expression and aggression is strong; the nonaggressive phenotype of V1B knockouts persists even on a more wild and aggressive genetic background (Caldwell and Young 2009).

Behavioral pharmacological data confirm the involvement of V1B receptors in aggression. For instance, the V1B antagonist SSR149415 reduces attack duration and frequency in resident male hamsters (Blanchard et al. 2005) and systemic pretreatment with SSR149415 reduced attack duration and defensive behaviors in mice, without impairing locomotor activity (Griebel et al. 2002, 2003). Various genetic markers for the V1B, but not V1A, receptors are reduced in blood and saliva samples from aggressive children (Zai et al. 2012). This suggests that the V1B receptor may modulate various forms of social aggression across species, while other components of the aggression circuit may regulate other forms of attack behavior (e.g., those necessary for the allocation of food) as shown by the results of Griebel et al. (2002), who studied defensive aggressive behaviors exhibited by mice in response to the presence of a rat.

The data on AVP receptor activation and blockade described above reveal little about where specific receptors are centrally active. Various studies have shown that activation of neurons within specific AVP receptor loci has differing effects on aggression and related behaviors. For example, in both hamsters and rats, scent (or flank) marking usually occurs during aggressive encounters (Adams 1976; Ferris et al. 1987), and is used to communicate social dominance (Ferris et al. 1987; Irvin et al. 1990). In hamsters, flank marking is enhanced after AVP microinjection into the AH and LS (Ferris et al. 1985; Irvin et al. 1990), and inhibited if the AVP injection is followed by a central injection of V1A antagonist (Ferris et al. 1985). These studies show that individual AVP receptors have specific effects on aggressive behavior, and demonstrate how microinjecion studies can target behavioral responses through stimulation of specific neural structures.

Aside from the regionally specific effects on behavior after exposure to agents that target AVP receptors, behaviors associated with aggression and AVP receptor activation are also highly dose-sensitive. Doses of V1A antagonists that inhibit flank marking when injected into specific brain regions or delivered more broadly via intracerebroventricular (IVC) injection also inhibit aggression in both hamsters and voles (Ferris and Potegal 1988; Gobrogge et al. 2009; Potegal and Ferris 1989; Winslow et al. 1993). One such study found that microinjection of the V1A antagonist, d(CH2)5Try(Me)AVP into the AH of large resident adult male hamsters produced a dose-dependent decrease in both offensive attacks and flank marking when presented with a smaller intruder hamster, as well as an increase in attack latency (Ferris and Potegal 1988). The V1A antagonist-specific effects on aggressive behavior were confirmed by the absence of any changes in social or sexually motivated behaviors (Ferris and Potegal 1988). More recent data show that bilateral intra-AH microinjection of a V1A antagonist reduces aggression toward novel females by sexually experienced, pair-bonded male voles (Gobrogge et al. 2009). Microinjection of AVP doses lower than those found to enhance flank marking in previous studies (Ferris et al. 1997) directly into the AH of aggressive male hamsters, as well as hamsters exposed to abnormally long daylight periods during their sleep/wake cycle, increases the number of attacks during resident-intruder testing, compared to vehicle-treated controls (Caldwell and Albers 2004; Ferris et al. 1997). Similarly, in sexually naive male voles, the bilateral microinjection of a high ($\sim$2 mM), but not a low (23 μM), concentration of AVP into the AH increases aggression toward novel females during an aggressive encounter (Gobrogge et al. 2009). The role of the AH V1A receptor in aggression control is further supported by data that show vector-enhancement of V1A receptor expression in the AH, but not the LS, increases the likelihood that sexually inexperienced voles will attack an unfamiliar female intruder, compared to controls (Gobrogge et al. 2009). At doses well above those that enhance hamster flank marking, microinjections of AVP into the MeA of castrated and gonadally intact rats decreased attack frequency, while microinjections into the LS of castrated rats decreased attack duration during aggression testing (Koolhaas et al. 1990, 1991). From these studies, it seems that behavioral responsiveness to low and high concentrations of AVP are inconsistent across species and also anatomical area. What is clear from these data is that the display of different agonistic behaviors is highly sensitive to AVP concentration, and may be modulated by the same brain regions across a variety of animals.

In light of the findings from studies that focus on the differing physiological and behavioral effects of artificially applied AVP and AVP ligands, it is not surprising that endogenous levels of AVP in specific brain regions have been shown to fluctuate during aggression tests. A recent microdialysis study found that tonic levels of AVP are elevated in the AH of highly aggressive AAS treated hamsters 30 min after a modified resident-intruder test (Melloni and Ricci 2010; Fig. 3). These findings are similar to those of a previous study that reported that both microdialysate AH AVP and aggressive responding were elevated in pair-bonded prairie voles when they were presented with an unfamiliar female (Gobrogge et al. 2009). Similarly, resident rats showed a positive correlation between aggressive behavior and AVP levels in the LS, and a negative correlation with AVP levels in the BNST (Veenema et al. 2010). Analogous to the results of the hamster studies that were mentioned previously, increased aggressive behavior duration can be blocked in highly aggressive rats before aggression testing by direct delivery of a V1A receptor antagonist into the LS (Veenema et al. 2010). Aggression duration in these same rats decreased following infusion of AVP into the BNST (Veenema et al. 2010). These studies indicate that the display of aggressive behaviors is highly sensitive to changes in the levels of AVP within the LS, BNST, and AH. More precisely, microdialysate measures of AVP fluctuations from hamster, vole, and rat studies suggest that increased AVP within the BNST suppresses aggressive behavior, while AVP elevations in the LS and AH positively correlate with enhanced aggression-like responding.

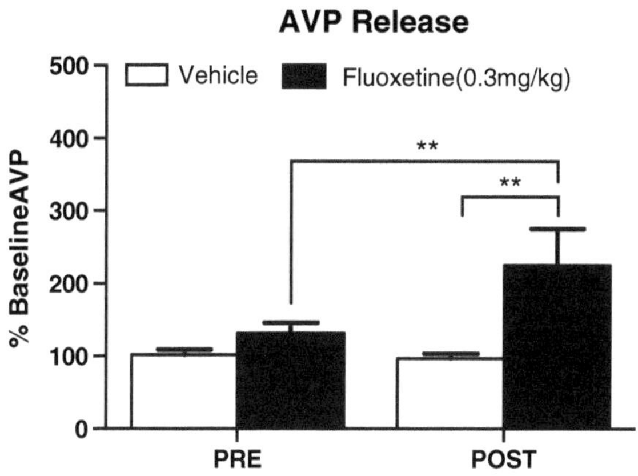

Fig. 3 Adolescent AAS exposure increases AVP release in the AH of pubertal Syrian hamsters. AH AVP release sampled using in vivo microdialysis every 30 min for 1 h prior to—(PRE) and following–(POST) exposure to intruders during resident/intruder testing in adolescent AAS- and vehicle-treated residents. Residents were separated from intruders by a wire mesh to prevent problems with the dialysis tether that often occur during testing. Even under these less than intense agonistic conditions, an increase in "intruder-evoked" release of AH AVP was seen in adolescent AAS-exposed residents (POST). ** $p < 0.01$

Considering the differential behavioral responsiveness between activation of neurons by AVP within the BNST, and also the LS and AH, it is interesting to note that functional and anatomical data have shown that putative AVP-containing afferents from the BNST innervate the LS (De Vries et al. 1985; De Vries and Buijs 1983; Wang et al. 1993). Lesion and immunocytochemical studies have shown that afferent and/or efferent connections exist between the LS and various hypothalamic areas including the lateral and anterior portions, as well as the NC and mSON (De Vries et al. 1985; De Vries and Buijs 1983; Deller et al. 1994; Ferris et al. 1990), areas that have all been implicated in the control of aggressive behavior. Combined with data that correlate shifts in AVP levels in specific rodent brain regions and aggressive behavior, localization data indicate that activation of AVP sensitive neurons in the BNST may modulate the activity of LS-mediated aggressive responding.

In addition to their anatomical interconnections and their influence over aggressive behavior, the BNST, LS, and AH, are also differentially affected by testosterone in a region-specific manner. For example, testosterone-depleted castrated adult male rats have far fewer immunoreactive AVP neuronal fibers projecting from the BNST to both the LS and anterior amygdala following gonadectomy than intact controls (De Vries et al. 1984, 1985). Immunocytochemical analysis revealed that testosterone replacement therapy in castrated rats often restored AVP fibers to control levels in both the LS and the anterior amygdala (De Vries et al. 1984). In other animals, however, the absence of testosterone did not affect the presence of putative AVP afferents extending from the suprachiasmatic nucleus into the PVN or the dorsomedial hypothalamus (De Vries et al. 1984; Hoorneman and Buijs 1982), thus indicating that the presence of testosterone may affect the expression of AVP pathways in a brain site-specific manner. The AVP-reducing effect of testosterone depletion has been shown to be developmentally sensitive in a study that examined the effects of castration and ovarectomy at birth. In their experiments, Wang et al. (1993) found that neonatally castrated male rats treated with testosterone replacement capsules after 3 months of age had a similar density of AVP neurons in the BNST and MeA, and a similar density of AVP fibers in the LS as ovarectomized female rats. In follow-up experiments, Wang et al. found that male rats castrated neonatally or 1 week following birth had fewer AVP cells in both the BNST and MeA, and also had fewer LS AVP fibers than male rats castrated 3 weeks or 3 months after birth. The temporal importance of postnatal castration was tested in neonatally castrated rats that were either untreated or treated with testosterone 1 week following birth. Testosterone treated animals had a greater density of AVP cells and fibers in the BNST and LS (respectively) than untreated animals suggesting that the presence of AVP structures in aggression areas is highly dependent on the interaction of AVP with testosterone during developmental time periods (Wang et al. 1993). Moreover, these studies suggest that the presence of AVP cells and fibers in some, but not all (as evidenced by the lack of an effect on hypothalamic AVP fibers after testosterone depletion) pre- and postsynaptic brain regions responsible for aggression control may be dependent on sex differences, and more importantly, the presence of testosterone.

In line with the hormone-sensitive nature of AVP neurons throughout the aggression circuit, sexually inexperienced male voles display less aggressive behavior toward unfamiliar females, have fewer V1A binding sites in the AH, but similar V1A binding in the MeA, BNST, and LS compared to pair-bonded voles (Gobrogge et al. 2009). Interestingly and counter-intuitively, serum testosterone levels of male prairie voles housed with female voles tend to be similar to that of individually housed males (Klein et al. 1997). Considering the influence that sexual experience has on V1A expression and aggressive behavior (Gobrogge et al. 2009), the lack of change in testosterone levels after mating suggests that hormone fluctuations do not factor into testosterone's interaction with AVP structures and levels in terms of rodent aggression. This is in line with hamster studies that have shown that castrated male hamsters that received both systemic testosterone treatments and also a microinjection of AVP directly into the VLH had shorter attack latencies, while AVP microinjection into the VLH of non-testosterone-treated castrates did not influence aggression control (Delville et al. 1996). These data suggest that control of aggressive behavior through neuroanatomical and neurochemical structures in the BNST, LS, MeA, and AH are sensitive to both the enhancement and blockade of AVP structures and the rise and fall of AVP and that this sensitivity is first and foremost dependent on the presence of testosterone.

Like testosterone, hormone-analogous synthetic derivatives (i.e., AAS) have been shown to affect the behavior-mediating sensitivity of neurons in AVP afferent and efferent locations. Hamsters treated with AAS within the developmental time period that parallels human adolescence (i.e., postnatal days 25–65; see: Miller et al. 1977; Schoenfeld and Leonard 1985; Vomachka et al. 1981; Festing 1972; Whitsett 1975) respond more aggressively than vehicle-treated conspecifics during resident-intruder encounters (for review see Melloni and Ricci 2010). Aggressive behavior is observed in these animals without any prior social experience or learning cues, thus making AAS exposure alone sufficient to generate a mature offensive aggressive phenotype. Studies using this animal model have revealed a great deal about the neurobiological mechanisms underlying both the AAS- and normal aggression circuits.

Anatomically, exposure to AAS during adolescent development alters AVP levels and AVP receptor expression in various locations throughout the hypothalamus. For instance, in the AH of the AAS-exposed hamster, there are significant increases in the density of AVP afferent fibers and terminals, AVP peptide content, and AVP release during an aggressive encounter (Delville et al. 1996; Melloni and Ricci 2010). Microinjection of an AVP antagonist into the AH of both AAS- and non-AAS-treated hamsters decreases levels of aggression intensity (i.e., bites and attacks; Harrison et al. 2000; Potegal and Ferris 1989), while direct application of AVP into the VLH of testosterone-dependent gonadectomized hamsters decreases attack latency (Delville et al. 1996). Along similar lines, in situ autoradiography data have revealed that V1A receptor binding is increased in the VLH and not the AH after adolescent AAS exposure in highly aggressive hamsters (DeLeon et al. 2002b). These data combined indicate that VLH neuronal V1A receptor activation may play a role in aggressive behavior initiation, while AH AVP receptors are at least partly responsible for regulating aggression intensity.

Various anatomical and functional studies have determined that the source of AVP to the AH region is magnocellular AVP neurons located along the dorsal surface of the optic chiasm in the mSON, and those in a small cluster centered directly within the AH proper in the NC (Ferris et al. 1989, 1990; Mahoney et al. 1990). During adolescent exposure to AAS, the population of AVP cells in presynaptic regions appears to remain constant (Harrison et al. 2000), though the density of AVP afferents in postsynaptic regions (e.g., the AH, LAH) increases (Grimes and Melloni 2002; Grimes et al. 2006, 2007; Harrison et al. 2000) along with the expression of AVP receptors (DeLeon et al. 2002b) and the release of AVP in aggression areas (Harrison et al. 2000). Thus AVP's role in aggressive responding, through neuroanatomical changes within various regions of the hypothalamus, is likely mediated by increased AVP receptor and afferent density, along with increased peptide concentration. Indeed, 20 days after cessation of AAS treatment, AVP afferent densities are reduced to control levels (Grimes et al. 2006). Not surprisingly, during this period of AAS withdrawal the return of AVP afferent density to normal levels is accompanied by a normalization of aggressive behavior (Grimes et al. 2006). All of these findings seem to suggest that many regulatory components of aggression are anatomically compartmentalized to specific regions of the hypothalamus (Harrison et al. 2000), and elevations in AVP, specifically in the hypothalamus, enhance aggressive responding.

In addition to AAS, various drugs of abuse alter AVP signaling and produce correlative changes in aggressive responding. For example, amphetamine affects aggressive responding across numerous contexts and species, including rats, mice, cats, squirrel monkeys, voles, and humans (Cherek et al. 1987; Gobrogge et al. 2009; Miczek and Tidey 1989; Steensland et al. 2005). In voles, subchronic amphetamine exposure produces aggressive responding that positively correlates with V1A expression, and can be blocked by local infusion of V1A antagonists (Gobrogge et al. 2009). As mentioned previously, chronic exposure to AAS or testosterone has been shown to enhance AVP signaling in rodents (Delville et al. 1996; Melloni and Ricci 2010), and accordingly, defensive aggressive behaviors in AAS-pretreated rats are enhanced after acute amphetamine exposure (Steensland et al. 2005). Cocaine shares similar binding targets with amphetamine (Horn 1990), and like amphetamine, its use has also been linked to alterations in AVP signaling and aggressive behavior. In animal models, chronic cocaine exposure during adulthood and adolescence dramatically elevates offensive, defensive, and maternal aggression (Blanchard and Blanchard 1999; Darmani et al. 1990; DeLeon et al. 2002a; Harrison et al. 2000; Jackson et al. 2005; Knyshevski et al. 2005a, b; Nephew and Febo 2010; Ricci et al. 2004), while also enhancing stimulated AVP release that can be blocked by microinjection of a V1A antagonist into the AH (Jackson et al. 2005). More recently, the recreational use and abuse of amphetamine analogs that contain cathiones and synthetic cathiones (commonly referred to as "bath salts") have become widespread (Fass et al. 2012; Gerona and Wu 2012) and have been linked to aggressive, erratic, and/or violent behavior in both animals and humans (Banjaw et al. 2006; Spiller et al. 2011). Although these drugs are believed to primarily produce their behavioral effects by interacting with catecholamine transporters (Baumann et al. 2013), some preclinical data combine with a handful of case studies suggest that AVP may be involved in cathione-induced maladaptive responding. For example, elevated levels of AVP induce excessive thirst in humans (Phillips et al. 1985), and excessive water intake has been reported in rats chronically exposed to cathione-like drugs (Nencini 1988). Similarly, case studies have indicated that emergency room patients that have consumed drugs containing synthetic cathiones and cathione derivatives often suffer from hyponaturmia (Ferris et al. 1985; Sammler et al. 2010; Wood et al. 2010), suggesting that cathione exposure may enhance AVP release in a manner similar to cocaine.

3 The Role of AVP and AVP/5HT Interactions in Aggressive Behavior

A number of behavioral pharmacology and neurobiology studies suggest a complex interaction between AVP and 5HT in the control of aggressive behavior. For instance, clinically prescribed drugs that affect AVP and 5HT neurotransmission also modulate aggression. Both chronic and acute exposure to the 5HT elevating drug fluoxetine increases both aggressive behavior and AVP afferent innervation in the LAH of the adolescent hamster (Ricci and Melloni 2012; Taravosh-Lahn et al. 2006). Interestingly, along with the influence they presumably have over AVP signaling and enhanced aggression, amphetamine, cocaine, cathiones, fluoxetine, and AAS are known to differentially affect 5HT structures. For example, long-term exposure to both AAS and cocaine decreases the expression of 5HT afferents in various aggression areas including the AH and MeA in the adolescent male hamster (DeLeon et al. 2002a; Grimes and Melloni 2002).

Conversely, chronic fluoxetine exposure enhances 5HT afferent innervation to many of the same locations where AAS and/or cocaine reduces them, namely the VLH and LAH (Ricci and Melloni 2012). AAS exposure affects the expression of 5HT1A and 5TH1B heteroreceptors in various postsynaptic aggression areas (Grimes and Melloni 2005; Ricci et al. 2006), and cocaine enhances the expression of excitatory 5HT3 receptors in the LAH, the CeA, VLH, and LS (Ricci et al. 2004), while chronic fluoxetine is known to desensitize 5HT1A and 5HT1B autoreceptors and increase postsynaptic 5HT release in various locations, including the hypothalamus (Dremencov et al. 2002).

Conversely, various other compounds affecting both AVP and 5HT systems can reduce aggressive behavior. Systemic administration of nicotine is known to increase circulating levels of AVP and central release of 5HT (Castro de Souza and Rocha E Silva 1977; Ribeiro et al. 1993), while reducing levels of aggressive behavior in humans (Cherek 1981; Clarke 1987). Like cathione-related substances, MDMA is known to induce hyponaturmia (Budisavljevic et al. 2003; Holden and Jackson 1996; Kessel 1994; Maxwell et al. 1993), and enhances AVP secretion (Henry et al. 1998), while also blocking the reuptake and metabolism of 5HT (Johnson et al. 1991; Leonardi and Azmitia 1994), thus enhancing 5HT release (Gough et al. 1991). Despite these neurochemical effects and its ability to enhance locomotor activity (Bexis and Docherty 2006; Spanos and Yamamoto 1989), MDMA is primarily associated with reduced aggressive behavior in laboratory animals (Kirilly et al. 2006; Navarro and Maldonado 1999). These effects of MDMA on 5HT release are in line with previous findings that elevations in fluoxetine-mediated elevations in 5HT are capable of blocking AVP-induced aggression (Ferris 1996; Ferris et al. 1997). Mechanistically, since drugs that enhance aggressive behaviors also enhance AVP signaling and reduce 5HT signaling (e.g., cocaine, amphetamine), the effects of MDMA fall in line with reports that 5HT depletion significantly increases aggressive behaviors without altering AVP structures (Adams et al. 1996; Keele 2001; Raleigh et al. 1980). Taken together, the substance exposure studies suggest that drugs that elevate both 5HT and AVP reduce aggressive responding, while drugs that elevate AVP and decrease 5HT enhance aggressive responding.

Despite a seemingly straightforward neurochemical association among AVP, 5HT, and behavior, complex structural changes are believed to underlie their role in the control of aggressive responding. Many regions of the brain associated with aggression are known to contain both 5HT- and AVP- ergic elements. For example, a modest density of V1B receptor mRNA has been detected in the VO PFx of the adult male rat (Vaccari et al. 1998), and direct microinjection of 5HT1B agonist into this region inhibits aggressive responding in mice (de Almeida et al. 2006). Similarly, the BNST and MeA are known to contain AVP cells (Caffé et al. 1987; Wang et al. 1993), as well as inhibitory 5HT receptors (Grimes and Melloni 2005; Guo and Rainnie 2010; Wright et al. 1995; Zhang et al. 1999). Despite these reports, numerous other studies suggest that the regulation of aggressive behaviors by functionally interacting elements of the 5HT and AVP systems is primarily localized within the AH.

AVP reaches the AH from cells originating in the mSON and NC (i.e., AH AVP; Ferris et al. 1989, 1990; Mahoney et al. 1990), while 5HT is delivered from afferents in the raphe nuclei (Delville et al. 2000). AH AVP neurons in mSON and NC have been shown to express inhibitory 5HT receptors (Melloni and Ricci 2010), and chronic exposure to AAS has been shown to reduce 5HT afferents that are otherwise in close apposition to these neurons in highly aggressive hamsters (Melloni and Ricci 2010). In these animals, 5HT afferents are reduced (Grimes and Melloni 2002), while AVP afferents are enhanced in close proximity to cells within the LAH region that are highly AVP and 5HT responsive (Ferris et al. 1997; Harrison et al. 2000), presumably as a function of the expression of 5HT1A, 5HT1B, and V1A receptors on these cells (Cooper et al. 2005; Ferris et al. 1997; Grimes and Melloni 2005; Melloni and Ricci 2010; Ricci et al. 2006). These anatomical data support various behavioral pharmacological findings that strongly suggest that, in the absence of drug treatment (e.g., cocaine, fluoxetine, AAS), 5HT mechanisms inhibit AVP neuronal activity that enhances aggressive behaviors by activating 5HT receptors expressed on both AH AVP neurons and AVP/5HT-sensitive neurons in the LAH.

An anatomical model illustrating the 5HT/AVP-mediated aggression circuit in the AH has been suggested (Ferris et al. 1997; Fig. 4) and was elaborated more recently in a review by Melloni and Ricci (2010) (Fig. 5). Various behavioral pharmacological data exist in support of this anatomical model. For instance, as mentioned previously, AVP microinjection into the AH region enhances aggression (Caldwell and Albers 2004; Delville et al. 1996; Ferris et al. 1997; Gobrogge et al. 2009). This AVP-induced enhancement can be inhibited by pretreatment with fluoxetine (Ferris 1996; Ferris et al. 1997), direct administration of 5HT1A agonists (Ferris et al. 1999), or co-microinfusion of AVP with a V1A antagonist (Gobrogge et al. 2009). Further, AVP release is enhanced during an aggressive encounter (Gobrogge et al. 2009; Melloni and Ricci 2010; Veenema et al. 2010) and is reduced in vitro after chronic high-dose fluoxetine exposure (Altemus et al. 1992). In contrast, 5HT levels drop in some postsynaptic regions during an aggressive encounter with a conspecific (Van Erp and Miczek 2000), and 5HT synthesis inhibition enhances aggressive responding (Adams et al. 1996; Keele 2001) and modulates AVP release (Carter and Murphy 1989; Iovino and Steardo 1985; Vacher et al. 2002, 2004). These data suggest that aggressive behavior is normally inhibited when 5HT interacts with receptors on both AH AVP and AVP/5HT-sensitive neurons in the LAH. Activation of these receptors inhibits the

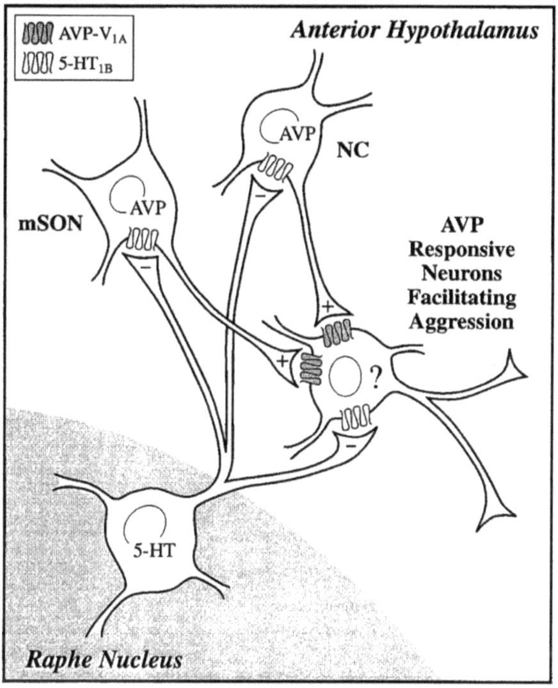

Fig. 4 The first model of the neurochemical regulation of offensive aggression from (Ferris et al. 1997) showing the hypothetical interaction between serotonin (*5HT*) with the arginine vasopressin (*AVP*) system in the anterior hypothalamus (*AH*). 5HT fibers originating from neurons in the medial and dorsal raphe nucleus innervate populations of AVP neurons in the AH localized to the medial supraoptic nucleus (*mSON*) and nucleus circularis (*NC*). Theses AVP neurons have been identified as potential sources of AVP innervation to the AH brain region involved in agonistic behavior. These AVP neurons together with 5HT neurons from the raphe nucleus impinge on neurons in the AH involved in the facilitation of aggression. The identity (?) of these postsynaptic neurons is unknown. 5HT is inhibitory (−), working through 5HT1 receptors, whereas AVP is excitatory (+), working through AVP V1A receptors

activity of AVP/5HT-sensitive cells in the LAH, and also any AH AVP neuronal activity, and subsequent AVP release, leading to the inhibition of aggressive behavior. When AVP levels are elevated, AVP facilitates aggression by binding excitatory V1A receptors expressed on 5HT/AVP-sensitive neurons in the LAH. Conversely, when fluoxetine or a 5HT1A agonist are administered, alone or prior to a microinjection of AVP, the elevated 5HT or the 5HT agonists bind 5HT receptors expressed by both AH AVP and AVP/5HT-sensitive neurons in the LAH. Both pharmacological and nonpharmacological activation of the 5HT receptors inhibits AVP signaling by AH AVP cells, while also presumably overriding any AVP-mediated excitatory postsynaptic potentials induced by activation of V1A receptors on AVP/5HT-sensitive cells, thus inhibiting the aggressive behavior-facilitating activity of these cells in the LAH.

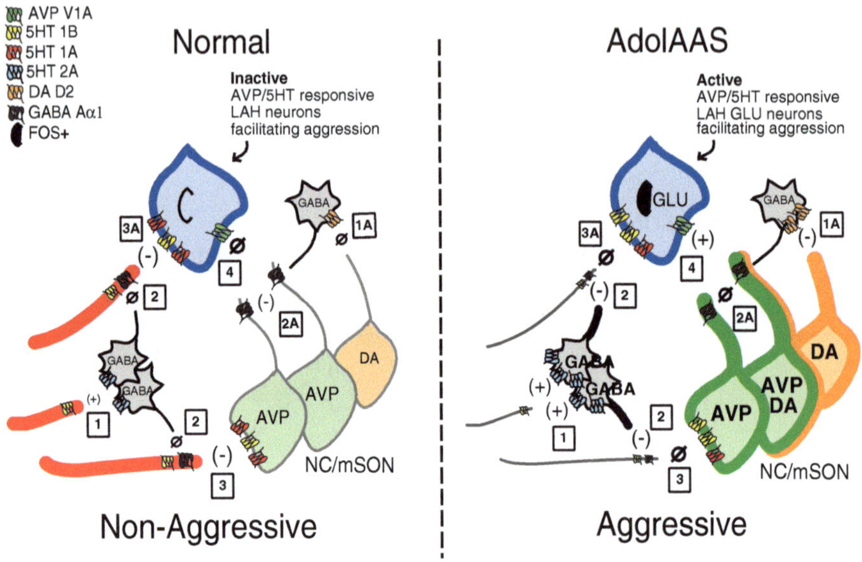

Fig. 5 A model of the anterior hypothalamic regulation of adolescent AAS offensive aggression. (Normal) This model shows the hypothetical interaction between the serotonin (*5HT*), arginine vasopressin (*AVP*), dopamine (*DA*), γ-amino-butyric acid (*GABA*), and the glutamate (*GLU*) systems in the latero-anterior hypothalamus (*LAH*). In the model, AVP is excitatory (+), working through AVP V1A receptors on AVP/5HT responsive (i.e., GLU) neurons located in the LAH sub-division of the AH, while 5HT is both inhibitory (−), working through 5HT1A/B receptors on AVP and non-AVP neurons in the AH, and excitatory (+), working through 5HT2A receptors on GABA neurons in the LAH. Also in the model DA is inhibitory (−), working through DA D2 receptors on GABA neurons in the LAH. In this case, we postulate that AVP and DA facilitate offensive aggression through activation of V1A receptors on AVP/5HT sensitive neurons and through DA D2 mediated inhibition of LAH GABA neurons. In this model, GABA is inhibitory (−), acting through GABA Aα1 receptors located on AVP and 5HT afferents in the LAH brain region. Thus, GABA's role in aggression may be dependent upon the nature and extent of its interaction with the AVP/5HT neural systems modulating offensive aggression. For instance, upon inhibiting AVP afferents in the LAH, GABA may suppress AH AVP release and offensive aggression, while inhibition of 5HT afferents in the LAH will function to suppress AH 5HT, also facilitating aggression. Lastly, the model proposes that GLU is excitatory (+), acting downstream of AVP, 5HT, DA, and GABA in other brain regions important for aggression control

This model also accounts for increased aggression induced by the inhibition of 5HT synthesis, either by 5HT depleting agents or by lesioning the 5HT nuclei, as mentioned throughout this chapter (Adams et al. 1996; Keele 2001; Vergnes et al. 1988; Vergnes and Kempf 1982). When tonic levels of 5HT are reduced, 5HT molecules fail to activate inhibitory 5HT1A and/or 5HT1B receptors expressed by AH AVP and AVP/5HT-sensitive cells in the LAH, thus leading to their disinhibition. The disinhibition of AH AVP neuronal activity enhances the release of AVP into the LAH, activating V1A receptors on AVP/5HT-sensitive cells, and leading to increased aggressive behavior.

While laboratory animal data exist as the backdrop for this model, evidence for similar functional interactions of the AVP and 5HT systems has been reported in humans. For example, it is well known that cerebro-spinal fluid concentrations of the 5HT metabolite 5-hydroxy-indole acetic acid (5-HIAA) are negatively correlated with aggressive tendencies, both in psychiatric patients and in nonpatient populations (Brown et al. 1979; Kruesi et al. 1990; Linnoila et al. 1983; Roy et al. 1988). Besides metabolic measures, fenfluramine-induced plasma prolactin secretion, which activates putative hypothalamic 5HT mechanisms, is believed to be an indicator of a normally functioning central 5HT system in healthy humans (Quattrone et al. 1983; Yatham and Steiner 1993). Numerous studies have shown that fenfluramine administration reduces prolactin concentrations in aggressive psychiatric patients, compared to healthy controls (Coccaro et al. 1989, 1998; Dolan et al. 2001; New et al. 2004; Soloff et al. 2003). One such study also found that, besides a diminished 5HT response, aggressive individuals also had heightened levels of AVP in their CSF (Coccaro et al. 1998). Similar to AVP, microinjection of Neuropeptide Y enhances AVP secretions in the SON (Willoughby and Blessing 1987) and decreases 5HT release in the ventromedial hypothalamus (Shimizu and Bray 1989), and thus it is interesting to note that Neuropeptide Y concentrations are elevated in the CSF of individuals with aggressive personality disorder (Coccaro et al. 2012). Indeed, as in many of the animal studies mentioned above, the hypothalamus appears to be an important site for the regulation of human aggression. Functional magnetic resonance imaging data have shown that the hypothalamus and other areas associated with aggression are activated when female participants are presented with pictures of “angry” human faces (Hermans et al. 2008). Recent case study data suggest that deep brain stimulation within the hypothalamus reduces self-aggressive acts, violent outbursts, and drug resistant aggression in mentally disabled persons (Franzini et al. 2010). One case study found that neuro-electric activity at the electrode stimulation site differed between an aggressive patient and a nonaggressive control (Rosa et al. 2012), suggesting that the basic neuronal physiology of aggressive individuals may be measurably altered compared to normal healthy individuals. Although more human studies are needed to confirm the translatability of the anatomical configuration from aggressive animal models to humans, the above data suggest that neuronal activity in the hypothalamus and interactions between the 5HT and AVP systems are involved in the regulation of aggressive behavior in humans.

4 Summary

The control of aggressive behavior is heavily dependent on the activity of the 5HT and AVP systems. Vast amounts of behavioral, pharmacological, neurophysiological, and neuroanatomical data suggest that 5HT and 5HT-sensitive structures are primarily responsible for inhibiting aggressive behavior associated with the activity of the AVP system. Moreover, animal and human data combined suggest that, due to the vast interconnections it has with various structures associated with aggression, the hypothalamus may serve as a significant site for AVP/5HT interactions and may be an important site of convergence in terms of the evolutionarily conserved anatomy which underlies aggressive responding across species.

References

Adamec RE (1990) Role of the amygdala and medial hypothalamus in spontaneous feline aggression and defense. Aggressive Behav 16(3–4):207–222

Adams CF, Liley NR, Gorzalka BB (1996) PCPA increases aggression in male firemouth cichlids. Pharmacology 53(5):328–330

Adams DB ((1976)) The relation of scent-marking, olfactory investigation, and specific postures in the isolation-induced fighting of rats. Behaviour 56(3–4–3):286–297

Aghajanian GK, Wang RY (1978) Physiology and pharmacology of central serotonergic neurons. In: Lipton MA, DiMascio A, Killam KF (eds) Psychopharmacology: a generation of progress. Raven Press, New York, pp 171–183

Albers HE, Bamshad M (1998) Role of vasopressin and oxytocin in the control of social behavior in Syrian hamsters (Mesocricetus auratus). In: Urban IJA, Burbach JPH, De Wied D (eds) Progress in brain research: advances in brain vasopressin, vol 119. Elsevier, New York, pp 395–408

Albers HE, Dean A, Karom MC, Smith D, Huhman KL (2006) Role of V1a vasopressin receptors in the control of aggression in Syrian hamsters. Brain Res 1073(16):425–430

Altemus M, Cizza G, Gold PW (1992) Chronic fluoxetine treatment reduces hypothalamic vasopressin secretion in vitro. Brain Res 593(2):311–313

Ansorge MS, Morelli E, Gingrich JA (2008) Inhibition of serotonin but not norepinephrine transport during development produces delayed, persistent perturbations of emotional behaviors in mice. J Neurosci 28(1):199–207

Ansorge MS, Zhou M, Lira A, Hen R, Gingrich JA (2004) Early-life blockade of the 5-HT transporter alters emotional behavior in adult mice. Science 306(5697):879–881

Antoni FA (1993) Vasopressinergic control of pituitary adrenocorticotropin secretion comes of age. Front Neuroendocrinol 14(2):76–122

Arnt J, Hyttel J (1989) Facilitation of 8-OHDPAT-induced forepaw treading of rats by the 5-HT2 agonist DOI. Eur J Pharmacol 161(1):45–51

Auerbach SB, Minzenberg MJ, Wilkinson LO (1989) Extracellular serotonin and 5-hydroxyindoleacetic acid in hypothalamus of the unanesthetized rat measured by in vivo dialysis coupled to high-performance liquid chromatography with electrochemical detection: dialysate serotonin reflects neuronal release. Brain Res 499(2):281–290

Azmitia EC, Segal M (1978) An autoradiographic analysis of the differential ascending projections of the dorsal and median raphe nuclei in the rat. J Comp Neurol 179(3):641–667

Azmitia EC, Whitaker-Azmitia PM (1995) Anatomy, cell biology, and plasticity of the serotonergic system. In: Bloom FE, Kupfer DJ (eds) Psychopharmacology: the fourth generation of progress: an official publication of the American College of Neuropsychopharmacology, 1st edn. Raven Press, New York, pp 443–449

Banjaw MY, Miczek K, Schmidt WJ (2006) Repeated Catha edulis oral administration enhances the baseline aggressive behavior in isolated rats. J Neural Transm 113(5):543–556

Barberis C, Mouillac B, Durroux T (1998) Structural bases of vasopressin/oxytocin receptor function. J Endocrinol 156(2):223–229

Barreto-Medeiros JM, Feitoza EG, Magalhaes K, Cabral-Filho JE, Manhaes-De-Castro FM, De-Castro CM, Manhaes-De-Castro R (2004) Malnutrition during brain growth spurt alters the effect of fluoxetine on aggressive behavior in adult rats. Nutr Neurosci 7(1):49–52

Baumann MH, Partilla JS, Lehner KR, Thorndike EB, Hoffman AF, Holy M, Rothman RB et al (2013) Powerful cocaine-like actions of 3,4-methylenedioxypyrovalerone (MDPV), a principal constituent of psychoactive "bath salts" products. Neuropsychopharmacol: Off Publ Am Coll Neuropsychopharmacol 38(4):552–562
Bell R, Donaldson C, Gracey D (1995) Differential effects of CGS 12066B and CP-94,253 on murine social and agonistic behaviour. Pharmacol Biochem Behav 52(1):7–16
Berlinger RW, Levinsky NG, Davidson DG, Eden M (1958) Dilution and concentration of the urine and the action of antidiuretic hormone. Am J Med 24(5):730–744
Berman ME, McCloskey MS, Fanning JR, Schumacher JA, Coccaro EF (2009) Serotonin augmentation reduces response to attack in aggressive individuals. Psychol Sci 20(6):714–720
Bester-Meredith JK, Young LJ, Marler CA (1999) Species differences in paternal behavior and aggression in peromyscus and their associations with vasopressin immunoreactivity and receptors. Horm Behav 36(1):25–38
Bexis S, Docherty JR (2006) Effects of MDMA, MDA and MDEA on blood pressure, heart rate, locomotor activity and body temperature in the rat involve alpha-adrenoceptors. Br J Pharmacol 147(8):926–934
Bielsky IF, Hu S-B, Szegda KL, Westphal H, Young LJ (2004) Profound impairment in social recognition and reduction in anxiety-like behavior in vasopressin V1a receptor knockout mice. Neuropsychopharmacology 29(3):483–493
Birnbaumer M (2000) Vasopressin receptors. Trends Endocrinol Metab 11(10):406–410
Bjork JM, Dougherty DM, Moeller FG, Swann AC (2000) Differential behavioral effects of plasma tryptophan depletion and loading in aggressive and nonaggressive men. Neuropsychopharmacology 22(4):357–369
Blader JC (2006) Pharmacotherapy and postdischarge outcomes of child inpatients admitted for aggressive behavior. J Clin Psychopharmacol 26(4):419–425
Blanchard DC, Blanchard RJ (1999) Cocaine potentiates defensive behaviors related to fear and anxiety. Neurosci Biobehav Rev 23(7):981–991
Blanchard RJ, Griebel G, Farrokhi C, Markham C, Yang M, Blanchard DC (2005) AVP V1b selective antagonist SSR149415 blocks aggressive behaviors in hamsters. Pharmacol Biochem Behav 80(1):189–194
Blier P, De Montigny C (1994) Current advances and trends in the treatment of depression. Trends Pharmacol Sci 15(7):220–226
Bluthé RM, Gheusi G, Dantzer R (1993) Gonadal steroids influence the involvement of arginine vasopressin in social recognition in mice. Psychoneuroendocrinology 18(4):323–335
Bond AJ (2005) Antidepressant treatments and human aggression. Eur J Pharmacol 526(1–3):218–225
Bouwknecht JA, Hijzen TH, Van Der Gugten J, Maes RA, Hen R, Olivier B (2001) Absence of 5-HT(1B) receptors is associated with impaired impulse control in male 5-HT(1B) knockout mice. Biol Psychiatry 49(7):557–568
Brown GL, Goodwin FK, Ballenger JC, Goyer PF, Major LF (1979) Aggression in humans correlates with cerebrospinal fluid amine metabolites. Psychiatry Res 1(2):131–139
Budisavljevic MN, Stewart L, Sahn SA, Ploth DW (2003) Hyponatremia associated with 3,4-methylenedioxymethylamphetamine ("ecstasy") abuse. Am J Med Sci 326(2):89–93
Caffé AR, Van Leeuwen FW, Luiten PG (1987) Vasopressin cells in the medial amygdala of the rat project to the lateral septum and ventral hippocampus. J Comp Neurol 261(2):237–252
Caldwell HK, Albers HE (2004) Effect of photoperiod on vasopressin-induced aggression in Syrian hamsters. Horm Behav 46(4):444–449
Caldwell HK, Young WS (2009) Persistence of reduced aggression in vasopressin 1b receptor knockout mice on a more "wild" background. Physiol Behav 97(1):131–134
Campbell M, Adams PB, Small AM, Kafantaris V, Silva RR, Shell J, Perry R et al (1995) Lithium in hospitalized aggressive children with conduct disorder: a double-blind and placebo-controlled study. J Am Acad Child Adolesc Psychiatry 34(4):445–453

Campbell M, Small AM, Green WH, Jennings SJ, Perry R, Bennett WG, Anderson L (1984) Behavioral efficacy of haloperidol and lithium carbonate. a comparison in hospitalized aggressive children with conduct disorder. Arch Gen Psychiatry 41(7):650–656

Campbell P, Ophir AG, Phelps SM (2009) Central vasopressin and oxytocin receptor distributions in two species of singing mice. J Comp Neurol 516(4):321–333

Carrillo M, Ricci LA, Schwartzer JJ, Melloni RH (2010) Immunohistochemical characterization of 5-HT(3A) receptors in the Syrian hamster forebrain. Brain Res 1329:67–81

Carter DA, Murphy D (1989) Independent regulation of neuropeptide mRNA level and poly(A) tail length. J Biol Chem 264(12):6601–6603

Castro de Souza E, Rocha E Silva M (1977) The release of vasopressin by nicotine: further studies on its site of action. J physiol 265(2):297–311

Centenaro LA, Vieira K, Zimmermann N, Miczek KA, Lucion AB, de Almeida RMM (2008) Social instigation and aggressive behavior in mice: role of 5-HT1A and 5-HT1B receptors in the prefrontal cortex. Psychopharmacology 201(2):237–248

Chamberlain B, Ervin FR, Pihl RO, Young SN (1987) The effect of raising or lowering tryptophan levels on aggression in vervet monkeys. Pharmacol Biochem Behav 28(4):503–510

Cherek DR (1981) Effects of smoking different doses of nicotine on human aggressive behavior. Psychopharmacology 75(4):339–345

Cherek DR, Steinberg JL, Kelly TH, Robinson D (1987) Effects of d-amphetamine on aggressive responding of normal male subjects. NIDA Res Monogr 21(3):257–265

Clarke PB (1987) Nicotine and smoking: a perspective from animal studies. Psychopharmacology 92(2):135–143

Cleare AJ, Bond AJ (1995) The effect of tryptophan depletion and enhancement on subjective and behavioural aggression in normal male subjects. Psychopharmacology 118(1):72–81

Clotfelter ED, O'Hare EP, McNitt MM, Carpenter RE, Summers CH (2007) Serotonin decreases aggression via 5-HT1A receptors in the fighting fish Betta splendens. Pharmacol Biochem Behav 87(2):222–231

Coccaro EF, Kavoussi RJ (1997) Fluoxetine and impulsive aggressive behavior in personality-disordered subjects. Arch Gen Psychiatry 54(12):1081–1088

Coccaro EF, Kavoussi RJ, Hauger RL, Cooper TB, Ferris CF (1998) Cerebrospinal fluid vasopressin levels: correlates with aggression and serotonin function in personality-disordered subjects. Arch Gen Psychiatry 55(8):708–714

Coccaro EF, Lee R, Liu T, Mathé AA (2012) Cerebrospinal fluid neuropeptide Y-like immunoreactivity correlates with impulsive aggression in human subjects. Biol Psychiatry 72(12):997–1003. doi:10.1016/j.biopsych.2012.07.029

Coccaro EF, Siever LJ, Klar HM, Maurer G, Cochrane K, Cooper TB, Mohs RC et al (1989) Serotonergic studies in patients with affective and personality disorders. Correlates with suicidal and impulsive aggressive behavior. Arch Gen Psychiatry 46(7):587–599

Cologer-Clifford A, Simon NG, Lu SF, Smoluk SA (1997) Serotonin agonist-induced decreases in intermale aggression are dependent on brain region and receptor subtype. Pharmacol Biochem Behav 58(2):425–430

Cooper MA, Karom M, Huhman KL, Albers HE (2005) Repeated agonistic encounters in hamsters modulate AVP V1a receptor binding. Horm Behav 48(5):545–551

Crabbe JC, Phillips TJ, Feller DJ, Hen R, Wenger CD, Lessov CN, Schafer GL (1996) Elevated alcohol consumption in null mutant mice lacking 5-HT1B serotonin receptors. Nat Genet 14(1):98–101

Cunningham MG, Connor CM, Zhang K, Benes FM (2005) Diminished serotonergic innervation of adult medial prefrontal cortex after 6-OHDA lesions in the newborn rat. Brain Res Dev Brain Res 157(2):124–131

Darmani NA, Hadfield MG, Carter WH, Martin BR (1990) Acute and chronic effects of cocaine on isolation-induced aggression in mice. Psychopharmacology 42(2):317–321

Datla KP, Bhattacharya SK (1990) Effect of selective monoamine oxidase A and B inhibitors on footshock induced aggression in paired rats. Indian J Exp Biol 28(8):742–745

Datla KP, Mitra SK, Bhattacharya SK (1991) Serotonergic modulation of footshock induced aggression in paired rats. Indian J Exp Biol 29(7):631–635
de Almeida RMM, Rosa MM, Santos DM, Saft DM, Benini Q, Miczek KA (2006) 5-HT(1B) receptors, ventral orbitofrontal cortex, and aggressive behavior in mice. Psychopharmacology 185(4):441–450
de Almeida RM, Lucion AB (1997) 8-OH-DPAT in the median raphe, dorsal periaqueductal gray and corticomedial amygdala nucleus decreases, but in the medial septal area it can increase maternal aggressive behavior in rats. Psychopharmacology 134(4):392–400
de Almeida RM, Nikulina EM, Faccidomo S, Fish EW, Miczek KA (2001) Zolmitriptan—a 5-HT1B/D agonist, alcohol, and aggression in mice. Psychopharmacology 157(2):131–141
de Almeida RMM, Miczek KA (2002) Aggression escalated by social instigation or by discontinuation of reinforcement ("frustration") in mice: inhibition by anpirtoline: a 5-HT1B receptor agonist. Neuropsychopharmacology 27(2):171–181
De Boer SF, Koolhaas JM (2005) 5-HT1A and 5-HT1B receptor agonists and aggression: a pharmacological challenge of the serotonin deficiency hypothesis. Eur J Pharmacol 526(1–3): 125–139
De Boer SF, Lesourd M, Mocaer E, Koolhaas JM (1999) Selective antiaggressive effects of alnespirone in resident-intruder test are mediated via 5-hydroxytryptamine1A receptors: a comparative pharmacological study with 8-hydroxy-2-dipropylaminotetralin, ipsapirone, buspirone, eltoprazine, and WAY-100635. J Pharmacol Exp Ther 288(3):1125–1133
De Montigny C, Chaput Y, Blier P (1990) Modification of serotonergic neuron properties by long-term treatment with serotonin reuptake blockers. J Clin Psychiatry 5(12 Supplement B):4–8
De Vries GJ, Buijs RM (1983) The origin of the vasopressinergic and oxytocinergic innervation of the rat brain with special reference to the lateral septum. Brain Res 273(2):307–317
De Vries GJ, Buijs RM, Sluiter AA (1984) Gonadal hormone actions on the morphology of the vasopressinergic innervation of the adult rat brain. Brain Res 298(1):141–145
De Vries GJ, Buijs RM, Swaab DF (1981) Ontogeny of the vasopressinergic neurons of the suprachiasmatic nucleus and their extrahypothalamic projections in the rat brain—presence of a sex difference in the lateral septum. Brain Res 218(1–2):67–78
De Vries GJ, Buijs RM, Van Leeuwen FW, Caffé AR, Swaab DF (1985) The vasopressinergic innervation of the brain in normal and castrated rats. J Comp Neurol 233(2):236–254
Deckel AW (1996) Behavioral changes in Anolis carolinensis following injection with fluoxetine. Behav Brain Res 78(2):175–182
DeLeon KR, Grimes JM, Connor DF, Melloni RH (2002a) Adolescent cocaine exposure and offensive aggression: involvement of serotonin neural signaling and innervation in male Syrian hamsters. Behav Brain Res 133(2):211–220
DeLeon KR, Grimes JM, Melloni RH (2002b) Repeated anabolic–androgenic steroid treatment during adolescence increases vasopressin V(1A) receptor binding in Syrian hamsters: correlation with offensive aggression. Horm Behav 42(2):182–191
Deller T, Leranth C, Frotscher M (1994) Reciprocal connections of lateral septal neurons and neurons in the lateral hypothalamus in the rat: a combined phaseolus vulgaris-leucoagglutinin and Fluoro-Gold immunocytochemical study. Neurosci Lett 168(1–2):119–122
Delville Y, De Vries GJ, Ferris CF (2000) Neural connections of the anterior hypothalamus and agonistic behavior in golden hamsters. Brain Behav Evol 55(2):53–76
Delville Y, Mansour KM, Ferris CF (1996) Testosterone facilitates aggression by modulating vasopressin receptors in the hypothalamus. Physiol Behav 60(1):25–29
DeNapoli JS, Dodman NH, Shuster L, Rand WM, Gross KL (2000) Effect of dietary protein content and tryptophan supplementation on dominance aggression, territorial aggression, and hyperactivity in dogs. J Am Vet Med Assoc 164(10):1042–1043
Dolan M, Anderson IM, Deakin JF (2001) Relationship between 5-HT function and impulsivity and aggression in male offenders with personality disorders. Br J Psychiatry 178(4):352–359
Dorrego MF, Canevaro L, Kuzis G, Sabe L, Starkstein SE (2002) A randomized, double-blind, crossover study of methylphenidate and lithium in adults with attention-deficit/hyperactivity disorder: preliminary findings. J Neuropsychiatry Clin Neurosci 14(3):289–295

Dorsa DM, Petracca FM, Baskin DG, Cornett LE (1984) Localization and characterization of vasopressin-binding sites in the amygdala of the rat brain. J Neurosci 4(7):1764–1770
Dremencov E, Gur E, Lerer B, Newman ME (2002) Effects of chronic antidepressants and electroconvulsive shock on serotonergic neurotransmission in the rat hippocampus. Prog Neuropsychopharmacol Biol Psychiatry 26(6):729–739
Drincic A, Robertson GL (1999) Treatment of diabetes insipidus. In: Mickle AW (ed) Hormone replacement therapy: contemporary endocrinology, 1st edn. Humana, New Jersey, pp 21–38
Duan X, Ju G (1998) The organization of chemically characterized afferents to the perivascular neuronal groups of the hypothalamic magnocellular neurosecretory system in the rat. Brain Res Bull 46(5):409–415
Dubois-Dauphin M, Pevet P, Tribollet E, Dreifuss JJ (1990) Vasopressin in the brain of the golden hamster: the distribution of vasopressin binding sites and of immunoreactivity to the vasopressin-related glycopeptide. J Comp Neurol 300(4):535–548
Dubois-Dauphin M, Tribollet E, Dreifuss JJ (1989) Distribution of neurohypophysial peptides in the guinea pig brain. I. An immunocytochemical study of the vasopressin-related glycopeptide. Brain Res 496(1–2):45–65
Dulawa SC, Holick KA, Gundersen B, Hen R (2004) Effects of chronic fluoxetine in animal models of anxiety and depression. Neuropsychopharmacology 29(7):1321–1330
El Mansari M, Sánchez C, Chouvet G, Renaud B, Haddjeri N (2005) Effects of acute and long-term administration of escitalopram and citalopram on serotonin neurotransmission: an in vivo electrophysiological study in rat brain. Neuropsychopharmacology 30(7):1269–1277
Engelmann M, Landgraf R (1994) Microdialysis administration of vasopressin into the septum improves social recognition in Brattleboro rats. Physiol Behav 55(1):145–149
Engelmann M, Wotjak CT, Neumann I, Ludwig M, Landgraf R (1996) Behavioral consequences of intracerebral vasopressin and oxytocin: focus on learning and memory. Neurosci Biobehav Rev 20(3):341–358
Faccidomo S, Bannai M, Miczek KA (2008) Escalated aggression after alcohol drinking in male mice: dorsal raphé and prefrontal cortex serotonin and 5-HT(1B) receptors. Neuropsychopharmacology 33(12):2888–2899
Fairbanks LA, Melega WP, Jorgensen MJ, Kaplan JR, McGuire MT (2001) Social impulsivity inversely associated with CSF 5-HIAA and fluoxetine exposure in vervet monkeys. Neuropsychopharmacology 24(4):370–378
Fass JA, Fass AD, Garcia AS (2012) Synthetic cathinones (bath salts): legal status and patterns of abuse. Ann Pharmacother 46(3):436–441. doi:10.1345/aph.1Q628
Fenton RA, Brønd L, Nielsen S, Praetorius J (2007) Cellular and subcellular distribution of the type-2 vasopressin receptor in the kidney. Am J Physiol Ren Physiol 293(3):748–760
Ferris CF (1996) Serotonin diminishes aggression by suppressing the activity of the vasopressin system. Ann N Y Acad Sci 794:98–103
Ferris CF, Axelson JF, Martin AM, Roberge LF (1989) Vasopressin immunoreactivity in the anterior hypothalamus is altered during the establishment of dominant/subordinate relationships between hamsters. Neuroscience 29(3):675–683
Ferris CF, Axelson JF, Shinto LH, Albers HE (1987) Scent marking and the maintenance of dominant/subordinate status in male golden hamsters. Physiol Behav 40(5):661–664
Ferris CF, Irvin RW, Potegal M, Axelson JF (1990) Kainic acid lesion of vasopressinergic neurons in the hypothalamus disrupts flank marking behavior in golden hamsters. J Neuroendocrinol 2(2):123–129
Ferris CF, Lu S-F, Messenger T, Guillon CD, Heindel N, Miller M, Koppel G et al (2006) Orally active vasopressin V1a receptor antagonist, SRX251, selectively blocks aggressive behavior. Pharmacol Biochem Behav 83(2):169–174
Ferris CF, Pollock J, Albers HE, Leeman SE (1985) Inhibition of flank-marking behavior in golden hamsters by microinjection of a vasopressin antagonist into the hypothalamus. Neurosci Lett 55(2):239–243
Ferris CF, Stolberg T, Delville Y (1999) Serotonin regulation of aggressive behavior in male golden hamsters (Mesocricetus auratus). Behav Neurosci 113(4):804–815

Ferris CF, Melloni RH, Koppel G, Perry KW, Fuller RW, Delville Y (1997) Vasopressin/serotonin interactions in the anterior hypothalamus control aggressive behavior in golden hamsters. J Neurosci 17(11):4331–4340

Ferris CF, Potegal M (1988) Vasopressin receptor blockade in the anterior hypothalamus suppresses aggression in hamsters. Physiol Behav 44(2):235–239

Festing MF (1972) Hamsters. In: University Federation for Animal Welfare (ed) The UFAW handbook on the care and management of laboratory animals, 4th edn. Williams and Wilkins, Baltimore, pp 242–256

Fish EW, Faccidomo S, Miczek KA (1999) Aggression heightened by alcohol or social instigation in mice: reduction by the 5-HT(1B) receptor agonist CP-94,253. Psychopharmacology 146(4):391–399

Fraga IC, Fregoneze JB, Carvalho FLQ, Dantas KB, Azevedo CS, Pinho CB, De Castro E, Silva E (2005) Acute fluoxetine administration differentially affects brain C-Fos expression in fasted and refed rats. Neuroscience 134(1):327–334

Franzini A, Messina G, Cordella R, Marras C, Broggi G (2010) Deep brain stimulation of the posteromedial hypothalamus: indications, long-term results, and neurophysiological considerations. Neurosurg Focus 29(2):E13

Gainer H (1998) Cell-specific gene expression in oxytocin and vasopressin magnocellular neurons. Adv Exp Med Biol 449:15–27

Gerona R, Wu A (2012) Bath salts. Clin Lab Med 32(3):415–427

Gerstberger R, Fahrenholz F (1989) Autoradiographic localization of V1 vasopressin binding sites in rat brain and kidney. Eur J Pharmacol 167(1):105–116

Giammanco S, Ernandes M, Lopez De Onate R, Paderni MA (1990) Short term diet of precooked corn meal almost lacking in tryptophan and interspecific rat-mouse aggressive behaviour. Archives Internationales de Physiologie et de Biochimie 98(1):23–26

Giri PR, Dave JR, Tabakoff B, Hoffman PL (1990) Arginine vasopressin induces the expression of c-fos in the mouse septum and hippocampus. Brain Res Mol Brain Res 7(2):131–137

Gobrogge KL, Liu Y, Young LJ, Wang Z (2009) Anterior hypothalamic vasopressin regulates pair bonding and drug-induced aggression in a monogamous rodent. Proc Nat Acad Sci USA 106(45):19144–19149

Gough B, Ali SF, Slikker W, Holson RR (1991) Acute effects of 3,4-methylenedioxymethamphetamine (MDMA) on monoamines in rat caudate. Pharmacol Biochem Behav 39(3): 619–623

Griebel G, Beeske S, Stahl S (2012) The vasopressin Vlb receptor antagonist SSR149415 in the treatment of major depressive and generalized anxiety disorders: results from 4 randomized, double-blind, placebo-controlled studies. J Clin Psychiatry 73(11):1403–1411

Griebel G, Simiand J, Serradeil-Le Gal C, Wagnon J, Pascal M, Scatton B, Maffrand J-P et al (2002) Anxiolytic- and antidepressant-like effects of the non-peptide vasopressin V1b receptor antagonist, SSR149415, suggest an innovative approach for the treatment of stress-related disorders. Proc Nat Acad Sci USA 99(9):6370–6375

Griebel G, Simiand J, Stemmelin J, Serradeil-Le GC, Steinberg R (2003) The vasopressin V1b receptor as a therapeutic target in stress-related disorders. Curr Drug Targets CNS Neurol Disord 2(3):191–200

Grimes JM, Melloni RH (2002) Serotonin modulates offensive attack in adolescent anabolic steroid-treated hamsters. Pharmacol Biochem Behav 73(3):713–721

Grimes JM, Melloni RH (2005) Serotonin-1B receptor activity and expression modulate the aggression-stimulating effects of adolescent anabolic steroid exposure in hamsters. Behav Neurosci 119(5):1184–1194

Grimes JM, Melloni RH (2006) Prolonged alterations in the serotonin neural system following the cessation of adolescent anabolic-androgenic steroid exposure in hamsters (Mesocricetus auratus). Behav Neurosci 120(6):1242–1251

Grimes JM, Ricci LA, Melloni RH (2006) Plasticity in anterior hypothalamic vasopressin correlates with aggression during anabolic-androgenic steroid withdrawal in hamsters. Behav Neurosci 120(1):115–124

Grimes JM, Ricci LA, Melloni RH (2007) Alterations in anterior hypothalamic vasopressin, but not serotonin, correlate with the temporal onset of aggressive behavior during adolescent anabolic-androgenic steroid exposure in hamsters (Mesocricetus auratus). Behav Neurosci 121(5):941–948

Guo JD, Rainnie DG (2010) Presynaptic 5-HT(1B) receptor-mediated serotonergic inhibition of glutamate transmission in the bed nucleus of the stria terminalis. Neuroscience 165(4):1390–1401

Harrison RJ, Connor DF, Nowak C, Nash K, Melloni RH (2000) Chronic anabolic-androgenic steroid treatment during adolescence increases anterior hypothalamic vasopressin and aggression in intact hamsters. Psychoneuroendocrinology 25(4):317–338

Henry JA, Fallon JK, Kicman AT, Hutt AJ, Cowan DA, Forsling M (1998) Low-dose MDMA ("ecstasy") induces vasopressin secretion. Lancet 351(9118):1784

Hermans EJ, Ramsey NF, Van Honk J (2008) Exogenous testosterone enhances responsiveness to social threat in the neural circuitry of social aggression in humans. Biol Psychiatry 63(3):263–270

Hernando F, Schoots O, Lolait SJ, Burbach JP (2001) Immunohistochemical localization of the vasopressin V1b receptor in the rat brain and pituitary gland: anatomical support for its involvement in the central effects of vasopressin. Endocrinology 142(4):1659–1668

Ho HP, Olsson M, Westberg L, Melke J, Eriksson E (2001) The serotonin reuptake inhibitor fluoxetine reduces sex steroid-related aggression in female rats: an animal model of premenstrual irritability? Neuropsychopharmacology 24(5):502–510

Holden R, Jackson MA (1996) Near-fatal hyponatraemic coma due to vasopressin over-secretion after "ecstasy" (3,4-MDMA). Lancet 347(9007):1052

Holmes CL, Walley KR, Chittock DR, Lehman T, Russell JA (2001) The effects of vasopressin on hemodynamics and renal function in severe septic shock: a case series. Intensive Care Med 27(8):1416–1421

Homberg JR, Schiepers OJG, Schoffelmeer ANM, Cuppen E, Vanderschuren LJMJ (2007) Acute and constitutive increases in central serotonin levels reduce social play behaviour in peri-adolescent rats. Psychopharmacology 195(2):175–182

Hoorneman EM, Buijs RM (1982) Vasopressin fiber pathways in the rat brain following suprachiasmatic nucleus lesioning. Brain Res 243(2):235–241

Horn AS (1990) Dopamine uptake: a review of progress in the last decade. Prog Neurobiol 34(5):387–400

Huang YY, Grailhe R, Arango V, Hen R, Mann JJ (1999) Relationship of psychopathology to the human serotonin1B genotype and receptor binding kinetics in postmortem brain tissue. Neuropsychopharmacology 21(2):238–246

Ibi D, Takuma K, Koike H, Mizoguchi H, Tsuritani K, Kuwahara Y, Kamei H et al (2008) Social isolation rearing-induced impairment of the hippocampal neurogenesis is associated with deficits in spatial memory and emotion-related behaviors in juvenile mice. J Neurochem 105(3):921–932

Invernizzi R, Belli S, Samanin R (1992) Citalopram's ability to increase the extracellular concentrations of serotonin in the dorsal raphe prevents the drug's effect in the frontal cortex. Brain Res 584(1–2):322–324

Invernizzi R, Bramante M, Samanin R (1994) Chronic treatment with citalopram facilitates the effect of a challenge dose on cortical serotonin output: role of presynaptic 5-HT1A receptors. Eur J Pharmacol 260(2–3):243–246

Invernizzi R, Bramante M, Samanin R (1996) Role of 5-HT1A receptors in the effects of acute and chronic fluoxetine on extracellular serotonin in the frontal cortex. Pharmacol Biochem Behav 54(1):143–147

Iovino M, Steardo L (1985) Effect of substances influencing brain serotonergic transmission on plasma vasopressin levels in the rat. Eur J Pharmacol 113(1):99–103

Irvin RW, Szot P, Dorsa DM, Potegal M, Ferris CF (1990) Vasopressin in the septal area of the golden hamster controls scent marking and grooming. Physiol Behav 48(5):693–699

Jackson D, Burns R, Trksak G, Simeone B, DeLeon KR, Connor DF, Harrison RJ et al (2005) Anterior hypothalamic vasopressin modulates the aggression-stimulating effects of adolescent cocaine exposure in Syrian hamsters. Neuroscience 133(3):635–646
Jarrell H, Hoffman JB, Kaplan JR, Berga S, Kinkead B, Wilson ME (2008) Polymorphisms in the serotonin reuptake transporter gene modify the consequences of social status on metabolic health in female rhesus monkeys. Physiol Behav 93(4–5):807–819
Johnson MP, Conarty PF, Nichols DE (1991) [3H]monoamine releasing and uptake inhibition properties of 3,4-methylenedioxymethamphetamine and p-chloroamphetamine analogues. Eur J Pharmacol 200(1):9–16
Joppa MA, Rowe RK, Meisel RL (1997) Effects of serotonin 1A or 1B receptor agonists on social aggression in male and female Syrian hamsters. Pharmacol Biochem Behav 58(2):349–353
Kania BF, Gralak MA, Wielgosz M (2012) Four-week fluoxetine (SSRI) exposure diminishes aggressive behaviour of male Siamese fighting fish (Betta splendens). J Behav Brain Sci 2(2):185–190
Keele NB (2001) Phenytoin inhibits isolation-induced aggression specifically in rats with low serotonin. NeuroReport 12(6):1107–1112
Kessel B (1994) Hyponatraemia after ingestion of ecstasy. BMJ (Clinical research ed.) 308(6925):414
Kirilly E, Benko A, Ferrington L, Ando RD, Kelly PAT, Bagdy G (2006) Acute and long-term effects of a single dose of MDMA on aggression in dark Agouti rats. Int J Neuropsychopharmacol Off Sci J Coll Int Neuropsychopharmacol CINP 9(1):63–76
Klein SL, Hairston JE, DeVries AC, Nelson RJ (1997) Social-environment and steroid-hormones affect species and sex-differences in immune function among voles. Horm Behav 32:30–39
Knyshevski I, Connor DF, Harrison RJ, Ricci LA, Melloni RH (2005a) Persistent activation of select forebrain regions in aggressive, adolescent cocaine-treated hamsters. Behav Brain Res 159(2):277–286
Knyshevski I, Ricci LA, McCann TE, Melloni RH (2005b) Serotonin type-1A receptors modulate adolescent, cocaine-induced offensive aggression in hamsters. Physiol Behav 85(2):167–176
Koolhaas JM, Everts H, De Ruiter AJ, De Boer SF, Bohus B (1998) Coping with stress in rats and mice: differential peptidergic modulation of the amygdala-lateral septum complex. Prog Brain Res 119:437–448
Koolhaas JM, Moor E, Hiemstra Y, Bohus B (1991) The testosterone-dependent vasopressinergic neurons in the medial amygdala and lateral septum: involvement in social behaviour of male rats. In: Jard S (ed) Vasopressin. John Libbey, Paris, pp 213–219
Koolhaas JM, Van den Brink THC, Roozendaal B, Boorsma F (1990) Medial amygdala and aggressive behavior: interaction between testosterone and vasopressin. Aggressive Behav 16(3–4):223–229
Kostowski W, Valzelli L, Kozak W, Bernasconi S (1984) Activity of desipramine, fluoxetine and nomifensine on spontaneous and p-CPA-induced muricidal aggression. Pharmacol Res Commun 16(3):265–271
Kruesi MJ, Rapoport JL, Hamburger S, Hibbs E, Potter WZ, Lenane M, Brown GL (1990) Cerebrospinal fluid monoamine metabolites, aggression, and impulsivity in disruptive behavior disorders of children and adolescents. Arch Gen Psychiatry 47(5):419–426
Kulikov AV, Osipova DV, Naumenko VS, Popova NK (2005) Association between Tph2 gene polymorphism, brain tryptophan hydroxylase activity and aggressiveness in mouse strains. Genes Brain Behav 4(8):482–485
Kuroda Y, Watanabe Y, McEwen BS (1994) Tianeptine decreases both serotonin transporter mRNA and binding sites in rat brain. Eur J Pharmacol 268(1):R3–R5
Lappalainen J, Long JC, Eggert M, Ozaki N, Robin RW, Brown GL, Goldman D (1998) Linkage of antisocial alcoholism to the serotonin 5-HT1B receptor gene in 2 populations. Arch Gen Psychiatry 55(11):989
Leonardi ET, Azmitia EC (1994) MDMA (ecstasy) inhibition of MAO type A and type B: comparisons with fenfluramine and fluoxetine (Prozac). Neuropsychopharmacology : Off Publ Am Coll Neuropsychopharmacol 10(4):231–238

Lesch KP, Merschdorf U (2000) Impulsivity, aggression, and serotonin: a molecular psychobiological perspective. Behav Sci law 18(5):581–604

Lewis DA (1990) The organization of chemically-identified neural systems in monkey prefrontal cortex: afferent systems. Prog Neuropsychopharmacol Biol Psychiatry 14(3):371–377

Linnoila M, Virkkunen M, Scheinin M, Nuutila A, Rimon R, Goodwin FK (1983) Low cerebrospinal fluid 5-hydroxyindoleacetic acid concentration differentiates impulsive from nonimpulsive violent behavior. Life Sci 33(26):2609–2614

Lolait SJ, O'Carroll AM, Mahan LC, Felder CC, Button DC, Young WS, Mezey E et al (1995) Extrapituitary expression of the rat V1b vasopressin receptor gene. Proc Nat Acad Sci USA 92(15):6783–6787

Lynn SE, Egar JM, Walker BG, Sperry TS, Ramenofsky M (2007) Fish on Prozac: a simple, noninvasive physiology laboratory investigating the mechanisms of aggressive behavior in Betta splendens. Adv Physiol Educ 31(4):358–363

Maciag D, Simpson KL, Coppinger D, Lu Y, Wang Y, Lin RCS, Paul IA (2006) Neonatal antidepressant exposure has lasting effects on behavior and serotonin circuitry. Neuropsychopharmacology 31(1):47–57

Mahoney PD, Koh ET, Irvin RW, Ferris CF (1990) Computer-aided mapping of vasopressin neurons in the hypothalamus of the male golden Hamster: evidence of magnocellular neurons that do not project to the neurohypophysis. J Neuroendocrinol 2(2):113–122

Maj J, Mogilnicka E, Klimek V, Kordecka-Magiera A (1981) Chronic treatment with antidepressants: protentiation of clonidine-induced aggression in mice via noradrenergic mechanism. J Neural Transm (Vienna, Austria: 1996) 52(3):189–197

Malay MB, Ashton RC, Landry DW, Townsend RN (1999) Low-dose vasopressin in the treatment of vasodilatory septic shock. J Trauma 47(4):699–705

Malone RP, Delaney MA, Luebbert JF, Cater J, Campbell M (2000) A double-blind placebo-controlled study of lithium in hospitalized aggressive children and adolescents with conduct disorder. Arch Gen Psychiatry 57(7):649–654

Mason WT, Ho YW, Hatton GI (1984) Axon collaterals of supraoptic neurones: anatomical and electrophysiological evidence for their existence in the lateral hypothalamus. Neuroscience 11(1):169–182

Matto V, Skrebuhhova T, Allikmets L (1998) The effect of antidepressants on rat aggressive behavior in the electric footshock and apomorphine-induced aggressiveness paradigms. Methods Find Exp Clin Pharmacol 20(4):329–337

Maxwell DL, Polkey MI, Henry JA (1993) Hyponatraemia and catatonic stupor after taking "ecstasy". BMJ (Clinical research ed) 307(6916):1399

Melloni RH, Connor DF, Hang PT, Harrison RJ, Ferris CF (1997) Anabolic-androgenic steroid exposure during adolescence and aggressive behavior in golden hamsters. Physiol Behav 61(3):359–364

Melloni RH, Ricci LA (2010) Adolescent exposure to anabolic/androgenic steroids and the neurobiology of offensive aggression: a hypothalamic neural model based on findings in pubertal Syrian hamsters. Horm Behav 58(1):177–191

Micev V, Lynch DM (1974) Effect of lithium on disturbed severely mentally retarded patients. Br J Psychiatry 125(584):110–111

Miczek KA, de Almeida RM (2001) Oral drug self-administration in the home cage of mice: alcohol-heightened aggression and inhibition by the 5-HT1B agonist anpirtoline. Psychopharmacology 157(4):421–429

Miczek KA, Tidey JW (1989) Amphetamines: aggressive and social behavior. NIDA Res Monogr 94:68–100

Miczek KA, Hussain S, Faccidomo S (1998) Alcohol-heightened aggression in mice: attenuation by 5-HT 1A receptor agonists. Psychopharmacology 139(1–2):160–168

Militante JD, Lombardini JB (2002) Treatment of hypertension with oral taurine: experimental and clinical studies. Amino Acids 23(4):381–393

Miller LL, Whitsett JM, Vandenbergh JG, Colby DR (1977) Physical and behavioral aspects of sexual maturation in male golden hamsters. J Comp Physiol Psychol 91(2):245–259

Mitchell PJ, Fletcher A, Redfern PH (1991) Is antidepressant efficacy revealed by drug-induced changes in rat behaviour exhibited during social interaction? Neurosci Biobehav Rev 15(4):539–544

Mocaer E, Rettori MC, Kamoun A (1988) Pharmacological antidepressive effects and tianeptine-induced 5-HT uptake increase. Clin Neuropharmacol 11(Suppl 2):S32–S42

Morales M, Bloom FE (1997) The 5-HT3 receptor is present in different subpopulations of GABAergic neurons in the rat telencephalon. J Neurosci 17(9):3157–3167

Mos J, Olivier B, Poth M, Van Oorschot R, Van Aken H (1993) The effects of dorsal raphe administration of eltoprazine, TFMPP and 8-OH-DPAT on resident intruder aggression in the rat. Eur J Pharmacol 238(2–3):411–415

Muehlenkamp F, Lucion A, Vogel WH (1995) Effects of selective serotonergic agonists on aggressive behavior in rats. Pharmacol Biochem Behav 50(4):671–674

Munro AD (1986) Effects of melatonin, serotonin, and naloxone on aggression in isolated cichlid fish (Aequidens pulcher). J Pineal Res 3(3):257–262

Navarro JF, Maldonado E (1999) Behavioral profile of 3,4-methylenedioxy-methamphetamine (MDMA) in agonistic encounters between male mice. Prog Neuropsychopharmacol Biol Psychiatry 23(2):327–334

Nencini P (1988) The role of opiate mechanisms in the development of tolerance to the anorectic effects of amphetamines. Pharmacol Biochem Behav 30(3):755–764

Nephew BC, Febo M (2010) Effect of cocaine sensitization prior to pregnancy on maternal care and aggression in the rat. Psychopharmacology 209(1):127–135

New AS, Trestman RF, Mitropoulou V, Goodman M, Koenigsberg HH, Silverman J, Siever LJ (2004) Low prolactin response to fenfluramine in impulsive aggression. J Psychiatr Res 38(3):223–230

Nielsen S, Frøkiaer J, Marples D, Kwon T-H, Agre P, Knepper MA (2002) Aquaporins in the kidney: from molecules to medicine. Physiol Rev 82(1):205–244

Olivier B, Mos J, Van Oorschot R, Hen R (1995) Serotonin receptors and animal models of aggressive behavior. Pharmacopsychiatry 28(Suppl 2):80–90

Olivier B (2004) Serotonin and aggression. Ann N Y Acad Sci 1036:382–392. doi:10.1196/annals.1330.022

Olivier B, Van Oorschot R (2005) 5-HT1B receptors and aggression: a review. Eur J Pharmacol 526(1–3):207–217

Ostrowski NL, Lolait SJ, Bradley DJ, O'Carroll AM, Brownstein MJ, Young WS (1992) Distribution of V1a and V2 vasopressin receptor messenger ribonucleic acids in rat liver, kidney, pituitary and brain. Endocrinology 131(1):533–535

Ostrowski NL, Lolait SJ, Young WS (1994) Cellular localization of vasopressin V1a receptor messenger ribonucleic acid in adult male rat brain, pineal, and brain vasculature. Endocrinology 135(4):1511–1528

Patel BM, Chittock DR, Russell JA, Walley KR (2002) Beneficial effects of short-term vasopressin infusion during severe septic shock. Anesthesiology 96(3):576–582

Pergola PE, Sved AF, Voogt JL, Alper RH (1993) Effect of serotonin on vasopressin release: a comparison to corticosterone, prolactin and renin. Neuroendocrinology 57(3):550–558

Perreault H, Semsar K, Godwin J (2003) Fluoxetine treatment decreases territorial aggression in a coral reef fish. Physiol Behav 79(4–5):719–724

Perry KW, Fuller RW (1993) Extracellular 5-hydroxytryptamine concentration in rat hypothalamus after administration of fluoxetine plus L-5-hydroxytryptophan. J Pharm Pharmacol 45(8):759–761

Phillips PA, Rolls BJ, Ledingham JG, Forsling ML, Morton JJ (1985) Osmotic thirst and vasopressin release in humans: a double-blind crossover study. Am J Physiol 248(6 Pt 2):R645–R650

Pinna G, Liskevych U, Doueiri MS, Costa E, Guidotti A (2003a) Antidepressants in doses that increase neurosteroid biosynthesis but fail to inhibit 5-HT reuptake reduce expression of aggression in socially isolated (SI) mice. Program No. 664.2. In 2003 Neuroscience Meeting Planner.

Pinna G, Costa E, Guidotti A (2006) Fluoxetine and norfluoxetine stereospecifically and selectively increase brain neurosteroid content at doses that are inactive on 5-HT reuptake. Psychopharmacology 186(3):362–372

Pinna G, Costa E, Guidotti A (2009) SSRIs act as selective brain steroidogenic stimulants (SBSSs) at low doses that are inactive on 5-HT reuptake. Curr Opin Pharmacol 9(1):24–30

Pinna G, Dong E, Matsumoto K, Costa E, Guidotti A (2003b) In socially isolated mice, the reversal of brain allopregnanolone down-regulation mediates the anti-aggressive action of fluoxetine. Proc Nat Acad Sci USA 100(4):2035–2040

Pitkow LJ, Sharer CA, Ren X, Insel TR, Terwilliger EF, Young LJ (2001) Facilitation of affiliation and pair-bond formation by vasopressin receptor gene transfer into the ventral forebrain of a monogamous vole. J Neurosci 21(18):7392–7396

Popa D, Léna C, Alexandre C, Adrien J (2008) Lasting syndrome of depression produced by reduction in serotonin uptake during postnatal development: evidence from sleep, stress, and behavior. J Neurosci 28(14):3546–3554

Popova NK, Naumenko VS, Plyusnina IZ (2007) Involvement of brain serotonin 5-HT1A receptors in genetic predisposition to aggressive behavior. Neurosci Behav Physiol 37(6):631–635

Potegal M, Ferris CF (1989) Intraspecific aggression in male hamsters is inhibited by intrahypothalamic vasopressin-receptor antagonist. Aggressive Behav 15(4):311–320

Puciłowski O, Płaźnik A, Kostowski W (1985) Aggressive behavior inhibition by serotonin and quipazine injected into the amygdala in the rat. Behav Neural Biol 43(1):58–68

Qu Y, Chang L, Klaff J, Seemann R, Rapoport SI (2003) Imaging brain phospholipase A2-mediated signal transduction in response to acute fluoxetine administration in unanesthetized rats. Neuropsychopharmacology 28(7):1219–1226

Quattrone A, Tedeschi G, Aguglia U, Scopacasa F, Direnzo GF, Annunziato L (1983) Prolactin secretion in man: a useful tool to evaluate the activity of drugs on central 5-hydroxytryptaminergic neurones. Studies with fenfluramine. Br J Clin Pharmacol 16(5):471–475

Raleigh MJ, Brammer GL, Yuwiler A, Flannery JW, McGuire MT, Geller E (1980) Serotonergic influences on the social behavior of vervet monkeys (Cercopithecus aethiops sabaeus). Exp Neurol 68(2):322–334

Ramboz S, Saudou F, Amara DA, Belzung C, Segu L, Misslin R, Buhot MC et al (1996) 5-HT1B receptor knock out—behavioral consequences. Behav Brain Res 73(1–2):305–312

Reist C, Nakamura K, Sagart E, Sokolski KN, Fujimoto KA (2003) Impulsive aggressive behavior: open-label treatment with citalopram. J Clin Psychiatry 59(1):676–680

Ribeiro EB, Bettiker RL, Bogdanov M, Wurtman RJ (1993) Effects of systemic nicotine on serotonin release in rat brain. Brain Res 621(2):311–318

Ricci LA, Grimes JM, Melloni RH (2004) Serotonin type 3 receptors modulate the aggression-stimulating effects of adolescent cocaine exposure in Syrian hamsters (Mesocricetus auratus). Behav Neurosci 118(5):1097–1110

Ricci LA, Knyshevski I, Melloni RH (2005) Serotonin type 3 receptors stimulate offensive aggression in Syrian hamsters. Behav Brain Res 156(1):19–29

Ricci LA, Melloni RH (2012) Repeated fluoxetine administration during adolescence stimulates aggressive behavior and alters serotonin and vasopressin neural development in hamsters. Behav Neurosci 126(5):640–653

Ricci LA, Rasakham K, Grimes JM, Melloni RH (2006) Serotonin-1A receptor activity and expression modulate adolescent anabolic/androgenic steroid-induced aggression in hamsters. Pharmacol Biochem Behav 85(1):1–11

Rilke O, Will K, Jähkel M, Oehler J (2001) Behavioral and neurochemical effects of anpirtoline and citalopram in isolated and group housed mice. Prog Neuropsychopharmacol Biol Psychiatry 25(5):1125–1144

Rittig S, Robertson GL, Siggaard C, Kovács L, Gregersen N, Nyborg J, Pedersen EB (1996) Identification of 13 new mutations in the vasopressin-neurophysin II gene in 17 kindreds with familial autosomal dominant neurohypophyseal diabetes insipidus. Am J Hum Genet 58(1):107–117

Roliński Z, Herbut M (1981) The role of the serotonergic system in foot shock-induced behavior in mice. Psychopharmacology 73(3):246–251
Rood BD, De Vries GJ (2011) Vasopressin innervation of the mouse (Mus musculus) brain and spinal cord. J Comp Neurol 519(12):2434–2474
Rosa M, Franzini A, Giannicola G, Messina G, Altamura AC, Priori A (2012) Hypothalamic oscillations in human pathological aggressiveness. Biol Psychiatry 72(12):e33–e35
Roth BL, Choudhary MS, Khan N, Uluer AZ (1997) High-affinity agonist binding is not sufficient for agonist efficacy at 5-hydroxytryptamine2A receptors: evidence in favor of a modified ternary complex model. J Pharmacol Exp Ther 280(2):576–583
Roy A, Adinoff B, Linnoila M (1988) Acting out hostility in normal volunteers: negative correlation with levels of 5HIAA in cerebrospinal fluid. Psychiatry Res 24(2):187–194
Sabol KE, Richards JB, Seiden LS (1992) Fluoxetine attenuates the DL-fenfluramine-induced increase in extracellular serotonin as measured by in vivo dialysis. Brain Res 585(1–2): 421–424
Saito M, Sugimoto T, Tahara A, Kawashima H (1995) Molecular cloning and characterization of rat V1b vasopressin receptor: evidence for its expression in extra-pituitary tissues. Biochem Biophys Res Commun 212(3):751–757
Salomé N, Stemmelin J, Cohen C, Griebel G (2006) Differential roles of amygdaloid nuclei in the anxiolytic- and antidepressant-like effects of the V1b receptor antagonist, SSR149415, in rats. Psychopharmacology 187(2):237–244
Sammler EM, Foley PL, Lauder GD, Wilson SJ, Goudie AR, O'Riordan JI (2010) A harmless high? Lancet 376(9742):742
Sánchez C, Arnt J, Hyttel J, Moltzen EK (1993) The role of serotonergic mechanisms in inhibition of isolation-induced aggression in male mice. Psychopharmacology 110(1–2): 53–59
Sánchez C, Hyttel J (1994) Isolation-induced aggression in mice: effects of 5-hydroxytryptamine uptake inhibitors and involvement of postsynaptic 5-HT1A receptors. Eur J Pharmacol 264(3):241–247
Saudou F, Amara DA, Dierich A, LeMeur M, Ramboz S, Segu L, Buhot MC et al (1994) Enhanced aggressive behavior in mice lacking 5-HT1B receptor. Science 265(5180): 1875–1878
Sawchenko PE, Swanson LW, Steinbusch HW, Verhofstad AA (1983) The distribution and cells of origin of serotonergic inputs to the paraventricular and supraoptic nuclei of the rat. Brain Res 277(2):355–360
Schoenfeld T, Leonard C (1985) Behavioral development in the syrian golden hamster. In: Siegel H (ed) The hamster: reproduction and behavior. Plenum Press, New York, pp 289–321
Schwartzer JJ, Ricci LA, Melloni RH (2009) Adolescent anabolic-androgenic steroid exposure alters lateral anterior hypothalamic serotonin-2A receptors in aggressive male hamsters. Behav Brain Res 199(2):257–262
Scordalakes EM, Rissman EF (2004) Aggression and arginine vasopressin immunoreactivity regulation by androgen receptor and estrogen receptor alpha. Genes Brain Behav 3(1):20–26
Semsar K, Kandel FL, Godwin J (2001) Manipulations of the AVT system shift social status and related courtship and aggressive behavior in the bluehead wrasse. Horm Behav 40(1):21–31
Share L (1988) Role of vasopressin in cardiovascular regulation. Physiol Rev 68(4):1248–1284
Sheard MH, Marini JL, Bridges CI, Wagner E (1976) The effect of lithium on impulsive aggressive behavior in man. Am J Psychiatry 133(12):1409–1413
Shimizu H, Bray GA (1989) Effects of neuropeptide Y on norepinephrine and serotonin metabolism in rat hypothalamus in vivo. Brain Res Bull 22(6):945–950
Shoaf SE, Carson R, Hommer D, Williams W, Higley JD, Schmall B, Herscovitch P et al (1998) Brain serotonin synthesis rates in rhesus monkeys determined by [11C]alpha-methyl-L-tryptophan and positron emission tomography compared to CSF 5-hydroxyindole-3-acetic acid concentrations. Neuropsychopharmacol: Off Publ Am Coll Neuropsychopharmacol 19(5):345–353

Sijbesma H, Schipper J, De Kloet ER, Mos J, Van Aken H, Olivier B (1991) Postsynaptic 5-HT1 receptors and offensive aggression in rats: a combined behavioural and autoradiographic study with eltoprazine. Pharmacol Biochem Behav 38(2):447–458

Soloff PH, Kelly TM, Strotmeyer SJ, Malone KM, Mann JJ (2003) Impulsivity, gender, and response to fenfluramine challenge in borderline personality disorder. Psychiatry Res 119(1–2):11–24

Spanos LJ, Yamamoto BK (1989) Acute and subchronic effects of methylenedioxymethamphetamine [(±)MDMA] on locomotion and serotonin syndrome behavior in the rat. Pharmacol Biochem Behav 32(4):835–840

Sperry TS, Moore IT, Meddle SL, Benowitz-Fredericks ZM, Wingfield JC (2005) Increased sensitivity of the serotonergic system during the breeding season in free-living American tree sparrows. Behav Brain Res 157(1):119–126

Sperry TS, Thompson CK, Wingfield JC (2003) Effects of acute treatment with 8-OH-DPAT and fluoxetine on aggressive behaviour in male song sparrows (Melospiza melodia morphna). J Neuroendocrinol 15(2):150–160

Spiller HA, Ryan ML, Weston RG, Jansen J (2011) Clinical experience with and analytical confirmation of "bath salts" and "legal highs" (synthetic cathinones) in the United States. Clin Toxicol 49(6):499–505

Sprouse J, Braselton J, Reynolds L, Clarke T, Rollema H (2001) Activation of postsynaptic 5-HT(1A) receptors by fluoxetine despite the loss of firing-dependent serotonergic input: electrophysiological and neurochemical studies. Synapse 41(1):49–57

Sprouse JS, Aghajanian GK (1987) Electrophysiological responses of serotoninergic dorsal raphe neurons to 5-HT1A and 5-HT1B agonists. Synapse 1(1):3–9

Steensland P, Hallberg M, Kindlundh A, Fahlke C, Nyberg F (2005) Amphetamine-induced aggression is enhanced in rats pre-treated with the anabolic androgenic steroid nandrolone decanoate. Steroids 70(3):199–204

Stemmelin J, Lukovic L, Salome N, Griebel G (2005) Evidence that the lateral septum is involved in the antidepressant-like effects of the vasopressin V1b receptor antagonist, SSR149415. Neuropsychopharmacology 30(1):35–42

Stribley JM, Carter CS (1999) Developmental exposure to vasopressin increases aggression in adult prairie voles. Proc Nat Acad Sci USA 96(22):12601–12604

Summers CH, Korzan WJ, Lukkes JL, Watt MJ, Forster GL, Øverli Ø, Höglund E et al (2005) Does serotonin influence aggression? Comparing regional activity before and during social interaction. Physiol Biochem Zool PBZ 78(5):679–694

Swanson LW, Kuypers HG (1980) The paraventricular nucleus of the hypothalamus: cytoarchitectonic subdivisions and organization of projections to the pituitary, dorsal vagal complex, and spinal cord as demonstrated by retrograde fluorescence double-labeling methods. J Comp Neurol 194(3):555–570

Taravosh-Lahn K, Bastida C, Delville Y (2006) Differential responsiveness to fluoxetine during puberty. Behav Neurosci 120(5):1084–1092

Thurmond JB, Lasley SM, Kramarcy NR, Brown JW (1979) Differential tolerance to dietary amino acid-induced changes in aggressive behavior and locomotor activity in mice. Psychopharmacology 66(3):301–308

Treiser SL, Cascio CS, O'Donohue TL, Thoa NB, Jacobowitz DM, Kellar KJ (1981) Lithium increases serotonin release and decreases serotonin receptors in the hippocampus. Science 213(4515):1529–1531

Tribollet E, Barberis C, Jard S, Dubois-Dauphin M, Dreifuss JJ (1988) Localization and pharmacological characterization of high affinity binding sites for vasopressin and oxytocin in the rat brain by light microscopic autoradiography. Brain Res 442(1):105–118

Tupin JP, Smith DB, Clanon TL, Kim LI, Nugent A, Groupe A (1973) The long-term use of lithium in aggressive prisoners. Compr Psychiatry 14(4):311–317

Vaccari C, Lolait SJ, Ostrowski NL (1998) Comparative distribution of vasopressin V1b and oxytocin receptor messenger ribonucleic acids in brain. Endocrinology 139(12):5015–5033

Vacher C-M, Calas A, Maltonti F, Hardin-Pouzet H (2004) Postnatal regulation by monoamines of vasopressin expression in the neuroendocrine hypothalamus of MAO-A-deficient mice. Eur J Neurosci 19(4):1110–1114

Vacher C-M, Frétier P, Créminon C, Calas A, Hardin-Pouzet H (2002) Activation by serotonin and noradrenaline of vasopressin and oxytocin expression in the mouse paraventricular and supraoptic nuclei. J Neurosci 22(5):1513–1522

Van Der Vegt BJ, Lieuwes N, Van De Wall EHEM, Kato K, Moya-Albiol L, Martínez-Sanchis S, De Boer SF et al (2003) Activation of serotonergic neurotransmission during the performance of aggressive behavior in rats. Behav Neurosci 117(4):667–674

Van Erp AM, Miczek KA (2000) Aggressive behavior, increased accumbal dopamine, and decreased cortical serotonin in rats. J Neurosci: Off J Soc Neurosci 20(24):9320–9325

Van Putten T, Sanders DG (1975) Lithium in treatment failures. J Nerv Ment Dis 161(4):255–264

Veenema AH, Beiderbeck DI, Lukas M, Neumann ID (2010) Distinct correlations of vasopressin release within the lateral septum and the bed nucleus of the stria terminalis with the display of intermale aggression. Horm Behav 58(2):273–281

Veiga CP, Miczek KA, Lucion AB, De Almeida RMM (2007) Effect of 5-HT1B receptor agonists injected into the prefrontal cortex on maternal aggression in rats. Braz J Med Biol Res 40(6):825–830

Vergnes M, Depaulis A, Boehrer A (1986) Parachlorophenylalanine-induced serotonin depletion increases offensive but not defensive aggression in male rats. Physiol Behav 36(4):653–658

Vergnes M, Depaulis A, Boehrer A, Kempf E (1988) Selective increase of offensive behavior in the rat following intrahypothalamic 5,7-DHT-induced serotonin depletion. Behav Brain Res 29(1–2):85–91

Vergnes M, Kempf E (1982) Effect of hypothalamic injections of 5,7-dihydroxytryptamine on elicitation of mouse-killing in rats. Behav Brain Res 5(4):387–397

Vertes RP (1991) A PHA-L analysis of ascending projections of the dorsal raphe nucleus in the rat. J Comp Neurol 313(4):643–668

Villalba C, Boyle PA, Caliguri EJ, De Vries GJ (1997) Effects of the selective serotonin reuptake inhibitor fluoxetine on social behaviors in male and female prairie voles (Microtus ochrogaster). Horm Behav 32(3):184–191

Vincent J-L, Su F (2008) Physiology and pathophysiology of the vasopressinergic system. Best Pract Res Clin Anaesthesiol 22(2):243–252

Vomachka AJ, Ruppert PH, Clemens LG, Greenwald GS (1981) Adult sexual behavior deficits and altered hormone levels in male hamsters given steroids during development. Physiol Behav 26(3):461–466

Wagner GC, Fisher H, Pole N, Borve T, Johnson SK (1993) Effects of monoaminergic agonists on alcohol-induced increases in mouse aggression. J Stud Alcohol Suppl 11:185–191

Wang Z, Bullock NA, De Vries GJ (1993) Sexual differentiation of vasopressin projections of the bed nucleus of the stria terminals and medial amygdaloid nucleus in rats. Endocrinology 132(6):2299–2306

Wersinger SR, Caldwell HK, Christiansen M, Young WS (2007a) Disruption of the vasopressin 1b receptor gene impairs the attack component of aggressive behavior in mice. Genes Brain Behav 6(7):653–660

Wersinger SR, Caldwell HK, Martinez L, Gold P, Hu S-B, Young WS (2007b) Vasopressin 1a receptor knockout mice have a subtle olfactory deficit but normal aggression. Genes Brain Behav 6(6):540–551

Wersinger SR, Ginns EI, O'Carroll A-M, Lolait SJ, Young WS (2002) Vasopressin V1b receptor knockout reduces aggressive behavior in male mice. Mol Psychiatry 7(9):975–984

Whale R, Quested DJ, Laver D, Harrison PJ, Cowen PJ (2000) Serotonin transporter (5-HTT) promoter genotype may influence the prolactin response to clomipramine. Psychopharmacology 150(1):120–122

Whitsett JM (1975) The development of aggressive and marking behavior in intact and castrated male hamsters. Horm Behav 6(1):47–57

Willoughby JO, Blessing WW (1987) Neuropeptide Y injected into the supraoptic nucleus causes secretion of vasopressin in the unanesthetized rat. Neurosci Lett 75(1):17–22

Winslow JT, Hastings N, Carter CS, Harbaugh CR, Insel TR (1993) A role for central vasopressin in pair bonding in monogamous prairie voles. Nature 365(6446):545–548

Wood DM, Davies S, Greene SL, Button J, Holt DW, Ramsey J, Dargan PI (2010) Case series of individuals with analytically confirmed acute mephedrone toxicity. Clin Toxicol 48(9): 324–327

Wright DE, Seroogy KB, Lundgren KH, Davis BM, Jennes L (1995) Comparative localization of serotonin1A, 1C, and 2 receptor subtype mRNAs in rat brain. J Comp Neurol 351(3):357–373

Yatham LN, Steiner M (1993) Neuroendocrine probes of serotonergic function: a critical review. Life Sci 53(6):447–463

Young LJ, Nilsen R, Waymire KG, MacGregor GR, Insel TR (1999) Increased affiliative response to vasopressin in mice expressing the V1a receptor from a monogamous vole. Nature 400(6746):766–768

Young WS, Li J, Wersinger SR, Palkovits M (2006) The vasopressin 1b receptor is prominent in the hippocampal area CA2 where it is unaffected by restraint stress or adrenalectomy. Neuroscience 143(4):1031–1039

Zai CC, Muir KE, Nowrouzi B, Shaikh SA, Choi E, Berall L, Trépanier M-O et al (2012) Possible genetic association between vasopressin receptor 1B and child aggression. Psychiatry Res 200(2–3):784–788

Zhang L, Hernández VS (2013) Synaptic innervation to rat hippocampus by vasopressin-immuno-positive fibres from the hypothalamic supraoptic and paraventricular nuclei. Neuroscience 228:139–162

Zhang L, Ma W, Barker JL, Rubinow DR (1999) Sex differences in expression of serotonin receptors (subtypes 1A and 2A) in rat brain: a possible role of testosterone. Neuroscience 94(1):251–259

Zimmermann M, Grabemann M, Mette C, Abdel-Hamid M, Uekermann J, Ueckermann J, Kraemer M et al (2012) The effects of acute tryptophan depletion on reactive aggression in adults with attention-deficit/hyperactivity disorder (ADHD) and healthy controls. PLoS ONE 7(3):e32023

Sex, Drugs, and Violence: Neuromodulation of Attachment and Conflict in Voles

Kyle L. Gobrogge

Abstract Prairie voles (*Microtus ochrogaster*) are a rodent species that display socially monogamous pair-bonds, a behavior illustrated by several types of social interactions such as mating-induced partner preference, selective aggression toward conspecific strangers, and bi-parental care. Therefore, this species has provided an excellent opportunity for the study of pair-bonding and its underlying neurochemical mechanisms. This chapter discusses the utility of this unique rodent in the study of attachment and conflict, and reviews recent findings illustrating the neuromodulatory mechanisms underlying mating-induced partner preference and aggression. Finally, implications of research using this animal model in human mental health are also discussed.

Keywords Addiction · Anterior hypothalamus · Dopamine · Drug abuse · Mating · Medial amygdala · Neural circuitry · Nucleus accumbens · Oxytocin · Social attachment · Vasopressin · Ventral tegmental area

Contents

K. L. Gobrogge (✉)
Department of Neurobiology, Harvard Medical School, 220 Longwood Avenue, Boston, MA 02115, USA
e-mail: kyle_gobrogge@hms.harvard.edu

Curr Topics Behav Neurosci (2014) 17: 229–264
DOI: 10.1007/7854_2013_264

Published Online: 4 December 2013

1 Introduction

Traditionally, the field of behavioral neuroscience has primarily used common laboratory rodents as animal model systems. Data from these studies provide valuable information for our better understanding of the neurobiological mechanisms underlying classic social behaviors including aggression, mating, and maternal care. However, previous research clearly demonstrates species differences in animals' social behaviors and associated neural mechanisms, and therefore caution should be exercised when data are extrapolated from one species to another. Furthermore, most commercially available laboratory animals do not readily display certain types of social behaviors and thus are not appropriate for some investigations. For example, laboratory rats and mice do not exhibit strong social bonds between mates, and males typically do not display paternal behavior or female-directed aggression.

Healthy social relationships are necessary for maintaining human mental health, yet we know very little about the brain microcircuitry involved in the formation and maintenance of pair-bonds. Many animal models have been used for investigation of the neurobiological mechanisms regulating the cognitive processes leading to the development of social behavior. However, social attachments require a complex suite of behaviors involving a number of cognitive functions including sensory processing, individual recognition, learning and memory, and motor output. Therefore, although the neurobiology of more traditional reproductive behaviors like courtship and aggression has been studied in a variety of animal species, we know far less about how multiple brain regions and neurochemical systems are interconnected to enable animals to form social memories and display pair-bonds.

Because mating naturally induces partner preference, aggression, and biparental care in the socially monogamous prairie vole (*Microtus ochrogaster*), this species represents a unique animal model for the study of neural mechanisms underlying social behavior associated with a monogamous life strategy (Carter et al. 1995; Insel et al. 1998; Wang and Aragona 2004; Young and Wang 2004; Young et al. 1998, 2008, 2011a). The following chapter begins by describing the prairie vole model and reviewing neuronal correlates, and circuitry underlying pair-bonding behavior focusing on the regulation of aggression by the neuropeptides arginine vasopressin (AVP) and oxytocin (OT), and the neurotransmitter dopamine (DA). The review ends by discussing the translational potential of pair-bonding studies in prairie voles for human research investigating the etiology of violence, with a particular emphasis on the neural underpinnings of drug-induced aggression.

2 Prairie Vole Model

Voles are a group of microtine rodents that are genetically and taxonomically similar, yet show remarkable differences in their life strategies and social behavior (Fig. 1; Young and Wang 2004; Young et al. 2008, 2011a). These animals, therefore, have provided excellent opportunities for comparative studies examining their social behaviors associated with different life strategies. For example, prairie (*Microtus ochrogaster*) and pine (*Microtus pinetorum)* voles are highly social and monogamous whereas meadow (*Microtus pennsylvanicus*) and montane (*Microtus montanus*) voles are asocial and promiscuous (Fig. 1; Jannett 1982; Dewsbury 1987; Insel and Hulihan 1995). In the laboratory prairie and pine voles are bi-parental, with both parents participating equally in caring for their young (Fig. 1). Meadow and montane voles are primarily maternal and males do not stay in the maternal nest (Fig. 1; McGuire and Novak 1984, 1986; Oliveras and Novak 1986). Following mating, prairie voles develop pair-bonds between mates, and males even display aggression toward conspecific strangers—behaviors that are not exhibited by promiscuous meadow or montane voles (Fig. 1; Insel et al. 1995; Lim et al. 2004). Interestingly, these vole species do not differ in their nonsocial behaviors (Tamarin 1985), further indicating associations between species-specific behavior and life strategy.

3 Pair-Bonding Behavior

Prairie voles have provided excellent opportunities for studies of bi-parental care and pair-bonding (Wang and Aragona 2004; Young and Wang 2004). However, this chapter will focus primarily on mating-induced partner preference and aggression. Field data indicate that the same male and female prairie vole pairs are often found together, throughout life, even after migrating to different locations (Getz et al. 1981). In the laboratory, pair-bonding can be reliably measured by two behavioral indices, classified as partner preference formation and selective aggression (Fig. 1; Getz and Hofmann 1986; Dewsbury, 1987; Carter and Getz 1993).

An animal's partner preference is tested in a three-chamber apparatus. Chambers are connected by plastic hollow tubes with photo beams mounted on top linked to a computer automatically recording the number of cage entries and time spent in each cage. During the 3-h test, a familiar partner, an animal that has social and/or sexual experience with the subject, and a conspecific stranger, an animal that has never encountered the subject, are tethered in their separate cages while the subject can move freely throughout the apparatus. A partner preference is defined as experimental animals spending significantly more time in side-by-side contact with their familiar partner versus a stranger (Figs. 2a and 4). It has been shown that 24 h of social cohabitation with ad libitum mating reliably induces partner preference formation in both male and female prairie voles (Figs. 2a and 4;

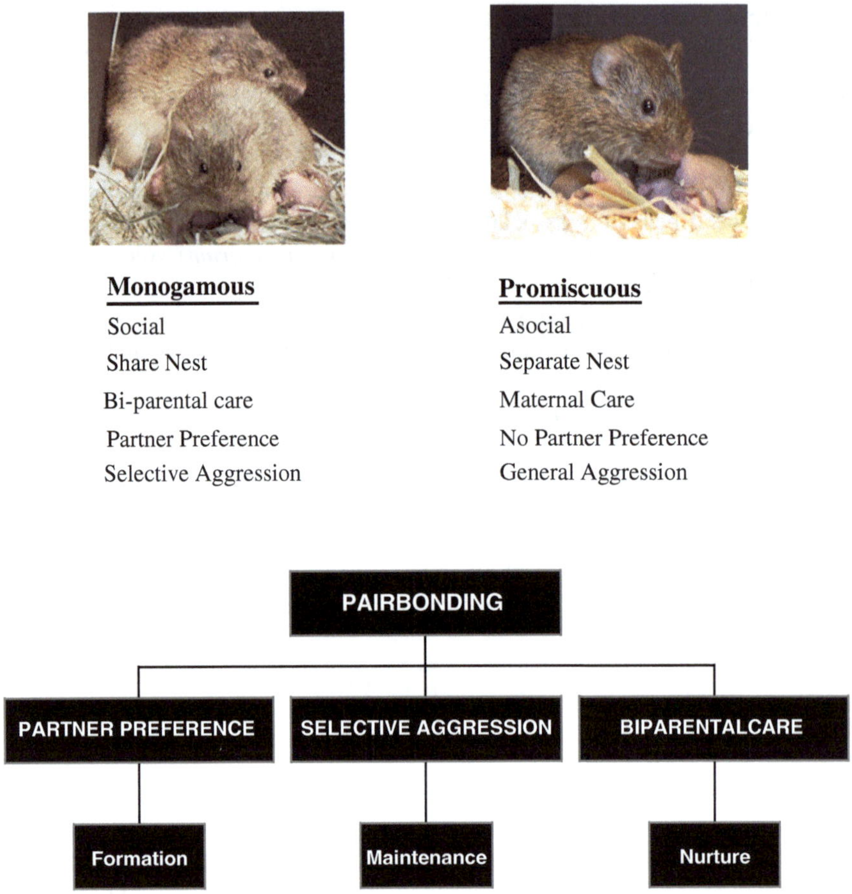

Fig. 1 Pair-Bonding Model. *Left* photomicrograph illustrates a pair-bonded male and female prairie vole protecting and huddling over their offspring. *Right* photomicrograph depicts a female meadow vole huddling over her offspring. A list of behaviors is provided to highlight species differences in social and mating strategies between prairie and meadow voles. The functional significance of pair-bonding is subdivided by behavioral phenotype and function

Winslow et al. 1993; Insel et al. 1995). It has also been shown that partner preference, once formed, is enduring. In male prairie voles, for example, partner preference persists 2 weeks after mating even in the absence of continued exposure to a familiar partner (Figs. 2b and 3d; Insel et al. 1995). Therefore, a 24-h mating paradigm has been used to examine if manipulation of central systems, such as blockade of a particular neurotransmitter receptor, can block mating-induced pair-bonding. Finally, in both male and female prairie voles, 6 h of social cohabitation without mating does not induce partner preference formation (Figs. 2a and 4a; Insel et al. 1995). This behavioral paradigm has been employed to examine if activation of a central system is sufficient to induce pair-bonding.

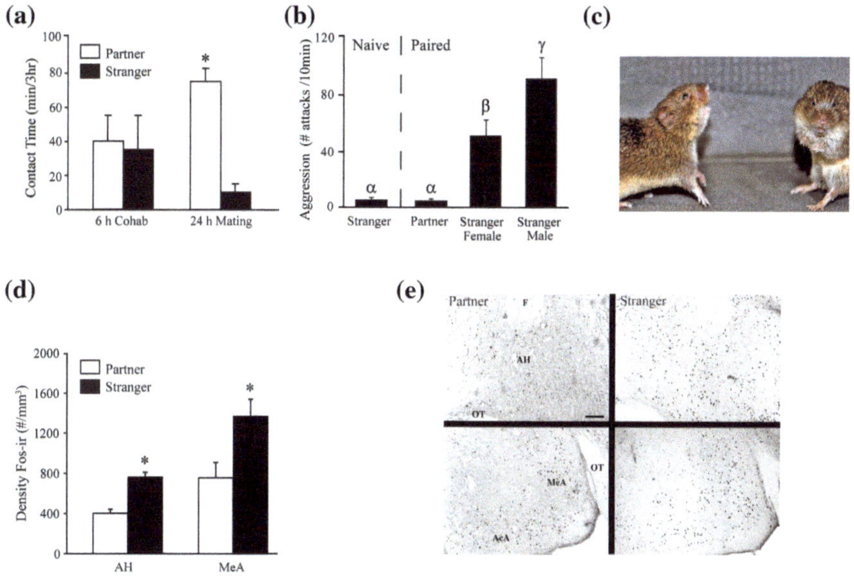

Fig. 2 Neuronal Activation Associated with Mating-Induced Aggression. **a** After 24, but not 6, hours of mating—male and female prairie voles display significantly more time in side-by-side physical contact with an opposite sex partner than with a stranger. **a** Sexually naïve male prairie voles (Naïve) do not display aggression toward a stranger female, whereas 2 weeks of sexual and social experience (Paired) induces selective aggression toward both male and female strangers but not toward familiar female partners. **c** Photomicrograph depicts a pair-bonded male prairie vole (*left*) preparing to attack a—sexually receptive—stranger female prairie vole (*right*). **d** Stereological estimates reveal a significantly higher density of Fos-ir cells in the anterior hypothalamus (AH) and medial amygdala (MeA) of pair-bonded males displaying aggression toward a stranger female (Stranger) compared to males displaying affiliation toward their familiar female partner (Partner). **e** Photomicrographs showing Fos-ir (dark nuclear staining) in the AH and MeA from pair-bonded males re exposed to their familiar female partner (Partner) or to a stranger female (Stranger). *F* fornix, *OT* optic tract. Bars indicate mean ± standard error of the mean. Bars with different Greek letters differ significantly from each other. $*p < 0.05$. Scale bar = 100 μm. Adapted from (Winslow et al. 1993; Wang et al. 1997a; Gobrogge et al. 2007; Aragona and Wang 2009; Gobrogge and Wang 2009, 2011; Young et al. 2011a)

Another behavioral index of pair-bonding is selective aggression, which is more prominent in male than in female prairie voles. It has been suggested that selective aggression is a behavioral trait associated with mate guarding that is important for the maintenance of monogamous pair-bonds (Carter et al. 1995). Experimental voles are tested for offensive aggression using the resident intruder paradigm, a well-characterized and ethologically valid model of offensive aggression (Miczek 1979). Briefly, an intruder of similar size and weight is introduced into the home cage of the experimental vole (resident) and their behavioral interactions are videotaped for 10 min (Winslow et al. 1993; Wang et al. 1997a; Aragona et al. 2006; Gobrogge et al. 2007, 2009). The resident is scored for specific and targeted attacks (Fig. 2b) including the frequency of lunges, bites, and chases as well as the

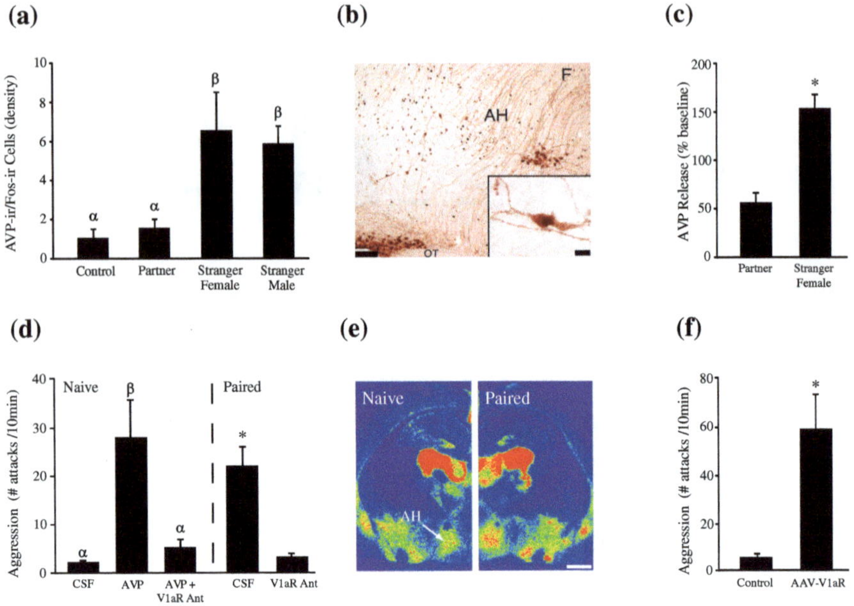

Fig. 3 Anterior hypothalamic vasopressin regulates pair-bonding-induced aggression. **a** Pair-bonded male prairie voles that display aggression toward female or male strangers exhibit a significantly higher density of AVP-ir/Fos-ir double-labeled cells in the AH relative to pair-bonded males re-exposed to their female partner or to males not exposed to any social stimulus (Control). **b** Photomicrograph of AVP-ir cell bodies and fibers (brown cytoplasmic staining), Fos-ir (dark nuclear staining), or AVP-ir/Fos-ir double-labeled cells (insert) in the anterior hypothalamus (AH). **c** In-vivo brain microdialysis reveals a significant increase in AH-AVP release in pair-bonded male prairie voles displaying aggression toward a stranger female compared to males re-exposed to their female partner. **d** Intra-AH AVP microinfusion, in sexually naïve males (Naïve), is sufficient to induce aggression toward a stranger female compared to control males infused with cerebral spinal fluid (CSF). AH-AVP-induced aggression is blocked with concurrent administration of AVP with a V1aR antagonist (V1aR Ant). Pair-bonded males (Paired) display aggression toward stranger females, which is blocked by intra-AH infusion of a V1aR Ant. **e** Pair-bonded males (Paired) exhibit a significantly higher density of AVP-V1a receptor (V1aR) binding in the AH relative to sexually naïve males (Naïve). **f** Sexually naïve males infused with an adeno-associated virus expressing the V1aR gene (AAV-V1aR) in the AH exhibit enhanced aggression toward stranger females relative to males infused with the Lac-Z gene (Control). *F* fornix, *OT* optic tract. Bars indicate mean ± standard error of the mean. Bars with different Greek letters differ significantly from each other. *$p < 0.05$. Scale bars = 100 μm, insert scale bar = 10 μm. Adapted from (Aragona et al. 2006; Gobrogge et al. 2007, 2009; Gobrogge and Wang 2011; Young et al. 2011a)

duration of affiliative side-by-side physical contact (Winslow et al. 1993) and time spent during anogenital investigation with a stranger male/female or partner (Gobrogge et al. 2007, 2009).

Selective aggression is associated with mating, as cohabitation in the absence of mating does not induce this behavior in male prairie voles (Fig. 2b; Winslow et al. 1993; Insel et al. 1995; Wang et al. 1997a). Selective aggression is also enduring

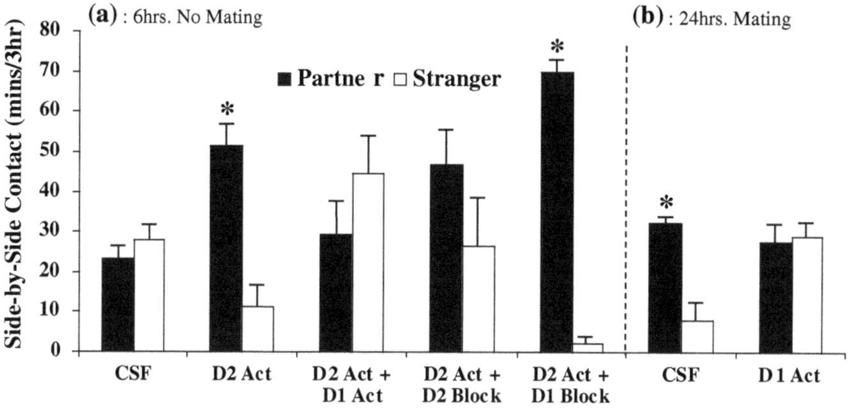

Fig. 4 Nucleus accumbens dopamine mediates partner preference formation. Bar graphs represent social affiliation behavior, across mating paradigms, of male prairie voles as a function of rostral nucleus accumbens shell (NAcc) dopamine pharmacology. **a** 6 h of nonsexual cohabitation has no effect on partner preference formation, which is not affected by intra-NAcc cerebral spinal fluid (CSF) injection. Dopamine D2 receptor activation (Act), in the absence of mating, induces partner preference formation which can be blocked (Block) by co-administration of a D1 agonist or D2 antagonist. A graded increase in the duration of side-by-side physical contact, with a familiar partner, is demonstrated by concurrent injection of a D2 agonist with D1 antagonist. **b** 24 h of mating and social cohabitation is sufficient to induce partner preference formation, which is not effected by intra-NAcc CSF microinfusion. Finally, dopamine D1 receptor activation (Act) is sufficient to block partner preference formation induced by 24 h of social and sexual contact. Bars indicate mean ± standard error of the mean. $*p < 0.05$. Adapted from (Aragona et al. 2006; Gobrogge et al. 2007, 2009; Gobrogge and Wang 2011; Young et al. 2011a)

(Insel et al. 1995; Aragona et al. 2006; Gobrogge et al. 2007, 2009) and lasts for at least 2 weeks after partner preference formation (Gobrogge et al. 2007, 2009) even in the absence of continuous exposure to a partner (Figs. 2b and 3d; Insel et al. 1995). Importantly, males display aggression not only toward conspecific males (Winslow et al. 1993; Insel et al. 1995; Wang et al. 1997a; Aragona et al. 2006; Gobrogge et al. 2007) but also toward sexually receptive females (Figs. 2b, c, 3d and 5f; Aragona et al. 2006; Gobrogge et al. 2007, 2009) and, thus, this selective aggression functions to maintain monogamous pair-bonds—as males reject potential female mates (Winslow et al. 1993; Wang et al. 1997a; Aragona et al. 2006; Gobrogge et al. 2007, 2009). Together, these reliably expressed and measurable agonistic behaviors make the prairie vole an excellent model for investigation of the neural mechanisms underlying naturally occurring aggression associated with monogamy. It should be mentioned that although female aggression has been less studied in voles, female prairie voles exhibit similar aggressive behavior as males and this behavior is influenced by a female's social and sexual experience (Bowler et al. 2002; Resendez et al. 2012).

4 Neuronal Correlates

There is a significant amount of ambiquity regarding which brain areas may be involved in the regulation of selective aggression, since overlapping brain areas are involved in several forms of social behavior (Newman 1999). Using a neuronal activation marker of an immediate early gene, *c-fos*, previous studies in voles have examined neuronal activity associated with aggression (Wang et al. 1997a), maternal (Katz et al. 1999) and paternal behavior (Kirkpatrick et al. 1994), mating (Curtis and Wang 2003; Lim and Young 2004), anxiety (Stowe et al. 2005), spatial learning (Kuptsov et al. 2005), chemosensory processing (Tubbiola and Wysocki 1997; Hairston et al. 2003), social experience (Cushing et al. 2003a; Kramer et al. 2006), or pharmacological challenges (Gingrich et al. 1997; Cushing et al. 2003b; Curtis and Wang 2005b). Within these studies, however, typically only one type of behavior was investigated. Little focus was aimed at examining other forms of social behavior, including affiliation or general social olfactory processing. Consequently, there is considerable overlap in brain-behavior relationships among these studies leading to ambiguity as to which brain areas regulate selective aggression. Nevertheless, in an early study, male prairie voles displayed aggression toward a male intruder following 24 h of mating, but not following 24 h of cohabitation with a female without mating (Wang et al. 1997a). However, despite their differences in sociosexual experience and in aggressive behavior, both types of male exposure led to equal levels of Fos-ir (immunoreactivity) expression in some brain areas, such as the bed nucleus of the stria terminalis (BNST). Males that mated for 24 h and displayed high levels of aggression toward either a male or a female intruder showed increased levels of Fos-ir expression in the medial amygdala (MeA), compared to males that cohabitated with a female without mating, implicating the MeA as a brain area associated with the display of mating-induced selective aggression (Figs. 2d and e).

In a more recent study, several brain areas including the BNST, medial preoptic area (MPOA), paraventricular nucleus (PVN), and lateral septum (LS) showed higher levels of Fos expression in pair-bonded males that had experienced an RIT compared to pair-bonded males not exposed to a social intruder (Gobrogge et al. 2007). However, no group differences in Fos expression across these brain areas were found among males that were exposed to different social stimuli or that displayed different patterns of social behavior, including aggression or affiliation toward intruders. These data suggest that the increased neuronal activation in these brain regions is probably a nonselective response to olfactory stimulation or general arousal associated with exposure to a conspecific. A unique pattern of Fos expression was found in the anterior hypothalamus (AH), in which exposure to a conspecific stranger, either male or female, induced a significant increase in AH-Fos compared to re-exposure to a familiar partner (Figs. 2d and e; Gobrogge et al. 2007). This increase in Fos staining may indicate a stimulus-specific response. The AH appears to be more responsive to chemosensory, tactile, and/or visual cues associated with conspecific strangers, but not familiar partners (Figs. 2d and e;

Gobrogge et al. 2007). These data indicate that the increased neuronal activation in the AH may be involved in aggressive behavior displayed by pair-bonded male prairie voles (Gobrogge et al. 2007). This notion is corroborated by previous research documenting a critical role of the hypothalamus in regulating aggression across several mammalian species. For example, the ventromedial hypothalamus (VMH) in mice (Choi et al. 2005; Lin et al. 2011) and AH in rats (Veening et al. 2005) is responsive to conspecific chemosensory cues, which elicit aggressive behavior. Electrical stimulation applied directly to the AH induces attack toward conspecifics in rats (Kruk 1991) and other animals (Albert and Walsh 1984; Siegel et al. 1999). Interestingly, in humans, surgical lesioning of the AH reduces physical violence (Sano et al. 1966; Ramamurthi 1988). In summary, data from vole studies demonstrate that activation of the MeA and AH is associated with the display of selective aggression (Wang et al. 1997a; Gobrogge et al. 2007).

5 Neurocircuitry

To dissect the neural circuitry underlying selective aggression, a series of monosynaptic tract-tracing experiments focusing on the AH and MeA were conducted. Intra-AH injections of an anterograde tracer, biotinylated dextran amine (BDA), resulted in BDA-ir staining in several brain regions. The AH projected to areas involved in processing chemosensory cues including the MeA; areas important for regulating social behavior including the LS, BNST, MPOA, ventromedial hypothalamus (VMH), and dorsal raphe (DR); and areas coordinating motor output, such as the periaqueductal gray (Gobrogge and Wang 2009). The AH also projected to brain areas implicated in evaluating incentive salience including the medial prefrontal cortex (mPFC), nucleus accumbens (NAcc), and ventral pallidum (VP), as well as areas involved in memory formation and consolidation including hippocampal regions CA3 and the dentate gyrus (Gobrogge and Wang 2009). Furthermore, site-specific injections of a retrograde tracer, fluorogold (FG), into the LS, NAcc, or MeA resulted in FG-ir staining in the AH, indicating reciprocal connections between the AH and regions involved in motivation and chemosensory communication (Gobrogge and Wang 2009).

Fos-ir staining was also used to assess how neuronal activation in this circuit was affected by the display of pair-bonding behavior. Males displaying aggression toward a stranger female showed a significantly higher density of Fos-ir in the AH and MeA relative to males displaying affiliation toward their female partner (Figs. 2d and e), replicating our previous findings (Wang et al. 1997a; Gobrogge et al. 2007). Interestingly, we identified a MeA-AH-LS circuit that was activated when males were displaying aggression and a DR-AH circuit that was recruited when males were displaying affiliation (Gobrogge and Wang 2009). The identification of these two neural circuits indicates a specific neuronal framework associated with the choice between courtship (DR-AH circuit) and aggression (MeA-AH-LS circuit) in pair-bonded male prairie voles.

6 Neuromodulation

Previous work has primarily focused on partner preference formation and documented a growing list of neurochemicals including oxytocin (OT), vasopressin (AVP), corticotropin releasing hormone, dopamine (DA), GABA, and glutamate as well as their interactions in the regulation of pair-bonding behavior (Williams et al. 1992, 1994; Winslow et al. 1993; Carter et al. 1995; DeVries et al. 1995; Wang et al. 1998, 1999; Gingrich et al. 2000; Liu et al. 2001; Aragona et al. 2003; Liu and Wang 2003; Lim et al. 2004; Lim and Young 2004; Curtis and Wang 2005a, b; Smeltzer et al. 2006). Importantly, data from several studies indicate a select subset of neurochemicals in the regulation of selective aggression.

6.1 Vasopressin/Oxytocin

Comparative approaches have been utilized in studies examining neuroendocrine mechanisms regulating pair-bonding behavior in social versus asocial vole species (Insel 2010). For example, studies have focused on examining the central AVP system, which has diverse forebrain projections, across monogamous and promiscuous vole species. AVP, a nine amino-acid neuropeptide, is an antidiuretic hormone and has been shown to stimulate three structurally distinct receptors, V1a, V1b, and V2, each activating very specific second messenger systems (Michell et al. 1979). Classically, AVP was first described as a primary homeostatic factor controlling kidney water reabsorption, blood volume/pressure, and vasodilatation in the peripheral nervous system. AVP and its receptors are widely expressed in the central nervous system (Thibonnier 1992) within specific brain regions (Johnson et al. 1993).

The V1a AVP receptor subtype (V1aR), in particular, has been extensively studied in the regulation of social behavior (Insel et al. 1994) including aggression (Winslow et al. 1993; Cooper et al. 2005; Albers et al. 2006; Ferris et al. 2006). V1aRs are directly coupled to stimulatory (s) Gq-11 proteins (Thibonnier et al. 1993). Stimulation of these G-proteins leads to activation of adenylate cyclase, cAMP, protein kinase C, and phospholipases C, A_2, and D (Raggenbass et al. 1991; Thibonnier 1992; Thibonnier et al. 1992, 1994) enhancing calcium influx through L-type calcium channels (Son and Brinton 2001). Such activation enhances learning and memory in the aging brain (Deyo et al. 1989; Yamada et al. 1996) via direct effects on gene expression (Murphy et al. 1991).

Because AVP regulates species-specific social behaviors such as aggression (Ferris et al. 1984; Ryding et al. 2008), it was initially hypothesized that the organization of central AVP systems may differ between monogamous and promiscuous vole species (Bamshad et al. 1993). To test this assumption, the distribution pattern of AVP cells, fibers, and receptors were mapped in the vole brain. In all vole species examined, AVP-ir neurons were found in several brain regions, including the PVN and supraoptic (SON) nucleus of the hypothalamus, the BNST, MeA, AH, and MPOA (Bamshad et al. 1993; Wang 1995, 1996; Gobrogge et al. 2007). AVP-ir fibers were localized in the LS, lateral habenular nucleus, diagonal band, BNST, MPOA, and MeA (Bamshad et al. 1993; Wang et al. 1996). Overall, AVP distribution patterns were highly conserved between monogamous and promiscuous vole species (Wang 1995; Wang et al. 1996).

Dramatic species differences in the distribution patterns and regional densities of V1aRs, however, were observed between vole species exhibiting different life strategies (Hammock and Young 2002). For example, prairie voles have higher densities of V1aRs in the BNST, VP, central, and basolateral nuclei of the amygdala, and accessory olfactory bulb whereas montane voles exhibit a higher density of V1aRs in the LS and mPFC (Insel et al. 1994; Wang et al. 1997c; Young et al. 1997; Lim et al. 2004; Smeltzer et al. 2006). Furthermore, prairie and pine voles exhibit similar patterns of V1aR binding, which differ from that of promiscuous meadow and montane voles (Insel et al. 1994; Lim et al. 2004). Species differences in V1aR distribution are stable across the lifespan (Wang et al. 1997b, c) and are receptor-specific, as no species differences are found in either the benzodiazepine or opiate receptor systems (Insel and Shapiro, 1992). In addition, monogamous prairie and promiscuous meadow voles differ in central AVP activity during mating and reproduction (Bamshad et al. 1993; Wang et al. 1994). Because of these anatomical and functional differences, central AVP was thought to underlie selective aggression in male prairie voles (Winslow et al. 1993). In summary, by examining the central AVP system, monogamous and promiscuous voles were found to differ in the distribution pattern (Insel et al. 1994; Smeltzer et al. 2006), 5′ promotor region sequence (Young et al. 1997), and microsatellite polymorphisms (Hammock and Young 2005) of AVP-V1a-type receptors in the brain. Similarly, prior work has shown that monogamous and promiscuous voles differed in brain distribution patterns of OT receptors (Insel and Shapiro 1992), DA neuroanatomy (Northcutt et al. 2007), and DA receptors (Aragona et al. 2006).

It is important to note that in some cases when multiple species were compared, clear differences in AVP and OT receptor distributions were found between species with different life strategies and sociosexual behavior (Insel and Shapiro 1992). It should also be noted that no species differences are found in other neurotransmitter systems, such as benzodiazepine and opioid receptors (Insel and Shapiro 1992), indicating neurochemical specificity of species differences in AVP, OT, and DA systems programming different mating strategies.

Among the neuropeptides underlying aggression (Miczek et al. 2007; Siever 2008), AVP and its homolog vasotocin have been found to regulate several forms of aggression across species (Riters and Panksepp 1997; Caldwell et al. 2008) and in diverse taxa (Goodson 2008; Backstrom and Winberg 2009). In humans, central AVP correlates with aggressive behavior (Coccaro et al. 1998) and mediates anger (Thompson et al. 2004). Thus, the central AVP system may have evolved to be primed by a wide variety of experiences to induce aggression, when appropriate, in social animals (Donaldson and Young 2008). Because central AVP underlies territorial aggression in other rodents (Ferris et al. 1984), it was proposed that AVP regulates mating-induced aggression in prairie voles. In a pharmacological study, injections of a V1aR antagonist into the lateral ventricle of male prairie voles blocked selective aggression induced by mating whereas injections of AVP induced aggression toward an intruder in the absence of mating (Winslow et al. 1993). These effects were neuropeptide-specific, as intracerebroventricular (ICV) infusion of an OT receptor antagonist had no effect on mating-induced aggression (Winslow et al. 1993). Importantly, developmental exposure to either AVP in male prairie voles (Stribley and Carter 1999) or OT in female prairie voles (Bales and Carter 2003) facilitates aggression in adulthood. Together, these data highlighted a critical role of neuropeptidergic regulation of aggression in prairie voles. However, the action site and release dynamics of neuropeptides in the regulation of selective aggression were unclear.

In more recent work, it was found that the display of selective aggression was associated with increased neuronal activation in the AH, specifically in neurons expressing AVP (Figs. 3a and b; Gobrogge et al. 2007). In a previous study in syrian hamsters (*Mesocricetus auratus*), aggression has also been shown to be associated with an increase in AVP-ir/Fos-ir double-labeled cells in the nucleus circularis (NC), medial SON (mSON), and surrounding areas ventral to the fornix in the AH (Delville et al. 2000). The vole data, combined with previous results from other species, suggests that the AH may be a brain area in which AVP regulates selective aggression. Indeed, AH-AVP has been implicated in various forms of aggressive behavior. In golden hamsters, for example, blockade of V1aRs in the AH diminished offensive aggression (Ferris and Potegal 1988; Caldwell and Albers 2004) whereas intra-AH administration of a V1aR agonist enhanced aggressive behavior (Ferris et al. 1997; Caldwell and Albers 2004). More recently, the density of V1aRs in the AH have been found to increase significantly in syrian hamsters after their display of offensive aggression (Albers et al. 2006).

Because selective aggression was associated with neuronal activation in the AH, specifically in AVP-containing neurons (Gobrogge et al. 2007), it was hypothesized that selective aggression is associated with AH-AVP release. In vivo brain microdialysis, with ELISA, was performed on male prairie voles that were pair-bonded for 2 weeks. Pair-bonded males displayed significantly higher levels of aggression toward novel females but more side-by-side affiliation with their familiar female partner (Gobrogge et al. 2009). ELISA analysis indicated that exposure to a stranger female, compared to a familiar partner, increased AH-AVP release (Fig. 3c), which is further confirmed by correlation analyses indicating that AH-AVP release was coupled positively with aggression and negatively with affiliation (Gobrogge et al. 2009). Moreover, intra-AH administration of AVP at a high (500 ng/side), but not a low (5 ng/side), dose in sexually naïve males induced aggression toward a novel female, and this effect was mediated by V1aR as concurrent administration of a tenfold higher dose of a V1aR antagonist (V1aR Ant) blocked AVP-induced aggression (Fig. 3d; Gobrogge et al. 2009). Furthermore, in pair-bonded males intra-AH infusions of a V1aR Ant (5 μg/side) blocked aggression and facilitated social affiliation toward novel females (Fig. 3d;

Gobrogge et al. 2009). Thus, AH-AVP is both necessary and sufficient to regulate mating-induced selective aggression in male prairie voles.

Monogamous male prairie voles encode several repetitive microsatellite DNA sequences in the promoter region of the V1aR gene that are not found in promiscuous male voles (Young 1999; Hammock and Young 2002, 2004). These genetic differences directly contribute to species differences in social organization (Pitkow et al. 2001; Landgraf et al. 2003; Lim et al. 2004). Furthermore, mice carrying a transgene encoding the prairie vole V1aR exhibit central V1aR patterns similar to prairie voles, and display enhanced social affiliation when injected with AVP (Young et al. 1999). Male voles given intra-VP injections of a virus expressing V1aR display enhanced partner preference in the absence of mating (Pitkow et al. 2001; Lim et al. 2004). Interestingly, individual variability in the genetic sequences encoding V1aRs has revealed remarkable within-species differences in sexual fidelity (Hammock and Young 2005; Ophir et al. 2008).

Because prairie voles carry varying lengths of DNA to program central V1aR distribution (Hammock and Young 2005), this genomic predisposition enables their brain to dynamically express V1aRs after sociosexual experience. This genetic loading distinguishes prairie voles from traditional laboratory rodents that lack this genetic make-up. Future work would benefit from comparing aggression levels between male prairie voles carrying long versus short versions of the promoter region encoding the V1aR gene (Gobrogge and Wang 2011). In humans, polymorphisms in the promoter region encoding V1aR are associated with differences in sociosexual bonding behaviors (Prichard et al. 2007, Walum et al. 2008) including altruism (Israel et al. 2008) and deficits in social communication observed in individuals with autistic spectrum disorders (Kim et al. 2002; Israel et al. 2008; Meyer-Lindenberg et al. 2009). To date, no study has examined potential associations between the molecular structure of the V1aR gene and patterns of aggression in humans (Gobrogge and Wang 2011). Thus, it would be interesting to investigate polymorphisms in the V1aR gene in patients with a lifetime history of pathological violence, as the V1aR system may harbor susceptibility genes underlying extreme forms of violence, increasing the prevalence of homicide and suicide in human populations (Gobrogge and Wang 2011).

Prior research has shown that the social environment has a significant impact on signaling and structural components of AVP systems in the brain. In marmoset monkeys, for example, prefrontal V1aR increases during fatherhood (Kozorovitskiy et al. 2006). In golden hamsters, several social and drug manipulations have been shown to directly alter the AH-AVP system to regulate offensive aggression (Ferris et al. 1989; Harrison et al. 2000b; Jackson et al. 2005; Grimes et al. 2007). In a previous study in male prairie voles, cohabitation with a female significantly increased the number of AVP mRNA-labeled cells in the BNST (Wang et al. 1994). Because social isolation increases the density of V1aRs in the AH to regulate offensive aggression in syrian hamsters (Albers et al. 2006), it was hypothesized that the density of AH-V1aR changes with pair-bonding experience to engage selective aggression. Pair-bonded males showed higher densities of V1aR binding, site-specifically, in the AH (Fig. 3e), with no change in OT receptor binding, demonstrating that pair-bonding experience induces neuroplasticity of V1aRs in a region- and receptor-specific manner (Gobrogge et al. 2009). To determine whether this increase in V1aR density in the AH following pair-bonding was directly related to the emergence of aggression toward novel females, prior work used viral vector mediated gene transfer to artificially elevate V1aR density in the AH. Males that received intra-AH microinfusions of the AAV-V1aR displayed higher levels of aggression toward a novel female compared to control males that received infusions of the LacZ-gene (Fig. 3f; Gobrogge et al. 2009). Similar viral vector mediated increases in V1aR expression in the VP in voles (Pitkow et al. 2001; Lim et al. 2004) and LS in mice (Bielsky et al. 2005)-enhanced affiliation. In male rats, inter-male aggression correlates with AVP release in the LS while AVP release in the BNST is inversely related to aggression levels (Veenema et al. 2010). Taken together, these data support the notion that region-specific AVP functioning regulates specific types of social behaviors, and multiple brain regions serve as a circuit in which AVP coordinates a range of adaptive behaviors important for reproductive success.

6.2 Dopamine

A detailed description of DA neuroanatomy can be found in earlier reviews (Zeiss 2005). Anatomically, central DA is divided into three distinct pathways: nigrostriatal, incertohypothalamic, and mesocorticolimbic. DA cell bodies projecting from the substantia nigra synapse in the dorsal striatum and comprise the nigrostriatal path (Swanson 1982). Incertohypothalamic paths extend from DA cell bodies of the A12-14 hypothalamic cell groups and project to the MPOA and PVN (Cheung et al. 1998). The mesocorticolimbic path represents DA cell bodies originating in the ventral tegmental area (VTA; Fig. 5c) projecting to the mPFC and NAcc (Swanson 1982; Carr and Sesack 2000). In addition, DA cells in the AH (Fig. 5b) project to forebrain areas including the striatum, LS, NAcc, and mPFC (Lindvall and Stenevi 1978, Maeda and Mogenson 1980; Alcaro et al. 2007).

DA preferentially binds to two families of receptors: D1-like and D2-like. Both types of DA receptors are found in the mPFC, NAcc, LS, and MeA (Boyson et al. 1986), and each of these brain regions are critical in regulating social behavior (Newman 1999). DA D1-like receptors consist of both D1 and D5 subtypes whereas D2-like DA receptors consist of D2, D3, and D4 subtypes. DA receptors are part of the large family of G-protein coupled receptors having seven transmembrane domains spanning intracellular and extracellular cell membrane surfaces. D1-like receptors are directly coupled to both stimulatory (s) $G\alpha$ and $G\alpha_{olf}$ proteins (Neve et al. 2004). Stimulation of these G-proteins leads to activation of adenylate cyclase, cAMP, and protein phosphatase-1 inhibitor DARP-32 (Neve et al. 2004). Conversely, D2-like receptors couple to inhibitory (i) $G\alpha$ proteins and, when activated, down-regulate adenylate cyclase, cAMP, and protein phosphatase-1 inhibitor DARP-32 (Neve et al. 2004). Phosphorylation of cAMP, with

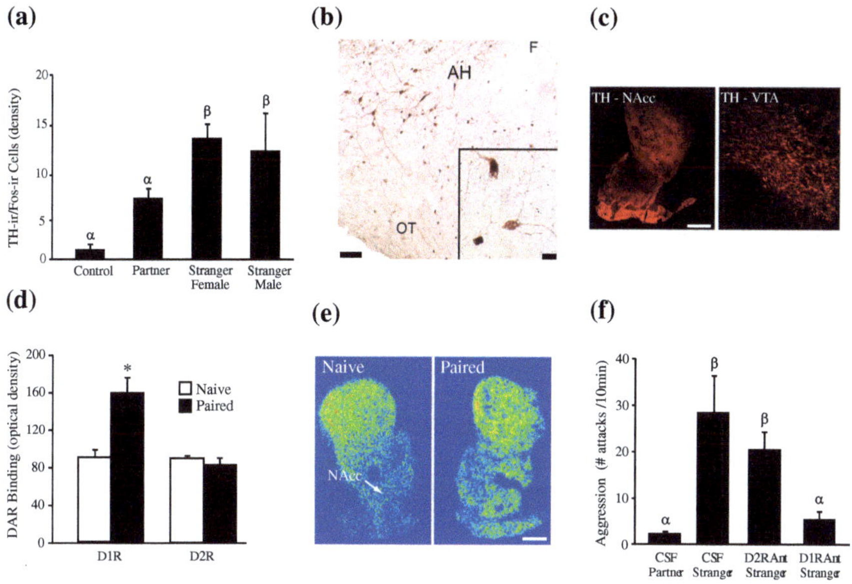

Fig. 5 Incertohypothalamic and mesocorticolimbic dopamine modulates selective aggression. **a** Pair-bonded male prairie voles that display aggression toward female or male strangers exhibit a significantly higher density of TH-ir/Fos-ir double-labeled cells in the AH relative to pair-bonded males re-exposed to their female partner or to males not exposed to any social stimulus (Control). **b** Photomicrograph of TH-ir cell bodies and fibers (brown cytoplasmic staining), Fos-ir (dark nuclear staining), or TH-ir/Fos-ir double labeled cells (insert) in the anterior hypothalamus (AH). **c** Photomicrograph of TH-ir fibers in the nucleus accumbens (NAcc) and neurons in the ventral tegmental area (VTA). **d** and **e** Pair-bonded males (Paired) exhibit a significantly higher density of DA D1-like (D1R), but not D2-like (D2R), receptor binding in the NAcc compared to sexually naïve males (Naïve). **f** Pair-bonded males, infused with CSF in the NAcc, display aggression toward a stranger female but not toward their female partner. Concurrent microinfusion of CSF with a DA D1R antagonist (D1R Ant), but not D2R antagonist (D2R Ant), in the NAcc is sufficient to attenuate pair-bonding-induced aggression toward a stranger female. *F* fornix, *OT* optic tract. Bars indicate mean ± standard error of the mean. Bars with different Greek letters differ significantly from each other. $*p < 0.05$. Scale bars = 100 μm, insert scale bar = 10 μm. Adapted from (Aragona et al. 2006; Gobrogge et al. 2007, 2009; Gobrogge and Wang 2011; Young et al. 2011a)

concurrent inhibition of protein phosphatase-1, stimulates other intracellular receptors, ion channels, and nuclear transcription factors (Neve et al. 2004). D1-receptor stimulation plays a critical role in calcium influx via L-type calcium channels, which is important for cellular long-term potentiation facilitating learning and memory in the aging brain (Deyo et al. 1989; Yamada et al. 1996) through direct influences on gene expression (Murphy et al. 1991). Detailed descriptions of DA-ergic intracellular signaling pathways can be found in other excellent reviews (Calon et al. 2000; Callier et al. 2003; Neve et al. 2004; Pollack 2004; Surmeier et al. 2007).

Because pair-bonding behavior represents a form of reward associative learning in which mesocorticolimbic DA plays a critical role, it is not surprising that central DA is involved in social memory (Young et al. 2011a). In prairie voles, DA producing cells and dendritic arbors are found in several brain regions important to social behavior, including the BNST, MPOA, PVN, AH (Fig. 5b), VTA (Fig. 5c), NAcc (Fig. 5c), striatum, MeA, and LS (Aragona et al. 2003; Lansing and Lonstein 2006; Gobrogge et al. 2007; Northcutt et al. 2007). Additionally, DA D1-type receptors are found in the NAcc (Figs. 5d and e), caudate putamen (CP) (Fig. 5e), and MeA with DA D2-type receptors being located in the NAcc (Fig. 5d), CP, MeA, substantia nigra, and VTA (Aragona et al. 2003, 2006).

Direct evidence of DA involvement in pair-bonding came from pharmacological experiments. Peripheral and intra-cerebro-ventricular (i.c.v) administration of a nonselective DA receptor antagonist, haloperidol, blocked mating-induced partner preference whereas administration of a nonselective DA receptor agonist, apomorphine, induced this behavior without mating in both male and female prairie voles (Wang et al. 1999; Gingrich et al. 2000; Aragona et al. 2003). In subsequent studies, DA in the NAcc was necessary and sufficient to regulate pair-bonding behavior. In both male and female prairie voles, mating induces an increase in DA release or DA turnover in the NAcc (Aragona et al. 2003; Curtis et al. 2003) while inhibiting DA receptors in the NAcc blocks partner preference formation induced by mating whereas activation of NAcc DA receptors induces this behavior without mating (Fig. 4a; Wang et al. 1999; Aragona et al. 2003). This effect also appears to be region-specific in the NAcc. DA in the rostral NAcc shell, but not caudal shell or core, plays a site-specific role in controlling the development of mating-induced partner preference formation in male prairie voles (Fig. 4; Aragona et al. 2003, 2006). Furthermore, data from female prairie voles indicate the importance of NAcc DA in partner preference formation, but not expression, as blockade of DA receptors prior to mating (which is essential for partner preference formation) blocked partner preference, while blockade after mating did not (Wang et al. 1999; Gingrich et al. 2000; Liu and Wang 2003). Intra-NAcc DA effects on partner preference formation are receptor-specific: activation of D2-type, but not D1-type, receptors facilitates partner preference formation whereas activation of D1-type receptors diminishes partner preference induced either by mating or by D2 receptor activation (Figs. 4a and b; Aragona et al. 2006). As described above, D1-type receptors are coupled to Gs proteins leading to activation of cAMP while D2 is coupled to Gi proteins inhibiting cAMP activity. Interestingly, prior work found that direct manipulation of the cAMP pathway in the NAcc has opposing effects on partner preference formation. Activation of the cAMP pathway blocks whereas inhibition of cAMP activity induces partner preference in male prairie voles (Aragona and Wang 2007). These data provide further evidence highlighting the intracellular mechanisms underlying receptor-specific DA-ergic regulation of the acquisition of partner preference social memory. Because mesocorticolimbic DA underlies partner preference formation (Wang et al. 1999; Gingrich et al. 2000; Aragona et al. 2003; Gobrogge et al. 2008; Aragona and Wang 2009), studies focused on examining the potential role of central DA regulating selective aggression (Gobrogge et al. 2008).

When sexually naïve male prairie voles are exposed to a novel female, DA is released in the NAcc and activation of D2-type receptors results in the formation of a partner preference (Fig. 4a). Thereafter, 2 weeks of pair-bonding experience induces neuroplastic changes, indicated by a significant increase in DA D1-type receptors (D1Rs), but not D2-type receptors (D2Rs), in the rostral NAcc shell (Figs. 5d and e) (Aragona et al. 2006). Interestingly, this upregulation of NAcc D1Rs is associated with increased selective aggression by male prairie voles toward conspecific strangers (males or females), and intra-NAcc blockade of D1Rs blocks selective aggression (Fig. 5f; Aragona et al. 2006). Males that cohabitated with a female for 24-h, with or without mating, did not exhibit an increase in D1Rs in the NAcc, supporting the hypothesis that increased densities of D1Rs, after pair-bonding, may directly contribute to selective aggression. To test this notion, pair-bonded male prairie voles were injected with a D1R antagonist into the NAcc. NAcc-D1R, not D2R, antagonism was sufficient to block selective aggression in pair-bonded male prairie voles (Fig. 5f; Aragona et al. 2006). Previous reviews have suggested that this form of aggression plays an important role in the maintenance of already established pair-bonds (Fig. 1) as bonded males aggressively reject conspecific male and female strangers that could serve as competitors or potential mates, respectively (Aragona et al. 2006; Gobrogge et al. 2007, 2009).

It should be noted that the NAcc is not the only brain area in which DA regulates pair-bonding behavior. For example, a species difference in DA receptor binding is found in the mPFC. Prairie voles have more D2-type, but less D1-type, receptor binding in the mPFC than meadow voles (Smeltzer et al. 2006). Manipulation of DA receptors in the mPFC also alters partner preference formation in female prairie voles, indicating a potential role of mPFC DA-ergic modulation of pair-bonding behavior (Wommack and Wang 2007). The VTA, which provides the major source of DA ergic axonal projections to the NAcc (Fig. 5c) and mPFC, is also involved in regulating pair-bonding behavior. Glutamate and GABA receptor blockade in the VTA, which alters DA activity in the NAcc, induces partner preference formation in the absence of mating in male prairie voles (Curtis and Wang 2005b).

In other work, both brief and extended cohabitation with unfamiliar conspecifics in female prairie voles significantly increased the number of DA-ergic cells in the BNST and MeA (Cavanaugh and Lonstein 2010), and blocking D2Rs during development decreased aggression-related behavior including infanticide in adult female, but not male, prairie voles (Hostetler et al. 2010). Furthermore, pair-bonded male prairie voles that displayed aggression toward either male or female intruders had a significantly higher density of cells in the AH that were double-labeled for tyrosine hydroxylase (TH, the rate limiting enzyme in DA biosynthesis)-ir and Fos-ir than males not exposed to a social stimulus or males that were re-exposed to their familiar female partner (Figs. 5a and b), implicating AH-DA involvement in selective aggression (Gobrogge et al. 2007).

6.3 Neurotransmitter Interactions

It is not surprising that complex social behaviors, such as pair-bonding, are regulated by multiple neuromodulators. In addition to AVP, OT, and DA, several other neurochemicals such as corticotrophin releasing factor (CRF), steroid hormones, GABA, glutamate, and opioids have also been implicated in pair-bonding behavior (Williams et al. 1992, 1994; Winslow et al. 1993; Carter et al. 1995; Wang et al. 1998; Liu et al. 2001; Aragona et al. 2003; Liu and Wang 2003; Lim and Young 2004; Curtis and Wang 2005b; Resendez et al. 2012).

Stress and stress hormones (e.g., glucocorticoids) have been found to interfere with pair-bonding in female prairie voles (DeVries et al. 1995). Peripheral administration of glucocorticoid and mineralcorticoid receptor antagonists facilitates partner preference formation, without mating, whereas concurrent central (i.c.v) administration of DA receptor antagonists blocks this behavior in female prairie voles, indicating an interaction between DA and glucocorticoids in partner preference formation (Curtis and Wang 2005a). As mentioned above, previous data also indicate a presynaptic interaction between DA and both glutamate and GABA receptors in partner preference formation in male prairie voles (Curtis and Wang 2005b).

Physical aggression is significantly more common in males than females, and these behavioral sex differences have been observed across many species (Gatewood et al. 2006). Research on the biological foundations of these sex differences has focused primarily on steroid hormones (Gatewood et al. 2006). Several studies have examined the role of androgens in the development of aggressive behavior, both organizationally (e.g., prenatal testosterone treatment) and activationally (e.g., postnatal testosterone treatment). Previous research has found that prenatal androgen exposure increases the behavioral expression of adult aggression (Vale et al. 1972; Michard-Vanhee 1988). Although organizational and activational influences of androgen on aggression have been noted, some inconsistent results have been reported. For example, castration in male rats (Koolhaas et al. 1990) and male prairie voles (Demas et al. 1999) has no affect on aggression. Thus, circulating testosterone cannot be the sole regulator of aggressive behavior in all rodent species including prairie voles. However, these findings do not rule out the possible effects of testosterone having organizational influences on other neurochemical systems in the brain which, together, may regulate aggression in adulthood. Thus, neurochemical-steroid hormone interactions underlying aggression have also been examined. For example, AVP administered directly into the MeA facilitates territorial aggression in male rats and is sufficient to block the effects of castration on reducing aggression (Koolhaas et al. 1990). Furthermore, castration (Bermond et al. 1982), but not ovariectomy (Kruk et al. 1984), decreases the excitability of neurons in the AH, blocking electrically-induced aggression, which in castrates can be reversed by testosterone treatment (Bermond et al. 1982). Therefore, circulating testosterone may be acting as a potent neuromodulator, interacting with neurochemicals like AVP to regulate aggression.

Additional evidence demonstrating steroid hormone-neurotransmitter interactions exists in research investigating central DA. For example, 75 % of TH-ir expressing cells in the hamster posterior MeA contain androgen receptors, express more DA-ir (i.e., do not co-label with DBH—dopamine beta hydroxylase), and are highly influenced by gonadal hormones compared to TH-ir cells found in the anterior MeA (Asmus et al. 1992; Asmus and Newman 1993). Interestingly, this same group of TH-ir cells is found in the posterior-dorsal MeA and BNST in male prairie voles, which appears to be influenced by testosterone (Northcutt et al. 2007) and activated after mating and social experience (Northcutt and Lonstein 2009). Together, these data suggest interactions between steroid hormones and central DA, in areas such as the MeA, in processing chemosensory cues related to social communication.

In addition to the effects of hormones and neurotransmitters underlying aggression, neuromodulators such as GABA and glutamate have also been shown to be involved in the display of agonistic behavior. For example, microinjection of a GABA antagonist (Adams et al. 1993; Roeling et al. 1993) with concurrent treatment of a glutamate agonist (Haller et al. 1998) in the AH, facilitates attack behavior in rodents, with higher doses having greater behavioral effects. Furthermore, reverse microdialysis infusion with a glutamate agonist and a GABA-A antagonist into the AH of rats after an agonistic encounter also facilitates aggression (Haller et al. 1998). More recent work has demonstrated site- (NAcc shell) and receptor-specific (kappa- but not mu-opioid) involvement in the regulation of selective aggression in both male and female prairie voles (Resendez et al. 2012).

7 Natural versus Artificial Reward

The data described above demonstrate that the mesocorticolimbic DA circuit is important in mediating social reward associated with pair-bonding in monogamous prairie voles. In fact, central DA has also been implicated in other types of natural reward including mating (Brackett et al. 1986; Pfaus et al. 1990; Meisel et al. 1993; Wersinger and Rissman 2000; Fisher et al. 2005; Hull and Dominguez 2006), food intake (Hernandez and Hoebel 1989; Clifton and Somerville 1994; Martel and Fantino 1996; Volkow et al. 2002; Pecina et al. 2003), and body weight homeostasis (Winn and Robbins 1985; Neill et al. 2002; Wang et al. 2002). In rats, for example, food seeking induces DA release in the NAcc, and released DA activates D1Rs to facilitate food seeking, illustrating receptor-specific DA-ergic regulation of behavior (Eyny and Horvitz 2003). Mating has been shown to induce DA release in several brain regions including the NAcc, CP, and MPOA in both male and female rats (Pfaus et al. 1990; Hull et al. 1993; Mas et al. 1995; Triemstra et al. 2005). In a series of elegant studies in male rats, mating was found to induce presynaptic DA release in the MPOA before and during copulation (Hull et al. 1993). Released DA then acts on post-synaptic DA receptors (Hull et al.

2004) and interacts with other neurochemicals, such as glutamate and GABA (Dominguez et al. 2006), to facilitate genital reflexes (e.g., erection, mounting, and ejaculation) and appetitive processing of the incentive value of potential mates (Hull and Dominguez 2007). Interestingly, previous tract tracing work shows that the MPOA has a direct anatomical and neurochemical connection with the mesocorticolimbic DA-ergic system (Maeda and Mogenson 1980) and can act synergistically with the VTA-NAcc circuit to regulate the motivational salience associated with mating (Moses et al. 1995).

The role of mesocorticolimbic DA in mediating actions of drugs of abuse has been well established. DA signaling has been implicated in processing a variety of abused drugs (Kosten 1996) including cocaine, amphetamine, nicotine, alcohol, marijuana, heroin, and morphine (Maldonado 2003; Nestler and Malenka 2004; Baler and Volkow 2006; Marsden 2006). The mesocorticolimbic DA pathway extending from the VTA to the NAcc has been consistently shown to play a significant role in behavioral patterns of drug addiction, including seeking, withdrawal, and relapse (Self et al. 1998; Koob 2000; Koob and Le Moal 2001; Yun et al. 2004). A detailed description of DA-ergic regulation of drugs of abuse is beyond the scope of the current chapter. However, in many cases, DA regulates drugs of abuse and natural reward in similar ways. For example, activation of D2Rs in the rostral NAcc shell facilitates cocaine-seeking behavior in rats, whereas activation of D1Rs in the same region diminishes it (Self et al. 1996). These opposing effects of NAcc D1(2) receptor involvement in drug seeking are similar to findings in prairie voles showing D2R activation facilitating and D1R activation preventing pair-bond formation (Aragona et al. 2006). Because the central DA system has evolved to mediate behavioral processes essential for survival, such as feeding and reproduction (Ikemoto and Panksepp 1999; Insel 2003), drugs of abuse usurp this neural circuitry which mediates associative learning reinforced by natural rewards, leaving an animal motivationally paralyzed to exclusively seek artificial drugs while neglecting social affiliation (Yamaguchi and Kandel 1985; Kaestner 1995), proper food intake (Ericsson et al. 1996; Chassler 1997), and parental behavior (Kandel et al. 1994).

Drug addiction is a significant mental health issue for many humans because drug abuse has such powerful control over social behavior essential for survival (Nesse and Berridge 1997; Kelley and Berridge 2002; Panksepp et al. 2002). Although the major focus on treatment has been pharmacological intervention targeted at neural circuitry known to mediate drug reward, the effects of social attachment on drug addiction should be evaluated especially since strong social attachments are likely associated with reduced addiction in humans (Havassy et al. 1995; Moos et al. 2002; Duncan et al. 2003). Unfortunately, studies of the interactions between pair-bonding and addiction are sparse partially due to the fact that traditional laboratory rats and mice show limited attachment behavior between adults. DA has been found to regulate pair-bonding in voles (Aragona et al. 2006) and drugs of abuse in other rodents (Self et al. 1996) in similar ways. A behavioral model was established to study drugs of abuse in prairie voles (Aragona et al. 2007).

In humans, substance abuse has been associated with weapon-related violence and homicide (Spunt et al. 1998; Madan et al., 2001; Hagelstam and Hakkanen 2006), intimate partner aggression, including partner-directed physical and psychological aggression (O'Farrell and Fals-Stewart 2000; Chermack et al. 2008), sexual abuse (El-Bassel et al. 2001), and child abuse (Haapasalo and Hamalainen 1996; Mokuau 2002; Walsh et al. 2003). Collectively, drug-related violence leads to family system dysfunction and incarceration (Krug et al. 2002), creating significant societal concerns. While aggression research in humans has provided valuable information regarding relationships between drug abuse and violence, animal models have been used to examine neural mechanisms underlying drug-induced aggression.

Drug use can override neurobiological programs to activate maladaptive forms of agonistic behavior, triggering inappropriate types of physical aggression (Swartz et al. 1998) such as domestic violence (Moore et al. 2008) and intimate partner homicide (Farooque et al. 2005). As a result, chronic drug abuse can cause permanent neural reorganization (Nestler and Aghajanian 1997; White and Kalivas 1998), impairing the adaptive social brain (Panksepp et al. 2002) and leading to the display of maladaptive social behavior (Wise 2002). Multiple studies have demonstrated that aggressive behavior may be escalated or reduced shortly after drug exposure, and that the direction of these effects depends on drug, dose, and individual differences between subjects.

Repeated exposure to many drugs of abuse during adolescence or adulthood persistently enhances agonistic behaviors, specifically those associated with offensive aggression. For example, syrian hamsters treated during adolescence with cocaine (Harrison et al. 2000a; Melloni et al. 2001; DeLeon et al. 2002a; Jackson et al. 2005; Knyshevski et al. 2005a, b) or anabolic–androgenic steroids (AAS) (Melloni and Ferris 1996; Melloni et al. 1997; Harrison et al. 2000b; DeLeon et al. 2002b) display enhanced offensive aggression in adulthood. Interestingly, these drug experiences reorganize AVP (Harrison et al. 2000b; Jackson et al. 2005; Grimes et al. 2007), DA (Ricci et al. 2009; Schwartzer et al. 2009); and GABA (Schwartzer et al. 2009) signaling in the AH. When compared with non-aggressive vehicle-treated control males, aggressive AAS-treated males exhibit significant neuroplastic changes in the AH including increased AVP-ir fiber density and AVP content (Harrison et al. 2000b), an increase in TH-ir cell and fiber density (Ricci et al. 2009), enhanced DA-D2R expression (Schwartzer et al. 2009), a higher number of GAD_{67}-ir cells (Schwartzer et al. 2009), and decreased $GABA_A$ receptor expression (Schwartzer et al. 2009). Furthermore, pharmacological blockade of D2 (Schwartzer and Melloni 2010a, b), but not D5 (Schwartzer and Melloni 2010b), DA receptors in the AH abolishes these effects. Together, these results suggest that AVP and DA signaling facilitates aggression by GABA inhibition in the AH of AAS-treated male golden hamsters.

As stated above, previous work has shown that exposure to drugs of abuse, such as cocaine, enhances male–male aggression by reorganizing the AH-AVP system in hamsters (Jackson et al. 2005). Therefore, it was suggested that amphetamine (AMPH), another commonly abused psychostimulant, would act in a similar fashion in affecting male-to-female aggression in prairie voles. Because recent data revealed that repeated AMPH treatment (1 mg/kg for three consecutive days) induced a conditioned place preference (Aragona et al. 2007) and blocked mating-induced partner preference (Fig. 6a; Liu et al. 2010), the following treatment regimen was used. To examine the selectivity of AMPH-induced aggression, males treated with saline or AMPH were tested for aggression toward an unfamiliar female or a familiar female (that had cohabitated with the male across a wire mesh screen for 24-h without mating). Compared with saline-treated controls, AMPH-treated males displayed significantly higher levels of aggression toward either familiar or unfamiliar females (Fig. 6b), indicating that AMPH exposure induces generalized aggression, rather than being selective to novel females. This AMPH treatment also increased the density of AVP-V1aR binding in the AH, but not MPOA, relative to saline control males (Fig. 6c). Furthermore, intra-AH infusions of the V1aR antagonist, but not CSF vehicle, diminished AMPH-induced aggression toward novel females (Fig. 6d). These data suggested that repeated exposure to AMPH can induce female-directed aggression and that this behavior is mediated by AH-AVP. Interestingly, these behavioral effects coincide with upregulation of D1Rs in the NAcc (Liu et al. 2010) and V1aRs in the AH (Gobrogge et al. 2009), both of which facilitate aggression toward novel females (Aragona et al. 2006; Gobrogge et al. 2009); indicating that drugs of abuse can hijack neuroplasticity that evolved to maintain monogamous pair-bonds. Finally, the prairie vole has also been established as an animal model for depression related to social separation (Grippo et al. 2007; Bosch et al. 2009). Furthermore, chronic metals ingestion—a potential model for autism—produces social avoidance in male, but not female, prairie voles exposed to unfamiliar same-sex conspecific strangers (Curtis et al. 2010), suggesting a developmental mechanism underlying the social deficits associated with autistic spectrum disorders in humans.

8 Summary and Future Directions

Pair-bonding behavior of prairie voles serves as an ideal model system for investigating the neural mechanisms of social decision-making. Prairie voles have provided an excellent animal model for studying the neurobiology of ethologically meaningful aggression associated with monogamous pair-bonds. Aggression can be easily manipulated under laboratory conditions and reliably expressed following mating and social cohabitation. Several brain areas (MeA, AH, and NAcc) work in a neural circuit to regulate selective aggression via AVP and DA. Commonly abused drugs, like AMPH, can usurp these neurochemical circuits to engage maladaptive social behavior impairing the formation and maintenance of monogamous pair-bonds.

Although male-to-male aggression has been studied in a variety of mammals, we know surprisingly little about male-to-female aggression and its underlying neuromechanisms. Interestingly, pair-bonded male prairie voles naturally display

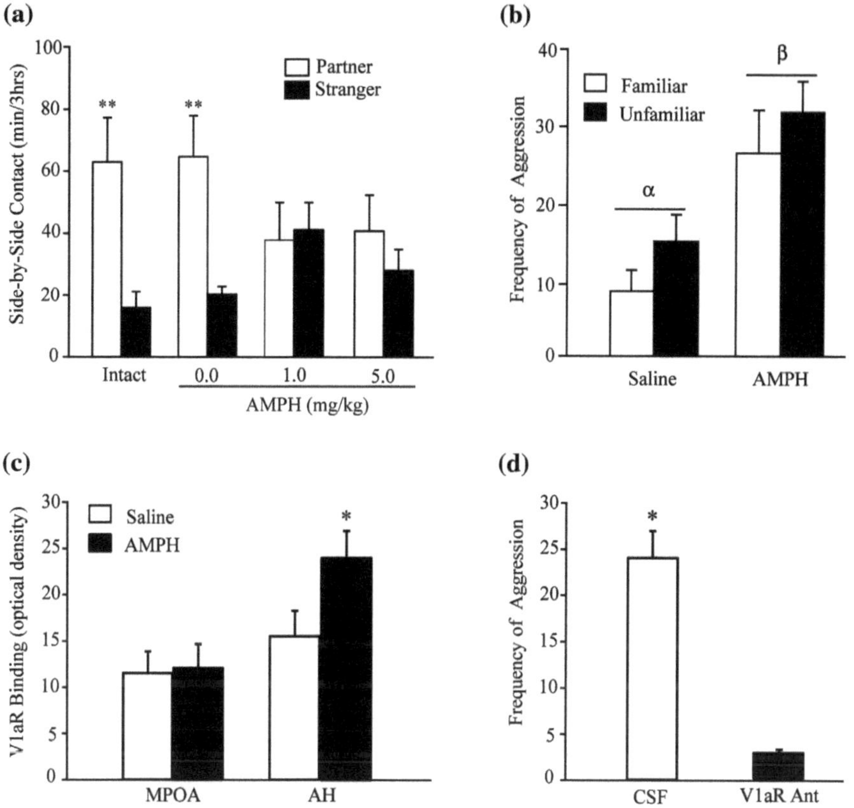

Fig. 6 Drug abuse impairs formation and maintenance of monogamous pair-bonds. **a** Pair-bonded (Intact) and saline-treated (0.0) male prairie voles, receiving 3-day—once daily; repeated injections, spend significantly more time in physical side-by-side contact with their familiar female partner than with an unfamiliar stranger female. Pair-bonded males injected (i.p.,—intraperitoneal) with 1.0 or 5.0 mg/kg amphetamine (AMPH) spend equal amounts of time in physical side-by-side contact with their female partner as with a stranger female. **b** and **e** Repeated AMPH exposure, in sexually naïve males, increases aggression toward both familiar and unfamiliar females and enhances the density of vasopressin (AVP) receptor (V1aR) expression in the anterior hypothalamus (AH) but not in the medial preoptic area (MPOA). **d** Site-specific microinfusion of an AVP-V1aR antagonist (V1aR Ant) into the AH, of males receiving 3-day repeated AMPH exposure (i.p.), significantly decreases AMPH-induced aggression toward novel females relative to males receiving intra-AH infusions of cerebral spinal fluid (CSF). Bars indicate mean ± standard error of the mean. Bars with different Greek letters differ significantly from each other. $^{*}p < 0.05$; $^{**}p < 0.01$. Adapted from (Gobrogge et al. 2009; Liu et al. 2010; Gobrogge and Wang 2011; Young et al. 2011a)

aggression toward conspecific females but not toward their female partner and, therefore, selective aggression allows investigation of the neurobiology of female-directed aggression. Data reviewed in this chapter demonstrate that this form of selective aggression is mediated by elevated AVP release and increased V1aR expression in the AH, which primes male prairie voles to respond aggressively to novel females. In addition, the same AH-AVP system mediates generalized, female-directed aggression induced by AMPH. Together with previous research from other animals (Ferris et al. 1989, 1997; Harrison et al. 2000b; Jackson et al. 2005; Veenema et al. 2006; Grimes et al. 2007), these data demonstrate a unique point of convergence in the mammalian brain (Choi et al. 2005; Motta et al. 2009). The AH-AVP system is highly conserved and functions to control different forms of aggression to maintain a wide range of resources important for reproductive success. These highly evolved neuropeptide systems appear to be extremely vulnerable to drugs of abuse, as hypothalamic AVP controls both naturally-occurring as well as drug-facilitated female-directed aggression, suggesting that psychostimulant drugs, like AMPH, are capable of switching adaptive (functional) forms of aggression (e.g., mate guarding) to aberrant (dysfunctional) forms of violent behavior (e.g., partner-directed aggression). Together, these data demonstrate the utility of the prairie vole model for evaluation of the effects of drug abuse on neural systems controlling adaptive forms of aggression, such as mate guarding behavior.

Because other neurochemicals, such as DA (Aragona et al. 2006) and serotonin (Villalba et al. 1997), also regulate selective aggression in prairie voles, and offensive aggression related to drug experience (Tidey and Miczek 1992) and AVP/5-HT interactions mediate aggression in other rodents (Ferris et al. 1997; Veenema et al. 2006), future studies should examine potential neurochemical interactions in the regulation of selective aggression. By understanding the basic neurobiology of pair-bonding behavior in prairie voles, we may eventually be able to better clarify the neural chemistry of mental health deficits associated with aberrations in social behavior in patients suffering from drug addiction or pathological violence.

9 Conclusions and Implications

In humans, clinical differences in social behavior underlie several forms of mental illness. AVP, OT, and DA are all involved in many human behaviors including social bonding (Bartels and Zeki 2004; Gonzaga et al. 2006; Marazziti et al. 2006), autism (Kendrick 2004), physical aggression (Pedersen 2004), and parenting (Swain et al. 2007). These systems work in concert with one-another to control complex social interactions. In prairie voles, these circuits have evolved to produce a monogamous life strategy. By understanding the basic neurobiology of pair-bonding behavior in prairie voles, we may eventually be able to describe the molecular and cellular mechanisms underlying mental health deficits associated with insecure attachment and conflict in humans. The data summarized in this chapter not only enhance our understanding of the neurochemical circuitry programming naturally occurring forms of agonistic behavior but may also reveal novel pharmaceutical treatments for escalated aggression and violence. For example, in recent studies, repeated treatment with the psychostimulant drug amphetamine (AMPH) induced aggression toward familiar female partners (Gobrogge et al. 2009) and impaired monogamous pair-bonds (Liu et al. 2010). Conversely, pair-bonding blocked the rewarding properties of AMPH-induced conditioning (Liu et al. 2011). Interestingly, this drug treatment coincides with upregulation of postsynaptic AVP-V1a-type receptors in the AHN (Gobrogge et al. 2009) and DA-D1-type receptors in the rostral NAcc shell (Liu et al. 2010). This neuroplastic remodeling is sufficient to enhance aggression directed toward both familiar (Gobrogge et al. 2009) and unfamiliar females (Aragona et al. 2006; Gobrogge et al. 2009). These findings together with current results identifying a neural mechanism regulating partner-directed aggression (Gobrogge and Wang 2009), indicate potential use of the prairie vole model for the study of the neurobiology of aggression associated with sex, drug abuse, and violence (Young et al. 2011b).

Acknowledgments The work reviewed in this chapter was supported by National Institutes of Health grants MHF31-79600 to KLG, MHR01-58616, DAR01-19627, & DAK02-23048 to ZXW, and NIH Program Training Grant T32 NS-07437.

Competing Financial Interests The author declares no competing financial interests.

References

Adams DB, Boudreau W, Cowan CW, Kokonowski C, Oberteuffer K, Yohay K (1993) Offense produced by chemical stimulation of the anterior hypothalamus of the rat. Physiol Behav 53:1127–1132

Albers EH, Dean A, Karom MC, Smith D, Huhman KL (2006) Role of V1a vasopressin receptors in the control of aggression in Syrian hamsters. Brain Res 1073–1074:425–430

Albert DJ, Walsh ML (1984) Neural systems and the inhibitory modulation of agonistic behavior: a comparison of mammalian species. Neurosci Biobehav Rev 8:5–24

Alcaro A, Huber R, Panksepp J (2007) Behavioral functions of the mesolimbic dopaminergic system: an affective neuroethological perspective. Brain Res Rev 56:283–321

Aragona BJ, Detwiler JM, Wang Z (2007) Amphetamine reward in the monogamous prairie vole. Neurosci Lett 418:190–194

Aragona BJ, Liu Y, Curtis JT, Stephan FK, Wang Z (2003) A critical role for nucleus accumbens dopamine in partner-preference formation in male prairie voles. J Neurosci 23:3483–3490

Aragona BJ, Liu Y, Yu YJ, Curtis JT, Detwiler JM, Insel TR, Wang Z (2006) Nucleus accumbens dopamine differentially mediates the formation and maintenance of monogamous pair bonds. Nat Neurosci 9:133–139

Aragona BJ, Wang Z (2007) Opposing regulation of pair bond formation by cAMP signaling within the nucleus accumbens shell. J Neurosci 27:13352–13356

Aragona BJ, Wang Z (2009) Dopamine regulation of social choice in a monogamous rodent species. Front Behav Neurosci 3:15

Asmus SE, Kincaid AE, Newman SW (1992) A species-specific population of tyrosine hydroxylase-immunoreactive neurons in the medial amygdaloid nucleus of the Syrian hamster. Brain Res 575:199–207

Asmus SE, Newman SW (1993) Tyrosine hydroxylase neurons in the male hamster chemosensory pathway contain androgen receptors and are influenced by gonadal hormones. J Comp Neurol 331:445–457

Backstrom T, Winberg S (2009) Arginine-vasotocin influence on aggressive behavior and dominance in rainbow trout. Physiol Behav 96:470–475
Baler RD, Volkow ND (2006) Drug addiction: the neurobiology of disrupted self-control. Trends Mol Med 12:559–566
Bales KL, Carter CS (2003) Sex differences and developmental effects of oxytocin on aggression and social behavior in prairie voles (*Microtus ochrogaster*). Horm Behav 44:178–184
Bamshad M, Novak MA, De Vries GJ (1993) Sex and species differences in the vasopressin innervation of sexually naive and parental prairie voles, *Microtus ochrogaster* and meadow voles, *Microtus pennsylvanicus*. J Neuroendocrinol 5:247–255
Bartels A, Zeki S (2004) The neural correlates of maternal and romantic love. Neuroimage 21:1155–1166
Bermond B, Mos J, Meelis W, van der Poel AM, Kruk MR (1982) Aggression induced by stimulation of the hypothalamus: effects of androgens. Pharmacol Biochem Behav 16:41–45
Bielsky IF, Hu SB, Ren X, Terwilliger EF, Young LJ (2005) The V1a vasopressin receptor is necessary and sufficient for normal social recognition: a gene replacement study. Neuron 47:503–513
Bosch OJ, Nair HP, Ahern TH, Neumann ID, Young LJ (2009) The CRF system mediates increased passive stress-coping behavior following the loss of a bonded partner in a monogamous rodent. Neuropsychopharmacology 34:1406–1415
Bowler CM, Cushing BS, Carter CS (2002) Social factors regulate female–female aggression and affiliation in prairie voles. Physiol Behav 76:559–566
Boyson SJ, McGonigle P, Molinoff PB (1986) Quantitative autoradiographic localization of the D1 and D2 subtypes of dopamine receptors in rat brain. J Neurosci 6:3177–3188
Brackett NL, Iuvone PM, Edwards DA (1986) Midbrain lesions, dopamine and male sexual behavior. Behav Brain Res 20:231–240
Caldwell HK, Albers HE (2004) Effect of photoperiod on vasopressin-induced aggression in Syrian hamsters. Horm Behav 46:444–449
Caldwell HK, Lee HJ, Macbeth AH, Young WS III (2008) Vasopressin: behavioral roles of an "original" neuropeptide. Prog Neurobiol 84:1–24
Callier S, Snapyan M, Le Crom S, Prou D, Vincent JD, Vernier P (2003) Evolution and cell biology of dopamine receptors in vertebrates. Biol Cell 95:489–502
Calon F, Hadj Tahar A, Blanchet PJ, Morissette M, Grondin R, Goulet M, Doucet JP, Robertson GS, Nestler E, Di Paolo T, Bedard PJ (2000) Dopamine-receptor stimulation: biobehavioral and biochemical consequences. Trends Neurosci 23:S92–100
Carr DB, Sesack SR (2000) Projections from the rat prefrontal cortex to the ventral tegmental area: target specificity in the synaptic associations with mesoaccumbens and mesocortical neurons. J Neurosci 20:3864–3873
Carter CS, De Vries AC, Getz LL (1995) Physiological substrates of mammalian monogamy: the prairie vole model. Neurosci Biobehav Rev 19:303–314
Carter CS, Getz LL (1993) Monogamy and the prairie vole. Sci Am 268:100–106
Cavanaugh BL, Lonstein JS (2010) Social novelty increases tyrosine hydroxylase immunoreactivity in the extended olfactory amygdala of female prairie voles. Physiol Behav 100:381–386
Chassler L (1997) Understanding anorexia nervosa and bulimia nervosa from an attachment perspective. Clin Soc Work J 25:407–423
Chermack ST, Murray RL, Walton MA, Booth BA, Wryobeck J, Blow FC (2008) Partner aggression among men and women in substance use disorder treatment: correlates of psychological and physical aggression and injury. Drug Alcohol Depend 98:35–44
Cheung S, Ballew JR, Moore KE, Lookingland KJ (1998) Contribution of dopamine neurons in the medial zona incerta to the innervation of the central nucleus of the amygdala, horizontal diagonal band of Broca and hypothalamic paraventricular nucleus. Brain Res 808:174–181
Choi GB, Dong HW, Murphy AJ, Valenzuela DM, Yancopoulos GD, Swanson LW, Anderson DJ (2005) Lhx6 delineates a pathway mediating innate reproductive behaviors from the amygdala to the hypothalamus. Neuron 46:647–660

Clifton PG, Somerville EM (1994) Disturbance of meal patterning following nucleus accumbens lesions in the rat. Brain Res 667:123–128

Coccaro EF, Kavoussi RJ, Hauger RL, Cooper TB, Ferris CF (1998) Cerebrospinal fluid vasopressin levels: correlates with aggression and serotonin function in personality-disordered subjects. Arch Gen Psychiatry 55:708–714

Cooper MA, Karom M, Huhman KL, Albers HE (2005) Repeated agonistic encounters in hamsters modulate AVP V1a receptor binding. Horm Behav 48:545–551

Curtis JT, Hood AN, Chen Y, Cobb GP, Wallace DR (2010) Chronic metals ingestion by prairie voles produces sex-specific deficits in social behavior: an animal model of autism. Behav Brain Res 213:42–49

Curtis JT, Stowe JR, Wang Z (2003) Differential effects of intraspecific interactions on the striatal dopamine system in social and non-social voles. Neuroscience 118:1165–1173

Curtis JT, Wang Z (2003) Forebrain c-fos expression under conditions conducive to pair bonding in female prairie voles (*Microtus ochrogaster*). Physiol Behav 80:95–101

Curtis JT, Wang Z (2005a) Glucocorticoid receptor involvement in pair bonding in female prairie voles: the effects of acute blockade and interactions with central dopamine reward systems. Neuroscience 134:369–376

Curtis JT, Wang Z (2005b) Ventral tegmental area involvement in pair bonding in male prairie voles. Physiol Behav 86:338–346

Cushing BS, Mogekwu N, Le WW, Hoffman GE, Carter CS (2003a) Cohabitation induced Fos immunoreactivity in the monogamous prairie vole. Brain Res 965:203–211

Cushing BS, Yamamoto Y, Hoffman GE, Carter CS (2003b) Central expression of c-Fos in neonatal male and female prairie voles in response to treatment with oxytocin. Brain Res Dev Brain Res 143:129–136

DeLeon KR, Grimes JM, Connor DF, Melloni RH Jr (2002a) Adolescent cocaine exposure and offensive aggression: involvement of serotonin neural signaling and innervation in male Syrian hamsters. Behav Brain Res 133:211–220

DeLeon KR, Grimes JM, Melloni RH Jr (2002b) Repeated anabolic-androgenic steroid treatment during adolescence increases vasopressin V(1A) receptor binding in Syrian hamsters: correlation with offensive aggression. Horm Behav 42:182–191

Delville Y, De Vries GJ, Ferris CF (2000) Neural connections of the anterior hypothalamus and agonistic behavior in golden hamsters. Brain Behav Evol 55:53–76

Demas GE, Moffatt CA, Drazen DL, Nelson RJ (1999) Castration does not inhibit aggressive behavior in adult male prairie voles (*Microtus ochrogaster*). Physiol Behav 66:59–62

DeVries AC, DeVries MB, Taymans S, Carter CS (1995) Modulation of pair bonding in female prairie voles (*Microtus ochrogaster*) by corticosterone. Proc Natl Acad Sci U S A 92:7744–7748

Dewsbury DA (1987) The comparative psychology of monogamy. Nebr Symp Motiv 35:1–50

Deyo RA, Straube KT, Disterhoft JF (1989) Nimodipine facilitates associative learning in aging rabbits. Science 243:809–811

Dominguez JM, Gil M, Hull EM (2006) Preoptic glutamate facilitates male sexual behavior. J Neurosci 26:1699–1703

Donaldson ZR, Young LJ (2008) Oxytocin, vasopressin, and the neurogenetics of sociality. Science 322:900–904

Duncan GJ, Wilkerson B, England P (2003) Cleaning up their act: the impact of marriage and cohabitation on licit and illicit drug use. Northwestern University, WP

El-Bassel N, Witte SS, Wada T, Gilbert L, Wallace J (2001) Correlates of partner violence among female street-based sex workers: substance abuse, history of childhood abuse, and HIV risks. AIDS Patient Care STDS 15:41–51

Ericsson M, Poston WS II, Foreyt JP (1996) Common biological pathways in eating disorders and obesity. Addict Behav 21:733–743

Eyny YS, Horvitz JC (2003) Opposing roles of D1 and D2 receptors in appetitive conditioning. J Neurosci 23:1584–1587

Farooque RS, Stout RG, Ernst FA (2005) Heterosexual intimate partner homicide: review of ten years of clinical experience. J Forensic Sci 50:648–651
Ferris CF, Albers HE, Wesolowski SM, Goldman BD, Luman SE (1984) Vasopressin injected into the hypothalamus triggers a stereotypic behavior in golden hamsters. Science 224:521–523
Ferris CF, Axelson JF, Martin AM, Roberge LF (1989) Vasopressin immunoreactivity in the anterior hypothalamus is altered during the establishment of dominant/subordinate relationships between hamsters. Neuroscience 29:675–683
Ferris CF, Lu SF, Messenger T, Guillon CD, Heindel N, Miller M, Koppel G, Robert Bruns F, Simon NG (2006) Orally active vasopressin V1a receptor antagonist, SRX251, selectively blocks aggressive behavior. Pharmacol Biochem Behav 83:169–174
Ferris CF, Melloni RH Jr, Koppel G, Perry KW, Fuller RW, Delville Y (1997) Vasopressin/serotonin interactions in the anterior hypothalamus control aggressive behavior in golden hamsters. J Neurosci 17:4331–4340
Ferris CF, Potegal M (1988) Vasopressin receptor blockade in the anterior hypothalamus suppresses aggression in hamsters. Physiol Behav 44:235–239
Fisher H, Aron A, Brown LL (2005) Romantic love: an fMRI study of a neural mechanism for mate choice. J Comp Neurol 493:58–62
Gatewood JD, Wills A, Shetty S, Xu J, Arnold AP, Burgoyne PS, Rissman EF (2006) Sex chromosome complement and gonadal sex influence aggressive and parental behaviors in mice. J Neurosci 26:2335–2342
Getz LL, Carter CS, Gavish L (1981) The mating system of the prairie vole, *Microtus ochrogaster*. Field and laboratory evidence for pair-bonding. Beh Ecol Sociobio 8:189–194
Getz LL, Hofmann JE (1986) Social organization in free-living prairie voles, *Microtus ochrogaster*. Beh Ecol Sociobio 18:275–282
Gingrich B, Liu Y, Cascio C, Wang Z, Insel TR (2000) Dopamine D2 receptors in the nucleus accumbens are important for social attachment in female prairie voles (*Microtus ochrogaster*). Behav Neurosci 114:173–183
Gingrich BS, Huot RL, Wang Z, Insel TR (1997) Differential Fos expression following microinjection of oxytocin or vasopressin in the prairie vole brain. Ann N Y Acad Sci 807:504–505
Gobrogge KL, Liu Y, Jia X, Wang Z (2007) Anterior hypothalamic neural activation and neurochemical associations with aggression in pair-bonded male prairie voles. J Comp Neurol 502:1109–1122
Gobrogge KL, Liu Y, Wang Z (2008) Dopamine regulation of pair bonding in monogamous prairie voles. In: Bridges R (ed) The neurobiology of the parental brain. Elsevier, San Diego, pp 345–358
Gobrogge KL, Liu Y, Young LJ, Wang Z (2009) Anterior hypothalamic vasopressin regulates pair-bonding and drug-induced aggression in a monogamous rodent. Proc Natl Acad Sci U S A 106:19144–19149
Gobrogge KL, Wang Z (2009) Neural circuitry underlying pair bonding-induced aggression in monogamous male prairie voles. In: Society for Neuroscience Abstracts #3774 Chicago, IL
Gobrogge KL, Wang Z (2011) Genetics of aggression in voles. Adv Genet 75:121–150
Gonzaga GC, Turner RA, Keltner D, Campos B, Altemus M (2006) Romantic love and sexual desire in close relationships. Emotion 6:163–179
Goodson JL (2008) Nonapeptides and the evolutionary patterning of sociality. Prog Brain Res 170:3–15
Grimes JM, Ricci LA, Melloni RH Jr (2007) Alterations in anterior hypothalamic vasopressin, but not serotonin, correlate with the temporal onset of aggressive behavior during adolescent anabolic-androgenic steroid exposure in hamsters (*Mesocricetus auratus*). Behav Neurosci 121:941–948
Grippo AJ, Cushing BS, Carter CS (2007) Depression-like behavior and stressor-induced neuroendocrine activation in female prairie voles exposed to chronic social isolation. Psychosom Med 69:149–157

Haapasalo J, Hamalainen T (1996) Childhood family problems and current psychiatric problems among young violent and property offenders. J Am Acad Child Adolesc Psychiatry 35:1394–1401

Hagelstam C, Hakkanen H (2006) Adolescent homicides in Finland: offence and offender characteristics. Forensic Sci Int 164:110–115

Hairston JE, Ball GF, Nelson RJ (2003) Photoperiodic and temporal influences on chemosensory induction of brain fos expression in female prairie voles. J Neuroendocrinol 15:161–172

Haller J, Abraham I, Zelena D, Juhasz G, Makara GB, Kruk MR (1998) Aggressive experience affects the sensitivity of neurons towards pharmacological treatment in the hypothalamic attack area. Behav Pharmacol 9:469–475

Hammock EA, Young LJ (2002) Variation in the vasopressin V1a receptor promoter and expression: implications for inter- and intraspecific variation in social behaviour. Eur J Neurosci 16:399–402

Hammock EA, Young LJ (2004) Functional microsatellite polymorphism associated with divergent social structure in vole species. Mol Biol Evol 21:1057–1063

Hammock EA, Young LJ (2005) Microsatellite instability generates diversity in brain and sociobehavioral traits. Science 308:1630–1634

Harrison RJ, Connor DF, Nowak C, Melloni RH Jr (2000a) Chronic low-dose cocaine treatment during adolescence facilitates aggression in hamsters. Physiol Behav 69:555–562

Harrison RJ, Connor DF, Nowak C, Nash K, Melloni RH Jr (2000b) Chronic anabolic-androgenic steroid treatment during adolescence increases anterior hypothalamic vasopressin and aggression in intact hamsters. Psychoneuroendocrinology 25:317–338

Havassy BE, Wasserman DA, Hall SM (1995) Social relationships and abstinence from cocaine in an American treatment sample. Addiction 90:699–710

Hernandez L, Hoebel BG (1989) Food intake and lateral hypothalamic self-stimulation covary after medial hypothalamic lesions or ventral midbrain 6-hydroxydopamine injections that cause obesity. Behav Neurosci 103:412–422

Hostetler CM, Harkey SL, Bales KL (2010) D2 antagonist during development decreases anxiety and infanticidal behavior in adult female prairie voles (*Microtus ochrogaster*). Behav Brain Res 210:127–130

Hull EM, Dominguez JM (2006) Getting his act together: roles of glutamate, nitric oxide, and dopamine in the medial preoptic area. Brain Res 1126:66–75

Hull EM, Dominguez JM (2007) Sexual behavior in male rodents. Horm Behav 52:45–55

Hull EM, Eaton RC, Moses J, Lorrain D (1993) Copulation increases dopamine activity in the medial preoptic area of male rats. Life Sci 52:935–940

Hull EM, Muschamp JW, Sato S (2004) Dopamine and serotonin: influences on male sexual behavior. Physiol Behav 83:291–307

Ikemoto S, Panksepp J (1999) The role of nucleus accumbens dopamine in motivated behavior: a unifying interpretation with special reference to reward-seeking. Brain Res Brain Res Rev 31:6–41

Insel TR (2003) Is social attachment an addictive disorder? Physiol Behav 79:351–357

Insel TR (2010) The challenge of translation in social neuroscience: a review of oxytocin, vasopressin, and affiliative behavior. Neuron 65:768–779

Insel TR, Hulihan TJ (1995) A gender-specific mechanism for pair bonding: oxytocin and partner preference formation in monogamous voles. Behav Neurosci 109:782–789

Insel TR, Preston S, Winslow JT (1995) Mating in the monogamous male: behavioral consequences. Physiol Behav 57:615–627

Insel TR, Shapiro LE (1992) Oxytocin receptor distribution reflects social organization in monogamous and polygamous voles. Proc Natl Acad Sci USA 89:5981–5985

Insel TR, Wang Z, Ferris CF (1994) Patterns of brain vasopressin receptor distribution associated with social organization in microtine rodents. J Neurosci 14:5381–5392

Insel TR, Winslow JT, Wang Z, Young LJ (1998) Oxytocin, vasopressin, and the neuroendocrine basis of pair bond formation. Adv Exp Med Biol 449:215–224

Israel S, Lerer E, Shalev I, Uzefovsky F, Reibold M, Bachner-Melman R, Granot R, Bornstein G, Knafo A, Yirmiya N, Ebstein RP (2008) Molecular genetic studies of the arginine vasopressin 1a receptor (AVPR1a) and the oxytocin receptor (OXTR) in human behaviour: from autism to altruism with some notes in between. Prog Brain Res 170:435–449

Jackson D, Burns R, Trksak G, Simeone B, DeLeon KR, Connor DF, Harrison RJ, Melloni RH Jr (2005) Anterior hypothalamic vasopressin modulates the aggression-stimulating effects of adolescent cocaine exposure in Syrian hamsters. Neuroscience 133:635–646

Jannett F (1982) Nesting patterns of adult voles, *Microtus montanus*, in field populations. J Mamm 63:495–498

Johnson AE, Audigier S, Rossi F, Jard S, Tribollet E, Barberis C (1993) Localization and characterization of vasopressin binding sites in the rat brain using an iodinated linear AVP antagonist. Brain Res 622:9–16

Kaestner R (1995) The effects of cocaine and marijuana use on marriage and marital stability. National bureau of economic research working paper no. 5038

Kandel DB, Rosenbaum E, Chen K (1994) Impact of maternal drug-use and life experiences on preadolescent children born to teenage mothers. J Marriage Fam 56:325–340

Katz LF, Ball GF, Nelson RJ (1999) Elevated Fos-like immunoreactivity in the brains of postpartum female prairie voles, *Microtus ochrogaster*. Cell Tissue Res 298:425–435

Kelley AE, Berridge KC (2002) The neuroscience of natural rewards: relevance to addictive drugs. J Neurosci 22:3306–3311

Kendrick KM (2004) The neurobiology of social bonds. J Neuroendocrinol 16:1007–1008

Kim SJ, Young LJ, Gonen D, Veenstra-VanderWeele J, Courchesne R, Courchesne E, Lord C, Leventhal BL, Cook EH Jr, Insel TR (2002) Transmission disequilibrium testing of arginine vasopressin receptor 1A (AVPR1A) polymorphisms in autism. Mol Psychiatry 7:503–507

Kirkpatrick B, Kim JW, Insel TR (1994) Limbic system fos expression associated with paternal behavior. Brain Res 658:112–118

Knyshevski I, Connor DF, Harrison RJ, Ricci LA, Melloni RH Jr (2005a) Persistent activation of select forebrain regions in aggressive, adolescent cocaine-treated hamsters. Behav Brain Res 159:277–286

Knyshevski I, Ricci LA, McCann TE, Melloni RH Jr (2005b) Serotonin type-1A receptors modulate adolescent, cocaine-induced offensive aggression in hamsters. Physiol Behav 85:167–176

Koob GF (2000) Neurobiology of addiction. Toward the development of new therapies. Ann N Y Acad Sci 909:170–185

Koob GF, Le Moal M (2001) Drug addiction, dysregulation of reward, and allostasis. Neuropsychopharmacology 24:97–129

Koolhaas JM, Van den Brink THC, Roozendaal B, Boorsma F (1990) Medial amygdala and aggressive behavior: interaction between testosterone and vasopressin. Aggressive Behav 16:223–229

Kosten TR (1996) Basic neurobiology of abused drugs. NIDA Res Monogr 162:54–56

Kozorovitskiy Y, Hughes M, Lee K, Gould E (2006) Fatherhood affects dendritic spines and vasopressin V1a receptors in the primate prefrontal cortex. Nat Neurosci 9:1094–1095

Kramer KM, Choe C, Carter CS, Cushing BS (2006) Developmental effects of oxytocin on neural activation and neuropeptide release in response to social stimuli. Horm Behav 49:206–214

Krug EG, Dahlberg LL, Mercy JA, Zwi AB, Lozito R (2002) World report on violence and health. World Health Organization, Geneva

Kruk MR (1991) Ethology and pharmacology of hypothalamic aggression in the rat. Neurosci Biobehav Rev 15:527–538

Kruk MR, Van der Laan CE, Mos J, Van der Poel AM, Meelis W, Olivier B (1984) Comparison of aggressive behaviour induced by electrical stimulation in the hypothalamus of male and female rats. Prog Brain Res 61:303–314

Kuptsov PA, Pleskacheva MG, Voronkova DN, Lipp HP, Anokhin KV (2005) Features of the c-Fos gene expression along the hippocampal rostro-caudal axis in common voles after rapid spatial learning. Zh Vyssh Nerv Deiat Im I P Pavlova 55:231–240

Landgraf R, Frank E, Aldag JM, Neumann ID, Sharer CA, Ren X, Terwilliger EF, Niwa M, Wigger A, Young LJ (2003) Viral vector-mediated gene transfer of the vole V1a vasopressin receptor in the rat septum: improved social discrimination and active social behaviour. Eur J Neurosci 18:403–411

Lansing SW, Lonstein JS (2006) Tyrosine hydroxylase-synthesizing cells in the hypothalamus of prairie voles (*Microtus ochrogaster*): sex differences in the anteroventral periventricular preoptic area and effects of adult gonadectomy or neonatal gonadal hormones. J Neurobiol 66:197–204

Lim MM, Wang Z, Olazabal DE, Ren X, Terwilliger EF, Young LJ (2004) Enhanced partner preference in a promiscuous species by manipulating the expression of a single gene. Nature 429:754–757

Lim MM, Young LJ (2004) Vasopressin-dependent neural circuits underlying pair bond formation in the monogamous prairie vole. Neuroscience 125:35–45

Lin D, Boyle MP, Dollar P, Lee H, Lein ES, Perona P, Anderson DJ (2011) Functional identification of an aggression locus in the mouse hypothalamus. Nature 470:221–226

Lindvall O, Stenevi U (1978) Dopamine and noradrenaline neurons projecting to the septal area in the rat. Cell Tissue Res 190:383–407

Liu Y, Aragona BJ, Young KA, Dietz DM, Kabbaj M, Mazei-Robison M, Nestler EJ, Wang Z (2010) Nucleus accumbens dopamine mediates amphetamine-induced impairment of social bonding in a monogamous rodent species. Proc Natl Acad Sci U S A 107:1217–1222

Liu Y, Curtis JT, Wang Z (2001) Vasopressin in the lateral septum regulates pair bond formation in male prairie voles (*Microtus ochrogaster*). Behav Neurosci 115:910–919

Liu Y, Wang Z (2003) Nucleus accumbens oxytocin and dopamine interact to regulate pair bond formation in female prairie voles. Neuroscience 121:537–544

Liu Y, Young KA, Curtis JT, Aragona BJ, Wang Z (2011) Social bonding decreases the rewarding properties of amphetamine through a dopamine D1 receptor-mediated mechanism. J Neurosci 31:7960–7966

Madan A, Beech DJ, Flint L (2001) Drugs, guns, and kids: the association between substance use and injury caused by interpersonal violence. J Pediatr Surg 36:440–442

Maeda H, Mogenson GJ (1980) An electrophysiological study of inputs to neurons of the ventral tegmental area from the nucleus accumbens and medial preoptic-anterior hypothalamic areas. Brain Res 197:365–377

Maldonado R (2003) The neurobiology of addiction. J Neural Transm Suppl 66:1–14

Marazziti D, Dell'Osso B, Baroni S, Mungai F, Catena M, Rucci P, Albanese F, Giannaccini G, Betti L, Fabbrini L, Italiani P, Del Debbio A, Lucacchini A, Dell'Osso L (2006) A relationship between oxytocin and anxiety of romantic attachment. Clin Pract Epidemol Ment Health 2:28

Marsden CA (2006) Dopamine: the rewarding years. Br J Pharmacol 147(Suppl 1):S136–S144

Martel P, Fantino M (1996) Influence of the amount of food ingested on mesolimbic dopaminergic system activity: a microdialysis study. Pharmacol Biochem Behav 55:297–302

Mas M, Fumero B, Gonzalez-Mora JL (1995) Voltammetric and microdialysis monitoring of brain monoamine neurotransmitter release during sociosexual interactions. Behav Brain Res 71:69–79

McGuire B, Novak M (1984) A comparison of maternal behavior in the meadow vole (*Microtus pennsylvanicus*), prairie vole (*M. ochrogaster*) and pine vole (*M. pinetorum*). Anim Behav 32:1132–1141

McGuire B, Novak M (1986) Parental care and its relation to social organization in the montane vole. J Mammal 67:305–311

Meisel RL, Camp DM, Robinson TE (1993) A microdialysis study of ventral striatal dopamine during sexual behavior in female Syrian hamsters. Behav Brain Res 55:151–157

Melloni RH Jr, Connor DF, Hang PT, Harrison RJ, Ferris CF (1997) Anabolic-androgenic steroid exposure during adolescence and aggressive behavior in golden hamsters. Physiol Behav 61:359–364

Melloni RH Jr, Connor DF, Todtenkopf MS, DeLeon KR, Sanyal P, Harrison RJ (2001) Repeated cocaine treatment activates flank marking in adolescent female hamsters. Physiol Behav 73:561–570

Melloni RH Jr, Ferris CF (1996) Adolescent anabolic steroid use and aggressive behavior in golden hamsters. Ann N Y Acad Sci 794:372–375

Meyer-Lindenberg A, Kolachana B, Gold B, Olsh A, Nicodemus KK, Mattay V, Dean M, Weinberger DR (2009) Genetic variants in AVPR1A linked to autism predict amygdala activation and personality traits in healthy humans. Mol Psychiatry 10:968–975

Michard-Vanhee C (1988) Aggressive behavior induced in female mice by an early single injection of testosterone is genotype dependent. Behav Genet 18:1–12

Michell RH, Kirk CJ, Billah MM (1979) Hormonal stimulation of phosphatidylinositol breakdown with particular reference to the hepatic effects of vasopressin. Biochem Soc Trans 7:861–865

Miczek KA (1979) A new test for aggression in rats without aversive stimulation: differential effects of d-amphetamine and cocaine. Psychopharmacology 60:253–259

Miczek KA, de Almeida RM, Kravitz EA, Rissman EF, de Boer SF, Raine A (2007) Neurobiology of escalated aggression and violence. J Neurosci 27:11803–11806

Mokuau N (2002) Culturally based interventions for substance use and child abuse among native Hawaiians. Public Health Rep 117(Suppl 1):S82–S87

Moore TM, Stuart GL, Meehan JC, Rhatigan DL, Hellmuth JC, Keen SM (2008) Drug abuse and aggression between intimate partners: a meta-analytic review. Clin Psychol Rev 28:247–274

Moos RH, Nichol AC, Moos BS (2002) Risk factors for symptom exacerbation among treated patients with substance use disorders. Addiction 97:75–85

Moses J, Loucks JA, Watson HL, Matuszewich L, Hull EM (1995) Dopaminergic drugs in the medial preoptic area and nucleus accumbens: effects on motor activity, sexual motivation, and sexual performance. Pharmacol Biochem Behav 51:681–686

Motta SC, Goto M, Gouveia FV, Baldo MV, Canteras NS, Swanson LW (2009) Dissecting the brain's fear system reveals the hypothalamus is critical for responding in subordinate conspecific intruders. Proc Natl Acad Sci U S A 106(12):4870–4875

Murphy TH, Worley PF, Baraban JM (1991) L-type voltage-sensitive calcium channels mediate synaptic activation of immediate early genes. Neuron 7:625–635

Neill DB, Fenton H, Justice JB Jr (2002) Increase in accumbal dopaminergic transmission correlates with response cost not reward of hypothalamic stimulation. Behav Brain Res 137:129–138

Nesse RM, Berridge KC (1997) Psychoactive drug use in evolutionary perspective. Science 278:63–66

Nestler EJ, Aghajanian GK (1997) Molecular and cellular basis of addiction. Science 278:58–63

Nestler EJ, Malenka RC (2004) The addicted brain. Sci Am 290:78–85

Neve KA, Seamans JK, Trantham-Davidson H (2004) Dopamine receptor signaling. J Recept Signal Transduct Res 24:165–205

Newman SW (1999) The medial extended amygdala in male reproductive behavior. A node in the mammalian social behavior network. Ann N Y Acad Sci 877:242–257

Northcutt KV, Lonstein JS (2009) Social contact elicits immediate-early gene expression in dopaminergic cells of the male prairie vole extended olfactory amygdala. Neuroscience 163:9–22

Northcutt KV, Wang Z, Lonstein JS (2007) Sex and species differences in tyrosine hydroxylase-synthesizing cells of the rodent olfactory extended amygdala. J Comp Neurol 500:103–115

O'Farrell TJ, Fals-Stewart W (2000) Behavioral couples therapy for alcoholism and drug abuse. J Subst Abuse Treat 18:51–54

Oliveras D, Novak M (1986) A comparison of paternal behavior in the meadow vole, *Microtus pennsylvanicus*, the pine vole, *Microtus pinetorum*, and prairie vole, *Microtus ochrogaster*. Anim Behav 34:519–526

Ophir AG, Wolff JO, Phelps SM (2008) Variation in neural V1aR predicts sexual fidelity and space use among male prairie voles in semi-natural settings. Proc Natl Acad Sci U S A 105:1249–1254

Panksepp J, Knutson B, Burgdorf J (2002) The role of brain emotional systems in addictions: a neuro-evolutionary perspective and new 'self-report' animal model. Addiction 97:459–469

Pecina S, Cagniard B, Berridge KC, Aldridge JW, Zhuang X (2003) Hyperdopaminergic mutant mice have higher "wanting" but not "liking" for sweet rewards. J Neurosci 23:9395–9402

Pedersen CA (2004) Biological aspects of social bonding and the roots of human violence. Ann N Y Acad Sci 1036:106–127

Pfaus JG, Damsma G, Nomikos GG, Wenkstern DG, Blaha CD, Phillips AG, Fibiger HC (1990) Sexual behavior enhances central dopamine transmission in the male rat. Brain Res 530:345–348

Pitkow LJ, Sharer CA, Ren X, Insel TR, Terwilliger EF, Young LJ (2001) Facilitation of affiliation and pair-bond formation by vasopressin receptor gene transfer into the ventral forebrain of a monogamous vole. J Neurosci 21:7392–7396

Pollack A (2004) Coactivation of D1 and D2 dopamine receptors: in marriage, a case of his, hers, and theirs. Sci STKE 2004:pe50

Prichard ZM, Mackinnon AJ, Jorm AF, Easteal S (2007) AVPR1A and OXTR polymorphisms are associated with sexual and reproductive behavioral phenotypes in humans. Mutation in brief no. 981. Online. Hum Mutat 28:1150

Raggenbass M, Goumaz M, Sermasi E, Tribollet E, Dreifuss JJ (1991) Vasopressin generates a persistent voltage-dependent sodium current in a mammalian motoneuron. J Neurosci 11:1609–1616

Ramamurthi B (1988) Stereotactic operation in behaviour disorders Amygdalotomy and hypothalamotomy. Acta Neurochir Suppl 44:152–157

Resendez SL, Kuhnmuench M, Krzywosinski T, Aragona BJ (2012) kappa-Opioid receptors within the nucleus accumbens shell mediate pair bond maintenance. J Neurosci 32:6771–6784

Ricci LA, Schwartzer JJ, Melloni RH Jr (2009) Alterations in the anterior hypothalamic dopamine system in aggressive adolescent AAS-treated hamsters. Horm Behav 55:348–355

Riters LV, Panksepp J (1997) Effects of vasotocin on aggressive behavior in male Japanese quail. Ann N Y Acad Sci 807:478–480

Roeling TA, Kruk MR, Schuurmans R, Veening JG (1993) Behavioural responses of bicucculline methiodide injections into the ventral hypothalamus of freely moving, socially interacting rats. Brain Res 615:121–127

Ryding E, Lindstrom M, Traskman-Bendz L (2008) The role of dopamine and serotonin in suicidal behaviour and aggression. Prog Brain Res 172:307–315

Sano K, Yoshioka M, Ogashiwa M, Ishijima B, Ohye C (1966) Postero-medial hypothalamotomy in the treatment of aggressive behaviors. Confin Neurol 27:164–167

Schwartzer JJ, Melloni RH Jr (2010a) Anterior hypothalamic dopamine D2 receptors modulate adolescent anabolic/androgenic steroid-induced offensive aggression in the Syrian hamster. Behav Pharmacol 21:314–322

Schwartzer JJ, Melloni RH Jr (2010b) Dopamine activity in the lateral anterior hypothalamus modulates AAS-induced aggression through D2 but not D5 receptors. Behav Neurosci 124:645–655

Schwartzer JJ, Ricci LA, Melloni RH Jr (2009) Interactions between the dopaminergic and GABAergic neural systems in the lateral anterior hypothalamus of aggressive AAS-treated hamsters. Behav Brain Res 203:15–22

Self DW, Barnhart WJ, Lehman DA, Nestler EJ (1996) Opposite modulation of cocaine-seeking behavior by D1- and D2-like dopamine receptor agonists. Science 271:1586–1589

Self DW, Genova LM, Hope BT, Barnhart WJ, Spencer JJ, Nestler EJ (1998) Involvement of cAMP-dependent protein kinase in the nucleus accumbens in cocaine self-administration and relapse of cocaine-seeking behavior. J Neurosci 18:1848–1859

Siegel A, Roeling TA, Gregg TR, Kruk MR (1999) Neuropharmacology of brain-stimulation-evoked aggression. Neurosci Biobehav Rev 23:359–389

Siever LJ (2008) Neurobiology of aggression and violence. Am J Psychiatry 165:429–442

Smeltzer MD, Curtis JT, Aragona BJ, Wang Z (2006) Dopamine, oxytocin, and vasopressin receptor binding in the medial prefrontal cortex of monogamous and promiscuous voles. Neurosci Lett 394:146–151

Son MC, Brinton RD (2001) Regulation and mechanism of L-type calcium channel activation via V1a vasopressin receptor activation in cultured cortical neurons. Neurobiol Learn Mem 76:388–402

Spunt B, Brownstein HH, Crimmins SM, Langley S, Spanjol K (1998) Alcohol-related homicides committed by women. J Psychoactive Drugs 30:33–43

Stowe JR, Liu Y, Curtis JT, Freeman ME, Wang Z (2005) Species differences in anxiety-related responses in male prairie and meadow voles: the effects of social isolation. Physiol Behav 86:369–378

Stribley JM, Carter CS (1999) Developmental exposure to vasopressin increases aggression in adult prairie voles. Proc Natl Acad Sci U S A 96:12601–12604

Surmeier DJ, Ding J, Day M, Wang Z, Shen W (2007) D1 and D2 dopamine-receptor modulation of striatal glutamatergic signaling in striatal medium spiny neurons. Trends Neurosci 30:228–235

Swain JE, Lorberbaum JP, Kose S, Strathearn L (2007) Brain basis of early parent-infant interactions: psychology, physiology, and in vivo functional neuroimaging studies. J Child Psychol Psychiatry 48:262–287

Swanson LW (1982) The projections of the ventral tegmental area and adjacent regions: a combined fluorescent retrograde tracer and immunofluorescence study in the rat. Brain Res Bull 9:321–353

Swartz MS, Swanson JW, Hiday VA, Borum R, Wagner HR, Burns BJ (1998) Violence and severe mental illness: the effects of substance abuse and nonadherence to medication. Am J Psychiatry 155:226–231

Tamarin R (1985) Biology of New World *Microtus*. Am Soc Mamm Spec Pub 8. Shippensburg, PA

Thibonnier M (1992) Signal transduction of V1-vascular vasopressin receptors. Regul Pept 38:1–11

Thibonnier M, Auzan C, Madhun Z, Wilkins P, Berti-Mattera L, Clauser E (1994) Molecular cloning, sequencing, and functional expression of a cDNA encoding the human V1a vasopressin receptor. J Biol Chem 269:3304–3310

Thibonnier M, Bayer AL, Simonson MS, Douglas JG (1992) Effects of amiloride analogues on AVP binding and activation of V1-receptor-expressing cells. Am J Physiol 262:E76–E86

Thibonnier M, Goraya T, Berti-Mattera L (1993) G protein coupling of human platelet V1 vascular vasopressin receptors. Am J Physiol 264:C1336–C1344

Thompson R, Gupta S, Miller K, Mills S, Orr S (2004) The effects of vasopressin on human facial responses related to social communication. Psychoneuroendocrinology 29:35–48

Tidey JW, Miczek KA (1992) Morphine withdrawal aggression: modification with D1 and D2 receptor agonists. Psychopharmacology 108:177–184

Triemstra JL, Nagatani S, Wood RI (2005) Chemosensory cues are essential for mating-induced dopamine release in MPOA of male Syrian hamsters. Neuropsychopharm 30:1436–1442

Tubbiola ML, Wysocki CJ (1997) FOS immunoreactivity after exposure to conspecific or heterospecific urine: where are chemosensory cues sorted? Physiol Behav 62:867–870

Vale JR, Ray D, Vale CA (1972) The interaction of genotype and exogenous neonatal androgen: agonistic behavior in female mice. Behav Biol 7:321–334

Veenema AH, Beiderbeck DI, Lukas M, Neumann ID (2010) Distinct correlations of vasopressin release within the lateral septum and the bed nucleus of the stria terminalis with the display of intermale aggression. Horm Behav 58:273–281

Veenema AH, Blume A, Niederle D, Buwalda B, Neumann ID (2006) Effects of early life stress on adult male aggression and hypothalamic vasopressin and serotonin. Eur J Neurosci 24:1711–1720

Veening JG, Coolen LM, de Jong TR, Joosten HW, de Boer SF, Koolhaas JM, Olivier B (2005) Do similar neural systems subserve aggressive and sexual behaviour in male rats? Insights from c-Fos and pharmacological studies. Eur J Pharmacol 526:226–239

Villalba C, Boyle PA, Caliguri EJ, De Vries GJ (1997) Effects of the selective serotonin reuptake inhibitor fluoxetine on social behaviors in male and female prairie voles (*Microtus ochrogaster*). Horm Behav 32:184–191

Volkow ND, Wang GJ, Fowler JS, Logan J, Jayne M, Franceschi D, Wong C, Gatley SJ, Gifford AN, Ding YS, Pappas N (2002) "Nonhedonic" food motivation in humans involves dopamine in the dorsal striatum and methylphenidate amplifies this effect. Synapse 44:175–180

Walsh C, MacMillan HL, Jamieson E (2003) The relationship between parental substance abuse and child maltreatment: findings from the Ontario Health Supplement. Child Abuse Negl 27:1409–1425

Walum H, Westberg L, Henningsson S, Neiderhiser JM, Reiss D, Igl W, Ganiban JM, Spotts EL, Pedersen NL, Eriksson E, Lichtenstein P (2008) Genetic variation in the vasopressin receptor 1a gene (AVPR1A) associates with pair-bonding behavior in humans. Proc Natl Acad Sci U S A 105:14153–14156

Wang GJ, Volkow ND, Fowler JS (2002) The role of dopamine in motivation for food in humans: implications for obesity. Expert Opin Ther Targets 6:601–609

Wang Z (1995) Species differences in the vasopressin-immunoreactive pathways in the bed nucleus of the stria terminalis and medial amygdaloid nucleus in prairie voles (*Microtus ochrogaster*) and meadow voles (*Microtus pennsylvanicus*). Behav Neurosci 109:305–311

Wang Z, Aragona BJ (2004) Neurochemical regulation of pair bonding in male prairie voles. Physiol Behav 83:319–328

Wang Z, Hulihan TJ, Insel TR (1997a) Sexual and social experience is associated with different patterns of behavior and neural activation in male prairie voles. Brain Res 767:321–332

Wang Z, Liu Y, Young LJ, Insel TR (1997b) Developmental changes in forebrain vasopressin receptor binding in prairie voles (*Microtus ochrogaster*) and montane voles (*Microtus montanus*). Ann N Y Acad Sci 807:510–513

Wang Z, Smith W, Major DE, De Vries GJ (1994) Sex and species differences in the effects of cohabitation on vasopressin messenger RNA expression in the bed nucleus of the stria terminalis in prairie voles (*Microtus ochrogaster*) and meadow voles (*Microtus pennsylvanicus*). Brain Res 650:212–218

Wang Z, Young LJ, De Vries GJ, Insel TR (1998) Voles and vasopressin: a review of molecular, cellular, and behavioral studies of pair bonding and paternal behaviors. Prog Brain Res 119:483–499

Wang Z, Yu G, Cascio C, Liu Y, Gingrich B, Insel TR (1999) Dopamine D2 receptor-mediated regulation of partner preferences in female prairie voles (*Microtus ochrogaster*): a mechanism for pair bonding? Behav Neurosci 113:602–611

Wang Z, Zhou L, Hulihan TJ, Insel TR (1996) Immunoreactivity of central vasopressin and oxytocin pathways in microtine rodents: a quantitative comparative study. J Comp Neurol 366:726–737

Wang Z, Young LJ, Liu Y, Insel TR (1997c) Species differences in vasopressin receptor binding are evident early in development: comparative anatomic studies in prairie and montane voles. J Comp Neurol 378:535–546

Wersinger SR, Rissman EF (2000) Dopamine activates masculine sexual behavior independent of the estrogen receptor alpha. J Neurosci 20:4248–4254

White FJ, Kalivas PW (1998) Neuroadaptations involved in amphetamine and cocaine addiction. Drug Alcohol Depend 51:141–153

Williams JR, Carter CS, Insel T (1992) Partner preference development in female prairie voles is facilitated by mating or the central infusion of oxytocin. Ann N Y Acad Sci 652:487–489

Williams JR, Insel TR, Harbaugh CR, Carter CS (1994) Oxytocin administered centrally facilitates formation of a partner preference in female prairie voles (*Microtus ochrogaster*). J Neuroendocrinol 6:247–250

Winn P, Robbins TW (1985) Comparative effects of infusions of 6-hydroxydopamine into nucleus accumbens and anterolateral hypothalamus induced by 6-hydroxydopamine on the response to dopamine agonists, body weight, locomotor activity and measures of exploration in the rat. Neuropharm 24:25–31

Winslow JT, Hastings N, Carter CS, Harbaugh CR, Insel TR (1993) A role for central vasopressin in pair bonding in monogamous prairie voles. Nature 365:545–548

Wise RA (2002) Brain reward circuitry: insights from unsensed incentives. Neuron 36:229–240

Wommack JC, Wang Z (2007) Dopamine in the prefrontal cortex regulates partner preference formation in female prairie voles. Poster P1.28, presented at the society for behavioral neuroendocrinology, Pacific Grove, CA

Yamada S, Uchida S, Ohkura T, Kimura R, Yamaguchi M, Suzuki M, Yamamoto M (1996) Alterations in calcium antagonist receptors and calcium content in senescent brain and attenuation by nimodipine and nicardipine. J Pharmacol Exp Ther 277:721–727

Yamaguchi K, Kandel DB (1985) On the resolution of role incompatibility – A life event history analysis of family roles and marijuana use. Am J Sociol 90:1284–1325

Young KA, Gobrogge KL, Liu Y, Wang Z (2011a) The neurobiology of pair bonding: Insights from a socially monogamous rodent. Front Neuroendocrinol 32:53–69

Young KA, Gobrogge KL, Wang Z (2011b) The role of mesocorticolimbic dopamine in regulating interactions between drugs of abuse and social behavior. Neurosci Biobehav Rev 35:498–515

Young KA, Liu Y, Wang Z (2008) The neurobiology of social attachment: A comparative approach to behavioral, neuroanatomical, and neurochemical studies. Comp Biochem Physiol C: Toxicol Pharmacol 148:401–410

Young LJ (1999) Frank A. Beach Award. Oxytocin and vasopressin receptors and species-typical social behaviors. Horm Behav 36:212–221

Young LJ, Nilsen R, Waymire KG, MacGregor GR, Insel TR (1999) Increased affiliative response to vasopressin in mice expressing the V1a receptor from a monogamous vole. Nature 400:766–768

Young LJ, Wang Z (2004) The neurobiology of pair bonding. Nat Neurosci 7:1048–1054

Young LJ, Wang Z, Insel TR (1998) Neuroendocrine bases of monogamy. Trends Neurosci 21:71–75

Young LJ, Winslow JT, Nilsen R, Insel TR (1997) Species differences in V1a receptor gene expression in monogamous and nonmonogamous voles: behavioral consequences. Behav Neurosci 111:599–605

Yun IA, Wakabayashi KT, Fields HL, Nicola SM (2004) The ventral tegmental area is required for the behavioral and nucleus accumbens neuronal firing responses to incentive cues. J Neurosci 24:2923–2933

Zeiss CJ (2005) Neuroanatomical phenotyping in the mouse: the dopaminergic system. Vet Pathol 42:753–773

Part II
Clinical Research

Gene–Environment Interactions in the Etiology of Human Violence

Manfred Laucht, Daniel Brandeis and Katrin Zohsel

Abstract This chapter reviews the current research on gene–environment interactions (G × E) with regard to human violence. Findings are summarized from both behavioral and molecular genetic studies that have investigated the interplay of genetic and environmental factors in terms of influencing violence-related behavior. Together, these studies reveal promising evidence that genetic factors combine with environmental influences to impact on the development of violent behavior and related phenotypes. G × E have been identified for a number of candidate genes implicated in violence. Moreover, the reviewed G × E were found to extend to a broad range of environmental characteristics, including both adverse and favorable conditions. As has been the case with other G × E research, findings have been mixed, with considerable heterogeneity between studies. Lack of replication together with serious methodological limitations remains a major challenge for drawing definitive conclusions about the nature of violence-related G × E. In order to fulfill its potential, it is recommended that future G × E research needs to shift its focus to dissecting the neural mechanisms and the underlying pathophysiological pathways by which genetic variation may influence differential susceptibility to environmental exposures.

Keywords Gene–environment interaction · Violence · Antisocial behavior · Aggression · Behavioral genetic studies · Molecular genetic studies

---

M. Laucht (✉) · D. Brandeis · K. Zohsel
Department of Child and Adolescent Psychiatry and Psychotherapy,
Central Institute of Mental Health, Medical Faculty Mannheim/Heidelberg University,
P.O. Box 122120, 68072 Mannheim, Germany
e-mail: manfred.laucht@zi-mannheim.de

M. Laucht
Department of Psychology, University of Potsdam, Potsdam, Germany

D. Brandeis
Department of Child and Adolescent Psychiatry, University of Zurich,
Zurich, Switzerland

Curr Topics Behav Neurosci (2014) 17: 267–295
DOI: 10.1007/7854_2013_260

Published Online: 21 December 2013

Contents

1 The Importance of Gene–Environment Interactions in the Etiology of Violence

Evidence from behavioral genetics suggests that violent behavior is moderately heritable, with about half of the variance being attributable to genetic influences in both males and females. In addition to a genetic predisposition, environmental factors account for the remaining half of the variance, mostly representing nonshared environmental influences. Current efforts to detect the specific genes that confer risk have, however, been challenged by the fact that identified common variants explain only a small fraction of the heritability of violence. The "mystery of the missing heritability" (Maher 2008), pertinent to most complex behaviors and disorders, may be related to the genetic architecture of these disorders and the selective pressure on variation impacting reproductive fitness, but also highlights the importance of investigating more sophisticated etiological models, such as those implicating G × E. According to this approach, different levels of environmental exposure could act to moderate the genetic risk in such a way that the genetic effect may only become apparent among individuals exposed to one environment and not among those exposed to another. Hence, G × E, defined as genetic effects that are conditional on environmental effects, or vice versa, could explain why some individuals carrying a susceptible genotype actually develop a phenotype while others remain unaffected. G × E research suggests a more plausible etiological model in which individual experiences and genetic make-up interact to produce a specific phenotype rather than a more simplistic model of independent effects of genes and environment.

Consequently, over the past decade, the study of G × E has emerged as an increasingly important research area that has moved the field beyond simple nature–nurture debates (Rutter et al. 2006). A growing body of evidence supports this approach highlighting its fruitfulness in studying the etiology of many psychiatric disorders and complex behaviors. Following the seminal work of Caspi et al. (2002, 2003, 2005), G × E involving a variety of genetic variants have been identified in humans with respect to a wide range of mental disorders (e.g., depression, anxiety, schizophrenia, ADHD, conduct disorder, and substance use disorder) implicating various pre-, peri-, and postnatal environmental influences (Uher and McGuffin 2008; Wermter et al. 2010). Numerous characteristics of family and social environment have been established as moderators of genetic influences on aggressive behavior in both twin/adoption studies and molecular genetic studies.

Although the G × E approach promises to provide new and intriguing insights for studying the etiology of antisocial behavior (Moffitt 2005), there is controversy regarding its usefulness given the notorious inconsistency of G × E findings (Caspi et al. 2010; Duncan and Keller 2011). A variety of weaknesses in terms of concept, methodology, and analytical methods have prompted critical appraisals (Eaves 2006; Munafo and Flint 2009). In particular, several methodological pitfalls have been discussed that challenge the investigation of G × E. Among these, multiple testing has long been recognized as a serious problem in genetic research. With regard to G × E, the availability of datasets which afford large numbers of subdivisions (due to different ways of defining genotype and environmental characteristics) offers numerous additional possibilities for data mining. A second, related issue of concern is that most studies of G × E are underpowered. The statistical power to detect an interaction depends upon a number of conditions, including study design, phenotype prevalence, distribution of genotypes, and environmental exposures. Given the likely small effects of any single G × E and the associated high risk of false positive results, this implies a critical need for replication. However, the rules on how to conduct a sound replication have become a matter of intense dispute (Duncan and Keller 2011). A third issue relates to the possibility of artefactual interactions produced by altered scaling, resulting from the fact that G × E vary greatly depending on the scale used to measure environmental exposure. A further problem is that an environment may not be independent of genetic factors, which requires the differentiation of G × E from gene–environment correlations (rGE), reflecting the nonrandom distribution of environments among genotypes (Moffitt et al. 2006; Rutter et al. 2006). In order to address these and other problems, a number of strategic recommendations have been proposed for future development of research into G × E (Moffitt 2005; Moffitt et al. 2006). Among others, these include the use of experimental designs to test for G × E and the need to adopt a developmental perspective in prospective studies.

In the following, the current evidence of G × E related to violence is reviewed. Findings are summarized from both behavioral and molecular genetic studies that have investigated the interplay of genetic and environmental factors influencing violent behavior and related phenotypes. Due to space restrictions, the review is limited to studies in humans, although it is important to note a burgeoning literature examining G × E in animal models of aggressive behavior (see elsewhere in this volume).

2 Behavioral Genetic Studies of G × E

This section comprises traditional genetic designs, particularly twin and adoption studies (see Tables 1 and 2 for an overview). In these designs, the genetic factor is not measured, but rather inferred, with genotype referring to an individual's entire

Table 1 Adoption studies of violence-related G × E

Study[a] (year)	Genetic factor (attribute of BP)	N	Age	Outcome	Results		
					G	E	G × E
Time in temporary care							
Crowe (1974)	Delinquency	98	25	ASPD, delinquency	+	–	(+)
Adverse adoptive home environment							
Cloninger et al. (1982)	Delinquency, alcohol abuse, low job status	862	23–43	Delinquency	+	+	(+)
Cadoret et al. (1983)	Antisocial behavior, alcohol abuse	515	n.r.	Adolescent antisocial behavior	(+)	+	+
Cadoret et al. (1995)	ASPD, alcohol, or drug abuse	197	18–47	CD symptoms, aggression, adult antisocial behavior	+	+	(+)
Psychiatric symptoms of adoptive parent							
Leve et al. (2010)	Delinquency, substance use, novelty seeking	348	0.75	Attention to frustrating events	+	+	+

ASPD antisocial personality disorder, *BP* biological parent, *CD* conduct disorder, *n. r.* not reported, + significant main/interaction effect, (+) trend or partial evidence

[a] First author

Table 2 Twin studies of violence-related G × E

Study[a] (year)	Design	N	Age	Outcome	Results		
					G	E	G × E
Childhood maltreatment							
Jaffee et al. (2005)	M	1116 T	5	CD symptoms, CD	+	+	+, (+)
Boutwell et al. (2011)	M	250 MZ, 1300 DZ	5	Antisocial behavior	+	+	m: +
Family environment							
Rowe et al. (1999)	V	1515 S	n.r.	Aggression	+	+	+, genetic influence ↘ with less family warmth
Button et al. (2005)	V	278 MZ, 378 DZ	5–18	Conduct problems	+	+	+, genetic influence ↘ with less family functioning
Feinberg et al. (2007)	V	93 MZ, 99 DZ, 528 S	9–18	Antisocial behavior			+, genetic influence ↘ with family warmth
Adverse peer relationships							
van Lier et al. (2007)	M	145 MZ, 212 DZ	6	Aggression	+	+	+
Button et al. (2007)	V	553 MZ, 558 DZ	11–18	Conduct problems	+	+	+, genetic influence ↗ with delinquent peers
Brendgen et al. (2008a)	V	406 T	7	Physical aggression	+	+	+, genetic influence ↗ with aggressive peers
Brendgen et al. (2008b)	M	506 T	6	Aggression	+	+	f: +
Other environments							
Tuvblad et al. (2006)	V	1133 T	16–17	Delinquency	+	+	+, genetic influence ↗ with socioeconomic status
Brendgen et al. (2011)	V	124 MZ, 93 DZ	7	Aggression	+	+	+, genetic influence ↗ with neg. teacher-relationship
Lamb et al. (2012)	V	8829 T	7–12	Externalizing behavior			+, genetic influence ↘ with different teacher

CD conduct disorder, *DZ* dizygotic twins, *f* females, *m* males, *M* mean levels approach, *MZ* monozygotic twins, *n. r.* not reported, *S* sibling pairs, *T* twins, *V* moderated variance components approach, + significant main/interaction effect, (+) trend or partial evidence

[a] First author

genetic make-up and not to particular genes. This may be regarded as advantageous, as the exact genetic underpinnings of violence are not yet fully understood and single genetic variants capture little variance. Although over one hundred adoption and twin studies of antisocial behavior exist (for a review, see Rhee and Waldman 2002), only a limited number of these has explicitly examined the question of G × E.

To study G × E effects using a behavioral genetic design, two methodological approaches are usually applied. The first examines whether the proportion of variance attributable to genetic effects is moderated by the exposure to an adverse environmental factor, for example, whether heritability of aggressive behavior differs under conditions of sensitive compared to harsh parenting (*variance components method*, see Thapar et al. 2007; Tuvblad and Baker 2011). Findings of this type of design are not easily interpretable on an individual level. A second method that is more easily comprehensible involves testing whether the probability of antisocial behavior is increased among individuals with a higher genetic risk compared to those with a lower genetic risk when exposed to an adverse environment (*mean levels approach*). In adoption studies, the individual "genetic risk" of an adoptee is defined by characteristics that are present or absent in a biological parent (e.g., delinquency, substance abuse, and antisocial personality disorder). In twin studies, the "genetic risk" is usually estimated by using the genetic relatedness within a sibling pair, which is 100 % in monozygotic and 50 % in dizygotic twins, and the phenotype of the "co-twin". In this respect, the highest genetic risk for antisocial behavior is attributed to a person whose monozygotic twin already displays aggressive behavior and the lowest genetic risk to a person with a monozygotic twin who shows no aggressive behavior. The estimated genetic risk of individuals with a dizygotic twin with or without aggressive symptoms would fall in between. From the two approaches described above, different patterns of results may emerge. For example, it is possible that under certain adverse environments, the mean level of antisocial behavior may be increased in individuals with a high genetic risk, while the overall variance explained by genetic factors may drop (Tuvblad and Baker 2011).

In the following, evidence for violence-related G × E will be summarized separately for environmental factors working within and outside of the family. To our knowledge, no studies conducted to date have addressed the impact of G × E involving the pre- or perinatal environment on antisocial behavior using a behavioral genetic design, e.g., by studying birth weight.

Environmental Factors within the Family. In the majority of studies on violence-related G × E, the focus has been on family environment. In early adoption studies carried out in the 1970s and 1980s in the USA and Scandinavia, the observation was made that adoptees with both a genetic risk and an adverse adoptive family environment had a much higher probability of antisocial behavior than would be predicted by simply adding both risk factors (Cadoret et al. 1983, 1995; Cloninger et al. 1982). For example, Cloninger et al. reported a criminality rate of 6.7 % in male Swedish adults with a low genetic risk exposed to an adverse adoptive family background, compared to 12.1 % in adoptees with a high genetic risk who were raised under favorable family conditions. When both risk factors were combined, a considerably increased criminality rate of 40 % was found (the interpretation of this finding was, however, limited by the small number of only 10 adoptees in this group). Studies carried out more recently tried to identify the specific environmental factors responsible for this effect. Potential candidates encompass parenting style, psychiatric symptoms of a parent, and physical punishment or maltreatment in the family.

Concerning parenting style, evidence for G × E was found in three samples of twins or other sibling pairs. Two studies (Button et al. 2005; Rowe et al. 1999) reported a decrease in the genetic influence on antisocial behavior in children and adolescents with lower levels of family functioning or parental warmth. This finding is consistent with the assumption that in favorable environments, biological factors are more likely to explain antisocial behavior, whereas in adverse environments, the risk of a negative outcome is enhanced regardless of biological factors, thus reducing the importance of genes. However, another carefully designed study (Feinberg et al. 2007) reported an opposite pattern. The authors supposed that, with high levels of parental warmth, the genetic liability to antisocial behavior may not be triggered. Such conflicting results are not easily reconciled, and similar inconsistency has been observed in other domains, e.g., in terms of cognitive abilities (Rutter et al. 2006).

Parental psychiatric symptoms as an environmental factor were addressed in a recent adoption study by Leve et al. (2010). In 9-month-old infants, attention to frustrating events was measured, which is regarded as a precursor of externalizing behavior. Children were found to show heightened attention to frustration when they were genetically at risk and had an adoptive mother with elevated anxious and depressive symptoms. Although the reported interaction only accounted for minor proportions of variance, this finding highlights the importance of early environmental factors. With regard to maltreatment, Jaffee et al. (2005) observed a significant increase of 24 % in the probability of diagnosable conduct disorder among 5-year-old siblings with a high genetic risk exposed to maltreatment, but an increase of only 2 % among maltreated children with a low genetic risk. This finding was replicated by Boutwell et al. (2011) for the use of physical punishment by the mother, with the interaction effect only reaching significance in boys. However, it cannot be completely ruled out that gene–environment correlations explain at least parts of those findings, as children with higher levels of genetic risk for antisocial behavior may also be exposed to more physical punishment as a consequence of their behavioral problems.

Environmental Factors Outside the Family. A small but increasing number of behavioral genetic studies have examined the influence of environmental factors outside of the family. Existing studies on G × E focus on peer victimization and deviant peer affiliation, teacher–child relationships, and socioeconomic conditions. In the field of peer relations, Brendgen et al. (2008b) reported that 6-year-old girls who were victimized by their peers were only prone to aggressive behavior when they additionally carried a high genetic risk for aggression. As, at least in young children, peer victimization seems to be independent of a genetic predisposition to aggressive behavior, this effect may be attributed to an interaction between genes and an environmental factor rather than to a correlation between these. Boys who were rejected by their peers, however, showed more aggressive behavior irrespective of their genetic risk for aggression. A consequence of being rejected by "normally" developing children might be affiliating with deviant peers, which then provides new opportunities to engage in antisocial activities. Twin studies have revealed that the contribution of genetic factors to aggression increases with higher levels of deviant behavior in the peer group of a child or adolescent (Brendgen et al. 2008a; Button et al. 2007). Moreover, aggressive friends enhanced the likelihood of antisocial behavior especially in children with a high genetic risk (van Lier et al. 2007).

Not only the peer group but also the school environment may constitute an environmental factor that interacts with the individual genetic risk for violence. Brendgen et al. (2011) reported a higher genetic impact on aggression when the teacher–child relationship was described as negative by the teacher. Conversely, a positive relationship with a teacher might be regarded as protective in children carrying a high genetic risk for aggressive behavior. Finally, a twin study on antisocial behavior reported an interaction between genetic factors and socioeconomic status (Tuvblad et al. 2006). Results suggested a higher genetic influence on antisocial behavior when neighborhood socioeconomic conditions were more advantaged. In less advantaged neighborhoods, however, adverse environmental conditions seemed to overlay the genetic contribution to antisocial behavior.

Summary and Limitations. In the limited number of available studies, considerable evidence is found for a role of G $\times$ E relating to violence. Summarizing studies that applied a mean levels approach, adverse environmental conditions within or outside of the family seemed to have little effect in individuals with a low genetic risk, but increased the probability of violent behavior when a genetic risk existed. With regard to studies that applied a variance component method, the picture was less consistent, with some describing an increasing genetic influence on antisocial behavior under more adverse and some under more favorable environmental conditions. In most studies, participants were children and adolescents, while G $\times$ E studies focusing on adult antisocial behavior were rare and the reported interaction effects were small (Cadoret et al. 1995; Cloninger et al. 1982). Possibly, environmental factors other than the ones studied so far play an additional role in adult antisocial behavior. Likewise, it is not yet clear to what extent G $\times$ E in antisocial behavior is moderated by sex.

For both adoption and twin studies on violence, general methodological limitations exist (Jaffee et al. 2012; Moffitt 2005). On the one hand, adoption studies are ideally suited to the study of G $\times$ E, as genetic and environmental influences may be disentangled and passive gene–environment correlations ruled out as a source of confounding, given that no selective placement has been conducted. On the other hand, results from adoption studies are limited in terms of generalizability, as adoptees commonly exhibit elevated rates of antisocial behavior. Moreover, adoption studies offer a restricted range of family environments, as adoptive families are usually selected to provide favorable rearing conditions.

Accordingly, G × E effects may be harder to detect (Moffitt 2005). Twins, in contrast, do not show enhanced prevalence rates of antisocial behavior compared to singletons. They do, however, exhibit other characteristics which may delimit generalizability, e.g., they suffer from more pre-and perinatal complications and have lower birth weights. An interactive effect of birth weight and a genetic factor on antisocial behavior has already been reported (Thapar et al. 2005). Hence, despite their usefulness for detecting G × E in antisocial behavior, adoption and twin studies should be supplemented by results from other genetic designs.

3 Molecular Genetic Studies of G × E

Behavioral genetic studies of G × E examine the degree to which aggregate genetic effects interact with environmental influences to predict variation in behavior. While these studies are unable to identify the specific genes involved, molecular genetic studies of G × E provide the possibility to uncover the contribution of single genetic markers. Traditionally, these studies have been conducted using the candidate gene approach, with genome-wide analysis to study G × E now ready to be launched (Thomas 2010). To date, six candidate genes implicated in the development of violence have been investigated in multiple studies in order to determine whether environmental factors differentially influence violent behaviors in individuals with different genotypes. While the vast majority of studies examined interactions involving single genes, a few studies tested for epistasis or applied composite measures of genetic effect that were based on a combination of variants from different genes (e.g., Beaver et al. 2010; Nobile et al. 2007). In addition, several recent studies, which are awaiting replication, reported evidence of G × E using new candidate genes, such as the gene coding for tryptophan hydroxylase 1 (TPH1), involved in the synthesis of serotonin (e.g., Bevilacqua et al. 2012; Cicchetti et al. 2012; Dick et al. 2011; Latendresse et al. 2011).

Monoamine Oxidase A Gene. One of the most influential studies examining G × E in violent behavior was conducted by Caspi et al. (2002). The authors investigated the interaction between a functional polymorphism in the MAOA gene (MAOA-LPR) encoding an enzyme that metabolizes monoamine neurotransmitters and early childhood maltreatment in the development of antisocial behavior. Using data of 442 male participants of an epidemiological cohort study, Caspi et al. reported consistent evidence for G × E. Maltreated individuals with a genotype conferring low levels of MAOA were more likely to develop antisocial problems, including conduct disorder, adult violent crime, and antisocial personality disorder, than carriers of the high activity genotype of MAOA with a history of maltreatment. Moreover, this G × E was found not to be due to either passive or evocative gene–environment correlation (rGE). Subsequent studies have replicated these findings in new cohorts and various age groups with additional measures of violence. Two earlier meta-analyses including five and eight studies, respectively, reported small to medium effect sizes for the MAOA by maltreatment interaction in conferring risk of antisocial behavior (Kim-Cohen et al. 2006, ES = 0.18; Taylor and Kim-Cohen 2007, ES = 0.17).

In the meantime, research aiming to replicate and extend the original findings of Caspi et al. has burgeoned. Currently, over 30 studies examining the MAOA G × E in the development of violence and related phenotypes have been published. Although findings have been mixed, as has been the case with other G × E research, a significant number of studies have confirmed the original findings, with a minority of studies reporting null (e.g., Prichard et al. 2008), or even opposite results (e.g., Tikkanen et al. 2010), so that the evidence from replication seems to be, overall, strong. The reasons for discrepancies are manifold and include, among others, variation in sample characteristics (age, gender, ethnicity), methodological features (study design, assessment), and definition of phenotype and environment studied, the implications of which for replicating G × E have been discussed at length (Caspi et al. 2010; Duncan and Keller 2011). Table 3 provides an overview of the major findings and the key characteristics of these studies. To this end, the studies are organized into groups based on the type of environment examined. The first group includes studies attempting to replicate the original finding by Caspi et al. in terms of childhood maltreatment, the second group focused on childhood adversity, a composite of multiple adversity indicators in the family such as low education, material deprivation, or poor parenting, and the third group examined other environmental factors affecting antisocial behavior, such as deviant peer affiliations.

One of the most rigorous tests of replication was carried out in a recent report by Fergusson et al. (2011a) with data of 398 males from a study that has strong methodological similarities to the original study by Caspi et al. in terms of geographical region, research design, and measurement methods. Using extensive analysis, Fergusson et al. were able to almost completely replicate the original findings of Caspi et al. for a series of antisocial outcomes spanning from adolescence to adulthood. This finding is of particular significance, as, at the same time, these authors were unable to confirm the even more extensively studied G × E landmark finding of Caspi et al. (2003) involving the serotonin transporter promoter polymorphism (5-HTTLPR) and life stress in conferring risk of depression (Fergusson et al. 2011b).

According to the current literature, the generality of the MAOA G × E extends to a wide range of *environments*. While the majority of studies replicated the original finding with regard to childhood maltreatment, a considerable number of investigations confirmed the interaction for a far broader spectrum of childhood adversities, encompassing, among others, death of a parent, material deprivation, or parental disengagement. In fact, the rate of nonreplications among these studies was even lower than among those attempting to replicate the original findings with regard to childhood maltreatment suggesting a more universal concept of environmental adversity to be effective in this interaction. Support for such a hypothesis comes from a number of studies which examined the MAOA G × E with regard to an even broader range of factors influencing the development of

Table 3 Molecular genetic studies of G × E involving MAOA

Study[a] (year)	N	Sex	Ethn.	Age	Sample	Outcome	Results G	Results E	Results G × E
Childhood maltreatment									
Caspi et al. (2002)	499	m	C	3–26	Com	CD, ASPD, violent offenses	–	+	+ L
Huang et al. (2004)	767	m, f	Mixed	37–38	Com	Aggression	–	+	–
Haberstick et al. (2005)	774	m	C	16–22	Com	Antisocial behavior	–	+	–
Huizinga et al. (2006)	277	m	C	37–43	Com	CD, ASPD, violence	–	+	–
Kim-Cohen et al. (2006)	975	m	C	7	Com	Antisocial behavior	+ L	+	+ L
Widom and Brzustowicz (2006)	631	m, f	C	41	CC	CD, ASPD, violence	–	+	+ L
Young et al. (2006)	347	m	Mixed	12–18	Clinical	CD	–	+	–
Sjoberg et al. (2007)	119	f	C	16–19	Com	Violence	–	+	–
Ducci et al. (2008)	291	f	NA	38	CC	ASPD	+ L	+	+ L
Prichard et al. (2008)	1,002	m	C	20–24	Com	Antisocial behavior	–	+	–
van der Vegt et al. (2009)	239	m	Mixed	10–15	Adoptee	Antisocial behavior	+ H	+	–
Weder et al. (2009)	114	m, f	Mixed	5–15	CC	Aggression	–	+	+ L
Beach et al. (2010)	538	m, f	C	45	Adoptee	ASPD	–	+	f: + L
Derringer et al. (2010)	841	m, f	C	25	Twins	Antisocial behavior, CD	–	+	+ L
Edwards et al. (2010)	250	m	C	6–22	Com	Aggression	–	+	–
Tikkanen et al. (2010)	174	m	C	33	Clinical	Violent offenses	–	+	+ H
Fergusson et al. (2011a)	398	m	C	19–30	Com	CD, violence	+ L	+	+ L
Aslund et al. (2011)	1,825	m, f	Mixed	17–18	Com	Violence	–	+	m: + L, f: + H
Philibert et al. (2011)	518	m, f	Mixed	45	CC	ASPD	–	+	f: + L
Cicchetti et al. (2012)	312	m	Mixed	10–12	CC	Antisocial behavior	–	+	+ L
Childhood adversity									
Foley et al. (2004)	514	m		8–17	Twins	CD	–	+	+ L

(continued)

Table 3 (continued)

Study[a] (year)	N	Sex	Ethn.	Age	Sample	Outcome	Results		
							G	E	G × E
Nilsson et al. (2006)	81	m	C	19	Com	Violence	–	+	+ L
Frazzetto et al. (2007)	235	m, f	C	31	CC	Aggression	–	+	m: + L
Reif et al. (2007)	169	m	C	34	Clinical	Violence	+ L	+	–
Kinnally et al. (2009)	159	f	C	39	Clinical	Aggression	–	+	+ L low care
Prom-Wormley et al. (2009)	721	f	C	8–17	Com	CD	+ L	+	+ H
Enoch et al. (2010)	6,129	m, f	C	4–7	Com	Conduct problems	–	+	+ L
Fergusson et al. (2012)	399	m	C	19–30	Com	Violence	+ L	+	+ L
Other environments									
Beaver et al. (2008)	1.994	m, f	C	18–27	Com	Desistance from delinquency	–	+	m: + L
McDermott et al. (2009)	78	m	Mixed	22	Com	Aggression (punishment)	+ L	+	+ L
Wakschlag et al. (2010)	176	m, f	C	15	Com	CD	–	+	m: + L, f: + H
Lee (2011)	672	m	C	12–20	Com	Aggression	–	+	+ H

m males, *f* females, *ethn.* ethnicity, *C* Caucasian, *NA* native American, *Com* community sample, *CC* case–control, *CD* conduct disorder, *ASPD* antisocial personality disorder, + significant main/interaction effect, with susceptible, *L* low activity genotype, *H* high activity genotype

[a] First author

violent behavior. For example, a most recent publication by Fergusson et al. (2012) revealed that the MAOA genotype interacted with multiple risk factors for antisocial behavior, including both personal and environmental factors, such as low childhood IQ and prenatal cigarette exposure. The latter finding is in accordance with another study by Wakschlag et al. (2010) which examined the interaction between MAOA genotype and maternal smoking during pregnancy in predicting adolescent conduct disorder using data from a pregnancy cohort with well-characterized exposure. Results revealed a sex specific pattern of G × E, with male carriers of the low activity genotype and female carriers of the high activity genotype being at increased risk for conduct disorder following prenatal smoke exposure.

Three further studies provided evidence of additional extensions of the G × E involving MAOA related to antisocial behavior. Using a sample of 672 male adolescents from the National Longitudinal Study of Adolescent Health (Add Health), Lee (2011) demonstrated a moderating effect of MAOA on the association between deviant peer affiliations, a potent risk factor for antisocial behavior, and aggression. Contrary to the hypotheses, a stronger impact of deviant peer affiliations on aggressive behavior was observed among individuals with the high activity genotype than among those carrying the low activity genotype. A study by Beaver et al. (2008) examined the interaction between MAOA and marital status, a well-replicated protective factor against involvement in antisocial behavior, in predicting desistance from delinquency. In a large sample of 1,555 young adults from Add Health, these authors reported that male offenders who had married were more likely to desist when carrying the low rather than the high activity genotype. The authors discussed their finding with regard to the concept of differential susceptibility (Belsky et al. 2009), which posits that the same genetic variants found to increase maladaptive behavior following exposure to adverse environments may also increase adaptive behavior under conditions of favorable environments (discussed more fully below). Finally, a most interesting study by McDermott et al. (2009) applied an experimental design to examine the role of MAOA in aggressive behavior. Using a "hot sauce" paradigm, the authors demonstrated that males with the low activity genotype displayed greater aggression in high provocation situations, while they did not differ from high activity males in low provocation situations.

Another conclusion from the current literature is that the MAOA G × E appears to hold independent of *age*. A number of investigations confirmed the interaction between MAOA genotype and child maltreatment with regard to antisocial behavior in studies of maltreated children (e.g., Weder et al. 2009), adolescents (e.g., Aslund et al. 2011), and adults (e.g., Ducci et al. 2008). In contrast, more inconsistency has been observed with regard to *gender*. While several studies replicated the original G × E in females (e.g., Weder et al. 2009), others reported effects in the reverse direction, with the high activity allele conferring risk of violent behavior in females exposed to maltreatment (e.g., Aslund et al. 2011), and others still found no interaction at all (e.g., Frazzetto et al. 2007). As a potential source of heterogeneity, differences in MAOA expression between males and females have been discussed. Since the MAOA gene is sex-linked and located on the X chromosome, females have two alleles of this gene, while males have only one. Typically, in females, only one allele is active due to random X inactivation. As this implies a degree of uncertainty regarding the relationship of MAOA genotype to enzyme activity, several studies excluded heterozygous females from the analysis. However, the process of X inactivation at the MAOA site is not yet fully understood (Pinsonneault et al. 2006).

Remarkably, all published studies used the MAOA-LPR polymorphism as a gene marker, thereby, facilitating comparison between studies. Additionally, a study by Ducci et al. (2008) tested the hypothesis that haplotypes covering the region of MAOA and the nearby monoamine oxidase B (MAOB) would outperform the MAOA-LPR locus in predicting G × E. The authors demonstrated that a haplotype-based analysis strengthened the association, while failing to ascertain a significant interaction for MAOB. Recently, Philibert et al. (2011) described a novel polymorphism located in the promoter region next to the classical variant that may be more influential in regulating MAOA expression, accounting for additional variance in the G × E.

Serotonin Transporter Gene. To date, about 13 studies have examined whether a functional polymorphism (5-HTTLPR) in the promoter region of the serotonin transporter gene moderates the effect of environmental factors on violent behavior and related phenotypes. Table 4 provides an overview of the major findings and the key characteristics of these studies. Compared to MAOA, findings are mixed to a greater extent. A number of studies reported G × E effects, suggesting that individuals carrying the S allele associated with reduced transporter availability were more likely to show higher levels of antisocial behavior when exposed to childhood adversity (e.g., Cicchetti et al. 2012). However, there is discussion as to whether this interaction applies irrespective of gender or ethnicity, given that several studies found the G × E to be confined to females (e.g., Li and Lee 2010) or to African–Americans (Douglas et al. 2011). Even more importantly, two studies reported an effect in the opposite direction, with carriers of the LL genotype displaying more violence under conditions of adversity (Aslund et al. 2012; Nobile et al. 2007). Notably, in both studies, measures of socioeconomic status (SES) were used as a proxy for environmental adversity. In contrast, in another study with SES-defined adversity, the SS genotype conferred a higher risk of callous-unemotional and narcissistic, but not impulsive features of psychopathy in adolescents from low as compared to high SES backgrounds (Sadeh et al. 2010). A further study using recent stress as a measure of adversity demonstrated a G × E involving 5-HTTLPR, which predicted an increase in aggressive behavior during adolescence among individuals with the S allele compared to those with the LL genotype when exposed to chronic but not to acute stress (Conway et al. 2012). In contrast, a study by Sonuga-Barke et al. (2009) demonstrated that males diagnosed with ADHD who carried the S allele were more responsive to the mother's expressed emotions, with elevated levels of conduct problems following exposure to maternal hostility and criticism. Finally, three studies examining the interaction of 5-HTTLPR with additional risk (deviant peer affiliations, prenatal

Table 4 Molecular genetic studies of G × E involving serotonergic genes (5-HTTLPR)

Study[a] (year)	N	Ethn.	Age	Sample	Outcome	Results		
						G	E	G × E
Childhood adversity								
Reif et al. (2007)	169	C	34	Clinical	Violence	–	+	+ S
Li and Lee (2010)	2,488	Mixed	16	Com	Antisocial behavior	–	+	f: + SS
Douglas et al. (2011)	1,381	Mixed	38	Clinical	ASPD	AA: + S	+	AA f: + SS
Simons et al. (2011)	505	AA	12–21	Com	Aggression	–	+	–
Cicchetti et al. (2012)	627	Mixed	10–12	CC	Antisocial behavior	–	+	+ SS
Socioeconomic status								
Nobile et al. (2007)	610	C	10–14	Com	Aggression	+ LL	+	+ LL
Sadeh et al. (2010)	180	Mixed	14	CC	Psychopathy	+ S	+	AA: + SS
Aslund et al. (2012)	1,467	C	17–18	Com	Violent delinquency	–	+	m: + LL, f: + S
Other environments								
Beaver et al. (2008)	1,994	C	18–27	Com	Desistance from delinquency	–	+	–
Langley et al. (2008)	131	C	9	Clinical	CD symptoms	–	+	–
Sonuga-Barke et al. (2009)	728	C	5–17	Clinical	CD	–	+	+ S
Vaughn (2009)	1,801	Mixed	18–26	Com	Violence	–	+	–
Conway et al. (2012)	381	C	15–20	Com	Aggression	–	+	+ S

m males, *f* females, *ethn.* ethnicity, *C* Caucasian, *AA* African American, *Com* community sample, *CC* case–control, *CD* conduct disorder, *ASPD* antisocial personality disorder, + significant main/interaction effect, with susceptible carriers of the S allele/SS/LL genotype

[a] First author

adversity) and protective factors (marital status) of violent behavior were unable to establish significant G × E (Beaver et al. 2007; Langley et al. 2008; Vaughn et al. 2009).

Dopamine Genes. About 20 studies have examined G × E on violence and related phenotypes with respect to dopamine genes (see Table 5 for an overview of the major findings). These encompass the dopamine receptor genes DRD2 and 4, the gene for the dopamine transporter (DAT) and the gene for the enzyme catechol-O-methytransferase (COMT). COMT has a key role in inactivating dopamine in the prefrontal cortex. Usually, the A1 allele of a 3′Taq1 polymorphism in DRD2, the 7-repeat (7r) allele of a 48-bp VNTR polymorphism in exon 3 of DRD4 (for another polymorphism, Martel et al. 2011), and the 10-repeat (10r) allele of a 40-bp VNTR polymorphism in DAT are considered as "risk alleles," which may be linked to a less efficient dopaminergic system (Pavlov et al. 2012). For COMT, homozygosity for the val allele of the Val158Met polymorphism is frequently regarded as risk genotype. The val allele is associated with higher enzymatic activity and, as a consequence, less prefrontal dopamine availability. However, higher levels of violent behavior have also been shown in carriers of the met allele, e.g., in male schizophrenic patients (Singh et al. 2012), possibly reflecting the fact that the impact of prefrontal dopamine variation on behavior is complex.

Interestingly, slightly different environments prevail in G × E studies on violent behavior involving dopamine genes when compared to MAOA and 5-HTTLPR. Thus, only one study examined a moderating effect of the COMT polymorphism on the impact of childhood maltreatment (Wagner et al. 2010). In contrast, several studies investigated whether dopamine genes moderated the effects of an adverse prenatal environment. Prenatal adversity may be inferred from a low birth weight or maternal smoking during pregnancy. Langley et al. (2008) found that ADHD children born with a low birth weight who were homozygous for the DAT 10r allele showed elevated levels of childhood conduct problems. A similar result was reported by Thapar et al. (2005) for ADHD children homozygous for the COMT val allele (though not replicated in a comparable study by Sengupta et al. 2006). Brennan et al. (2011) found no interaction between COMT genotype and birth weight in a community sample of young adults. Participants, however, were described as more aggressive when they were homozygous carriers of the COMT val allele and their mothers had been smoking during pregnancy. An enhanced sensitivity to the adverse effects of prenatal smoke exposure with regard to ADHD and ODD symptoms was also demonstrated for children homozygous for the DAT 10r allele (Becker et al. 2008, Kahn et al. 2003).

Another environmental factor often investigated in dopamine-related G × E studies is parenting, which, however, may not necessarily be independent of genotype (e.g., Propper et al. 2007). Most relevant work focused on the DRD4 polymorphism. A much cited study by Bakermans-Kranenburg and van Ijzendoorn (2006) showed that only toddlers who were exposed to insensitive maternal care and, at the same time, were carriers of the DRD4 7r allele were prone to

Table 5 Molecular genetic studies of G × E involving dopaminergic genes (DRD2, DRD4, DAT, COMT)

Study[a] (year)	N	Ethn.	Age	Sample	Outcome	Results		
						G	E	G × E
Prenatal adversity								
Kahn et al. (2003)	161	Mixed	5	Com	ODD symptoms	DAT: –	–	+ 10r/10r
Thapar et al. (2005)	240	C	5–14	Clinical	CD symptoms	COMT: + valval	+	+ valval
Sengupta et al. (2006)	191	Mixed	6–12	Clinical	CD, CD symptoms	COMT: –	–	–
Becker et al. (2008)	305	C	15	Com	ODD/CD symptoms	DAT: –	m: +	m: + 10r/10r
Langley et al. (2008)	266	n.r.	9	Clinical	ODD/CD symptoms	DRD4 DAT: n.r.	n.r.	(+) 7r + 10r/10r
Brennan et al. (2011)	430	C	15, 20	Com	Aggression	COMT: –	+	+ valval
Rearing environment								
Bakermans-Kranenburg and van Ijzendoorn (2006)	47	n.r.	3	Twin	Externalizing behavior	DRD4: –	–	+ 7r
Propper et al. (2007)	169	Mixed	2.5	Com	Externalizing behavior	DRD4: –	+	AA: + 2–6r neg.
Bakermans-Kranenburg et al. (2008)	157	C	1–3	Com	Decrease in externalizing behavior post intervention	DRD4: –	–	+ 7r
DeLisi et al. (2008)	324	Mixed	18–26	Com	Age at first police contact/arrest	DRD2: – DRD4: + 7r	n.r.	+ A1 –/+ 7r
Sonuga-Barke et al. (2009)	728	C	5–17	Clinical	Conduct problems	DAT: – DRD4: –	+	+ 9r/10r, (9r/9r) –
Lahey et al. (2011)	162	Mixed	9–14	CC	CD symptoms	DAT: –	–	+ 9r/9r neg.
Martel et al. (2011)	548	Mixed	6–18	CC	ODD symptoms	DRD4 120 bp: –	+	+ long/long
Socioeconomic status								
Nobile et al. (2007)	607	C	10–14	Com	Aggression	DRD4: +	+	+ 6–8r
Nobile et al. (2010)	575	C	10–14	Com	ODD/CD symptoms	COMT: –	+	–

(continued)

Table 5 (continued)

Study[a] (year)	N	Ethn.	Age	Sample	Outcome	Results		
						G	E	G × E
Other environments								
Beaver et al. (2008)	1,994	Mixed	18–27	Com	Desistance from delinquency	DAT: –	+	n.r.
						DRD2: –		m: + A1
						DRD4: + 7r		m: + 7r
Guo et al. (2008)	1,130	Mixed	12–23	twin	Delinquency	DAT: + 10r	+	n.r.
						DRD2: + A1/A2		+A1/A2
Beaver et al. (2009)	814	Mixed	12–18	Com	Delinquency	DRD2: –	+	+ A1
Vaughn et al. (2009)	1,801	Mixed	18–26	Com	Violence	DAT: –	+	+ 10r
Wagner et al. (2010)	112	C	n.r.	Clinical	Impulsive aggression	COMT: –	+ neg.	+ valval neg.

m males, *ethn.* ethnicity, *C* Caucasian, *AA* African American, *Com* community sample, *CC* case–control, *CD* conduct disorder, *neg.* negatively related to the outcome, *n. r.* not reported, *ODD* oppositional defiant disorder, + significant main/interaction effect, (+) trend or partial evidence

[a] First author

externalizing behavior. In carriers of other DRD4 alleles, maternal insensitivity had no effect. Most interestingly, toddlers possessing a DRD4 7r allele also profited most from a parent intervention program aimed at reducing externalizing behavior (Bakermans-Kranenburg et al. 2008). This and similar findings gave rise to the hypothesis that the so-called "risk alleles" of dopamine receptor and transporter gene polymorphisms generally enhance the sensitivity to environmental conditions (Bakermans-Kranenburg and van Ijzendoorn 2011). Hence, a higher risk for violent behavior may result from exposure to adverse, but less violent behavior from exposure to favorable conditions. Support for this so-called differential susceptibility hypothesis has been found for various environments: For example, young adults possessing a DRD2 A1 or DRD4 7r allele showed a later crime onset than participants with other genotypes when they reported a close relationship to their mother (DeLisi et al. 2008). As well, they had a higher desistance from delinquency after marriage (Beaver et al. 2008). In contrast, more aggression was observed in young DRD4 7r carriers growing up in socioeconomic disadvantaged conditions (Nobile et al. 2007).

Summary and Limitations. Overall, molecular genetic studies have provided significant evidence in support of violence-related G × E in humans. The majority of studies reviewed above found at least some evidence to suggest a G × E effect, with the interaction involving MAOA being the best studied and confirmed. Of the 32 studies investigating MAOA, 23 revealed evidence in support of G × E. Although inconsistent results concerning the susceptible genotype were reported, in most cases, carriers of the low activity allele were found to be at higher risk. With respect to the other genes examined, evidence was more mixed. Findings of studies investigating 5-HTTLPR are conflicting and cannot be considered as replications. The literature is marked by a wide variation in the definition of environmental factors with little overlap between studies. Further research is essential in order to clarify a possible moderating role of 5-HTTLPR in the development of violent behavior. Findings of studies examining G × E involving dopamine genes suggest a role in moderating prenatal adversity and rearing conditions, but a replication of results is scarce.

A salient feature that characterizes molecular genetic research of G × E is the substantial heterogeneity among samples, data collection strategies, environmental factors, and outcome measures used to conduct studies of G × E. In particular, studies differ considerably with respect to the modeling and measurement of environmental exposures. A wide range of exposures have been investigated using various assessment methods. However, a systematic analysis of which environments are likely candidates to reliably and validly examine G × E effects on violent behavior is overdue. Recently, Jaffee et al. (2012) provided a comprehensive evaluation of relevant environments revealing that, for the most part, the suggested environmental candidates were not specific to violence, but conferred risk of a range of adverse health outcomes. Together with the inconsistent patterns of results, this ambiguity makes it difficult to interpret and summarize findings and understand the nature of G × E effects related to violent behavior.

4 Directions for Future Research

Although progress has been made in increasing our understanding of the role attributed to G × E in the development of violence, many issues remain to be resolved. A major problem of molecular genetic research into G × E is that interactions involving single genes prove difficult to replicate due to small effect sizes (Duncan and Keller 2011). An attractive opportunity to extend traditional approaches is to investigate the ways in which multiple genes interact with multiple environmental factors to influence individual susceptibility to violence. In the current post-GWAS era, the advent of *genome–environment-wide association studies* promises to provide the methodological basis upon which to simultaneously examine the impact of multiple genetic and environmental factors and their interplay (Aschard et al. 2012).

Another avenue for developing research in the field is to examine G × E affecting *endophenotypes* in order to uncover the complex biological mechanisms underlying G × E. As discussed elsewhere in this volume, the endophenotypes related to human violence and their genetic risk variants involve structure and function of limbic and cortical regions such as amygdala, hippocampus, and medial prefrontal as well as cingulate cortex (Buckholtz and Meyer-Lindenberg 2008), plus anterior temporal and rostral prefrontal areas in antisocial offenders with psychopathy (Gregory et al. 2012). Similar structural deviance is already evident in children and adolescents with conduct disorder (Huebner et al. 2008) or callous-unemotional traits (Viding and McCrory 2012). Neurophysiological endophenotypes of violence include altered oscillatory resting EEG asymmetries due to reduced right frontal activation (Keune et al. 2012), attenuations of attentional P300 activity as central markers (Gao and Raine 2009), and psychophysiological deviations with hypoarousal and reduced variability of cardiac and skin conductance as autonomous system measures (Lorber 2004; Ortiz and Raine 2004).

Recently, a number of studies have employed neurophysiology and neuroimaging to investigate G × E (Hyde et al. 2011). Although few of these studies have focused on violence-related genes and endophenotypes, they directly confirmed that environmental factors stratified by genotype impacted on brain activity and morphology in regions associated with violence. The findings of G × E with regard to amygdala reactivity to affective faces were mixed, revealing interactive effects between childhood adversity and the FKBP5 gene encoding a glucocorticoid receptor (GR) regulating co-chaperone shown to alter GR sensitivity resulting in increased amygdala activity in carriers of the risk allele (White et al. 2012), but additive effects for the serotonin transporter (Walsh et al. 2012). Structural MR evidence also points to additive effects of risks due to the serotonin transporter variant and those due to methamphetamine abuse related to aggression. For the gene encoding the brain-derived neurotrophic factor BDNF, the met risk variant was found to interact with elevated early life stress to produce reduced amygdala volumes, higher cardiac reactivity (Gatt et al. 2009), enhanced right frontal EEG reactivity (Gatt et al. 2010), and reduced subgenual anterior cingulate cortex (ACC) volumes (Gerritsen et al. 2012). Starting from this encouraging evidence, future research should aim to further elucidate the specific neural correlates of G × E that affect violent behavior and render individuals susceptible to adverse environmental effects.

Endophenotypes (including some psychophysiological markers) have been criticized for not being associated with more robust gene effects than the phenotypes (Flint and Munafo 2007). However, more specific neurophysiological markers of impulsivity such as the NoGo P300 component, which has proven useful for uncovering genetic mechanisms (Baehne et al. 2009; Dresler et al. 2010; Reif et al. 2009) and genetic epistasis (Heinzel et al. 2012) of inhibitory response control, still remain to be tested for G × E. Although gene main effects are no prerequisite for detecting G × E, the use of neuroimaging-based endophenotypes to isolate effects on task-related core brain regions and networks appears most promising. The success of this specific high resolution approach is illustrated by the meta-analytic finding of robust but opposite COMT effects on brain activation in emotional and executive tasks obtained in functional neuroimaging studies (Mier et al. 2010). Most of the less specific but reliable markers of impaired brain and autonomic function in violence such as reduced neural activity to targets (amplitudes of the P300 component), reduced skin conductance, and heart rate variability still remain to be examined for G × E. Initial findings suggest that the endophenotype approach may be important to dissect genetic pathways to aggression, as genetic influences on psychophysiological (cardiac) function appeared to mediate the effects on aggression in a longitudinal twin study (Baker et al. 2009).

Several mechanisms have been discussed as determining the transduction of environmental influences into changes in brain physiology and morphology. Among these, a major role has been attributed to epigenetic regulation (Meaney 2010; Roth and Sweatt 2011). *Epigenetics* refers to modifications to the genome that occur without altering the DNA sequence. Such modifications include chemical marks that regulate the transcription of the genome. Ample evidence of such "environmental programming" has demonstrated that environmental conditions, particularly early in life, can directly alter the epigenetic state of the genome leading to changes in gene expression and neural function that persist into adulthood. Epigenetic research into violence will provide exciting new insights into G × E mechanisms (Tremblay and Szyf 2010). For example, epigenetic differences in DNA methylation due to childhood aggression have recently been linked to lower 5-HT expression in orbitofrontal cortex using peripheral methylation measures in blood samples (Wang et al. 2012). Although methylation is unlikely to affect single genes, epigenetic research may complicate the current picture through new G × E findings of the expression of new genomic regions that were not previously implicated in violence. Alternatively, epigenetic findings may also clarify mechanisms underlying established G × E interactions, particularly, when combined with longitudinal research on the course of G × E across development.

The majority of G × E findings reported above have been interpreted within a *diathesis-stress* framework, suggesting that an underlying genetic predisposition increases the probability of adverse conditions to foster violent behavior. In recent years, this concept has been challenged by a model that conceptualizes G × E as *differential susceptibility* to environmental influences (Belsky et al. 2007; Belsky and Pluess 2009). According to this perspective, individuals vary in their plasticity or susceptibility to environmental exposure, i.e., due to their particular genetic make-up, some individuals may be more responsive than others to environmental conditions regardless of whether these are positive or negative. Hence, such individuals will suffer particularly under conditions of adversity, but will simultaneously show the utmost benefit from exposure to favorable environments. Several studies reviewed above provide illustrative examples of differential susceptibility, indicating that groups with genetic risk that were more likely to display violent behavior when exposed to adversity actually developed less violence in the absence thereof (Kinnally et al. 2009; Simons et al. 2011). Future studies that encompass the full range of the environmental spectrum are needed to investigate whether the effects of genes implicated in violence-related G × E studies are directional or increase variability in response to the environment.

For the most part, the reviewed G × E were not *specific* to violence, but extended to a broader range of antisocial behavior, including conduct disorder, antisocial personality disorder, delinquency, and psychopathy. Moreover, most of the identified G × E did not differentiate between different types of violent behavior, such as reactive (impulsive) versus proactive (instrumental) aggression. One exception to this is MAOA, where both human and animal studies point to a functional role in impulsive-aggressive behavior. Although the mechanism of interaction is still unclear, Buckholtz and Meyer-Lindenberg (2008) suggested that a reduction in social evaluation and emotional regulation abilities in individuals low in MAOA activity may enhance the effects of adverse environments. More careful consideration of the specific behavioral outcomes associated with violence-related G × E will help to further uncover the mediating processes linking genes and environments to aggressive phenotypes.

To conclude, the study of G × E in violence offers prospects for a number of exciting possibilities (Moffitt et al. 2005). It may stimulate progress in basic neuroscience by providing new insights into the pathophysiology of violence-related disorders. Understanding G × E mechanisms may also provide useful hints with regard to the prevention of and intervention in these disorders. New findings in G × E may advance the development of individual treatment strategies and give new impulses for the field of pharmacogenetics. However, in order to fulfill its potential, future G × E research needs to shift its focus to dissecting the neural mechanisms and the underlying pathophysiological pathways by which genetic variation may influence differential susceptibility to environmental exposures.

References

Aschard H, Lutz S, Maus B et al (2012) Challenges and opportunities in genome-wide environmental interaction (GWEI) studies. Hum Genet 131:1591–1613

Aslund C, Comasco E, Nordquist N et al (2012) Self-reported family socioeconomic status, the 5-HTTLPR genotype, and delinquent behavior in a community-based adolescent population. Aggress Behav. doi:10.1002/ab.21451

Aslund C, Nordquist N, Comasco E et al (2011) Maltreatment, MAOA, and delinquency: sex differences in gene-environment interaction in a large population-based cohort of adolescents. Behav Genet 41:262–272

Baehne CG, Ehlis AC, Plichta MM et al (2009) Tph2 gene variants modulate response control processes in adult ADHD patients and healthy individuals. Mol Psychiatry 14:1032–1039

Baker LA, Tuvblad C, Reynolds C et al (2009) Resting heart rate and the development of antisocial behavior from age 9 to 14: genetic and environmental influences. Dev Psychopathol 21:939–960

Bakermans-Kranenburg MJ, van Ijzendoorn MH (2006) Gene–environment interaction of the dopamine D4 receptor (DRD4) and observed maternal insensitivity predicting externalizing behavior in preschoolers. Dev Psychobiol 48:406–409

Bakermans-Kranenburg MJ, van Ijzendoorn MH (2011) Differential susceptibility to rearing environment depending on dopamine-related genes: new evidence and a meta-analysis. Dev Psychopathol 23:39–52

Bakermans-Kranenburg MJ, van Ijzendoorn MH, Pijlman FT et al (2008) Experimental evidence for differential susceptibility: dopamine D4 receptor polymorphism (DRD4 VNTR) moderates intervention effects on toddlers' externalizing behavior in a randomized controlled trial. Dev Psychol 44:293–300

Beach SR, Brody GH, Gunter TD et al (2010) Child maltreatment moderates the association of MAOA with symptoms of depression and antisocial personality disorder. J Fam Psychol 24:12–20

Beaver KM, Gibson CL, Jennings WG et al (2009) A gene × environment interaction between DRD2 and religiosity in the prediction of adolescent delinquent involvement in a sample of males. Biodemography Soc Biol 55:71–81

Beaver KM, Sak A, Vaske J et al (2010) Genetic risk, parent-child relations, and antisocial phenotypes in a sample of African–American males. Psychiatry Res 175:160–164

Beaver KM, Wright JP, DeLisi M et al (2007) Evidence of a gene × environment interaction in the creation of victimization: results from a longitudinal sample of adolescents. Int J Offender Ther Comp Criminol 51:620–645

Beaver KM, Wright JP, DeLisi M et al (2008) Desistance from delinquency: the marriage effect revisited and extended. Soc Sci Res 37:736–752

Becker K, El Faddagh M, Schmidt MH et al (2008) Interaction of dopamine transporter genotype with prenatal smoke exposure on ADHD symptoms. J Pediatr 152:263–269

Belsky J, Bakermans-Kranenburg MJ, van Ijzendoorn MH (2007) For better and for worse: differential susceptibility to environmental influences. Curr Dir Psychol Sci 16:300–304

Belsky J, Jonassaint C, Pluess M et al (2009) Vulnerability genes or plasticity genes? Mol Psychiatry 14:746–754

Belsky J, Pluess M (2009) Beyond diathesis stress: differential susceptibility to environmental influences. Psychol Bull 135:885–908

Bevilacqua L, Carli V, Sarchiapone M et al (2012) Interaction between FKBP5 and childhood trauma and risk of aggressive behavior. Arch Gen Psychiatry 69:62–70

Boutwell BB, Franklin CA, Barnes JC et al (2011) Physical punishment and childhood aggression: the role of gender and gene–environment interplay. Aggress Behav 37:559–568

Brendgen M, Boivin M, Dionne G et al (2011) Gene–environment processes linking aggression, peer victimization, and the teacher–child relationship. Child Dev 82:2021–2036

Brendgen M, Boivin M, Vitaro F et al (2008a) Linkages between children's and their friends' social and physical aggression: evidence for a gene–environment interaction? Child Dev 79:13–29

Brendgen M, Boivin M, Vitaro F et al (2008b) Gene-environment interaction between peer victimization and child aggression. Dev Psychopathol 20:455–471

Brennan PA, Hammen C, Sylvers P et al (2011) Interactions between the COMT Val108/158Met polymorphism and maternal prenatal smoking predict aggressive behavior outcomes. Biol Psychol 87:99–105

Buckholtz JW, Meyer-Lindenberg A (2008) MAOA and the neurogenetic architecture of human aggression. Trends Neurosci 31:120–129

Button TM, Corley RP, Rhee SH et al (2007) Delinquent peer affiliation and conduct problems: a twin study. J Abnorm Psychol 116:554–564

Button TM, Scourfield J, Martin N et al (2005) Family dysfunction interacts with genes in the causation of antisocial symptoms. Behav Genet 35:115–120

Cadoret RJ, Cain CA, Crowe RR (1983) Evidence for gene-environment interaction in the development of adolescent antisocial behavior. Behav Genet 13:301–310

Cadoret RJ, Yates WR, Troughton E et al (1995) Genetic–environmental interaction in the genesis of aggressivity and conduct disorders. Arch Gen Psychiatry 52:916–924

Caspi A, Hariri AR, Holmes A et al (2010) Genetic sensitivity to the environment: the case of the serotonin transporter gene and its implications for studying complex diseases and traits. Am J Psychiatry 167:509–527

Caspi A, McClay J, Moffitt TE et al (2002) Role of genotype in the cycle of violence in maltreated children. Science 297:851–854

Caspi A, Moffitt TE, Cannon M et al (2005) Moderation of the effect of adolescent-onset cannabis use on adult psychosis by a functional polymorphism in the catechol-O-methyltransferase gene: longitudinal evidence of a gene × environment interaction. Biol Psychiatry 57:1117–1127

Caspi A, Sugden K, Moffitt TE et al (2003) Influence of life stress on depression: moderation by a polymorphism in the 5-HTT gene. Science 301:386–389

Cicchetti D, Rogosch FA, Thibodeau EL (2012) The effects of child maltreatment on early signs of antisocial behavior: genetic moderation by tryptophan hydroxylase, serotonin transporter, and monoamine oxidase A genes. Dev Psychopathol 24:907–928

Cloninger CR, Sigvardsson S, Bohman M et al (1982) Predisposition to petty criminality in Swedish adoptees. II. Cross-fostering analysis of gene–environment interaction. Arch Gen Psychiatry 39:1242–1247

Conway CC, Keenan-Miller D, Hammen C et al (2012) Coaction of stress and serotonin transporter genotype in predicting aggression at the transition to adulthood. J Clin Child Adolesc Psychol 41:53–63

Crowe RR (1974) An adoption study of antisocial personality. Arch Gen Psychiatry 31:785–791

DeLisi M, Beaver KM, Wright JP et al (2008) The etiology of criminal onset: The enduring salience of nature and nurture. J Crim Justice 36:217–223

Derringer J, Krueger RF, Irons DE et al (2010) Harsh discipline, childhood sexual assault, and MAOA genotype: an investigation of main and interactive effects on diverse clinical externalizing outcomes. Behav Genet 40:639–648

Dick DM, Meyers JL, Latendresse SJ et al (2011) CHRM2, parental monitoring, and adolescent externalizing behavior: evidence for gene–environment interaction. Psychol Sci 22:481–489

Douglas K, Chan G, Gelernter J et al (2011) 5-HTTLPR as a potential moderator of the effects of adverse childhood experiences on risk of antisocial personality disorder. Psychiatr Genet 21:240–248

Dresler T, Ehlis AC, Heinzel S et al (2010) Dopamine transporter (SLC6A3) genotype impacts neurophysiological correlates of cognitive response control in an adult sample of patients with ADHD. Neuropsychopharmacology 35:2193–2202

Ducci F, Enoch MA, Hodgkinson C et al (2008) Interaction between a functional MAOA locus and childhood sexual abuse predicts alcoholism and antisocial personality disorder in adult women. Mol Psychiatry 13:334–347

Duncan LE, Keller MC (2011) A critical review of the first 10 years of candidate gene-by-environment interaction research in psychiatry. Am J Psychiatry 168:1041–1049

Eaves LJ (2006) Genotype × environment interaction in psychopathology: fact or artifact? Twin Res Hum Genet 9:1–8

Edwards AC, Dodge KA, Latendresse SJ et al (2010) MAOA-uVNTR and early physical discipline interact to influence delinquent behavior. J Child Psychol Psychiatry 51:679–687

Enoch MA, Steer CD, Newman TK et al (2010) Early life stress, MAOA, and gene–environment interactions predict behavioral disinhibition in children. Genes Brain Behav 9:65–74

Feinberg ME, Button TM, Neiderhiser JM et al (2007) Parenting and adolescent antisocial behavior and depression: evidence of genotype × parenting environment interaction. Arch Gen Psychiatry 64:457–465

Fergusson DM, Boden JM, Horwood LJ et al (2012) Moderating role of the MAOA genotype in antisocial behaviour. Br J Psychiatry 200:116–123

Fergusson DM, Boden JM, Horwood LJ et al (2011a) MAOA, abuse exposure and antisocial behaviour: 30-year longitudinal study. Br J Psychiatry 198:457–463

Fergusson DM, Horwood LJ, Miller AL et al (2011b) Life stress, 5-HTTLPR and mental disorder: findings from a 30-year longitudinal study. Br J Psychiatry 198:129–135

Flint J, Munafo MR (2007) The endophenotype concept in psychiatric genetics. Psychol Med 37:163–180

Foley DL, Eaves LJ, Wormley B et al (2004) Childhood adversity, monoamine oxidase a genotype, and risk for conduct disorder. Arch Gen Psychiatry 61:738–744

Frazzetto G, Di Lorenzo G, Carola V et al (2007) Early trauma and increased risk for physical aggression during adulthood: the moderating role of MAOA genotype. PLoS ONE 2:e486

Gao Y, Raine A (2009) P3 event-related potential impairments in antisocial and psychopathic individuals: a meta-analysis. Biol Psychol 82:199–210

Gatt JM, Nemeroff CB, Dobson-Stone C et al (2009) Interactions between BDNF Val66Met polymorphism and early life stress predict brain and arousal pathways to syndromal depression and anxiety. Mol Psychiatry 14:681–695

Gatt JM, Nemeroff CB, Schofield PR et al (2010) Early life stress combined with serotonin 3A receptor and brain-derived neurotrophic factor valine 66 to methionine genotypes impacts emotional brain and arousal correlates of risk for depression. Biol Psychiatry 68:818–824

Gerritsen L, Tendolkar I, Franke B et al (2012) BDNF Val66Met genotype modulates the effect of childhood adversity on subgenual anterior cingulate cortex volume in healthy subjects. Mol Psychiatry 17:597–603

Gregory S, ffytche D, Simmons A et al (2012) The antisocial brain: psychopathy matters. Arch Gen Psychiatry 69:962–972

Guo G, Roettger ME, Cai T (2008) The integration of genetic propensities into social-control models of delinquency and violence among male youths. Am Soc Rev 73:543–568

Haberstick BC, Lessem JM, Hopfer CJ et al (2005) Monoamine oxidase A (MAOA) and antisocial behaviors in the presence of childhood and adolescent maltreatment. Am J Med Genet B Neuropsychiatr Genet 135B:59–64

Heinzel S, Dresler T, Baehne CG et al (2012) COMT × DRD4 epistasis impacts prefrontal cortex function underlying response control. Cereb Cortex 23:1453–1462

Huang YY, Cate SP, Battistuzzi C et al (2004) An association between a functional polymorphism in the monoamine oxidase a gene promoter, impulsive traits and early abuse experiences. Neuropsychopharmacology 29:1498–1505

Huebner T, Vloet TD, Marx I et al (2008) Morphometric brain abnormalities in boys with conduct disorder. J Am Acad Child Adolesc Psychiatry 47:540–547

Huizinga D, Haberstick BC, Smolen A et al (2006) Childhood maltreatment, subsequent antisocial behavior, and the role of monoamine oxidase A genotype. Biol Psychiatry 60:677–683

Hyde LW, Bogdan R, Hariri AR (2011) Understanding risk for psychopathology through imaging gene–environment interactions. Trends Cogn Sci 15:417–427
Jaffee SR, Caspi A, Moffitt TE et al (2005) Nature X nurture: genetic vulnerabilities interact with physical maltreatment to promote conduct problems. Dev Psychopathol 17:67–84
Jaffee SR, Strait LB, Odgers CL (2012) From correlates to causes: can quasi-experimental studies and statistical innovations bring us closer to identifying the causes of antisocial behavior? Psychol Bull 138:272–295
Kahn RS, Khoury J, Nichols WC et al (2003) Role of dopamine transporter genotype and maternal prenatal smoking in childhood hyperactive-impulsive, inattentive, and oppositional behaviors. J Pediatr 143:104–110
Keune PM, van der Heiden L, Varkuti B et al (2012) Prefrontal brain asymmetry and aggression in imprisoned violent offenders. Neurosci Lett 515:191–195
Kim-Cohen J, Caspi A, Taylor A et al (2006) MAOA, maltreatment, and gene-environment interaction predicting children's mental health: new evidence and a meta-analysis. Mol Psychiatry 11:903–913
Kinnally EL, Huang YY, Haverly R et al (2009) Parental care moderates the influence of MAOA-uVNTR genotype and childhood stressors on trait impulsivity and aggression in adult women. Psychiatr Genet 19:126–133
Lahey BB, Rathouz PJ, Lee SS et al (2011) Interactions between early parenting and a polymorphism of the child's dopamine transporter gene in predicting future child conduct disorder symptoms. J Abnorm Psychol 120:33–45
Lamb DJ, Middeldorp CM, van Beijsterveldt CE et al (2012) Gene–environment interaction in teacher-rated internalizing and externalizing problem behavior in 7- to 12-year-old twins. J Child Psychol Psychiatry 53:818–825
Langley K, Turic D, Rice F et al (2008) Testing for gene × environment interaction effects in attention deficit hyperactivity disorder and associated antisocial behavior. Am J Med Genet B Neuropsychiatr Genet 147B:49–53
Latendresse SJ, Bates JE, Goodnight JA et al (2011) Differential susceptibility to adolescent externalizing trajectories: examining the interplay between CHRM2 and peer group antisocial behavior. Child Dev 82:1797–1814
Lee SS (2011) Deviant peer affiliation and antisocial behavior: interaction with Monoamine Oxidase A (MAOA) genotype. J Abnorm Child Psychol 39:321–332
Leve LD, Kerr DC, Shaw D et al (2010) Infant pathways to externalizing behavior: evidence of genotype × environment interaction. Child Dev 81:340–356
Li JJ, Lee SS (2010) Latent class analysis of antisocial behavior: interaction of serotonin transporter genotype and maltreatment. J Abnorm Child Psychol 38:789–801
Lorber MF (2004) Psychophysiology of aggression, psychopathy, and conduct problems: a meta-analysis. Psychol Bull 130:531–552
Maher B (2008) Personal genomes: the case of the missing heritability. Nature 456:18–21
Martel MM, Nikolas M, Jernigan K et al (2011) The dopamine receptor D4 gene (DRD4) moderates family environmental effects on ADHD. J Abnorm Child Psychol 39:1–10
McDermott R, Tingley D, Cowden J et al (2009) Monoamine oxidase A gene (MAOA) predicts behavioral aggression following provocation. Proc Natl Acad Sci USA 106:2118–2123
Meaney MJ (2010) Epigenetics and the biological definition of gene × environment interactions. Child Dev 81:41–79
Mier D, Kirsch P, Meyer-Lindenberg A (2010) Neural substrates of pleiotropic action of genetic variation in COMT: a meta-analysis. Mol Psychiatry 15:918–927
Moffitt TE (2005) The new look of behavioral genetics in developmental psychopathology: gene–environment interplay in antisocial behaviors. Psychol Bull 131:533–554
Moffitt TE, Caspi A, Rutter M (2005) Strategy for investigating interactions between measured genes and measured environments. Arch Gen Psychiatry 62:473–481
Moffitt TE, Caspi A, Rutter M (2006) Measured gene–environment interactions in psychopathology. Perspect Psychol Sci 1:5–27

Munafo MR, Flint J (2009) Replication and heterogeneity in gene × environment interaction studies. Int J Neuropsychopharmacol 12:727–729

Nilsson KW, Sjoberg RL, Damberg M et al (2006) Role of monoamine oxidase A genotype and psychosocial factors in male adolescent criminal activity. Biol Psychiatry 59:121–127

Nobile M, Giorda R, Marino C et al (2007) Socioeconomic status mediates the genetic contribution of the dopamine receptor D4 and serotonin transporter linked promoter region repeat polymorphisms to externalization in preadolescence. Dev Psychopathol 19:1147–1160

Nobile M, Rusconi M, Bellina M et al (2010) COMT Val158Met polymorphism and socioeconomic status interact to predict attention deficit/hyperactivity problems in children aged 10–14. Eur Child Adolesc Psychiatry 19:549–557

Ortiz J, Raine A (2004) Heart rate level and antisocial behavior in children and adolescents: a meta-analysis. J Am Acad Child Adolesc Psychiatry 43:154–162

Pavlov KA, Chistiakov DA, Chekhonin VP (2012) Genetic determinants of aggression and impulsivity in humans. J Appl Genet 53:61–82

Philibert RA, Wernett P, Plume J et al (2011) Gene environment interactions with a novel variable Monoamine Oxidase A transcriptional enhancer are associated with antisocial personality disorder. Biol Psychol 87:366–371

Pinsonneault JK, Papp AC, Sadee W (2006) Allelic mRNA expression of X-linked monoamine oxidase a (MAOA) in human brain: dissection of epigenetic and genetic factors. Hum Mol Genet 15:2636–2649

Prichard Z, Mackinnon A, Jorm AF et al (2008) No evidence for interaction between MAOA and childhood adversity for antisocial behavior. Am J Med Genet B Neuropsychiatr Genet 147B:228–232

Prom-Wormley EC, Eaves LJ, Foley DL et al (2009) Monoamine oxidase A and childhood adversity as risk factors for conduct disorder in females. Psychol Med 39:579–590

Propper C, Willoughby M, Halpern CT et al (2007) Parenting quality, DRD4, and the prediction of externalizing and internalizing behaviors in early childhood. Dev Psychobiol 49:619–632

Reif A, Jacob CP, Rujescu D et al (2009) Influence of functional variant of neuronal nitric oxide synthase on impulsive behaviors in humans. Arch Gen Psychiatry 66:41–50

Reif A, Rosler M, Freitag CM et al (2007) Nature and nurture predispose to violent behavior: serotonergic genes and adverse childhood environment. Neuropsychopharmacology 32:2375–2383

Rhee SH, Waldman ID (2002) Genetic and environmental influences on antisocial behavior: a meta analysis of twin and adoption studies. Psychol Bull 128:490–529

Roth TL, Sweatt JD (2011) Epigenetic marking of the BDNF gene by early-life adverse experiences. Horm Behav 59:315–320

Rowe DC, Almeida DM, Jacobson KC (1999) School context and genetic influences on aggression in adolescence. Psychol Sci 10:277–280

Rutter M, Moffitt TE, Caspi A (2006) Gene–environment interplay and psychopathology: multiple varieties but real effects. J Child Psychol Psychiatry 47:226–261

Sadeh N, Javdani S, Jackson JJ et al (2010) Serotonin transporter gene associations with psychopathic traits in youth vary as a function of socioeconomic resources. J Abnorm Psychol 119:604–609

Sengupta SM, Grizenko N, Schmitz N et al (2006) COMT Val108/158Met gene variant, birth weight, and conduct disorder in children with ADHD. J Am Acad Child Adolesc Psychiatry 45:1363–1369

Simons RL, Lei MK, Beach SR et al (2011) Social environmental variation, plasticity genes, and aggression: evidence for the differential susceptibility hypothesis. Am Sociol Rev 76:833–912

Singh JP, Volavka J, Czobor P et al (2012) A meta-analysis of the Val158Met COMT polymorphism and violent behavior in schizophrenia. PLoS ONE 7:e43423

Sjoberg RL, Nilsson KW, Wargelius HL et al (2007) Adolescent girls and criminal activity: role of MAOA-LPR genotype and psychosocial factors. Am J Med Genet B Neuropsychiatr Genet 144B:159–164

Sonuga-Barke EJ, Oades RD, Psychogiou L et al (2009) Dopamine and serotonin transporter genotypes moderate sensitivity to maternal expressed emotion: the case of conduct and emotional problems in attention deficit/hyperactivity disorder. J Child Psychol Psychiatry 50:1052–1063
Taylor A, Kim-Cohen J (2007) Meta-analysis of gene–environment interactions in developmental psychopathology. Dev Psychopathol 19:1029–1037
Thapar A, Harold G, Rice F et al (2007) The contribution of gene–environment interaction to psychopathology. Dev Psychopathol 19:989–1004
Thapar A, Langley K, Fowler T et al (2005) Catechol O-methyltransferase gene variant and birth weight predict early-onset antisocial behavior in children with attention-deficit/hyperactivity disorder. Arch Gen Psychiatry 62:1275–1278
Thomas D (2010) Gene–environment-wide association studies: emerging approaches. Nat Rev Genet 11:259–272
Tikkanen R, Ducci F, Goldman D et al (2010) MAOA alters the effects of heavy drinking and childhood physical abuse on risk for severe impulsive acts of violence among alcoholic violent offenders. Alcohol Clin Exp Res 34:853–860
Tremblay RE, Szyf M (2010) Developmental origins of chronic physical aggression and epigenetics. Epigenomics 2:495–499
Tuvblad C, Baker LA (2011) Human aggression across the lifespan: genetic propensities and environmental moderators. Adv Genet 75:171–214
Tuvblad C, Grann M, Lichtenstein P (2006) Heritability for adolescent antisocial behavior differs with socioeconomic status: gene–environment interaction. J Child Psychol Psychiatry 47:734–743
Uher R, McGuffin P (2008) The moderation by the serotonin transporter gene of environmental adversity in the aetiology of mental illness: review and methodological analysis. Mol Psychiatry 13:131–146
van der Vegt EJ, Oostra BA, Arias-Vasquez A et al (2009) High activity of monoamine oxidase A is associated with externalizing behaviour in maltreated and nonmaltreated adoptees. Psychiatr Genet 19:209–211
van Lier P, Boivin M, Dionne G et al (2007) Kindergarten children's genetic vulnerabilities interact with friends' aggression to promote children's own aggression. J Am Acad Child Adolesc Psychiatry 46:1080–1087
Vaughn M, DeLisi M, Beaver KM et al (2009) DAT1 and 5-HTT are associated with pathological criminal behavior in a nationally representative sample of youth. Crim Just Behav 36:1103–1114
Viding E, McCrory EJ (2012) Genetic and neurocognitive contributions to the development of psychopathy. Dev Psychopathol 24:969–983
Wagner S, Baskaya O, Anicker NJ et al (2010) The catechol o-methyltransferase (COMT) val(158)met polymorphism modulates the association of serious life events (SLE) and impulsive aggression in female patients with borderline personality disorder (BPD). Acta Psychiatr Scand 122:110–117
Wakschlag LS, Kistner EO, Pine DS et al (2010) Interaction of prenatal exposure to cigarettes and MAOA genotype in pathways to youth antisocial behavior. Mol Psychiatry 15:928–937
Walsh ND, Dalgleish T, Dunn VJ et al (2012) 5-HTTLPR-environment interplay and its effects on neural reactivity in adolescents. Neuroimage 63:1670–1680
Wang D, Szyf M, Benkelfat C et al (2012) Peripheral SLC6A4 DNA methylation is associated with in vivo measures of human brain serotonin synthesis and childhood physical aggression. PLoS ONE 7:e39501
Weder N, Yang BZ, Douglas-Palumberi H et al (2009) MAOA genotype, maltreatment, and aggressive behavior: the changing impact of genotype at varying levels of trauma. Biol Psychiatry 65:417–424
Wermter AK, Laucht M, Schimmelmann BG et al (2010) From nature versus nurture, via nature and nurture, to gene × environment interaction in mental disorders. Eur Child Adolesc Psychiatry 19:199–210

White MG, Bogdan R, Fisher PM et al (2012) FKBP5 and emotional neglect interact to predict individual differences in amygdala reactivity. Genes Brain Behav 11:869–878

Widom CS, Brzustowicz LM (2006) MAOA and the "cycle of violence": childhood abuse and neglect, MAOA genotype, and risk for violent and antisocial behavior. Biol Psychiatry 60:684–689

Young SE, Smolen A, Hewitt JK et al (2006) Interaction between MAO-A genotype and maltreatment in the risk for conduct disorder: failure to confirm in adolescent patients. Am J Psychiatry 163:1019–1025

Neurobiological Mechanisms for Impulsive-Aggression: The Role of MAOA

Hayley M. Dorfman, Andreas Meyer-Lindenberg and Joshua W. Buckholtz

Abstract Aggression may be present across a large part of the spectrum of psychopathology, and underlies costly criminal antisocial behaviors. Human aggression is a complex and underspecified construct, confounding scientific discovery. Nevertheless, some biologically tractable subtypes are apparent, and one in particular—impulsive (reactive) aggression—appears to account for many facets of aggression-related dysfunction in psychiatric illness. Impulsive-aggression is significantly heritable, suggesting genetic transmission. However, the specific neurobiological mechanisms that mediate genetic risk for impulsive-aggression remain unclear. Here, we review extant data on the genetics and neurobiology of individual differences in impulsive-aggression, with particular attention to the role of genetic variation in Monoamine Oxidase A (MAOA) and its impact on serotonergic signaling within corticolimbic circuitry.

Keywords Aggression · Violence · Corticolimbic · Amygdala · Genetic · MAOA · Prefrontal

Contents

H. M. Dorfman · J. W. Buckholtz
Department of Psychology, Harvard University, Cambridge, MA, USA

A. Meyer-Lindenberg
Central Institute of Mental Health, Mannheim, Germany

J. W. Buckholtz (✉)
Center for Brain Science, Harvard University, Cambridge, MA, USA
e-mail: joshuabuckholtz@fas.harvard.edu

Curr Topics Behav Neurosci (2014) 17: 297–313
DOI: 10.1007/7854_2013_272

Published Online: 28 January 2014

In 2011, over 1 million violent crimes were committed in the United States alone (Baltic 2011). While the physical and emotional consequences imposed on the victims of these crimes are severe, violent crime has larger social costs as well. Taking into account the aggregate costs of medical treatment, incarceration, and lost productivity, it is estimated that the US spends approximately 3.3 % of their gross domestic product (GDP) dealing with the aftermath of violent crimes (Waters et al. 2004). In Germany, the corresponding figure is 6.5 % of GDP. The staggering costs of antisocial aggression are all the more astounding given our relative ignorance of individual-level causal mechanisms. This is due in part to our limited understanding of the repertoire of aggressive behaviors in humans.

Other species have detailed taxonomies for aggressive behavior, but the lack of a consensus human analog hinders research progress. In animals, ethologists often distinguish between "functional" and "pathological" aggression, and this distinction may be useful here. Functional aggression is contextually appropriate and constrained by species-specific rules and rituals; the ritualized violence of professional football or hockey perhaps represents a human analog of this functional aggression. By contrast, pathological aggressive behaviors are out of context, unstructured, and do not adhere to norms that would otherwise serve to mitigate their damaging impact to both victim and community. Drilling down further, human pathological aggression can be further subtyped into reactive-impulsive and proactive-instrumental variants, mirroring ethological distinctions between defensive and predatory aggression (Krueger and South 2009; Valzelli and Bernasconi 1979; Valzelli et al. 1981a). These two forms of aggression are associated with discrete social, cognitive, and biological profiles, and likely have partially unique genetic diatheses (Valzelli et al. 1981a, b; Vergnes and Kempf 1981; Stark et al. 1985). Reactive-impulsive aggression appears to be the more common form, and impulsive-aggressive symptoms are common to numerous psychiatric conditions, including antisocial personality disorder, borderline personality disorder, psychopathy, and bipolar disorder. Overlap in symptom expression among these disorders suggests that common aggression-linked genetic and neurobiological alterations may account for some of their shared genetic variance (Krueger and South 2009). Uncovering the neurobiological underpinnings of aggression could thus expose a transdiagnostic risk mechanism that contributes to dysfunction across this large spectrum of psychopathology (Buckholtz and Meyer-Lindenberg 2012).

1 The Causal Architecture of Human Aggression

Though human aggression is complex, heterogeneous, and multifactorial, there is compelling evidence that genetic factors play a significant role in predisposing aggressive traits and behaviors. Work by Farrington and others provided some of the first clear empirical evidence for a familial clustering of crime within communities, validating a long-held folk intuition that crime "breeds true" (Farrington et al. 2001). These studies show that the majority of criminal offenses within a given community are committed by a small number of families in that community. The familial nature of antisocial behavior suggests that heritable factors may comprise a vehicle for its intergeneration transmission, and subsequent twin studies confirmed that genetic factors account for a moderate proportion of the population variance in antisocial traits and behaviors.

We note that while antisociality and aggression are related they are not synonymous; importantly, twin and family studies have also provided evidence for a genetic structure of aggression. Aggression-specific phenotypes show evidence of heritability in mice (Miczek et al. 2001) and nonhuman primates (Suomi 2003; Fairbanks et al. 2004) as do aggression-linked personality traits and aggressive symptoms in human adolescents and adults (Gelhorn et al. 2005, 2006; Coccaro et al. 1997). Of relevance to the distinction between antisocial behavior and aggression made above, aggressive antisocial behavior shows significantly higher heritability than does nonaggressive antisocial behavior (Eley et al. 2003; Button et al. 2004). Genetic contributions to aggression appear early (van Beijsterveldt et al. 2004) and are stable throughout childhood (Hudziak et al. 2003; Baker et al. 2007), and genetic factors account for variance in the transition from childhood to adolescent aggression (Valzelli and Bernasconi 1979; Beitchman et al. 2004; Jacob et al. 2005). Intriguingly, work in preadolescent boys suggests partially distinct genetic risk factors for reactive (impulsive) and proactive (instrumental) aggression (Baker et al. 2008). Other data in adults suggest genetic bases for other observed aggression subtypes, such as physical versus nonphysical (Yeh et al. 2010). Taken together, work to date supports the notion that genetic factors play an important role in susceptibility to aggressive behavior. However, none of the specific genetic variants involved were known until Han Brunner's landmark identification of *MAOA* in the early 1990s.

2 MAOA: The Original "Violence Gene"

The very first candidate gene for antisocial behavior, *MAOA*, was identified during molecular genetic analysis of a large, multigenerational, and notoriously violent, Dutch kindred (Brunner et al. 1993a). As the syndrome appeared to be transmitted via an X-linked mode of inheritance and was accompanied by mild mental retardation, Brunner's linkage analyses of this family focused on the X-chromosome.

The location of the linkage peak, combined with evidence of aberrant monoamine metabolism, implicated the *MAOA* locus on Xp11.23-11.4. Sequencing at this locus revealed a nonsense point mutation in exon 8 of the *MAOA* gene (C936T). Studies of MAO-A activity in fibroblasts donated by affected males confirmed the functional impact of the mutation, which can be considered to produce, for all intents and purposes, a human *MAOA* knockout (Brunner et al. 1993a, b).

MAOA encodes the mitochondrial catabolic enzyme monoamine oxidase A (MAO-A), which catalyzes the oxidative deamination of biogenic amines. MAO-A is one of two MAO isoforms (the other being MAOA-B); the genes encoding both map to adjacent sites on chromosome Xp11.23 (Grimsby et al. 1991). The monoamine oxidases are localized to the outer mitochondrial membrane in the presynaptic terminal of monoamine projection neurons (MAOA) (Arai et al. 2002; Westlund et al. 1993) and in astrocytes (MAO-A and MAO-B) (Westlund et al. 1988; Levitt et al. 1982) where they are positioned to regulate both the amount of intracellular substrate available for release and the degree of extrasynaptic monoamine inactivation. Functional genetic variation in *MAOA* is therefore likely to disrupt signaling at monoaminergic synapses throughout the brain (Shih and Thompson 1999).

Both MAO-A and MAO-B are comprised of 15 exons and demonstrate matching exon–intron organization, suggesting their derivation from the same ancestral gene by a duplication event (Grimsby et al. 1991). However, the two isoforms show significant divergence in regional expression and activity patterns (Jahng et al. 1997; Westlund et al. 1985; Willoughby et al. 1988; Fowler et al. 1997; Saura et al. 1996a, b). Other differences between the two isoforms are potentially instructive for understanding the *MAOA*-aggression linkage. First, while dopamine (DA) is a good substrate for both isoforms of MAO, MAO-A has higher affinity for serotonin (5-HT) and norepinephrine (NE). The relative importance MAO-A for regulating 5-HT and NE (i.e., vs. DA) is underscored by the finding that *MAOB* knockout mice show selective increases in brain 5HT and NE compared to wild-type, while DA is unaffected (Cases et al. 1995). Moreover, the neurochemical, neuromorphological and behavioral phenotypes produced by genetic deletion of *MAOB* are quite distinct from those of *MAOA* knockout mice. The explanation for the relative importance of MAO-A over MAO-B may pertain to the divergent developmental trajectories of MAO-A and MAO-B expression. MAO-A expression precedes that of MAO-B, and is the more critical for regulating monoamine catabolism ante-partum. While MAO-A is present at adult levels at birth, MAO-B appears only postnatally and subsequently exhibits a striking increase (Nicotra et al. 2004; Benedetti 1992; Tsang et al. 1986).

While the C936T nonsense mutation characterized by Brunner is fortunately rare, being a private mutation held only, as far as we know, by members of that specific Dutch family, many common polymorphic variants have been identified in the *MAOA* gene region. Of these, the most studied is a variable number of tandem repeats (VNTR) polymorphism located in the promoter region of the gene, 1.2 kilobase pairs upstream of its start site. Initially discovered by Sabol and colleagues, the *MAOA* upstream VNTR (*MAOA* u-VNTR) comprises 30 bp sequence that is repeated 2, 3, 3.5, 4, or 5 times (Sabol and Hamer 1998). 3, 3.5, and 4 repeat alleles are common, while 2 and 5 repeats are more rare. Repeat length has been shown to impact transcription and enzyme activity in vitro, with higher expression linked to 3.5 or 4 repeats (therefore termed *MAOA-H* or "high" alleles) relative to 3 repeats (likewise termed an *MAOA-L* or "low" allele) (Mertins et al. 2011; Garpenstrand et al. 2001; Daruna and Kent 1976; Syagailo et al. 2001). However, contradictory findings do exist, and the functional significance of the 2 and 5 repeats is still somewhat controversial (Balciuniene et al. 2002; Cirulli and Goldstein 2007).

3 The Role of MAOA in Aggression: Preclinical Work

A wealth of evidence implicates MAO-A as an important biological regulator of impulsive-aggressive behavior. Genetic deletion studies in rodents largely recapitulate the findings of Brunner and colleagues, with *MAOA* knockouts showing dramatically heightened reactive aggression, poor fear learning, and increased forebrain 5-HT and NE levels (Cases et al. 1995; Scott et al. 2008; Popova et al. 2001; Godar et al. 2010). Of note, a naturally occurring nonsense mutation ($MAOA^{A863T}$) has also been identified in mice, providing a closer parallel to the genetic anomaly described by Brunner and colleagues. Mutant allele carriers exhibit high levels of reactive aggression, including an increased frequency of fighting behaviors, amplified tail rattling, and reduced attack latency (Scott et al. 2008). Intriguingly, these animals also display aberrant exploratory behaviors: while they are slower to investigate objects and familiar environments, they show much quicker unconditioned escape and fear responses to discrete threat stimuli and threatening contexts (Godar et al. 2010). These deficits are not accompanied by significant visual, locomotor, or olfactory abnormalities, implying a selective impairment in emotional reactivity and risk assessment (Godar et al. 2010).

Molecular and behavioral pharmacology studies converge with the findings highlighted above, and suggest that, although the enzymatic activity of the gene product extends to several monoamines, the relationship between *MAOA* and aggression is specifically mediated by genetically determined dysregulation in serotonin signaling pathways. Murine *MAOA* knockouts show a significant increase in serotonin that is particularly dramatic early in life, with mutant pups showing a 9-fold increase compared to wild-type. Changes in DA and NE levels, while apparent, are much smaller in magnitude (Cases et al. 1995). Moreover, blunting postsynaptic serotonin signaling via the 5HT2A antagonist ketanserin reduces aggressive behavior in these animals (Shih and Chen 1999). Neuromorphological findings in *MAOA* knockouts also implicate excess serotonin signaling during development as a potential pathomechanism. *MAOA* deficient mice show cytoarchitectonic changes in sensory cortex, which are replicated by pharmacological attenuation of MAO-A by clorgyline. This pharmacologically induced neuromorphological phenocopy is developmentally specific; clorgyline administration replicates the knockout phenotype only when administered during a critical developmental window. Changes in neuronal morphology (e.g., dendritic size, length, branching) have also been observed in the orbitofrontal cortex and basolateral amygdala of *MAOA* knockout mice (Bortolato et al. 2011). Finally, pharmacological depletion of serotonin, but not of catecholamines, rescues the phenotype in *MAOA* deficient animals. Of note, this rescue is only observed during a specific developmental window (Cases et al. 1996), which is consistent with other work showing developmentally specific effects of aberrant serotonin signaling in producing adult affective illness (Ansorge et al. 2004, 2007; Gingrich et al. 2003; Gross and Hen 2004). The observation of sensitive periods during development suggests that there may be critical windows during youth and early adolescence for mitigating the impact of genetic risk (de Boer et al. 2009; Tost et al. 2010).

4 The Role of MAOA in Aggression: Clinical Data

A body of behavioral genetic data provides qualified support for the notion that lowered *MAOA* expression (e.g., in *MAOA*-L carriers) predisposes disinhibitory psychopathology. Impulsive-aggressive symptoms have been linked to reduced MAO-A activity in platelets and brain (Davis et al. 1983; Perris et al. 1984; af Klinteberg et al. 1987; Mukasa et al. 1990; Belfrage et al. 1992; Prochazka et al. 2003; Skondras et al. 2004; Eklund et al. 2005; Alia-Klein et al. 2008), and numerous associations to *MAOA-L* have been reported for substance abuse, (Contini et al. 2006; Saito et al. 2002; Parsian et al. 2003) antisocial personality disorder, and other clinical diagnoses related to pathological aggression (Mertins et al. 2011; Reti et al. 2011; Williams et al. 2009). Of note, associations have also been seen using quantitative behavioral phenotypes indexing impulsive-aggression. Gallardo-Pujol and colleagues found that *MAOA-L* allele carriers were substantially more aggressive in the point subtraction aggression paradigm than their *MAOA-H* counterparts, an effect that was magnified by provocations such as social exclusion and aversive acoustical stimuli (Gallardo-Pujol et al. 2013; Kuepper et al. 2013). Similarly, *MAOA-L* participants were significantly more likely than high-expressing allele carriers to administer physical punishment (hot sauce) to an opponent when they believed that 80 % of their earnings had been taken in a power-to-take task (McDermott et al. 2009). Low *MAOA* expression has also been linked to decreased cooperation in economic tasks, with *MAOA-L* males making lower contributions to community funds in public goods games (Mertins et al. 2011). Despite the positive findings reported above, support for *MAOA-L* as a causal variant for commonly used clinical diagnostic categories is, on the whole, mixed. For example, some groups have reported enhanced impulsive-aggression (Beitchman et al. 2004), elevated risk for Cluster B personality disorders (Jacob et al. 2005), lower CSF 5-HIAA levels (Jönsson et al. 2000) and blunted serotontegic responsivity in *MAOA-H*, rather than *MAOA-L* allele carriers

(Beitchman et al. 2004; Manuck et al. 2000, 2002; Doudet et al. 1995). Associations to the *MAOA-H* allele are particularly common in female samples, and in the presence of specific environmental factors (e.g. prenatal nicotine exposure; see below) (Sjöberg et al. 2008; Wakschlag et al. 2010).

5 Neurobiological Mechanisms for Genetic Risk

Prior functional imaging work in patients with impulsive-aggressive symptoms, including adults diagnosed with intermittent explosive disorder (Coccaro et al. 2007; Best et al. 2002), adolescents with conduct disorder (Stadler et al. 2007; Sterzer et al. 2005), and convicted murderers (Raine et al. 1994, 1997, 1998), have demonstrated amygdala hyperactivity and reduced prefrontal engagement, particularly during emotion processing tasks. This is especially interesting in light of preclinical research suggesting that amygdala and medial prefrontal cortex are wired together as a negative regulatory circuit (Amaral and Price 1984; Carmichael and Price 1995, 1996; Ghashghaei and Barbas 2002; Ongür and Price 2000; Stefanacci and Amaral 2002; Cavada et al. 2000). Moreover, lesion studies support the notion that disrupting this circuit produces deficits in emotion regulation and impulse control (Anderson et al. 1999; Goursaud and Bachevalier 2007; Machado and Bachevalier 2006; Butter and Snyder 1972; Blair and Cipolotti 2000; Clark and Manes 2004). Together, these data indicate that genetic risk for impulsive-aggression may exert its effects by disrupting corticolimbic circuits for emotional arousal and regulation.

5.1 Brain Structure

One of the first studies to implicate *MAOA* in the development of corticolimbic circuitry was a morphometric analysis of women with X-chromosome abnormalities (e.g., Turner syndrome). Good (2003) found that such women had increased gray matter volume in both amygdala and orbitofrontal cortex. Notably, these structural changes were accompanied by deficits in facial emotion recognition, particularly for fearful facial expressions (Good 2003). Subsequent studies have shown that *MAOA-L* allele carriers show pronounced structural reductions in limbic and paralimbic regions involved in socio-affective processing. For example, Meyer-Lindenberg and colleagues (Meyer-Lindenberg et al. 2006) reported a reduction in gray matter volume for *MAOA-L* allele carriers within the amygdala, insula, and cingulated gyrus, as well as an increase in lateral orbitofrontal volume (for males only) (Meyer-Lindenberg et al. 2006). These findings have been independently replicated (for cingulate and orbitofrontal cortex) (Alia-Klein et al. 2011; Kalin et al. 2008; Roiser et al. 2009; Cerasa et al. 2010).

5.2 Brain Function

If *MAOA* genetic variation accounts for some of the variance in risk for impulsive-aggressive behavior, risk allele carriers should show information processing differences in brain circuits subserving emotion regulation and inhibitory control. This would be consistent with a model of risk wherein multiple genes, each of small effect size, impact susceptibility in a nondeterministic fashion by exerting a relatively deleterious influence over neural circuits that are associated with symptom-specific aspects of cognition. Supporting this notion, functional MRI studies in *MAOA-L* carriers show differences in frontoparietal and corticolimbic circuit function during tasks that index impulse control, affective arousal, and emotional memory. Early work in this area found decreased anterior cingulate and ventrolateral prefrontal cortex activation during a combined conflict-resolution and response inhibition in *MAOA-L* subjects (Fan et al. 2003; Passamonti et al. 2008). Building on this work, we (Meyer-Lindenberg et al. 2006) used a battery of tasks to measure the impact of *MAOA* genetic variation on affective responsiveness, emotional memory, and impulse control. Across these tasks, *MAOA-L* carriers showed a pattern of enhanced subcortical (amygdala, hippocampus, insula) activation in the context of reduced cortical (ACC, orbitofrontal cortex) recruitment (Meyer-Lindenberg et al. 2006).

While the work described above was performed in healthy volunteers, imaging genetic studies in patient samples provide further evidence of a role for *MAOA* in impulsive-aggression and related traits. For example, Nymberg and colleagues used fMRI in a sample of adolescent ADHD patients to measure the impact of *MAOA* genetic variation on neural circuits for response inhibition and reward learning. They found a significant relationship between ADHD symptoms and frontostriatal activation during Stop-Signal and Monetary Incentive Delay tasks, which was moderated by *MAOA* genotype. Specifically, carriers of a putatively low-expressing allele at an intronic SNP (rs12843268) had reduced activation in ventral striatum and inferior frontal gyrus during the reward and response inhibition tasks, respectively. These findings suggest that alterations in motivation and inhibitory control in ADHD may be driven, in part, by *MAOA* genetic variability (Nymberg et al. 2013). Other work suggests that alterations in a frontoparietal circuit for working memory mediates the relationship between *MAOA* genetic variation and externalizing symptoms in adolescents (Ziermans et al. 2012).

5.3 Brain Connectivity

Collectively, work to date suggests that low-expressing *MAOA* variants dysregulate corticolimbic circuitry for affective arousal and inhibitory control. In particular, *MAOA-L* carriers show a pattern of exaggerated amygdala activation in the context of reduced regulatory control by prefrontal cortex—especially anterior cingulate and inferior frontal gyrus. Given the wealth of data suggesting that these cortical regions exert a negative regulatory influence over amygdala (Rebsam et al. 2002; Salichon et al. 2001)—along with evidence suggesting that this circuit is disordered in patients with impulsive aggressive symptoms (Coccaro et al. 2011)—we and others have sought to investigate the impact of *MAOA* genetic variation on functional coupling within this network. Using functional connectivity analyses in a sample of healthy volunteers, we have shown that *MAOA-L* carriers demonstrate aberrant coupling between amygdala and ventromedial prefrontal cortex (vmPFC) during emotion tasks (Buckholtz and Meyer-Lindenberg 2008). Of note, connections between amygdala and vmPFC are sparse (Ghashghaei and Barbas 2002; Ongür and Price 2000; Ghashghaei et al. 2007), suggesting the involvement of another region with strong anatomical connections to both; subsequent analyses identified perigenual anterior cingulate as playing this mediating role. These findings, which have been replicated by others (Dannlowski et al. 2009) support the hypothesis that *MAOA* variation disrupts the functional coupling of amygdala with vmPFC to induce emotion regulation deficits.

6 GxE Interactions

As noted in preceding sections, inconsistent associations make claims for a direct linkage between the *MAOA* uVNTR and impulsive-aggression less than compelling. However, a more consistent picture emerges when environmental influences are taken into full account. A landmark study by (Caspi et al. 2002) found that early life maltreatment significantly potentiated the relationship between *MAOA* genotype and adult antisocial behavior. Specifically, *MAOA-L* men who were previously exposed to early life abuse exhibited significantly higher levels of violent behavior compared to *MAOA-H* men (Caspi et al. 2002). These data have been consistently replicated in both men and women (Foley et al. 2004; Huang et al. 2004; Nilsson et al. 2006; Kim-Cohen et al. 2006; Widom and Brzustowicz 2006; Frazzetto et al. 2007; Reif et al. 2007; Ducci et al. 2006, 2008; Kinnally et al. 2009; Enoch et al. 2010; Edwards et al. 2010; Fergusson et al. 2011; Derringer et al. 2010; Weder et al. 2013; Haberstick et al. 2013), making this one of the best supported GxE observations in the entire literature of psychiatric genetics.

Subsequent work in rodents provides some mechanistic insight into the nature of this gene-by-environment interaction. Animals who experienced prepubertal stress exhibited an increase in basal metabolic rate in the amygdala, which was accompanied by decreased activity in orbitofrontal cortex. Connectivity analyses revealed a negative association between the medial orbital cortex and amygdala activity which was disrupted in stressed animals, providing further evidence for corticolimbic circuit dysfunction in impulsive-aggression (Márquez et al. 2013). Critically, stressed rats also showed a significant increase in prefrontal *MAOA* expression, along with enhanced aggression and anxiety (Márquez et al. 2013). On the surface, these findings would appear to contradict prior work linking aggression to lower *MAOA* expression. However, some have suggested that both extremely high and extremely low *MAOA* expression can contribute to impulsive-aggression, particularly during developmental sensitive periods (de Boer et al. 2009; Tost et al. 2010). Consistent with this, clorgyline administration reduced aggression in stressed-exposed animals while enhancing aggression in control animals. Together, these data suggest that environmental factors that predispose aggression may do so, in part, by modulating *MAOA* expression during developmental critical periods.

Intriguingly, GxE interactions with *MAOA* do not appear to be limited to early life stress, as *MAOA* genotype appears to potentiate the deleterious impact of a range of adverse environmental exposures. For example, prenatal exposure to cigarette smoke—which contains compounds that affect *MAOA* expression—magnifies the impact of the *MAOA-L* allele on risk for conduct disorder symptoms (Wakschlag et al. 2010). Importantly, this effect appears to be independent from the impact of parenting style and childhood stress exposure (Wakschlag et al. 2010). Also of note, *MAOA* appears to modulate the impact of adverse environmental influences in adulthood as well as childhood and adolescence. Alia-Klein and colleagues have shown that the deleterious effects of chronic cocaine use on impulsivity and emotion regulation are potentiated in *MAOA-L* carriers (Verdejo-García 2013), an effect mediated through corticolimbic circuits (e.g., orbitofrontal cortex and hippocampus) (Alia-Klein et al. 2011). This is consistent with the notion that excess serotonin signaling during development sensitizes or labilizes a core corticolimbic circuit for emotional arousal and regulation (Buckholtz and Meyer-Lindenberg 2008), making it more vulnerable to a wide range of adverse environmental exposures throughout the lifespan.

7 Conclusion

In this review, we highlight evidence that functional genetic variation in *MAOA* predisposes risk for impulsive-aggression by disrupting corticolimbic circuitry for affective arousal and regulation. Both preclinical and human work to date strongly suggests that these effects are due to dysregulated serotonin signaling during a critical developmentally sensitive period. However, it is worth noting here that impulsive-aggression is etiologically complex, with serotonergic alterations comprising but one causal factor. In particular, social neuropeptides such as vasopressin and oxytocin are also likely to play an critical role in mediating impulsive-aggression given their known involvement in social behavior generally and aggression regulation specifically (Ferris et al. 2005; Heinrichs and Domes 2008). Crucially, serotonin-sensitive corticolimbic circuitry related to emotion regulation and social cognition also seems to be influenced by variability in human neuropeptide signaling (Tost et al. 2010; Meyer-Lindenberg et al. 2009). Finally, it is imperative to reiterate that, considered in isolation, effect sizes for single genetic variants such as *MAOA* u-VNTR are small. Thus, while it may be tempting to think of *MAOA* as a "violence" gene, the majority of individuals who harbor common risk alleles at this locus would not meet diagnostic criteria for impulsive-aggressive disorders. One can think of such alleles as exerting a small influence on the development of a complex system, which may serve to magnify, unmask, or protect against the impact of other influences (e.g., other variants at other genetic loci, environmental exposures) on this system. To that end, large scale, multimodal imaging studies of the interplay between polygenic risk profiles and a variety of environment factors will be key to unraveling the manner by which individual genetic background, modified by experience, translates into individual behavior (Buckholtz and Meyer-Lindenberg 2012).

References

af Klinteberg B, Schalling D, Edman G, Oreland L, Asberg M (1987) Personality correlates of platelet monoamine oxidase (MAO) activity in female and male subjects. Neuropsychobiology 18:89–96

Alia-Klein N et al (2008) Brain monoamine oxidase A activity predicts trait aggression. J Neurosci 28:5099–5104

Alia-Klein N et al (2011) Gene x disease interaction on orbitofrontal gray matter in cocaine addiction. Arch Gen Psychiatry 68:283–294

Amaral DG, Price JL (1984) Amygdalo-cortical projections in the monkey (Macaca fascicularis). J Comp Neurol 230:465–496

Anderson SW, Bechara A, Damasio H, Tranel D, Damasio AR (1999) Impairment of social and moral behavior related to early damage in human prefrontal cortex. Nat Neurosci 2:1032–1037

Ansorge MS, Hen R, Gingrich JA (2007) Neurodevelopmental origins of depressive disorders. Curr Opin Pharmacol 7:8–17

Ansorge MS, Zhou M, Lira A, Hen R, Gingrich JA (2004) Early-life blockade of the 5-HT transporter alters emotional behavior in adult mice. Science 306:879–881

Arai R et al (2002) Differential subcellular location of mitochondria in rat serotonergic neurons depends on the presence and the absence of monoamine oxidase type B. Neuroscience 114:825–835

Baker LA, Jacobson KC, Raine A, Lozano DI, Bezdjian S (2007) Genetic and environmental bases of childhood antisocial behavior: a multi-informant twin study. J Abnorm Psychol 116:219–235

Baker LA, Raine A, Liu J, Jacobson KC (2008) Differential genetic and environmental influences on reactive and proactive aggression in children. J Abnorm Child Psychol 36:1265–1278

Balciuniene J, Emilsson L, Oreland L, Pettersson U, Jazin E (2002) Investigation of the functional effect of monoamine oxidase polymorphisms in human brain. Hum Genet 110:1–7

Baltic S (2011) Crime in the United States 2011—Google books

Beitchman JH, Mik HM, Ehtesham S, Douglas L, Kennedy JL (2004) MAOA and persistent, pervasive childhood aggression. Mol Psychiatry 9:546–547

Belfrage H, Lidberg L, Oreland L (1992) Platelet monoamine oxidase activity in mentally disordered violent offenders. Acta Psychiatr Scand 85:218–221

Best M, Williams JM, Coccaro EF (2002) Evidence for a dysfunctional prefrontal circuit in patients with an impulsive aggressive disorder. Proc Natl Acad Sci USA 99:8448–8453

Blair RJ, Cipolotti L (2000) Impaired social response reversal. A case of 'acquired sociopathy'. Brain 123(Pt 6):1122–1141

Bortolato M et al (2011) Social deficits and perseverative behaviors, but not overt aggression, in MAO-A hypomorphic mice. Neuropsychopharmacology 36:2674–2688

Brunner HG et al (1993b) X-linked borderline mental retardation with prominent behavioral disturbance: phenotype, genetic localization, and evidence for disturbed monoamine metabolism. Am J Hum Genet 52:1032–1039

Brunner HG, Nelen M, Breakefield XO, Ropers HH, van Oost BA (1993a) Abnormal behavior associated with a point mutation in the structural gene for monoamine oxidase A. Science 262:578–580

Buckholtz JW, Meyer-Lindenberg A (2008) MAOA and the neurogenetic architecture of human aggression. Trends Neurosci 31:120–129

Buckholtz JW, Meyer-Lindenberg A (2012) Psychopathology and the human: toward a transdiagnostic model of risk for mental illness. Neuron 74:990–1004

Butter CM, Snyder DR (1972) Alterations in aversive and aggressive behaviors following orbital frontal lesions in rhesus monkeys. Acta Neurobiol Exp (Wars) 32:525–565

Button TMM, Scourfield J, Martin N, McGuffin P (2004) Do aggressive and non-aggressive antisocial behaviors in adolescents result from the same genetic and environmental effects? Am J Med Genet B Neuropsychiatr Genet 129B:59–63

Carmichael ST, Price JL (1995) Limbic connections of the orbital and medial prefrontal cortex in macaque monkeys. J Comp Neurol 363:615–641

Carmichael ST, Price JL (1996) Connectional networks within the orbital and medial prefrontal cortex of macaque monkeys. J Comp Neurol 371:179–207

Cases O et al (1995) Aggressive behavior and altered amounts of brain serotonin and norepinephrine in mice lacking MAOA. Science 268:1763–1766

Cases O et al (1996) Lack of barrels in the somatosensory cortex of monoamine oxidase A-deficient mice: role of a serotonin excess during the critical period. Neuron 16:297–307

Caspi A et al (2002) Role of genotype in the cycle of violence in maltreated children. Science 297:851–854

Cavada C, Compañy T, Tejedor J, Cruz-Rizzolo RJ, Reinoso-Suárez F (2000) The anatomical connections of the macaque monkey orbitofrontal cortex. A review. Cereb Cortex 10:220–242

Cerasa A et al (2010) Morphological correlates of MAO A VNTR polymorphism: new evidence from cortical thickness measurement. Behav Brain Res 211:118–124

Cirulli ET, Goldstein DB (2007) In vitro assays fail to predict in vivo effects of regulatory polymorphisms. Hum Mol Genet 16:1931–1939

Clark L, Manes F (2004) Social and emotional decision-making following frontal lobe injury. Neurocase 10:398–403

Coccaro EF, Kavoussi RJ, Cooper TB, Hauger RL (1997) Central serotonin activity and aggression: inverse relationship with prolactin response to d-fenfluramine, but not CSF 5-HIAA concentration, in human subjects. Am J Psychiatry 154:1430–1435

Coccaro EF, McCloskey MS, Fitzgerald DA, Phan KL (2007) Amygdala and orbitofrontal reactivity to social threat in individuals with impulsive aggression. Biol Psychiatry 62:168–178

Coccaro EF, Sripada CS, Yanowitch RN, Phan KL (2011) Corticolimbic function in impulsive aggressive behavior. Biol Psychiatry 69:1153–1159

Contini V, Marques FZC, Garcia CED, Hutz MH, Bau CHD (2006) MAOA-uVNTR polymorphism in a Brazilian sample: further support for the association with impulsive behaviors and alcohol dependence. Am J Med Genet B Neuropsychiatr Genet 141B:305–308

Dannlowski U et al (2009) Reduced amygdala-prefrontal coupling in major depression: association with MAOA genotype and illness severity. Int J Neuropsychopharmacol 12:11–22

Daruna JH, Kent EW (1976) Comparison of regional serotonin levels and turnover in the brain of naturally high and low aggressive rats. Brain Res 101:489–501

Davis BA, Yu PH, Boulton AA, Wormith JS, Addington D (1983) Correlative relationship between biochemical activity and aggressive behaviour. Prog Neuropsychopharmacol Biol Psychiatry 7:529–535

de Boer SF, Caramaschi D, Natarajan D, Koolhaas JM (2009) The vicious cycle towards violence: focus on the negative feedback mechanisms of brain serotonin neurotransmission. Front Behav Neurosci 3:52
Derringer J, Krueger RF, Irons DE, Iacono WG (2010) Harsh discipline, childhood sexual assault, and MAOA genotype: an investigation of main and interactive effects on diverse clinical externalizing outcomes. Behav Genet 40:639–648
Doudet D et al (1995) Cerebral glucose metabolism, CSF 5-HIAA levels, and aggressive behavior in rhesus monkeys. Am J Psychiatry 152:1782–1787
Ducci F et al (2006) A functional polymorphism in the MAOA gene promoter (MAOA-LPR) predicts central dopamine function and body mass index. Mol Psychiatry 11:858–866
Ducci F et al (2008) Interaction between a functional MAOA locus and childhood sexual abuse predicts alcoholism and antisocial personality disorder in adult women. Mol Psychiatry 13:334–347
Edwards AC et al (2010) MAOA-uVNTR and early physical discipline interact to influence delinquent behavior. J Child Psychol Psychiatry 51:679–687
Eklund J, Alm PO, af Klinteberg B (2005) Monoamine oxidase activity and tri-iodothyronine level in violent offenders with early behavioural problems. Neuropsychobiology 52:122–129
Eley TC, Lichtenstein P, Moffitt TE (2003) A longitudinal behavioral genetic analysis of the etiology of aggressive and nonaggressive antisocial behavior. Dev Psychopathol 15:383–402
Enoch M-A, Steer CD, Newman TK, Gibson N, Goldman D (2010) Early life stress, MAOA, and gene-environment interactions predict behavioral disinhibition in children. Genes Brain Behav 9:65–74
Fairbanks LA et al (2004) Genetic contributions to social impulsivity and aggressiveness in vervet monkeys. Biol Psychiatry 55:642–647
Fan J, Fossella J, Sommer T, Wu Y, Posner MI (2003) Mapping the genetic variation of executive attention onto brain activity. Proc Natl Acad Sci USA 100:7406–7411
Farrington DP, Jolliffe D, Loeber R, Stouthamer-Loeber M, Kalb LM (2001) The concentration of offenders in families, and family criminality in the prediction of boys' delinquency. J Adolesc 24:579–596
Fergusson DM, Boden JM, Horwood LJ, Miller AL, Kennedy MA (2011) MAOA, abuse exposure and antisocial behaviour: 30-year longitudinal study. Br J Psychiatry 198:457–463
Ferris CF (2005) Vasopressin/oxytocin and aggression. Novartis Found. Symp 268:190–8–discussion 198–200– 242–53
Foley DL et al (2004) Childhood adversity, monoamine oxidase a genotype, and risk for conduct disorder. Arch Gen Psychiatry 61:738–744
Fowler JS et al (1997) Age-related increases in brain monoamine oxidase B in living healthy human subjects. Neurobiol Aging 18:431–435
Frazzetto G et al (2007) Early trauma and increased risk for physical aggression during adulthood: the moderating role of MAOA genotype. PLoS ONE 2:e486
Gallardo-Pujol D, Andrés-Pueyo A, Maydeu-Olivares A (2013) MAOA genotype, social exclusion and aggression: an experimental test of a gene-environment interaction. Genes Brain Behav 12:140–145
Garpenstrand H, Annas P, Ekblom J, Oreland L, Fredrikson M (2001) Human fear conditioning is related to dopaminergic and serotonergic biological markers. Behav Neurosci 115:358–364
Gelhorn HL et al (2005) Genetic and environmental influences on conduct disorder: symptom, domain and full-scale analyses. J Child Psychol Psychiatry 46:580–591
Gelhorn HL et al (2006) Common and specific genetic influences on aggressive and nonaggressive conduct disorder domains. J Am Acad Child Adolesc Psychiatry 45:570–577
Ghashghaei HT, Barbas H (2002) Pathways for emotion: interactions of prefrontal and anterior temporal pathways in the amygdala of the rhesus monkey. Neuroscience 115:1261–1279
Ghashghaei HT, Hilgetag CC, Barbas H (2007) Sequence of information processing for emotions based on the anatomic dialogue between prefrontal cortex and amygdala. NeuroImage 34:905–923

Gingrich JA, Ansorge MS, Merker R, Weisstaub N, Zhou M (2003) New lessons from knockout mice: the role of serotonin during development and its possible contribution to the origins of neuropsychiatric disorders. CNS Spectr 8:572–577
Godar SC et al (2010) Maladaptive defensive behaviours in monoamine oxidase A-deficient mice. Int J Neuropsychopharmacol 14:1195–1207
Good CD (2003) Dosage-sensitive X-linked locus influences the development of amygdala and orbitofrontal cortex, and fear recognition in humans. Brain 126:2431–2446
Goursaud A-PS, Bachevalier J (2007) Social attachment in juvenile monkeys with neonatal lesion of the hippocampus, amygdala and orbital frontal cortex. Behav Brain Res 176:75–93
Grimsby J, Chen K, Wang LJ, Lan NC, Shih JC (1991) Human monoamine oxidase A and B genes exhibit identical exon-intron organization
Gross C, Hen R (2004) Genetic and environmental factors interact to influence anxiety. Neurotox Res 6:493–501
Haberstick BC et al (2013) MAOA genotype, childhood maltreatment, and their interaction in the etiology of adult antisocial behaviors. Biol Psychiatry. doi:10.1016/j.biopsych.2013.03.028
Heinrichs M, Domes G (2008) Neuropeptides and social behaviour: effects of oxytocin and vasopressin in humans. Prog Brain Res 170:337–350
Huang Y–Y et al (2004) An association between a functional polymorphism in the monoamine oxidase a gene promoter, impulsive traits and early abuse experiences. Neuropsychopharmacology 29:1498–1505
Hudziak JJ et al (2003) Individual differences in aggression: genetic analyses by age, gender, and informant in 3-, 7-, and 10-year-old Dutch twins. Behav Genet 33:575–589
Jacob CP et al (2005) Cluster B personality disorders are associated with allelic variation of monoamine oxidase A activity. Neuropsychopharmacology 30:1711–1718
Jahng JW et al (1997) Localization of monoamine oxidase A and B mRNA in the rat brain by in situ hybridization. Synapse 25:30–36
Jönsson EG et al (2000) A promoter polymorphism in the monoamine oxidase A gene and its relationships to monoamine metabolite concentrations in CSF of healthy volunteers. J Psychiatr Res 34:239–244
Kalin NH et al (2008) The serotonin transporter genotype is associated with intermediate brain phenotypes that depend on the context of eliciting stressor. Mol Psychiatry 13:1021–1027
Kim-Cohen J et al (2006) MAOA, maltreatment, and gene-environment interaction predicting children's mental health: new evidence and a meta-analysis. Mol Psychiatry 11:903–913
Kinnally EL et al (2009) Parental care moderates the influence of MAOA-uVNTR genotype and childhood stressors on trait impulsivity and aggression in adult women. Psychiatr Genet 19:126–133
Krueger RF, South SC (2009) Externalizing disorders: cluster 5 of the proposed meta-structure for DSM-V and ICD-11. Psychol Med 39:2061–2070
Kuepper Y, Grant P, Wielpuetz C, Hennig J (2013) MAOA-uVNTR genotype predicts interindividual differences in experimental aggressiveness as a function of the degree of provocation. Behav Brain Res 247:73–78
Levitt P, Pintar JE, Breakefield XO (1982) Immunocytochemical demonstration of monoamine oxidase B in brain astrocytes and serotonergic neurons. Proc Natl Acad Sci USA 79:6385–6389
Machado CJ, Bachevalier J (2006) The impact of selective amygdala, orbital frontal cortex, or hippocampal formation lesions on established social relationships in rhesus monkeys (Macaca mulatta). Behav Neurosci 120:761–786
Manuck SB, Flory JD, Ferrell RE, Mann JJ, Muldoon MF (2000) A regulatory polymorphism of the monoamine oxidase-A gene may be associated with variability in aggression, impulsivity, and central nervous system serotonergic responsivity. Psychiatry Res 95:9–23
Manuck SB, Flory JD, Muldoon MF, Ferrell RE (2002) Central nervous system serotonergic responsivity and aggressive disposition in men. Physiol Behav 77:705–709

Márquez C et al (2013) Peripuberty stress leads to abnormal aggression, altered amygdala and orbitofrontal reactivity and increased prefrontal MAOA gene expression. Transl Psychiatry 3:e216

McDermott R, Tingley D, Cowden J, Frazzetto G, Johnson DDP (2009) Monoamine oxidase A gene (MAOA) predicts behavioral aggression following provocation. Proc Nat Acad Sci 106:2118–2123

Mertins V, Schote AB, Hoffeld W, Griessmair M, Meyer J (2011) Genetic susceptibility for individual cooperation preferences: the role of monoamine oxidase A gene (MAOA) in the voluntary provision of public goods. PLoS ONE 6:e20959

Meyer-Lindenberg A et al (2006) Neural mechanisms of genetic risk for impulsivity and violence in humans. Proc Natl Acad Sci USA 103:6269–6274

Meyer-Lindenberg A et al (2009) Genetic variants in AVPR1A linked to autism predict amygdala activation and personality traits in healthy humans. Mol Psychiatry 14:968–975

Miczek KA, Maxson SC, Fish EW, Faccidomo S (2001) Aggressive behavioral phenotypes in mice. Behav Brain Res 125:167–181

Mukasa H, Nakamura J, Yamada S, Inoue M, Nakazawa Y (1990) Platelet monoamine oxidase activity and personality traits in alcoholics and methamphetamine dependents. Drug Alcohol Depend 26:251–254

Nicotra A, Pierucci F, Parvez H, Senatori O (2004) Monoamine oxidase expression during development and aging. Neurotoxicology 25:155–165

Nilsson KW et al (2006) Role of monoamine oxidase A genotype and psychosocial factors in male adolescent criminal activity. Biol Psychiatry 59:121–127

Nymberg C et al (2013) Neural mechanisms of attention-deficit/hyperactivity disorder symptoms are stratified by MAOA genotype. Biol Psychiatry. doi:10.1016/j.biopsych.2013.03.027

Ongür D, Price JL (2000) The organization of networks within the orbital and medial prefrontal cortex of rats, monkeys and humans. Cereb Cortex 10:206–219

Parsian A, Cloninger CR, Sinha R, Zhang ZH (2003) Functional variation in promoter region of monoamine oxidase A and subtypes of alcoholism: haplotype analysis. Am J Med Genet B Neuropsychiatr Genet 117B:46–50

Passamonti L et al (2008) Genetically dependent modulation of serotonergic inactivation in the human prefrontal cortex. NeuroImage 40:1264–1273

Perris C, Eisemann M, Knorring von L, Oreland L, Perris H (1984) Personality traits and monoamine oxidase activity in platelets in depressed patients. Neuropsychobiology 12:201–205

Popova NK et al (2001) Behavioral characteristics of mice with genetic knockout of monoamine oxidase type A. Neurosci Behav Physiol 31:597–602

Prochazka H, Anderberg UM, Oreland L, Knorring LV, Agren H (2003) Self-rated aggression related to serum testosterone and platelet MAO activity in female patients with the fibromyalgia syndrome. World J Biol Psychiatry 4:35–41

Raine A et al (1994) Selective reductions in prefrontal glucose metabolism in murderers. Biol Psychiatry 36:365–373

Raine A et al (1998) Reduced prefrontal and increased subcortical brain functioning assessed using positron emission tomography in predatory and affective murderers. Behav Sci Law 16:319–332

Raine A, Buchsbaum M, LaCasse L (1997) Brain abnormalities in murderers indicated by positron emission tomography. Biol Psychiatry 42:495–508

Rebsam A, Seif I, Gaspar P (2002) Refinement of thalamocortical arbors and emergence of barrel domains in the primary somatosensory cortex: a study of normal and monoamine oxidase a knock-out mice. J Neurosci 22:8541–8552

Reif A et al (2007) Nature and nurture predispose to violent behavior: serotonergic genes and adverse childhood environment. Neuropsychopharmacology 32:2375–2383

Reti IM et al (2011) Monoamine oxidase A regulates antisocial personality in whites with no history of physical abuse. Compr Psychiatry 52:188–194

Roiser JP et al (2009) A genetically mediated bias in decision making driven by failure of amygdala control. J Neurosci 29:5985–5991
Sabol S, Hamer D (1998) A functional polymorphism in the monoamine oxidase A gene promoter
Saito T et al (2002) Analysis of monoamine oxidase A (MAOA) promoter polymorphism in Finnish male alcoholics. Psychiatry Res 109:113–119
Salichon N et al (2001) Excessive activation of serotonin (5-HT) 1B receptors disrupts the formation of sensory maps in monoamine oxidase a and 5-ht transporter knock-out mice. J Neurosci 21:884–896
Saura J et al (1996a) Localization of monoamine oxidases in human peripheral tissues. Life Sci 59:1341–1349
Saura J et al (1996b) Molecular neuroanatomy of human monoamine oxidases A and B revealed by quantitative enzyme radioautography and in situ hybridization histochemistry. Neuroscience 70:755–774
Scott AL, Bortolato M, Chen K, Shih JC (2008) Novel monoamine oxidase A knock out mice with human-like spontaneous mutation. NeuroReport 19:739–743
Shih J, Thompson R (1999) Monoamine oxidase in neuropsychiatry and behavior. Am J Hum Genet 65:593–598
Shih JC, Chen K (1999) MAO-A and -B gene knock-out mice exhibit distinctly different behavior. Neurobiology (Bp) 7:235–246
Sjöberg RL et al (2008) A non-additive interaction of a functional MAO-A VNTR and testosterone predicts antisocial behavior. Neuropsychopharmacology 33:425–430
Skondras M, Markianos M, Botsis A, Bistolaki E, Christodoulou G (2004) Platelet monoamine oxidase activity and psychometric correlates in male violent offenders imprisoned for homicide or other violent acts. Eur Arch Psychiatry Clin Neurosci 254:380–386
Stadler C et al (2007) Reduced anterior cingulate activation in aggressive children and adolescents during affective stimulation: association with temperament traits. J Psychiatr Res 41:410–417
Stark P, Fuller RW, Wong DT (1985) The pharmacologic profile of fluoxetine. J Clin Psychiatry 46:7–13
Stefanacci L, Amaral DG (2002) Some observations on cortical inputs to the macaque monkey amygdala: an anterograde tracing study. J Comp Neurol 451:301–323
Sterzer P, Stadler C, Krebs A, Kleinschmidt A, Poustka F (2005) Abnormal neural responses to emotional visual stimuli in adolescents with conduct disorder. Biol Psychiatry 57:7–15
Strolin Benedetti M, Dostert P, Tipton KF (1992) Developmental aspects of the monoamine-degrading enzyme monoamine oxidase. Dev Pharmacol Ther 18:191–200
Suomi SJ (2003) Gene-environment interactions and the neurobiology of social conflict. Ann N Y Acad Sci 1008:132–139
Syagailo YV et al (2001) Association analysis of the functional monoamine oxidase A gene promoter polymorphism in psychiatric disorders. Am J Med Genet 105:168–171
Tost H et al (2010) A common allele in the oxytocin receptor gene (OXTR) impacts prosocial temperament and human hypothalamic-limbic structure and function. Proc Nat Acad Sci 107:13936–13941
Tsang D, Ho KP, Wen HL (1986) Ontogenesis of multiple forms of monoamine oxidase in rat brain regions and liver. Dev Neurosci 8:243–250
Valzelli L, Bernasconi S (1979) Aggressiveness by isolation and brain serotonin turnover changes in different strains of mice. Neuropsychobiology 5:129–135
Valzelli L, Bernasconi S, Dalessandro M (1981b) Effect of tryptophan administration on spontaneous and P-CPA-induced muricidal aggression in laboratory rats. Pharmacol Res Commun 13:891–897
Valzelli L, Garattini S, Bernasconi S, Sala A (1981a) Neurochemical correlates of muricidal behavior in rats. Neuropsychobiology 7:172–178
van Beijsterveldt CEM, Verhulst FC, Molenaar PCM, Boomsma DI (2004) The genetic basis of problem behavior in 5-year-old Dutch twin pairs. Behav Genet 34:229–242

Verdejo-García A, Albein-Urios N, Molina E, Ching-López A, Martínez-González JM, Gutiérrez B (2013) A MAOA gene*cocaine severity interaction on impulsivity and neuropsychological measures of orbitofrontal dysfunction: Preliminary results. Drug Alcohol Depend 133(1):287–290. doi:10.1016/j.drugalcdep.2013.04.031

Vergnes M, Kempf E (1981) Tryptophan deprivation: effects on mouse-killing and reactivity in the rat. Pharmacol Biochem Behav 14(Suppl 1):19–23

Wakschlag LS et al (2010) Interaction of prenatal exposure to cigarettes and MAOA genotype in pathways to youth antisocial behavior. Mol Psychiatry 15:928–937

Waters H, Hyder AA, Rajkotia Y, Basu S, Rehwinkel JA (2004) The economic dimensions of interpersonal violence/editorial committee: Hugh Waters...[et al.]

Weder N et al (2013) MAOA genotype, maltreatment, and aggressive behavior: the changing impact of genotype at varying levels of trauma. Biol Psychiatry 65:417–424

Westlund KN, Denney RM, Kochersperger LM, Rose RM, Abell CW (1985) Distinct monoamine oxidase A and B populations in primate brain. Science 230:181–183

Westlund KN, Denney RM, Rose RM, Abell CW (1988) Localization of distinct monoamine oxidase A and monoamine oxidase B cell populations in human brainstem. Neuroscience 25:439–456

Westlund KN, Krakower TJ, Kwan SW, Abell CW (1993) Intracellular distribution of monoamine oxidase A in selected regions of rat and monkey brain and spinal cord. Brain Res 612:221–230

Widom CS, Brzustowicz LM (2006) MAOA and the 'cycle of violence:' childhood abuse and neglect, MAOA genotype, and risk for violent and antisocial behavior. BPS 60:684–689

Williams LM et al (2009) A polymorphism of the MAOA gene is associated with emotional brain markers and personality traits on an antisocial index. Neuropsychopharmacology 34:1797–1809

Willoughby J et al (1988) Monoamine oxidase activity and distribution in marmoset brain: implications for MPTP toxicity. Neurosci Lett 90:100–106

Yeh MT, Coccaro EF, Jacobson KC (2010) Multivariate behavior genetic analyses of aggressive behavior subtypes. Behav Genet 40:603–617

Ziermans T et al (2012) Working memory brain activity and capacity link MAOA polymorphism to aggressive behavior during development. Transl Psychiatry 2:e85

Early Development of Physical Aggression and Early Risk Factors for Chronic Physical Aggression in Humans

Richard E. Tremblay

Abstract This chapter describes the state of knowledge on the development of physical aggression from early childhood to adulthood, the long term outcomes of chronic physical aggression during childhood and the risk factors for chronic physical aggression. Unraveling the development of physical aggression is important to understand when and why humans start using physical aggression, to understand why some humans suffer from chronic physical aggression and to understand how to prevent the development of this disorder which causes much distress to the aggressors and their victims. The study of the developmental origins of aggression also sheds light on the reasons why situational prevention of aggression is important at all ages and in all cultures.

Keywords Physical aggression · Development · Risk factors · Prevention

Contents

R. E. Tremblay (✉)
School of Public Health, Physiotherapy and Population Science,
University College Dublin, Dublin, Ireland
e-mail: tremblar@grip.umontreal.ca

R. E. Tremblay
Departments of Pediatrics and Psychology, University of Montreal, Montreal, Canada

Curr Topics Behav Neurosci (2014) 17: 315–327
DOI: 10.1007/7854_2013_262

Published Online: 19 February 2014

1 Introduction

The development of physical aggression throughout early childhood has only very recently been studied with large population samples. This lack of attention to the developmental origins of physical aggression during early childhood appears to be the result of a long-held belief that humans start to use physical aggression after early childhood as a result of exposition to violent behaviour (e.g. Bandura 1977; Lefkowitz et al. 1977; Zimbardo 2007).

A good summary of this social learning perspective can be found in the 1993 report of the US Academy of Science Panel on Understanding Violent Behaviour: "Modern Psychological perspectives emphasize that aggressive and violent behaviors are learned responses to frustration, that they can also be learned as instruments for achieving goals, and that the learning occurs by observing models of such behavior. Such models may be observed in the family, among peers, elsewhere in the neighborhood, through the mass media…" (Reiss and Roth 1993). The 2002 World Health Organization report on violence was still more explicit on the timing of physical aggression onset: "The majority of young people who become violent are adolescent-limited offenders who, in fact, show little or no evidence of high levels of aggression or other problem behaviours during their childhood." (Krug et al. 2002).

Physical aggression is a crucial component of human's behavioural heritage. Our ancestors needed to be skilled in the art of physical aggression to eat, to defend themselves against predators, to compete for mating, to protect their brood and to acquire resources. However, like all other social animals, humans need to learn to use aggression sparingly because physically aggressive encounters can be fatal, and lack of self-control among social animals leads to social exclusion (Boivin et al. 2005; Suomi 2005). The relative absence of fear from being murdered is recent. Historical analyses of homicide rates indicate that physical violence has systematically and substantially decreased among European citizens over the past 500 years (Eisner 2003). Homicides in European cities decreased from 40 to 1 per 100,000 citizens per year. Looking further back from an evolutionary perspective, the estimated rate of "homicide" among our closest non-human relatives, chimpanzees, is 261 per 100,000 (Wrangham et al. 2006).

2 Definition of Physical Aggression

It is important to closely examine the operational definition of physical aggression in human aggression studies because there are three major definition problems. First, there has been a tendency to confuse disruptive and antisocial behaviour with physically aggressive behaviour (Tremblay 2000). Physical aggressions can be defined as disruptive and antisocial, but disruptive and antisocial behaviour is not necessarily physically aggressive. Second, many have argued that intent to harm is required to conclude that behaviour is aggressive. This "intent" criterion has significantly limited the study of the early development of aggressive behaviour because it was assumed that young children cannot aggress since they cannot intend to hurt others. After a 6-year-old boy gunned down a 6-year-old girl in their classroom, one of the frequently asked questions was: "did the boy really understand what he had done?" The intent question is interesting in itself, however it is a different question from the fact that infants physically aggress when angry or when they want to take something away from someone. Research on aggression among mice, rats and monkeys has not been inhibited by the intentionality issue, while most human aggression investigators kept away, until very recently from studying aggression in infants and toddlers. The intent criterion is a problem not only for human infants and non-human animals. Behaviour driven by anger and fear is often not under the control of one's will, even during adulthood. Many, if not most of the physically aggressive behaviours following intense frustration are impulsive behaviours that were not "planned". Third, some investigators include a severity dimension to the definition of physical aggression. This issue is essentially related to the consequences of the aggression for the victim. Pushing and slapping generally have less serious consequences than choking and stabbing. However, from a developmental perspective, one needs to take into account that age from birth to adolescence is positively correlated with increase in physical strength, cognitive development and unsupervised access to objects that can be used as weapons. A punch from a 1.8 m, 90 kg 17-year-old has, potentially, more serious consequences than a punch from the same individual when he was 24-month-old. A boy looking for revenge will have easier access to a knife or a gun at 12 than at 3 years of age. Thus severity of physical aggression generally increases with age, but this does not mean that a 16-year-old who is committing the most serious physical aggressions for his age was not committing the most serious physical aggressions for his age when he was 2 years old.

The simplest way to define "physical aggression" is to follow the ethological approach and list the physical aggressions in agonistic encounters at a given age (Blurton-Jones 1972; Restoin et al. 1985). However, it is difficult to make a complete list of the multitude of means humans invented to physically hurt other humans. Examples of the most basic physical aggressions should be sufficient to make the concept clear: hitting, slapping, kicking, biting, pushing, grabbing, pulling, shoving, beating, twisting and chocking. Some scales use terms such as fighting and bullying to summarise these behaviours. Threatening to physically aggress, use of objects and weapons to aggress is also included in the definition used by ethologists.

Chronic physical aggression (CPA) can be defined as a tendency to use physical aggressions significantly more frequently than the large majority of a birth cohort over many years. Thus, repeated assessments over many years (longitudinal studies) of random samples of a population birth cohort are needed to estimate the prevalence of CPA during specific developmental periods. Such studies provide an opportunity to assess the different developmental trajectories of physical aggression in a population and estimate the proportion of individuals on a CPA trajectory (Nagin and Tremblay 1999).

In contrast, occasional physical aggression (OPA) can be defined as the use of physical aggression at a rate that corresponds to the rate of a relatively large number of members in a population cohort. As will be described below, the rate of physical aggression for the majority of members in a birth cohort varies substantially with age and subgroups of OPA can be identified. However, risk factors that discriminate between CPA and OPA are easier to identify than risk factors that discriminate subgroups of OPA.

3 Development of Physical Aggression During Early Childhood

Studies of physical aggression during infancy have clearly shown that humans start to use physical aggression towards the end of the first year after birth when they have acquired the motor coordination to push, pull, hit, kick, etc. (Alink et al. 2006; Hay et al. 2011; Tremblay et al. 1999). Figure 1 illustrates the results of the physical aggression developmental trajectory analyses from 17 to 60 months with a population birth cohort (Côté et al. 2007). We can see that half of the children are in the middle trajectory of physical aggression frequency, a third are on a low trajectory, while 17 % are on a high trajectory. Based on the definition already given, we can consider the high trajectory as a CPA trajectory and the two others as OPA trajectories.

These analyses are based on prospective-repeated assessments of physical aggressions reported by mothers over 4 years. From this perspective, developmental trajectories should be a better estimate of a chronic behaviour problem than an assessment at a given point of time, even if that assessment attempts to reconstruct past behaviour. Longitudinal data have shown that within a year mothers do not recall the age of onset of their children's physical aggressions (Tremblay 2000). In a clinical study of boys between 7 and 12 years of age, the mean age of physical aggression onset reported by parents was 6.75 years (Frick et al. 1993). Retrospective information collected in the Pittsburgh Youth Study (Loeber and Hay 1997; Loeber and Stouthamer-Loeber 1998) compared to prospective data is a good example of the problem with retrospective dating of physical aggression onset. The subjects ($N = 503$) represented the Pittsburgh public schools' male 8th graders and were close to 14 years old (mean age = 13.8; $SD = 0.80$) at the first data collection. The cumulative age of onset of physical aggressions reported by the mothers and the boys at that first data collection indicated that by age 5 years less than 5 % of the boys had initiated use of physical aggressions and almost no one had initiated fighting. In sharp contrast, the prospective data represented in Fig. 1 on physical aggression from 17 months after birth indicate that children who do not initiate physical aggression before 3 years

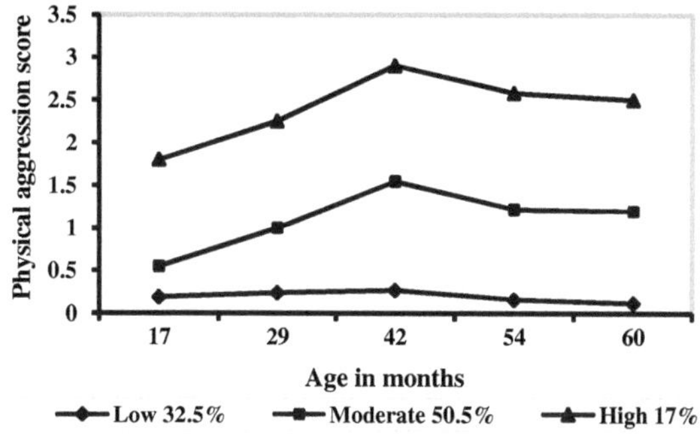

Fig. 1 Developmental trajectories of physical aggression from 17 to 60 months (Côté et al. 2007 AGP)

of age are extremely rare. These prospective studies suggest that the peak frequency in use of physical aggression for most humans is somewhere between 2 and 4 years of age (see Fig. 2 and NICHD Early Child Care Research Network 2004). The recall problem suggests that retrospective assessments of children or adolescents cannot identify the age of onset and developmental trajectories of physical aggression use or of CPA. Hence, the conclusion reached by the World Health Organization report on Violence (Krug et al. 2002) cited above needs to be seriously amended.

4 Developmental Trajectories of Physical Aggression After Early Childhood

The developmental trajectories of physical aggression after early childhood have now been studied in many different cultures. From these studies, we can expect between 7 and 11 % of elementary school children on a CPA trajectory (Broidy et al. 2003; Campbell et al. 2010; Nagin and Tremblay 1999). That percentage tends to be higher for preschool children (Côté et al. 2007; Tremblay et al. 2004) and lower for adolescents (Brame et al. 2001). This decrease in CPA cases with age corresponds to the general decrease in frequency of physical aggression with age, after the peak in early childhood (see Fig. 2 from Tremblay and Côté 2009).

Most children use physical aggression during the preschool years, but most children also learn to use alternatives to physical aggression with age, and this applies to a number of chronic cases during early childhood and preadolescence (Nagin and Tremblay 1999). In fact there is good evidence that the learning process to gain control over physical aggression continues throughout adulthood. A longitudinal study of juvenile delinquents up to old age showed that the number of their violent offenses decreased with age (Sampson and Laub 2003; see also Sweeten et al. 2013).

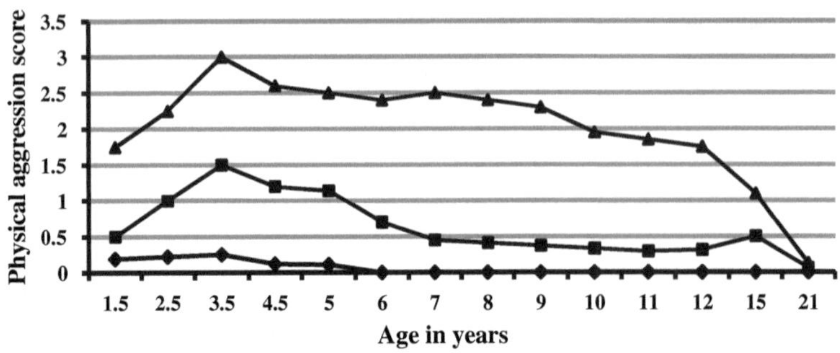

Fig. 2 Trajectories of physical aggression from infancy to adulthood (Tremblay and Côté 2009)

Crime records from the middle ages to modern times suggest that this phenomenon is not new. The likelihood of committing a homicide and most other crimes has always decreased from late adolescence and early adulthood to old age (Eisner 2003; Quetelet 1984). Trajectories of physical aggression covering different age periods (early childhood to childhood, childhood to adolescence, adolescence to adulthood) also indicate that CPA very rarely onsets after early childhood (Barker et al. 2007; NICHD Early Child Care Research Network 2004; van Lier et al. 2009).

5 Outcomes of Chronic Physical Aggression Trajectories During Childhood

Longitudinal studies of physical aggression trajectories during childhood have been used to study how well the trajectories predict future outcomes such as school performance, social adjustment, mental health and violent behaviour. The first longitudinal study to describe developmental trajectories of physical aggression from school entry to adolescence (Nagin and Tremblay 1999) reported that boys on a teacher-rated trajectory of frequent physical aggression from 6 to 15 years of age were at highest risk of self-reported violence as well as other forms of delinquency at 17 years of age, even after having controlled for hyperactivity and oppositional behaviour. The chronically aggressive boys were also at highest risk of school drop-out. A study which used six longitudinal cohorts from Canada, New Zealand and the USA (Broidy et al. 2003) reached the same conclusion for male adolescent violent delinquency, but not for female adolescent violent delinquency. The authors attributed the sex difference in prediction to the fact that the prevalence of female adolescent violent delinquency was too low. However, a later analysis of one of the female samples (Fontaine et al. 2008) reported that elementary school girls who were on a CPA trajectory combined with a chronic hyperactivity trajectory were more likely than others to report physical and psychological aggression towards intimate partners by age 21 years. They were also more likely to report early pregnancy, welfare assistance, nicotine use problems and low educational attainment. A more recent analysis of a population sample of males and females (Pingault et al. 2013) reported that the 9.5 % of children on a high physical aggression trajectory between 6 and 12 years, according to mother and teacher rating, represented 28.2 % of all those who had a criminal record by age 24 years. In addition, they represented 45.9 % of all recorded criminal charges and 57.4 % of the violence charges. Therefore, children on a high trajectory of physical aggression during elementary school are not only more likely to have a criminal record but also to have more criminal charges. There is evidence that the criminal outcomes of childhood physical aggression during adolescence and adulthood are preceded by a large range of negative social and academic outcomes by the end of elementary school for boys and girls (Campbell et al. 2010).

6 Early Risk Factors of Chronic Physical Aggression and Putative Mechanisms

Sex of the individual is one of the most important risk factor for CPA. When children start using physical aggression at the end of the first year after birth there are no significant difference in frequency of physical aggressions between boys and girls (Hay et al. 2011), however the differences appear soon after and increase until adolescence (e.g. Côté 2007). Males between 10 and 15 years of age are close to 20 times (OR = 18.84) more at risk than females of being on a CPA trajectory (van Lier et al. 2009).

Twin studies have become important tools to understand the contributions of environmental and genetic factors in the development of human characteristics, including aggression. However, to date there appears to be only one longitudinal study that used a large sample of twins from infancy onwards to study the contributions of genetic and environmental factors in the development of physical aggression. The study reported that 19 months after birth 58 % of the variance in frequency of physical aggression rated by mothers could be attributed to genetic contributions and 42 % to common environmental contributions (Dionne et al. 2003). At 72 months after birth, physical aggression was rated by kindergarten teachers. Genetic contribution to variance in frequency was then estimated to be 66 % with 34 % attributed to common environmental factors (van Lier et al. 2007).

Although these results suggest a substantial contribution to physical aggression by genetic factors to the use of physical aggression from infancy to school entry, environmental factors are also very important. The developmental trajectories of physical aggression described above indicate that the environmental conditions are essential to learn alternatives to physical aggression during early childhood. Studies of physical aggression trajectories during early childhood with singletons have identified the following types of environmental risk factors: (a) Maternal characteristics, including life style and mental health, (b) family characteristics,

(c) maternal parenting, (d) child characteristics (Campbell et al. 2010; Côté et al. 2006; Hay et al. 2011; NICHD Early Child Care Research Network 2004; Tremblay et al. 2004).

Maternal and family characteristics are key for planning preventive interventions because they can be used to identify pregnant women at risk of having children on a CPA trajectory (e.g. Olds et al. 1998). The maternal characteristics associated with CPA include mothers' young age at birth of their child, mothers' smoking during pregnancy, mothers' antisocial behaviour during adolescence, mothers' depression and mothers' low level of education. Family characteristics included low income, family dysfunction and the presence of siblings. High-risk maternal parenting behaviour includes mother's hostile-coercive-harsh parenting and lack of sensitivity.

It is important to note that these studies of environmental risk factors were not done in the context of genetically informative designs (e.g. twin studies or sibling studies), hence we do not know to what extent the significant environmental risk factors are correlated or interact with genetic factors (Plomin 1994; Szyf et al. 2009). Nonetheless, the environmental risk factors identified by these studies can be used to identify at risk groups for preventive experiments. Such experiments are useful to test the effectiveness of the interventions as well as test causal hypotheses (Schwartz et al. 1981; Tremblay 2003). Maternal and family characteristics are especially key for early preventive interventions because they can be used to identify at-risk pregnant women (e.g. Olds et al. 1998). The challenge for the future is to integrate preventive experiments within genetically informative longitudinal studies of risk factors.

The discovery of environmental effects on gene expression (epigenetics) is providing the tolls to meet this challenge (Tremblay and Szyf 2010; Tremblay 2010). The term "epigenetic" refers to the mechanisms which programme genes and can change gene function without modifying gene sequence, mainly through changes in DNA methylation and chromatin structure. This programming is responsive to environmental effects, especially during foetal and early post natal development. Thus, environments can impact phenotypes through their chemical impact on programming of gene function (Mill and Petronis 2008; Szyf et al. 2009).

To grasp the potential contribution of epigenetics for understanding the mechanisms involved in the development of CPA it is important to understand the difference between the traditional gene-environment interaction story and the epigenetic story. Key differences can be understood by comparing two studies which attempt to explain the effects of a maltreating environment on development. In the gene-environment statistical interaction approach (Caspi et al. 2002) males brought up in a "maltreating" environment were observed to be more at risk of violent behaviour if they had a short rather than a long allele on the promoter region of the MAOA gene. In this example, the maltreating environment does not physically interact with the MAOA gene. The interaction is statistical, i.e. obtained from a 2 × 2 table comparing four groups created from two variables: maltreating-not maltreating and short-long MAOA allele. It is presumed that individuals who inherited either short or long alleles react differently to the maltreating environment because their neural system functions differently and such differences are due to MAOA activity.

Epigenetic studies have a different approach; they focus specifically on the physical effects of the environment on gene expression at a given moment during development. The classic example for effects of neglectful environments comes from an experimental study of maternal behaviour in rats which showed that rat pups insufficiently licked by their mothers in the days following birth (i.e. neglected) have increased methylation of the gene encoding the Glucocorticoid receptor in the hippocampus, resulting in reduced expression (Weaver et al. 2004, 2006). The study further showed that this gene methylation effect had downstream effects on the hypothalamic–pituitary–adrenal axis which regulates stress responses in the body.

Epigenetic mechanisms are especially important because they provide a powerful explanation for early maternal and family effects on the development of physical and mental health problems, including CPA. Furthermore, DNA methylation changes over time can be used as markers of environmental effects during development, including assessment of preventive and corrective intervention effects. The above discussion of early risk factors showed that trajectories of CPA are specifically related to maternal characteristics: maternal age at first pregnancy, history of behaviour problems, education, smoking, depression, coercive parenting, etc. This can easily be understood from the traditional environmental perspective: a poor early environment has an impact on the developing foetus and infant. Mother characteristics turn out to be more important risk factors than father characteristics because the former carry the child in their womb during foetal life and are more involved in care giving during early childhood. However, the exact bio-psycho-social mechanisms linking poor quality environment to disorganised behaviour remain unclear.

The epigenetic story provides a basic mechanism that has the advantage of being parsimonious, testable and promising for prevention. The most fascinating aspect of this mechanism is that it provides an environmentally based explanation of intergenerational transmission for physical and mental disorders which involves genes but is not genetically transmitted. These mechanisms are still far from being clearly understood, but they provide a challenging alternative perspective to the traditional gene versus environment and gene–environment interaction hypotheses (McGowan et al. 2008, 2009; Rutten and Mill 2009).

The first epigenetic study of children with CPA (Wang et al. 2012) used subjects from the longitudinal study of boys living in low socio-economic environments described above (Nagin and Tremblay 1999; Tremblay et al. 1994). Blood was collected from two groups of the boys when they were 27 years old to assess DNA methylation patterns (in monocytes and T cells) of the serotonin transporter (5-HT). The first group included boys who were on a high trajectory of physical aggression during childhood. The second group included boys who were on a normal trajectory of physical aggression. Brain imaging of serotonin synthesis was also obtained from the same boys around age 27 years. We found that CPA during childhood was associated with increased DNA methylation in specific CpG sites in monocytes and

T cells of the serotonin transporter gene. Interestingly, we also found associations between measures of serotonin synthesis in the brain and differential DNA methylation in T cells and monocytes in the same CpG sites that revealed association with CPA during childhood. These findings are the first evidence of the association between environmentally related differences in DNA methylation in white blood cells and in vivo measures of 5-HT in the living human brain.

Experimental studies with rats and monkeys which started at birth (e.g. Provençal et al. 2012; Weaver et al. 2004) are suggesting that the associations between DNA methylation and aggression observed in humans were caused by perinatal environmental effects. To confirm this hypothesis we will need two types of studies. The longitudinal studies approach would be to assess DNA methylation at birth and follow the children up to adulthood to confirm the differences observed in adults. However, this would take a long time and would offer only correlational evidence. The shorter and more rigorous approach would be the use of experimental preventive interventions starting during the perinatal period. The maternal and family risk factors described above can be used to identify high-risk pregnant women and randomly allocate them to preventive interventions that have been shown to have long-term impacts on children's behaviour development (e.g. Eckenrode et al. 2010). By comparing the DNA methylation profiles at birth and throughout development of children in the experimental and control groups, we will be able to confirm if inadequate prenatal environments impact methylation patterns and if these impacts can be reversed by pre- and early postnatal preventive interventions (Tremblay 2010).

7 Conclusions

From the available data on the development of physical aggression between early childhood and adulthood, we can conclude that: (a) the vast majority of humans have used physical aggression; (b) the vast majority of humans also learn with age to use other means of solving problems than the use of physical aggression; (c) some need more time than others to learn; (d) females learn more quickly than males; (e) by adolescence not much more than 5 % of males can be considered cases of CPA, while female cases are exceptional; (f) most of the CPA cases during adolescence were on a CPA trajectory since early childhood; (g) since most humans used physical aggression and learned alternatives during childhood, most adults are at risk of using physical aggression if in a given situation it appears to be the "best" solution to the problem. Thus, all societies need to have effective situational prevention of physical aggression as well as effective early preventive interventions of CPA. To advance our understanding of the pre-, peri- and early postnatal mechanisms leading to CPA, we need preventive experiments that will manipulate environmental conditions and measure their effects at the DNA expression level, at the metabolomic level, at the neurodevelopmental level and at the behavioural level.

References

Alink LR et al (2006) The early childhood aggression curve: development of physical aggression in 10- to 50-months-old children. Child Dev 77:954–966

Bandura A (1977) Social learning theory. Prentice Hall Inc., Englewood Cliffs

Barker ED et al (2007) Developmental trajectories of male physical violence and theft: relations to neurocognitive performance. Arch Gen Psychiatry 64:592–599

Blurton-Jones N (1972) Categories of child–child interactions. In: Blurton-Jones N (ed) Ethological studies of child behaviour. Cambridge University Press, New York, pp 97–128

Boivin M et al (2005) Peer relationships and the development of aggressive behavior in early childhood. In: Tremblay RE, Hartup WW, Archer J (eds) Developmental origins of aggression. Guilford Press, New York, pp 376–397

Brame B et al (2001) Developmental trajectories of physical aggression from school entry to late adolescence. J Child Psychol Psychiatry 42:503–512

Broidy LM et al (2003) Developmental trajectories of childhood disruptive behaviors and adolescent delinquency: a six site, cross national study. Dev Psychol 39:222–245

Campbell SB et al (2010) Predictors and sequelae of trajectories of physical aggression in school-age boys and girls. Dev Psychopathol 22:133–150

Caspi A et al (2002) Role of genotype in the cycle of violence in maltreated children. Science 297:851–854

Côté SM (2007) Sex differences in physical and indirect aggression: a developmental perspective. Eur J Crim Policy Res 13:183–200

Côté SM et al (2006) The development of physical aggression from toddlerhood to pre-adolescence: a nation wide longitudinal study of Canadian children. J Abnorm Child Psychol 34:71–85

Côté SM et al (2007) The role of maternal education and non-maternal care services in the prevention of children's physical aggression. Arch Gen Psychiatry 64:1305–1312

Dionne G et al (2003) Physical aggression and expressive vocabulary in 19 month-old twins. Dev Psychol 39:261–273

Eckenrode J et al (2010) Long-term effects of prenatal and infancy nurse home visitation on the life course of youths: 19-year follow-up of a randomized trial. Arch Pediatr Adolesc Med 164:9–15

Eisner M (2003) Long-term historical trends in violent crime. In: Tonry M (ed) Crime and justice: a review of research, vol 30. University of Chicago Press, Chicago, pp 83–142

Fontaine N et al (2008) Girls' hyperactivity and physical aggression during childhood and adjustment problems in early adulthood. Arch Gen Psychiatry 65:320–328

Frick PJ et al (1993) Oppositional defiant disorder and conduct disorder: a meta-analytic review of factor-analyses and cross-validation in a clinic sample. Clin Psychol Rev 13:319–340

Hay DF et al (2011) Known risk factors for violence predict 12-month-old infants' aggressiveness with peers. Psychol Sci 22:1205–1211

Krug EG et al (2002) World report on violence and health. www.who.int/violence_injury_prevention/violence/world_report/wrvh1/en, accessed

Lefkowitz MM et al (1977) Growing up to be violent: a longitudinal study of the development of aggression. Pergamon Press, New York

Loeber R, Hay DF (1997) Key issues in the development of aggression and violence from childhood to early adulthood. Annu Rev Psychol 48:371–410

Loeber R, Stouthamer-Loeber M (1998) Development of juvenile aggression and violence. Some common misconceptions and controversies. Am Psychol 53:242–259

McGowan PO et al (2008) Diet and the epigenetic (re)programming of phenotypic differences in behavior. Brain Res 1237:12–24

McGowan PO et al (2009) Epigenetic regulation of the glucocorticoid receptor in human brain associates with childhood abuse. Nat Neurosci 12:342–348

Mill J, Petronis A (2008) Pre- and peri-natal environmental risks for attention-deficit hyperactivity disorder (ADHD): the potential role of epigenetic processes in mediating susceptibility. J Child Psychol Psychiatry 49:1020–1030
Nagin D, Tremblay RE (1999) Trajectories of boys' physical aggression, opposition, and hyperactivity on the path to physically violent and nonviolent juvenile delinquency. Child Dev 70:1181–1196
NICHD Early Child Care Research Network (2004) Trajectories of physical aggression from toddlerhood to middle school: predictors, correlates, and outcomes. SRCD Monogr 69:1–146
Olds D et al (1998) Long-term effects of nurse home visitation on children's criminal and antisocial behavior: 15-year follow-up of a randomized controlled trial. J Am Med Assoc 280:1238–1244
Pingault J-B et al (2013) Childhood hyperactivity, physical aggression and criminality: a 19 year prospective population-based study. PLoS One 8:1–7
Plomin R (1994) Genetics and experience: the interplay between nature and nurture. Sage Publications, Thousand Oaks
Provençal N et al (2012) The signature of maternal rearing in the methylome in rhesus macaque prefrontal cortex and Tcells. J Neurosci 32:15626–15642
Quetelet A (1984) Research on the propensity for crime at different ages, trans. S.F. Sylvester, Anderson, Cincinnati (Original work published in 1833)
Reiss AJ, Roth JA (eds) (1993) Understanding and preventing violence. National Academy Press, Washington
Restoin A et al (1985) Chronologie des comportements de communication et profils de comportement chez le jeune enfant. In: Tremblay RE, Provost MA, Strayer FF (eds) Ethologie et développement de l'enfant. Editions Stock/Laurence Pernoud, Paris, pp 93–130
Rutten BPF, Mill J (2009) Epigenetic mediation of environmental influences in major psychotic disorders. Schizophr Bull 35:1045–1056
Sampson RJ, Laub JH (2003) Life-course desisters? Trajectories of crime among delinquent boys followed to age 70. Criminology 41:555–592
Schwartz D et al (1981) Clinical trials. Academic Press, London
Suomi SJ (2005) Genetic and environmental factors influencing the expression of impulsive aggression and serotonergic functioning in Rhesus monkeys. In: Tremblay RE, Hartup WW, Archer J (eds) Developmental origins of aggression. Guilford Press, New York, pp 63–82
Sweeten G et al (2013) Age and the Explanation of Crime, Revisited [Published online]. J Youth Adolescenc 1–18
Szyf M et al (2009) Epigenetics and behaviour. In: Tremblay RE, van Aken MAG, Koops W (eds) Development and prevention of behaviour problems: from genes to social policy. Psychology Press, Sussex, pp 25–59
Tremblay RE (2000) The origins of youth violence (ISUMA). Can J Policy Res 1:19–24
Tremblay RE (2003) Why socialization fails? The case of chronic physical aggression. In: Lahey BB, Moffitt TE, Caspi A (eds) Causes of conduct disorder and juvenile delinquency. Guilford Publications, New York, pp 182–224
Tremblay RE (2010) Developmental origins of disruptive behaviour problems: the 'original sin' hypothesis, epigenetics and their consequences for prevention. J Child Psychol Psychiatry 51:341–367
Tremblay RE, Côté SM (2009) Development of sex differences in physical aggression: the maternal link to epigenetic mechanisms [comment]. Behav Brain Sci 32:290–291
Tremblay RE, Szyf M (2010) Developmental origins of chronic physical aggression and epigenetics. Epigenomics 2:495–499
Tremblay RE et al (1994) Predicting early-onset of male antisocial-behavior from preschool behavior. Arch Gen Psychiatry 51:732–739
Tremblay RE et al (1999) The search for the age of "onset" of physical aggression: Rousseau and Bandura revisited. Crim Behav Mental Health 9:8–23
Tremblay RE et al (2004) Physical aggression during early childhood: trajectories and predictors. Pediatrics 114:e43–e50

van Lier P et al (2007) Kindergarten children's genetic vulnerabilities interact with friends' aggression to promote children's own aggression. J Am Acad Child Adolesc Psychiatry 46:1080–1087

van Lier PAC et al (2009) Developmental links between trajectories of physical violence, vandalism, theft, and alcohol-drug use from childhood to adolescence. J Abnorm Child Psychol 37:481–492

Wang D et al (2012) Peripheral SLC6A4 DNA methylation is associated with in vivo measures of human brain serotonin synthesis and childhood physical aggression. PLoS ONE 7:1–8

Weaver ICG et al (2006) Maternal care effects on the hippocampal transcriptome and anxiety-mediated behaviors in the offspring that are reversible in adulthood. Proc Nat Acad Sci USA 103:3480–3485

Weaver ICG et al (2004) Epigenetic programming by maternal behavior. Nat Neurosci 7:847–854

Wrangham RW et al (2006) Comparative rates of violence in chimpanzees and humans. Primates 47:14–26

Zimbardo PG (2007) The Lucifer Effect: understanding how good people turn evil. Random House, New York

Violence Among People with Schizophrenia: Phenotypes and Neurobiology

Sheilagh Hodgins, Magdalena J. Piatosa and Boris Schiffer

Abstract People with schizophrenia are at increased risk, as compared to the general population, to acquire convictions for violent crimes and homicide. They also show elevated levels of aggressive behaviour. While psychotic symptoms explain aggressive behaviour that is common during acute episodes, they do not explain such behaviour at other stages of illness or prior to illness onset. Three distinct phenotypes have been identified: individuals with a childhood onset of conduct disorder who display antisocial and aggressive behaviour both before and after schizophrenia onset; individuals with no history of conduct problems who begin engaging in aggressive behaviour as illness onsets; and individuals who after many years of illness engage in a severe physical assault. Little is known about the aetiology of the three types of offenders and about the neural mechanisms that initiate and maintain these behaviours. We hypothesize that schizophrenia preceded by conduct disorder is associated with a combination of genes conferring vulnerability for both disorders and altering the effects of environmental factors on the brain, and thereby, with a distinct pattern of neural development. Some

To be published in K.Miczek & A. Meyer-Lindenberg (Eds.), Neuroscience of Aggression. Heidelberg: Springer.

S. Hodgins (✉)
Département de Psychiatrie, Université de Montréal, Montréal, Canada
e-mail: s.hodgins@kcl.ac.uk

S. Hodgins
Institute of Psychiatry, King's College London, London, UK

M. J. Piatosa
Institute of Psychology, Justus-Liebig-Universität-Giessen, Giessen, Germany
e-mail: Magdalena.J.Piatosa@psychol.uni-giessen.de

B. Schiffer
Division of Forensic Psychiatry,
Department of Psychiatry, Psychotherapy and Preventive Medicine,
LWL-University Hospital Bochum, Bochum, Germany
e-mail: boris.schiffer@rub.de

Curr Topics Behav Neurosci (2014) 17: 329–368
DOI: 10.1007/7854_2013_259

Published Online: 7 December 2013

evidence is available to support this hypothesis. By contrast, offending among adults with schizophrenia who have no history of such behaviour prior to illness may result from the changes in the brain that occur as illness onsets, and that are further altered by comorbid conditions such as substance misuse, or by the progressive changes in the brain through adulthood that may result from the illness and from the use of antipsychotic medications.

Keywords Schizophrenia · Violence · Phenotypes

Contents

This chapter begins by reviewing the epidemiological evidence about crime and aggressive behaviour[1] among persons with schizophrenia and children and adolescents who are developing schizophrenia. Next the associations, and the lack

[1] Throughout this chapter we use the term aggressive behaviour to refer to physically aggressive behaviour against another person. In some cases such behaviour leads to criminal prosecution and convictions for violent crimes. The term violence, commonly used in the relevant scientific literature, is also used to describe physically aggressive behaviour towards others.

of associations, between psychotic symptoms and aggressive behaviour are discussed. The following section of the chapter presents a typology of persons with schizophrenia who engage in violence. A diversion is then taken to briefly review the evidence on the neural mechanisms associated with aggressive behaviour among persons without severe mental illness. In the subsequent sections, this evidence provides a framework for reviewing the small number of studies that have examined neural structures and functions among people with schizophrenia who display aggressive behaviour. Finally, propositions for future studies are explored.

1 Violent Criminal Offending Among Persons with Schizophrenia

There is now robust evidence demonstrating that both men and women with schizophrenia[2] are at elevated risk, as compared to the general population, to be convicted of non-violent criminal offences, at higher risk to be convicted of violent criminal offences, and at even higher risk to be convicted of homicide (Wallace et al. 2004). For example, we examined a birth cohort composed of all the 358,180 persons born in Denmark from 1944 through 1947 followed until they were in their mid-forties. We excluded those who had died or emigrated before the end of the follow-up period. The official criminal records of cohort members who had been admitted to a psychiatric ward at least once with a discharge diagnosis of schizophrenia were compared to those with no psychiatric admissions. The risk of a violent crime was elevated 4.6 (3.8–5.6) times among the men and 23.2 (14.4–37.4) times among the women with schizophrenia as compared to those with no admissions to a psychiatric ward (Brennan et al. 2000). Similar elevations in risk have been documented among persons with schizophrenia identified in other birth and populations cohorts (Wallace et al. 2004; Arseneault et al. 2000; Hodgins 1992; Tiihonen et al. 1997). While fewer women than men, with and without schizophrenia, are convicted of crimes, schizophrenia confers a greater risk for offending among women than among men.

This association between schizophrenia and violent offending is robust: it has been observed by different research teams which recruited samples from countries with different cultures and health and justice systems, and which measured the association of schizophrenia and offending using different experimental designs including longitudinal investigations of birth and population cohorts, comparisons of people with schizophrenia and their neighbours, and diagnostic studies of random samples of convicted offenders (Hodgins and Janson 2002). There is no evidence to suggest that the elevated rates of violent offending among persons with schizophrenia as compared to the general population result from discrimination on

[2] The term schizophrenia is used to refer to both schizophrenia and schizo-affective disorder.

the part of the criminal justice systems (Hodgins and Janson 2002) in the different countries where these investigations have been conducted. It is important to note that these convictions result from crimes that have been committed in the community and not on inpatient psychiatric wards, where aggressive behaviour towards others is rather common but rarely leads to criminal prosecution.

Results of the epidemiological investigations are consistent in showing that the proportions of persons with schizophrenia who commit crimes vary from one study to another, while the elevations in risk among those with schizophrenia as compared to the general population are similar (Hodgins 1998). In countries with high rates of violent crimes, proportionately more individuals with schizophrenia have convictions for violent crimes than in a country with lower rates of violent crimes.[3] This observation suggests that at least some of the factors that contribute to violent crime in the general population, such as the availability of guns, also influence violent crime among persons with schizophrenia. The proportions of persons with schizophrenia who acquire convictions for violent crimes also varies by time period reflecting differences in policies regarding diversion of mentally ill persons accused of crimes from the justice system to the health system (Hodgins and Janson 2002). Most violent offences committed by persons with schizophrenia are assaults. While homicides attract much attention from the media, they are rare. In some countries, all persons accused of homicide undergo thorough psychiatric evaluations prior to trial. These evaluations have been used to estimate the proportion of homicides that are committed by individuals with schizophrenia and estimates vary from 6 to 28 % indicating substantial variation from one country to another and within countries by time period (Hodgins and Janson 2002). Because the number of homicides varies greatly from one country to another while the prevalence of schizophrenia is relatively stable at just less than 1 %, the numbers of persons with schizophrenia who commit homicides differ across countries. Thus, the proportion of persons with schizophrenia who engage in violence varies by place and time period, but evidence shows that it always remains higher than the proportion of offenders in the general population.

Preventing crime by people with schizophrenia would reduce the homicide rate anywhere from 6 to 28 % depending on the country, but also reduce the rate of other forms of violence. For example, in the Danish cohort described above, 2.2 % of the men had severe mental illness and they committed 8.4 % of the physically aggressive sex offences, 9.0 % of the non-physically aggressive sex offences and in all they comprised 8.1 % of the sex offenders (Alden et al. 2007). Assuming treatment would have removed this disease-associated risk, more than 8 % of all sex offences in Denmark in this period could have been prevented. Adult mental health services provide care for a subgroup of individuals with schizophrenia who present high levels of violent behaviour. For example, we recently assessed a

[3] Epidemiological studies comparing criminality among persons with and without schizophrenia produce valid results only when conducted in countries that have national registers of both health care and criminal convictions.

sample of 205 inpatients with severe mental illness from a UK inner city mental health trust, most of whom suffered from schizophrenia. The patients were, on average, in their late-thirties and more than 80 % had previously required inpatient care. Official criminal records indicated that 46.7 % of the men and 16.5 % of the women had at least one conviction for a violent crime, and on average, the violent offenders had each been convicted of more than two crimes. The 82 men with criminal records had committed 1,792 crimes and the 23 female offenders had committed 458 crimes (Hodgins et al. 2007). The risks of conviction in this UK inner city sample of patients, as compared to the general UK population, were similar to those observed in comparisons of severely mentally ill inpatient and general population samples in Sweden and Denmark. This subgroup of patients with a history of violence presents a challenge to mental health services. Some of these patients are transferred to forensic inpatient services that have dramatically increased in capacity in several European countries (Priebe et al. 2005). Most beds in such forensic psychiatric hospitals are occupied by men with schizophrenia (Hodgins and Müller-Isberner 2004).

Among people with schizophrenia who acquire a criminal conviction, as many as 72 % already had at least one conviction prior to the onset of schizophrenia (Wallace et al. 2004). There are few studies documenting the proportions of patients who acquire criminal convictions prior to onset of schizophrenia. A recent study in Denmark, where the age of criminal responsibility is 15, used national registers to document criminality among all persons born in 1963 or later and diagnosed with schizophrenia by 1999. Thirty-seven per cent of the men and 7 % of the women were found to have a criminal conviction prior to contact with mental health services (Munkner et al. 2003). In studies of clinical samples of patients with schizophrenia, the proportions with a criminal conviction prior to first admission or contact with services are lower, for example, 20 % in Berne, Switzerland in 1985–1987 (Modestin and Ammann 1996) and 16 % in the Edinburgh High Risk Study (Humphreys et al. 1994). We examined a representative sample of individuals within one geographic area in the UK who received treatment for a first episode of psychosis. One-third of the men and 10 % of the women had a conviction for a criminal offence and 20 % of the men and 5 % of the women had at least one conviction for a violent offence (Hodgins et al. 2011). Generally, across such studies, the proportions of people with schizophrenia who are convicted of criminal offences prior to first treatment for schizophrenia are higher in studies that use national register data not requiring individual informed consent (Wallace et al. 2004; Munkner et al. 2003, 2009) than in studies of clinical samples, in studies of samples born more recently (Coid et al. 1993; Fazel et al. 2009), among men as compared to women (Munkner et al. 2003, 2009), and in countries where the age of criminal responsibility is 15 years (Munkner et al. 2003, 2009). Looking at the same phenomenon from the opposite point of view, another Danish study showed a relatively high rate of subsequent schizophrenia among juvenile delinquents (Gosden et al. 2005).

1.1 Schizophrenia Is also Associated with Elevated Rates of Aggressive Behaviour See footnote 2 Towards Others

An interview protocol, the MacArthur Community Violence Instrument, has been developed specifically to assess aggressive behaviour and victimisation among persons with mental disorders. Patients and collateral informants independently complete an interview (Steadman et al. 1998). Studies have shown that severely mentally ill patients report aggressive incidents at only a slightly lower rate than do collaterals (Steadman et al. 1998). We used this instrument to examine aggressive behaviour in the sample of UK inpatients with severe mental illness, most of whom had schizophrenia, described above. Rates of aggressive behaviour were high: 47.1 % of the men and 21.2 % of the women reported having engaged in at least one serious assault sometime in their life, 49.2 % of the men and 38.8 % of the women reported having engaged in at least one act of physical aggression against another person in the previous 6 months, and 21.7 % of the men and 18.8 % of the women reported engaging in a life-threatening act of violence against another person in the previous 6 months (Hodgins et al. 2007). A study of a large sample of outpatients with severe mental illness found that one-in-five assaulted another person over a 2-year period (Walsh et al. 2001). Similarly, other studies of samples of outpatients with schizophrenia report rates of physically aggressive behaviour towards others ranging from 8 to 77 % during the previous 6 months and rates of life-threatening violence towards others ranging from 0 to 40 % (Hodgins et al. 2007). In a meta-analysis of nine studies, pooled estimates of the proportion of patients with first-episode psychosis committing serious violence were 16.6 % and severe violence 0.6 (Large and Nielssen 2011). As noted previously, rates of convictions for violent crimes are lower among women than men with schizophrenia, but several studies suggest that the prevalence of aggressive behaviour is similar (Hodgins et al. 2007; Walsh et al. 2001; Dean et al. 2006). Importantly, the correlates of violent offending and of physically aggressive behaviour towards others when patients are living in the community are similar. One of the important consequences of violent behaviour among persons with schizophrenia is elevated rates of victimisation, and particularly violent victimisation (Hodgins et al. 2007; Teplin et al. 2005; Silver et al. 2005; Walsh et al. 2003).

1.2 Conclusion

The extant literature provides robust evidence that individuals who develop schizophrenia are at elevated risk for violent criminality as compared to the general population. Most, but not all, offenders with schizophrenia have a history of aggressive behaviour and/or criminal offending prior to first contact with mental health services for psychosis. Among adolescents engaging in antisocial and criminal behaviours, a disproportionate number develop schizophrenia. Further, many individuals with schizophrenia engage in physical aggression towards others.

2 Psychotic Symptoms and Aggressive Behaviour

In general population samples, the presence of psychotic symptoms has been associated with a five-fold increase in the risk of assaulting another person (Mojtabai 2006). In a study of 18,104 Japanese adolescents, psychotic-like experiences were associated with an increased risk for violent behaviour towards others (Kinoshita et al. 2011). Yet the results of studies of the association of positive symptoms of psychosis and violent behaviour among persons with schizophrenia are contradictory (Appelbaum et al. 2000; Bjørkly 2002a, b). The lack of consistency in the evidence results primarily from methodological features of studies including retrospective assessment of symptoms and failure to take account of other factors such as previous violence, childhood behaviour problems and intoxication that are known to be associated with violence (Appelbaum et al. 2000; Bjørkly 2002a, b; Hodgins et al. 2003). There is some evidence that when positive symptoms are accompanied by depression or distress, the risk of aggressive behaviour is elevated, even after controlling for a previous history of antisocial behaviour and current and past substance misuse (Hodgins et al. 2003; Crocker et al. 2005).

The evidence is consistent, however, in showing that the correlates of violent behaviour during acute episodes of psychosis among patients hospitalised in a psychiatric ward differ from the correlates of violent behaviour that occurs in the community. Many acutely psychotic patients who are in most cases admitted involuntarily to psychiatric wards behave aggressively. The aggressive behaviour declines rapidly in the days following admission. Aggressive incidents on an acute psychiatric ward are as common among women as men, and unlike aggressive behaviour that occurs in the community they are strongly associated with confusion and thought disorder (Krakowski and Czobor 2004; Steinert 2002). Thus, positive psychotic symptoms constitute the principal factor associated with aggressive behaviour during an acute episode. Also, some studies show that a long duration of untreated psychosis increases the risk of aggressive behaviour (Hodgins et al. 2011; Large and Nielssen 2011). Very high levels of positive symptoms may be associated with aggressive behaviour because they lower the threshold for threat or increase insensitivity to others. By contrast, once the florid psychotic symptoms are reduced other factors such as male sex, young age, threat control-override symptoms and depression, a history of conduct problems, previous aggressive behaviour and illicit drug use are associated with aggressive behaviour (Hodgins and Riaz 2011).

Among people with schizophrenia who are not experiencing an acute episode of psychosis, the lack of association between aggressive behaviour and positive symptoms of psychosis is not surprising if considered in the light of evidence that psychosocial functioning generally is not associated with positive symptoms. Aggressive behaviour reflects a lack of interpersonal skills, or simply one aspect of psychosocial functioning. Interpersonal skills, community activities and work skills are most strongly associated with performance on neuropsychological tests, negative symptoms and depression, and not with positive symptoms (Bowie et al. 2006). We assessed psychosocial functioning among 248 men with schizophrenia living in the community. Psychosocial functioning was defined to include measures of independent living, occupational functioning, social and leisure activities, and the absence of aggressive behaviour, substance misuse and self-harm. We found that psychosocial functioning was associated with two static predictors—level of education, past diagnoses of substance misuse disorders—and three current predictors—depression http://ajp.psychiatryonline.org/index.dtl, non-compliance with antipsychotic medication, and experiences of physical victimisation (Hodgins et al. 2009). The results of this study confirmed findings from previous investigations showing that factors other than positive symptoms are most strongly associated with real-life functioning among persons with schizophrenia, as well as the need for adherence to antipsychotic medication.

2.1 Conclusion

When levels of positive psychotic symptoms are high, aggressive behaviour is common and only associated with these symptoms. When levels of positive symptoms are lower, other factors such as previous history of conduct problems and substance use are associated with aggressive behaviour. Individuals without schizophrenia who report psychotic-like experiences are at elevated risk of violent crime. Taken together, the evidence suggests that the neural anomalies that lead to psychosis render individuals vulnerable for aggressive behaviour.

3 Furthering Understanding of Violent Behaviour Among Persons with Schizophrenia

Many studies have compared people with schizophrenia with and without criminal convictions or with and without aggressive behaviour. These studies thereby failed to take account of the consistent evidence showing that among people with schizophrenia, as in the general population, violent offenders constitute a population that is heterogeneous with respect to age of onset and persistence of violent behaviour and, it is hypothesized, with respect to the neurobiological mechanisms that initiate and maintain the aggressive behaviour. Much epidemiological evidence converges in identifying three distinct types of individuals with schizophrenia who engage in aggressive behaviour towards others: Type I—those who display antisocial and aggressive behaviour from a young age that remains stable into adulthood undisturbed by the onset of schizophrenia; Type II—those with no history of conduct problems in childhood and adolescence who begin engaging in aggressive behaviour as the illness onsets; and Type III—a small group with no history of antisocial or aggressive behaviour who engage in one act of serious violence in the third or fourth decade of life.

4 Type I: Schizophrenia Preceded by Conduct Disorder (SZ + CD)

Early start offenders with schizophrenia have been defined in different ways—those with a conviction for violence prior to illness onset, those with a history of childhood conduct problems and those who meet criteria for Conduct Disorder (CD) prior to age 15. As compared to other offenders with schizophrenia, those with a childhood history of conduct problems defined in any of these ways, are convicted for more non-violent and violent crimes (Crocker et al. 2005; Fulwiler and Ruthazer 1999; Mueser et al. 2006), commit a more diverse array of crimes (Hodgins 2004), and have criminal histories similar to those of non-mentally ill offenders who also have a childhood history of conduct problems (Hodgins and Côté 1993; Schug et al. 2007). In addition, almost all display a pattern of substance misuse going back to early adolescence (Fulwiler et al. 1997; Moran and Hodgins 2004; Tengström et al. 2001; Mueser et al. 1999).

CD and conduct problems are precursors of schizophrenia. For example, in a prospective investigation that followed a Dunedin, New Zealand birth cohort to age 26, 40 % of the cohort members who developed schizophreniform disorders had displayed CD prior to age 15 (Kim-Cohen et al. 2003). In clinical samples of adults with schizophrenia, the prevalence of CD is lower. The CD modules of the Structured Clinical Interview for DSM-III-R and IV were designed to diagnose CD prior to age 15 among adults. We have used this interview protocol, in some studies supplemented by information from family members, school, social service and justice files, to diagnose CD among adults with schizophrenia. In most samples, the prevalence is approximately 20 % among both women and men (Kim-Cohen et al. 2003), but for example in the UK sample of inpatients described above, CD prior to age 15 characterised 42.0 % of the men and 22.4 % of the women (Hodgins et al. 2008). While these samples of patients with schizophrenia were recruited in general psychiatric services, among patients in forensic services the prevalence of CD is higher, and among those incarcerated it is further elevated (Hodgins et al. 1998).

CD and conduct problems are not only precursors of schizophrenia, they are more common among people who develop schizophrenia than in the general population. Recent estimates of the prevalence of CD prior to age 15 range from 7.5 % among boys and 3.9 % among girls in a large UK cohort (Green et al. 2004), to 12.0 % among boys and 7.1 % among girls in a US community sample (Nock et al. 2006). This more recent evidence is consistent with older findings from a US investigation of a population sample of 20,000 adults, in which higher rates of CD were observed among persons with, than without, schizophrenia (Robins et al. 1991), and the number of CD symptoms prior to age 15 was positively and linearly associated with the likelihood of developing schizophrenia (Robins 1993). The accumulated evidence indicates that the prevalence of CD is higher among people who develop schizophrenia than in the general population, and that among men and women with schizophrenia the prevalence is similar while in the general population the prevalence is much higher among males. Thus, not only is the proportion of offenders greater in the population of persons who develop schizophrenia than in the general population, so is the proportion of offenders who display an early onset and stable pattern of antisocial behaviour.

We conducted an investigation to examine the association of CD prior to age 15 and crime and aggressive behaviour among a sample of 248 men with schizophrenia who were, on average, aged 39.8 years old at the time of the study (Hodgins et al. 2005). They were assessed in the 2 weeks before discharge from hospital using multiple sources of information including complete criminal records. Fifty-two (21 %) of these men met criteria for CD prior to age 15. Incident Rate Ratios (IRR) were calculated to estimate the association between CD and the number of convictions for violent crimes. A diagnosis of CD prior to age 15 was associated with an increase of 2.29 (1.31–4.03) in the number of convictions for violent crimes after controlling for lifetime diagnoses of alcohol and/or drug abuse and/or dependence. As shown in Fig. 1, each CD symptom present before the age of 15 was associated with a 1.15 (1.06–1.25) increase in the number of convictions for violent crimes, again after controlling for diagnoses of substance misuse disorders. CD and CD symptoms were also associated with the number of convictions for non-violent crimes as depicted in Fig. 1.

We have replicated these results in the sample of UK inpatients with severe mental illness described above (Hodgins et al. 2007). In this study, after controlling for sex, age, current alcohol and drug use, CD prior to age 15 was associated with a two-fold increase in the number of convictions for violent crimes. Again, after controlling for sex, age and substance misuse, each CD symptom present before age 15 was associated with a slight increase in the number of violent crimes (odds ratio 1.16, 1.01–1.35). Both CD diagnosis and the number of CD symptoms were associated with the number of convictions for non-violent crimes. Importantly, no sex differences in the associations of CD and later offending were detected. The results of these two studies indicate that among individuals who develop schizophrenia, as among those who do not, CD in childhood is a precursor of criminal offending in adulthood. They concur with

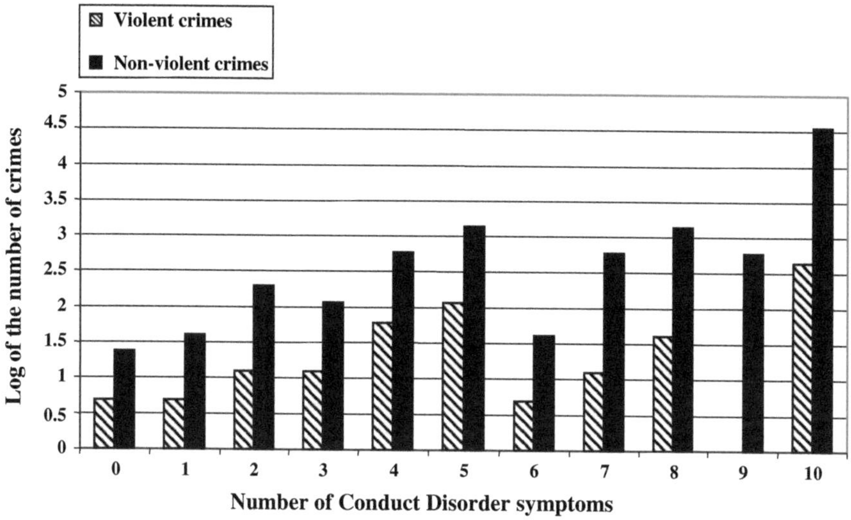

Fig. 1 Number of non-violent and violent crimes as a function of the number of Conduct Disorder symptoms among men with schizophrenia

results from other studies that used different definitions of childhood conduct problems (Fulwiler and Ruthazer 1999; Rice and Harris 1995; Tengström et al. 2004).

Both in the general population (Moffitt and Caspi 2001) and among people with schizophrenia, CD is a precursor of aggressive behaviour, as well as violent crime. The MacArthur Community Violence Instrument (Steadman et al. 1998) has been used in three studies to assess aggressive behaviour among people with schizophrenia and its association with Conduct Disorder. In the sample of 248 men with schizophrenia described above, the diagnosis of CD and the number of CD symptoms were associated with an increased risk of aggressive behavior, after taking account of lifetime diagnosis of alcohol abuse and/or dependence, alcohol and drug use during the follow-up period, depot medication or medication compliance and obligatory care. The diagnosis of CD did not predict aggressive behavior, however, once we controlled for lifetime diagnosis of drug abuse and/or dependence diagnosis. Each CD symptom increased the risk of aggressive behavior by a factor of 1.2, and this remained significant after controlling for lifetime diagnoses of alcohol and drug use disorders, self-reported alcohol and drug use, drug detected in urine or hair or a refusal to provide a sample, depot medication or self-reported compliance, and a court order to comply with treatment (Hodgins et al. 2005). In the UK sample of severely mentally ill men and women in their late-thirties described above, we found that after controlling for age, sex and current substance misuse, CD diagnosis prior to age 15 was associated with an increased odds of aggressive behaviour towards others in the previous 6 months (odds ratio 2.66, 1.24–5.68), as was each CD symptom (odds ratio 1.29,

1.11–1.50) (Hodgins et al. 2008). We also undertook analyses with this sample to further investigate the role of psychotic symptoms in four 6-month periods when patients were living in the community. Symptoms were measured at the beginning of each 6-month period and at the end of the period both collateral informants and patients reported on aggressive behaviour that had occurred in the previous 6 months. In regression models, we first entered the presence or absence of Antisocial Personality Disorder (ASPD) (the adult diagnosis that requires the presence of CD prior to age 15), then current substance misuse measured both by self-report and hair and urine analyses, and then numbers of positive and negative symptoms and changes in symptoms. In the final model, the risk of aggressive behaviour was increased six times by a diagnosis of APD, and one and one-half times by the number of positive symptoms, by the change in the level of positive symptoms and by the presence of anxiety (Hodgins et al. 2003; Hodgins 2008).

Similar findings emerged from analyses of baseline data collected for a large trial of medications in the US. While this study used the same instrument to assess aggressive behaviour as did our studies, unlike our study described above symptoms were not measured prospectively but at the same time as the aggressive behaviour. Two or more CD symptoms were found to be associated with aggressive behaviour in the previous 6 months after controlling for numerous other factors, and, as in our studies, there was no association with substance misuse after taking account of childhood conduct problems. While positive symptoms and depression were positively associated with life-threatening violence, negative symptoms were protective. A combination of above median scores for positive symptoms and below median scores for negative symptoms was associated with a three-fold increase in risk of life-threatening violence (Swanson et al. 2006).

Thus, the accumulated evidence indicates that CD is a precursor of schizophrenia in a minority of cases, that CD is more common among individuals who develop schizophrenia than in the general population and that among adults with schizophrenia, a diagnosis of CD and the number of CD symptoms present prior to age 15 are associated with violent behaviour through adulthood. While some studies do show that positive symptoms are associated with aggressive behaviour even after taking account of CD, the existing evidence indicates that those with CD are not distinguished from other patients with schizophrenia by symptom profiles (Moran and Hodgins 2004). Based on a comparison of the results of a small number of studies, the strength of the association between CD in childhood and aggressive behaviour in adulthood appears to be similar among individuals who develop schizophrenia (Hodgins et al. 2005, 2008; Swanson et al. 2006) as it is among those who do not (Loeber et al. 2005; Simonoff et al. 2004). In addition, the number of CD symptoms prior to age 15 is linearly associated with the development of ASPD in adulthood among persons with schizophrenia, as it is in the general population (Robins et al. 1991).

There is little prospective research on individuals who present conduct problems in childhood and who subsequently develop schizophrenia (SZ + CD). An important finding has emerged from the prospective longitudinal investigation of a birth cohort in Dunedin, New Zealand. As previously noted, and consistent with other epidemiological evidence, the risk of violence was elevated among cohort members who developed schizophreniform disorder by age 26. This association was partially explained by the presence of aggressive behaviour at ages 7, 9 and 11 and psychotic-like experiences at age 11 (Arseneault et al. 2003). In a cohort of twins aged 12 years, those reporting psychotic-like experiences also reported significantly more antisocial behaviour, depression and anxiety (Polanczyk et al. 2010). Most other relevant findings on the childhood characteristics of individuals with SZ + CD derive from studies of clinical samples of adults in which data on childhood are collected retrospectively from multiple sources, some objective—school, social service and juvenile justice records, and some subjective—reports from patients, parents, older siblings. For example, in our sample of 248 men with schizophrenia, we compared those with and without CD prior to age 15. More of those with, than without, CD obtained lower than average marks in elementary school, failed to graduate from secondary school, and prior to age 18 abused substances, experienced physical abuse and were institutionalised (Hodgins et al. 2005). The results of other similar studies concur (Fulwiler et al. 1997; Tengström et al. 2001; Schanda et al. 1992).

4.1 Factors Contributing to the Development of Schizophrenia Preceded by Conduct Disorder

Nothing is known definitively about the aetiology of SZ + CD (SZ + CD), but the extant literature allows us to propose hypotheses that can be tested in future studies. Both schizophrenia (Weinberger 1987) and CD (Stadler et al. 2010; Blair et al. 2006) are neurodevelopmental disorders. In both disorders, from conception onwards, combinations of genes, in addition and in interaction with environmental events, are thought to modify brain structure and function. Thus, we reason that when schizophrenia develops in parallel with CD, neurodevelopment will be distinct from both that associated with schizophrenia and that associated with CD, but will include features that are common to each disorder and others that differ across the two disorders.

4.1.1 Heritability

Twin studies have shown that CD that persists into adulthood is moderately heritable (Bornovalova et al. 2010; Krueger et al. 2002; Slutske et al. 1998; Young et al. 2000) with shared variance in childhood and unique variance in adulthood playing important roles (Burt et al. 2005; Burt 2009). The inherited vulnerability is for early onset and persistent antisocial behaviour and substance misuse. The effects of the genes conferring this vulnerability are increased by addition or interaction with environmental stressors such as maltreatment, poor relationships with mother and/or father, stressful life events, poor academic achievement and antisocial peers (Hicks et al. 2009). Further, the associations of these environmental pathogens with genetic vulnerabilities vary by type of maltreatment, dose, duration and developmental stage at exposure (Cicchetti et al. 2012). While heritability estimates for this pattern of antisocial behaviour and substance misuse that onsets early in childhood and remains stable across the life span are similar among males and females, the timing of genetic and environmental effects and the specific genetic variants that are involved differ (Javdani et al. 2011).

It is possible that individuals who develop SZ + CD simply inherit the multiple genes associated with both disorders and additionally are exposed to the environmental risk factors associated with both disorders. This is unlikely, however, given that CD is more common among people with schizophrenia than in the general population and that among people with schizophrenia the prevalence among men and women is similar while in the general population it is much higher in males than females. These epidemiological findings suggest, rather, that children with a vulnerability for schizophrenia are more vulnerable than other children to develop CD. This vulnerability could be conferred by genes, gene-to-gene interactions, gene-environment interactions, and/or environmental factors altering the expression of genes, and/or environmental factors.

4.1.2 Genes

Variants of the monoamine oxidase A (MAOA) gene (Taylor and Kim-Cohen 2007), the serotonin transporter (5-HTLPPR), tryptophan hydroxylase (Cicchetti et al. 2012), catechol O-methyltransferase (COMT) genes (Caspi et al. 2008) have each been linked to persistent violent behaviour. As will be described in other chapters in this volume (Leucht and coworkers), the low activity variant of the MAOA gene has been associated with conduct problems in childhood and persistent violent behaviour in adulthood among males (Taylor and Kim-Cohen 2007). Further, in studies of males recruited from the general population, this low activity variant of MAOA was associated with changes in orbitofrontal cortex (OFC) volume, amygdala and hippocampus hyper-reactivity during aversive recall, and impaired cingulate activation during cognitive inhibition. While the limbic system was reduced in grey matter (GM) volume, it was hyper-reactive to angry faces and emotionally negative stimuli with indications that cortical down-regulation was limited (Meyer-Lindenberg 2011). The authors speculated that a dysregulated and hyper-reactive amygdala response contributed to both anxiety and violence, with reactive violence resulting only when higher control centres such as the OFC and caudal cingulate are impaired. The structural and functional brain anomalies observed in males carrying the low activity variant of the MAOA gene were not observed in females (Meyer-Lindenberg 2011). This is consistent with a small number of studies showing that the high activity variant of MAOA (Sjöberg et al. 2007; Nilsson et al. 2006; McGrath et al. 2012) is associated with antisocial behaviour in females. In another study however, both males and females carrying the low activity variant of the MAO gene reported higher trait aggression and showed significantly greater activity in the dorsal anterior cingulate cortex (ACC) in response to social exclusion. Analyses revealed that both dorsal ACC activity and trait interpersonal hypersensitivity mediated the relationship between MAOA and trait aggression, suggesting that the relationship between MAOA and aggression might be due, in part, to a heightened sensitivity to negative socio-emotional experience (Eisenberger et al. 2007).

4.1.3 Obstetrical Complications

To our knowledge there is only one study of obstetrical complications and offending in schizophrenia. In a Swedish birth cohort, we found that among males, but not females, who developed a major mental disorder (most were cases of schizophrenia) complications in the neonatal period increased the risk of childhood conduct problems and offending three-fold. Neither pregnancy nor birth complications, inadequate parenting, or SES of the family of origin increased the risk for offending (Hodgins et al. 2002). Importantly, neonatal complications did not characterise males without major mental disorders who presented a similar pattern of childhood conduct problems and offending. Again this evidence suggests that environmental factors have distinct effects on individuals developing SZ + CD due to a distinct genetic diathesis (Hodgins et al. 2001).

4.1.4 Physical Maltreatment

Children who will subsequently develop schizophrenia (Read et al. 2005; Morgan and Fisher 2006; Bendall et al. 2008), like those with CD (Stouthamer-Loeber et al. 2001; Widom 1989; Murray and Farrington 2010), often experience physical maltreatment during childhood. Maltreatment in childhood may alter the functioning of the hypothalamic pituitary adrenal (HPA) axis and the methylation of glucocorticoid genes thereby modifying the expression of these genes throughout life (McGowan et al. 2009), and lead to hyper-reactivity of the limbic system to perceived threat (Dannlowski et al. 2012). Anomalies in hippocampal volumes are observed among adults with schizophrenia and their siblings. Risk genes for schizophrenia are associated with hippocampal volumes (Honea et al. 2008; Goldman et al. 2008). Further, evidence is accumulating to suggest that these anomalies precede illness onset (Cullen et al. 2012). Childhood maltreatment has been shown to lead to similar reductions in GM volumes of the hippocampus in adults, suggesting that they result from chronic stress reactions within the HPA axis (McCrory et al. 2010). Thus, physical maltreatment may affect children developing schizophrenia differently, or more severely, because of the neural alterations associated with the developing illness. Physical abuse is not only perpetrated by parents or adults close to children, but also by peers. Both children reporting psychotic-like-experiences (Arseneault et al. 2011; Schreier et al. 2009) and those with CD are victims of bullying and physical attacks from other children (Arseneault et al. 2010). There are almost no studies of the early childhood development of individuals who subsequently succumb to schizophrenia. Prospective studies highlight the role of inconsistent parenting and poor attachment to parents and teachers as factors contributing to CD (Odgers et al. 2012). Approximately 40 % of parents of children who will subsequently develop schizophrenia present schizophrenia or a spectrum disorder (Kety 1987). A spectrum disorder, for example, the most common schizotypal personality disorder, is characterised by distrust of others, inappropriate or constricted affect, lack of close friends or confidants, excessive anxiety and odd behaviour. Such an individual would be unlikely to provide a warm relationship, nor parenting with appropriate rewards and sanctions, and would not model appropriate ways for coping with stress. A child with a propensity to overreact to stress, to be easily frightened, would not learn to dampen these fears in the family environment, and might even learn to increase them. Among adults with schizophrenia, physical victimisation continues to be prevalent; for example in a UK sample we studied, more than one-third of the patients reported being physically assaulted in the 6 months prior to interview. Among both males and females, the best predictors of victimisation were the patient's own aggressive behaviour both past and current (Dick 2010). These repeated experiences of victimisation may maintain hyperactivation of neural circuits to threat that began in childhood as a consequence of maltreatment.

4.1.5 Learning Not to Be Aggressive

Aggressive behaviour is most common among toddlers and the great majority of children learn not to be physically aggressive prior to school entry (Petitclerc and Tremblay 2009). A small group, however, continue to display high levels of aggressive behaviour across the life span (Broidy et al. 2003). Children who are developing schizophrenia already at age 2, show motor abnormalities such as delays in walking and talking and specific neurological soft signs (Walker et al. 1994), and in the subsequent years there is further evidence of motor deficiencies, neurological signs, receptive language deficits (Cannon et al. 2002; Cannon and Clarke 2005) and lower than average IQ (Dickson et al. 2012). Further, recent evidence strongly suggests that these children experience psychotic-like experiences by mid-childhood (Laurens et al. 2007; Kelleher and Cannon 2011). Perhaps, one or several of these impairments limits these children from learning not be physically aggressive and to resolve interpersonal conflicts using other strategies, for example, negotiating. Before age 13 children who will subsequently develop schizophrenia present lower than average IQ (Dickson et al. 2012), as do children with CD. Lower than average IQ may index difficulties in using language to interact with others and reduced cognitive control over emotionally charged responses.

4.1.6 Difficulty in Recognizing Facial Expressions

Children with conduct problems present marked abnormalities in the recognition of emotions in the faces of others, including seeing neutral faces as threatening (Dadds et al. 2006; Frick and Marsee 2006). Adults with schizophrenia display deficits in the recognition of emotions in the faces of others (Schneider et al. 2006) that are present at illness onset and in the prodromal phase (Addington et al. 2008). Studies have also shown that this deficit characterises many of their healthy siblings leading to the suggestion that it may signal the presence of one or more susceptibility genes for schizophrenia (Bediou et al. 2007). Children with CD have been shown to perceive neutral faces as threatening, while children who present both conduct problems and callous traits fail to perceive sadness in the faces of others. Similarly, antisocial adults who engage in reactive aggression have been found to be hyper-responsive to both neutral (Bobes et al. 2012) and angry faces while antisocial adults with psychopathic traits fail to identify sadness in the faces of others. It is easy to imagine that in early childhood inaccurate reading of the facial expressions of others would have profound consequences for the child's understanding of his/her environment but also for his/her behaviour. Adults communicate much information to children via facial expressions, including information regarding inappropriate behaviour. A failure to understand these communications might thereby contribute to conduct problems. The failure to recognise anger in the face of adults would deprive the child of information that their behaviour was inappropriate, the failure to recognise fear in others would be associated with a lack of empathy and understanding of others (Blair et al. 2006), perceiving threat in neutral faces would lead to unnecessary anxiety, while a general failure to recognise emotions would be associated with problems with peers (D'Acremont and van der Linden 2007). Among adults with schizophrenia, poor recognition of emotions in faces, especially fear and anger (Weiss et al. 2006), has been associated with past violent behaviour (Frommann et al. 2009; Fullam and Dolan 2006).

4.1.7 Cannabis Use

The results of prospective longitudinal studies have consistently shown that heavy cannabis use in early adolescence is one of the causal factors for schizophrenia (Arseneault et al. 2002; McGrath et al. 2010). Among children who are vulnerable to schizophrenia, cannabis further alters the developing brain as it transitions to a disordered state. There is a great deal of evidence showing that children with CD are exposed earlier and begin abusing drugs earlier than other children (Robins and McEvoy 1990). In a sample of individuals experiencing their first episode of psychosis, we observed that the number of CD symptoms prior to age 14 was associated with lifetime cannabis use and with use of cannabis prior to age 14 (Malcolm et al. 2011). These results suggest that CD increases the risk of early use of cannabis which in turn increases the risk of schizophrenia among vulnerable individuals. Both carriers of the C/C variant of AKT1 gene (Di Forti et al. 2012) and the val/val alleles of COMT have been shown to have an elevated risk of schizophrenia if they used cannabis daily in early adolescence, although attempts to replicate these findings have been mixed.

4.2 Conclusion

CD is a precursor to schizophrenia in as many as 40 % of cases and is associated with an early onset lifelong pattern of antisocial and aggressive behaviour and criminality. Such a pattern of behaviour is known to be moderately heritable and a small bit of evidence shows elevated rates of criminality and substance misuse in relatives of individuals with SZ + CD. We hypothesize that these individuals carry a distinct combination of genes that render them vulnerable to both schizophrenia and CD and alter the ways in which they react to environmental factors. Thus, individuals with SZ + CD differ both from those with schizophrenia and no CD (SZ-CD) and individuals with CD as to genetic and environmental factors contributing to their disorders. The evidence that CD is more common among people with, than without, schizophrenia suggests that the vulnerability to schizophrenia increases the risk of CD, but whether this occurs via genes, environmental factors or a combination of both is presently unknown. The motor, cognitive, and emotional deficits and the associated anomalies of brain functioning, that characterise children who are developing schizophrenia somehow increase the likelihood of CD. In other ways, however, the presence of CD among those vulnerable for schizophrenia may lead to behaviours such as daily use of cannabis that increase the risk of the onset of schizophrenia.

5 Type II: Offenders with Schizophrenia and No Childhood History of Antisocial Behaviour Who Begin Offending at Illness Onset

Some offenders with schizophrenia present no history of antisocial or aggressive behaviour prior to illness onset. Thereafter, they repeatedly engage in aggressive behaviour towards others. Such a pattern of aggressive behaviour that onsets in adulthood is very rare, as almost all persistently violent offenders have a long history of antisocial and aggressive behaviour stretching back to childhood (Loeber and Stouthammer-Loeber 1998; Moffitt 1993). In the sample of 248 men with schizophrenia described previously in this chapter, the offenders who did not have a childhood history of antisocial behaviour were compared with those with childhood conduct problems. The two groups included similar proportions with at least one conviction for a violent crime, but on average, SZ-CD had acquired fewer convictions for violent crimes and many fewer convictions for non-violent crimes. Importantly, however, a greater proportion of patients with no antisocial behaviour prior to illness onset, had been convicted for a homicide (23.9 %) as compared to patients with prior CD (10.4 %) ($X^2(N = 186) = 3.99$, $p = 0.046$) (Hodgins 2008). As would be expected, patients with no history of antisocial behaviour prior to illness onset were significantly older at first conviction for a violent crime than SZ + CD.

There are few studies of patients with schizophrenia, no history of antisocial behaviour prior to illness onset, who repeatedly engage in aggressive behaviour towards others. Some evidence (Mueser et al. 1999, 2006) suggests that they may be particularly vulnerable to illicit drug use, and that drug use may be directly associated with their violent behaviour. In the study described above, we found no differences in the prevalence of lifetime diagnoses of alcohol abuse and/or dependence or drug abuse and/or dependence among patients with and without antisocial behaviour prior to illness onset. Similarly, in our study of the UK inpatient sample, we found no difference in levels of substance misuse between those with and without a CD. The most important study describing this type of offender with schizophrenia was conducted in Denmark using national health and criminal registers (Munkner et al. 2005). All Danes born after 1963 who acquired a diagnosis of schizophrenia by 1999 were included in the cohort. First those who died, emigrated from Denmark, disappeared, or were diagnosed before age 15, the legal age for criminal responsibility in Denmark, were excluded, leaving 4619 persons with schizophrenia. Among them, 422 committed violent offences and 818 non-violent offences before first contact with psychiatric services, 241 committed a first non-violent offence and 144 a violent offence after first contact with services but before the diagnosis of schizophrenia. The 3,138 individuals who had not committed any crimes by the time they were diagnosed with schizophrenia, were, on average, 23.8-years old at diagnosis. Of them, 396 (13 %) were convicted of a non-violent crime and 215 of a violent crime during the follow-up period that lasted from 6 months to 20 years with a median of 5.4 years. Being convicted of a crime was associated with a younger age at diagnosis, male sex, and having a substance use disorder before or at the time of being diagnosed with schizophrenia.

Drug use by patients with schizophrenia puts them at risk for violence in several different ways. With impaired social cognition, such patients make contact with antisocial offenders who congregate in high crime areas with high levels of violence in order to purchase illicit drugs and use their limited financial resources that may subsequently impact on nutrition. The use of either alcohol and/or illicit drugs further contributes to limiting their psychosocial functioning generally and to increasing their cognitive impairments. It usually also limits their engagement with services and their compliance with medication. Further, the consequences of different substances on an already malfunctioning brain, and most specifically on the fragile dopaminergic and serotonergic systems, may lower the threshold for engaging in aggressive behaviour. Understanding of the links between substance misuse and violence among persons with schizophrenia with no history of antisocial behaviour prior to illness onset is a priority for future research.

While substance use by persons with schizophrenia has attracted a great deal of attention, no studies to our knowledge have examined the links between the progressive brain changes that characterise the illness and aggressive behaviour. Further, studies have not been undertaken to understand if individual reactions to these brain changes, and/or to illness onset, are linked to violence. Patients with schizophrenia who repeatedly engage in aggressive behaviour in the absence of a prior history of such behaviour represent a distinct sub-group. Further research is needed to identify the processes leading to the aggressive behaviour and treatments to end it.

5.1 Conclusion

Little is known about individuals with no history of conduct problems who begin engaging in aggressive behaviour when they become ill. We hypothesize that misuse of substances, changes in brain structure and functioning associated with illness onset and substance misuse, and the individual's reaction to these changes increase the risk of aggressive behaviour.

6 Type III Offenders with Schizophrenia

These are usually men in their late-thirties with chronic schizophrenia and no history of antisocial or aggressive behaviour who kill, or try to kill, someone, often a caregiver. Among men with schizophrenia, the number of convictions for violent offences decreases with the age at first conviction (Hodgins 2004). But this is not true of homicide. Homicides are often committed by patients with schizophrenia who have no history of violence or of antisocial behaviour (Beaudoin et al. 1993; Erb et al. 2001). We know of only one study that has examined this type of offender. Using the sample of 248 men with schizophrenia described above, we found that scores for the items from the Psychopathy Checklist Revised (Hare 2003) that assess the affective trait, callousness, were associated with violent offending and with negative symptoms, but not with CD or substance misuse (Sunak 2006). Similarly, in a study of persistently aggressive inpatients, three distinct groups were identified, one of which was characterised by a lack of remorse (Nolan et al. 2003). We hypothesize that among patients with chronic schizophrenia, callousness may result from progressive brain changes and increase the risk for aggressive behaviour.

6.1 Conclusion

Thus, the age of onset and persistence of aggressive behaviour distinguishes three types of offenders with schizophrenia. The differences in the patterns of aggressive behaviour and associated characteristics of the three phenotypes and the evidence of heritability of CD and of increased antisocial behaviour among relatives of those with SZ + CD suggest that the three phenotypes may carry distinct combinations of susceptibility genes. This would mean that each phenotype reacts to environmental factors from conception onwards in a distinct manner and presents distinct abnormalities of neural development.

7 A Diversion: The Neurobiology of Aggressive Behaviour

Much research into the neurobiological mechanisms underlying aggressive behaviour has distinguished between reactive, emotionally charged, aggressive behaviour and instrumental aggressive behaviour defined as aggressive acts that are undertaken to achieve a goal such as obtaining money, avenging a previous insult or injury, or punishing. As discussed in more detail elsewhere in this volume, instrumental aggression characterises individuals who in addition to an early onset stable pattern of antisocial behaviour, are distinguished, even in childhood, by the presence of traits of callousness and a lack of empathy. Importantly, early onset and persistent patterns of reactive and instrumental aggression characterise individuals who display an early onset and persistence of other antisocial behaviours, attitudes and ways of thinking. This means that from early childhood onwards they fail to follow rules and laws, cheat, lie and show a profound disrespect for the rights of others. Those who additionally present high levels of the personality traits of psychopathy exhibit in addition to instrumental aggression, a capacity to manipulate others, and to identify potential victims (Wheeler et al. 2009). While conduct problems that persist into adulthood are moderately heritable, when accompanied by the trait of callousness, heritability is high (Viding et al. 2008), as it is for instrumental aggression (Tuvblad et al. 2009).

While most research has focused on those with a childhood onset of persistent antisocial and aggressive behaviour that remains stable across the life span, the traits that characterise these individuals—antisociality, impoverished affective responding operationalised as a lack of remorse, shallow affect, callous/lacking empathy and a failure to accept responsibility for one's own action, and an interpersonal style characterised by lying, manipulation and a grandiose sense of self-worth—vary dimensionally in the general population. The interpersonal trait is the one aspect of psychopathy that is specifically associated with instrumental aggressive behaviour (Laurell et al. 2010; Vitacco et al. 2006) and a recent study suggests that it is also specifically associated with deficient amygdala response to fearful faces (Carré et al. 2012). Previous evidence and theorising, however, consistently linked the deficient amygdala responses to fearful faces to the affective trait of psychopathy (Blair 1995, 2010). The antisocial characteristics of psychopathy, essentially an early onset persistent pattern of antisocial and aggressive behaviour, are associated with reactive aggression (Hall et al. 2004; Coccaro et al. 2007) and with heightened amygdala responses to angry faces (Carré et al. 2012; Sterzer and Stadler 2009). The antisocial characteristics of psychopathy are also associated with a diminished responsiveness to reward in the ventral striatum (Carré et al. 2012). Psychopathy is associated with increased risk taking and sensation seeking but the syndrome is not associated with impulsivity which characterises individuals who engage in reactive aggression and persistent antisocial behaviour.

Thus, much research shows that among humans, differences in genetic liability and anomalies of brain structure and functioning, cognitive functioning, emotional processing and personality traits are observed in individuals who show persistent reactive aggression from childhood onwards and those who exhibit instrumental aggression from childhood onwards.

8 Neural Mechanisms of Aggressive Behaviour Among Persons with Schizophrenia[4]

The neural mechanisms that have been found to underlie reactive and instrumental aggressive behaviour that onsets in childhood and persists through adulthood, and the similar mechanisms associated with dimensional traits of antisociality, callousness, lying and manipulation in general population samples, may underlie aggressive behaviour in people who are developing, or who have schizophrenia. However, to date no studies have examined whether aggressive behaviour among persons with schizophrenia is reactive or instrumental. Among persons with SZ + CD, neurodevelopment may well be distinct from both that among SZ-CD and CD, and the causes of the structural and functional anomalies underlying aggressive behaviour are hypothesized to differ. Among persons with schizophrenia whose aggressive behaviour onsets with the illness, the neural anomalies underlying aggressive behaviour may develop as a consequence of changes in the brain during transition to illness, substance misuse and medications, that either lower the threat threshold or increase callousness, or both. Among the small group who engage in one severe violent act after many years of chronic illness, changes in the brain are hypothesized to increase callousness.

[4] Studies of brain structure and function that compare individuals with schizophrenia who committed a violent crime to those with no criminal record are not reviewed. By lumping together individuals who have a history of aggressive behaviour prior to illness onset, those with multiple incidents of aggression towards others since illness onset, those who have committed one serious incident of violence in their lives, these studies fail to acknowledge the heterogeneity of violent offenders with schizophrenia. Other studies focus on patients with schizophrenia with persistent aggressive behaviour that keeps them hospitalised (Hoptman et al. 2002, 2005, 2006, 2010). Other studies examine violent offenders with schizophrenia recruited from institutions where placement was decided by the court and influenced by the media, and based largely on the characteristics of one incident of violence and of the victim (Barkataki et al. 2006; Puri et al. 2008; Kumari et al. 2009, 2012; Narayan et al. 2007) and compare them to schizophrenia patients with no criminal records. Other studies select patients based on the type of offence (Schug et al. 2010).

8.1 Neural Mechanisms Associated with Aggressive Behaviour Among People with SZ + CD

Initial studies and reviews led to the hypothesis that men with SZ + CD presented better cognitive performance, particularly executive skills, than men with SZ-CD (Naudts and Hodgins 2006; Joyal et al. 2003). For example, two studies examined male offenders with schizophrenia: one compared those with and without comorbid ASPD, and another compared those with and without high psychopathy scores (Rasmussen et al. 1995). Both included small samples and reported fewer neuropsychological deficits among the antisocial participants in tests tapping the dorsolateral prefrontal and OFC functions. However, subsequent studies have not concurred. We examined a complete cohort of individuals consulting psychiatric services with a first episode of psychosis in one large geographic catchment area in the United Kingdom. Men who already had convictions for crimes prior to a first episode of psychosis were presumed to have presented CD earlier in childhood or adolescence. Those with violent convictions prior to illness onset, as compared to the non-offenders, performed significantly more poorly on tests of verbal memory, visual-spatial perception and organisation, and verbal intelligence, and in one of two tests on processing speed. Violent offenders obtained, on average, similar scores as the non-offenders on tests of working memory and executive function. They obtained higher scores than non-offenders in one test on executive function, the Raven's CPM—Set B test, but this difference was not significant (Hodgins et al. 2011).

In a subsequent study, we compared three groups of men: SZ + CD, SZ-CD and healthy. Those with SZ + CD obtained IQ scores approximately one-half of a standard deviation lower than those with SZ-CD, and made significantly more errors on the Wisconsin Card Sort Test, made fewer correct solutions on the Tower of London task although this difference was not significant, and made more errors on the Go/NoGo test, while performance on the Trail Making Test, and Working Memory visuo-spatial memory and verbal memory subtests were similar.

Functional MRI (fMRI) studies of violent offenders with schizophrenia observed decreased activation in the OFC and basal cortex during a Go/NoGo task and increased activity in the motor, premotor and anterior cingulate regions among those with ASPD (Joyal et al. 2007) and attenuated amygdala activation to fearful faces among those with high psychopathy scores (Dolan and Fullam 2009). Additionally, OFC responses were negatively correlated with both lifestyle and antisocial facets of psychopathy in the entire sample during processing of sad faces.

We recently completed the first structural MRI study, comparing four groups of men, SZ + CD, SZ-CD, CD and healthy, matched on IQ and level of education (Schiffer et al. 2012). The SZ + CD and SZ-CD men were similar as to age of onset and duration of illness, scores for positive and negative psychotic symptoms, and dose of medication measured in chlorpromazine equivalent units. Yet, relative to men with SZ-CD, those with SZ + CD displayed increased GM volumes in the hypothalamus, left putamen, right cuneus/precuneus, and right inferior parietal cortex and no regions of decreased GM volumes after controlling for age and self-reports of lifelong alcohol and drug use. The SZ + CD and CD men were similar as to mean number of CD symptoms prior to age 15, scores for lifelong aggression, numbers of criminal convictions and proportions with a diagnosis of ASPD. Relative to men with CD, those with SZ + CD exhibited no regions in which GM volumes were increased, and volume decreases in the inferior, medial and superior temporal lobes (BA20,21,22,38), the temporo-parietal junction (BA19,39,40), parahippocampal gyrus dorsolateral prefrontal cortex (BA9,46), frontopolar regions (BA10) and the thalamus, again after controlling for age and self-reports of lifelong alcohol and drug use.

Conjunction analyses indicated that relative to the healthy men, the SZ + CD and CD groups exhibited similarly increased volumes in the hypothalamus, the right inferior parietal cortex (BA40) and left superior parietal cortex (BA5,7). Further, the GM volumes of the hypothalamus were significantly correlated with scores for fighting and assault and the number of aggressive CD symptoms. To our knowledge this is the first study to report abnormal GM volumes of the hypothalamus in antisocial males and consequently these findings require replication. Few studies of humans have reported hypothalamic abnormalities associated with aggression. Lower metabolism in the right hypothalamus was observed in violent men with high LHA scores (George et al. 2004) and children with gelastic seizures and hypothalamic hamartomas present elevated rates of CD (Weissenberger et al. 2001). In rodents and cats, stimulation of the hypothalamus leads to aggressive behavior (Lin et al. 2011; Herman et al. 2008). The amygdala-hypothalamus-periaqueductal grey system within the midbrain is thought to mediate reactive aggression in response to a real or perceived threat (Panksepp 1998; Gregg and Siegel 2001; Blair 2012). This system is organised hierarchically such that aggression evoked by stimulation of the periaqueductal gray is not dependent on the functional integrity of the amygdala (Panksepp 1998; Gregg and Siegel 2001; Blair 2012). Elevated scores for lifelong aggression have been positively correlated with cerespinal fluid arginine vasopressin (Coccaro et al. 1998), negatively correlated with brain serotonin levels (Coccaro et al. 2011) and 5-HT transporter platelet binding sites (Coccaro et al. 2010), reduced cerebral spinal fluid oxytocin (Lee et al. 2009), and as noted with reduced metabolism in the right hypothalamus (George et al. 2004). Thus, among men with SZ + CD and those with CD aggressive behavior may be associated with dysfunction at the bottom of this neural circuit (Blair 2004; Lotze et al. 2007).

Supporting this notion, intranasal administration of vasopressin compared to placebo has been shown to selectively reduce the ability to infer mental states (particularly negative emotions) in male but not female eyes in healthy men (Uzefovsky et al. 2012). The ability to infer and comprehend the mental states of self and other, often referred to as mentalizing or theory of mind (TOM), has been proposed as an important construct in the understanding of violence (Addy et al. 2007; Blair et al. 2004, 2006; Covell and Scalora 2002; Hare 2006; Levinson and Fonagy 2004). This ability, which includes both cognitive (i.e., reasoning about beliefs) and affective (i.e., reasoning about emotions) components (Abu-Akel 2003; Baron-Cohen 1995; Blair 2005; Brothers and Ring 1992; Mitchell 1998), is considered a prerequisite for inter- and intra-relational adjustment, including affecting regulatory processes and impulse control (Weiss et al. 2006; Allen et al. 2008; Baumeister and Heatherton 1996), as well as the ability to empathize and feel guilt (Blair 2005). More specifically, it is suggested that understanding the emotional mental states of others reduces the propensity to engage in violence (Fullam and Dolan 2006; Eisenberg and Fabes 1990; Miller and Eisenberg 1988; Tangney 1991). However, investigations of ToM among violent offenders with schizophrenia are rare (Bo et al. 2011) and nothing is known about mentalizing abilities and their neural underpinnings in the three sub-types of offenders with schizophrenia.

In our study, the SZ + CD and CD men were similar as to increased hypothalamic volumes and also increased volumes of the right inferior parietal cortex, and left superior parietal cortex. Spatial impairments are evident by age 3 among males whose antisocial and criminal behavior persists into adulthood (Raine et al. 2005), structural abnormalities of the parietal lobes have been reported in boys with CD (Hyatt et al. 2012; de Brito et al. 2008), and functional abnormalities in male offenders with high psychopathy scores (Koenigs et al. 2011; Birbaumer et al. 2005). The right hemisphere is dominant in the first months and years of life, and atypical development, it has been hypothesized, impairs the child's recognition of mother's facial expressions and consequent mother–child interactions (Raine et al. 2002).

Altered serotonergic functioning is associated with reactive aggressive behavior (Davidson et al. 2000) and impulsive-antisocial temperament is associated with excess neurochemical and functional engagement of the mesolimbic dopamine system in response to reward (Buckholtz et al. 2010). Research in rodents suggests that mesolimbic dopamine is critical for the expression of aggression (Lewis et al. 1994; Couppis and Kennedy 2008; Rodriguiz et al. 2004) and that genetic manipulations that reduce striatal dopamine clearance increase aggressive behavior (Rodriguiz et al. 2004). Increased striatal dopamine synthesis capacity predates the onset of schizophrenia (Howes et al. 2011). Youth with externalising pathology exhibit activation in the caudate to reward and to non-reward, while healthy youth show activations in the striatum only to reward (Gatzke-Kopp et al. 2009). Another study reported that youth with psychopathic traits displayed increased activations in the caudate during punished errors while healthy youths did not (Finger et al. 2008). Thus, abnormalities of dopamine function in the striatum are associated with both psychosis and CD and CD coupled with callousness (Fig. 2).

In our study, the SZ + CD participants, but not the CD participants, displayed larger cerebellum volumes than the healthy men. Similar increases have been reported in boys with conduct problems and callous traits (de Brito et al. 2008) and male offenders with ASPD (Tiihonen et al. 2008). As antipsychotic medications are associated with reduced cerebellar volumes (Ho et al. 2011), the greater volumes

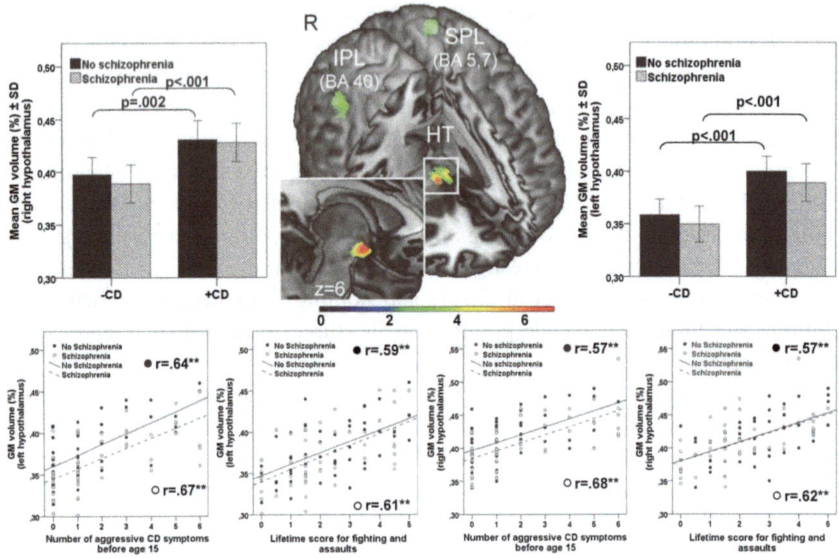

Fig. 2 Grey matter volumes associated with conduct disorder and aggressive behaviour among men with schizophrenia and conduct disorder and men with conduct disorder

observed among the SZ + CD men may have been even greater prior to treatment. The cerebellum develops late and is particularly vulnerable to environmental insults (Tiemeier et al. 2010; ten Donkelaar et al. 2003; Caspi et al. 2003).

8.1.1 Conclusion

While there are few studies, emerging evidence suggests that men with SZ + CD show lower verbal IQ, and distinctive anomalies of cognition, brain structure and function as compared to men with SZ-CD, with CD and those with no mental disorders and criminality.

8.2 Neural Mechanisms Associated with Aggressive Behaviour Among People with Schizophrenia with No History of Aggressive Behaviour Prior to Illness Onset

Aggressive behaviour is as stable across the lifespan as IQ (Olweus 1979). When an adult with no history of such behaviour begins engaging in aggressive behaviour it suggests a change in brain functioning. Among individuals with no history of antisocial or aggressive behaviours who repeatedly engage in aggressive behaviour while transitioning to, or after, the first full blown episode of psychosis, the changes in the brain (Ziermans et al. 2012) that underlie the psychosis may either alter the threshold for aggressive behaviour or increase callousness towards others.

These individuals may worsen their own condition as they transition towards illness by using alcohol and/or drugs. These substances, like most environmental factors, impact individuals with schizophrenia differently than healthy individuals. For example, similar abuse of alcohol leads to greater reductions in GM volumes among men with schizophrenia than healthy men (Mathalon et al. 2003). Antipsychotic medications may lead to changes in neural structures that increase the likelihood of aggressive behaviour. For example, people with schizophrenia exhibit increased GM volumes in most structures forming the dorsal striatum including the caudate, putamen and globus pallidus. These morphological anomalies show no evidence of heritability (Goldman et al. 2008) but rather appear to result from antipsychotic medication (Honea et al. 2008). Men with high levels of psychopathic traits have been found to present similarly increased striatum volumes (Glenn et al. 2010).

Aggressive behaviour that suddenly erupts in an individual with chronic schizophrenia may result from progressive brain changes associated with illness. Prospective studies that repeatedly image individuals with schizophrenia show evidence of progressive GM volume changes over time (Ellison-Wright et al. 2008) that are greater than those observed among healthy adults through the same age period. While there is evidence that antipsychotic medications are responsible, at least in part, for all or some of these changes (Ho et al. 2011), other factors such as substance misuse, obesity and lifestyle may also contribute to these changes (Meyer-Lindenberg et al. 2006).

9 Conclusions

There is a wealth of evidence showing that adults with schizophrenia and children and adolescents developing schizophrenia are at increased risk, compared to the general population, to engage in aggressive behavior towards others and to acquire convictions for violent crimes. While positive symptoms do explain aggressive that commonly occurs in acute episodes of psychosis, they do not explain the aggressive behavior in other stages of illness or before illness onset. Much evidence identifies three distinct phenotypes, one with CD prior to illness onset, one whose aggressive behavior onsets with illness, and a rare, third type, who engage in violence after several decades of illness. To date, few studies have examined neural mechanisms of these phenotypes of individuals with schizophrenia who engage in aggressive behavior. Further, the aggressive behavior has not been classified as reactive or instrumental. Several structural and functional anomalies characterise both people with schizophrenia and with antisocial behaviour. In addition to abnormalities of the hypothalamus, parietal corticies, striatum, decreased GM volumes in the left middle temporal gyrus, BA 21, have been observed in adults with schizophrenia as compared to their siblings (Honea et al. 2008) and among persistent violent offenders with the syndrome of psychopathy (Gregory et al. 2012). Thus, understanding the neural mechanisms underlying aggressive behavior among people with schizophrenia is presently very limited. Consequently, this chapter includes much speculation about both aetiological factors and neural mechanisms associated with aggressive behavior of these three phenotypes, but also proposed hypotheses that could be tested in future research. Prospective longitudinal studies following children characterised by known antecedents of schizophrenia (Cullen et al. 2012; Laurens et al. 2007, 2010) are needed to identify the mechanisms associated with aggressive behaviour in this population. Such studies offer the possibility of taking account of neural changes underlying transition to illness and the consequences on the brain and behaviour of both the changes and antipsychotic medication. Studies are urgently needed to determine whether antipsychotic medications contribute to aggressive behaviour either alone or in combination with alcohol and/or illicit drugs. Almost all of the research reviewed involved males and consequently very little is known about aggressive behavior among women with schizophrenia.

References

Abu-Akel A (2003) A neurobiological mapping of theory of mind. Brain Res Brain Res Rev 43:29–40. doi:10.1016/S0165-0173(03)00190-5

Addington J, Penn D, Woods SW, Addington D, Perkins DO (2008) Facial affect recognition in individuals at clinical high risk for psychosis. Br J Psychiatry 192:67–68. doi:10.1192/bjp.bp.107.039784

Addy K, Shannon K, Brookfield K (2007) Theory of mind function, motor empathy, emotional empathy and schizophrenia: a single case study. J Forensic Psychiat Psychol 18:293–306. doi:10.1080/09670870701292746

Alden A, Brennan P, Hodgins S, Mednick S (2007) Psychotic disorders and sex offending in a Danish birth cohort. Arch Gen Psychiatry 64:1251–1258. doi:10.1001/archpsyc.64.11.1251

Allen JG, Fonagy P, Bateman A (2008) Mentalizing in clinical practice. American Psychiatric Pub, Arlington

Appelbaum PS, Robbins PC, Monahan J (2000) Violence and delusions: data from the MacArthur violence risk assessment study. Am J Psychiatry 157:566–572. doi:10.1176/appi.ajp.157.4.566

Arseneault L, Moffitt TE, Caspi A, Taylor PJ, Silva PA (2000) Mental disorders and violence in a total birth cohort: results from the Dunedin study. Arch Gen Psychiatry 57:979–986. doi:10.1001/archpsyc.57.10.979

Arseneault L, Cannon M, Poulton R, Murray R, Caspi A, Moffitt TE (2002) Cannabis use in adolescence and risk for adult psychosis: longitudinal prospective study. BMJ 325:1212–1213. doi:10.1136/bmj.325 7374.1212

Arseneault L, Cannon M, Murray R, Poulton R, Caspi A, Moffitt TE (2003) Childhood origins of violent behaviour in adults with schizophreniform disorder. Br J Psychiatry 183:520–525

Arseneault L, Bowes L, Shakoor S (2010) Bullying victimization in youths and mental health problems: 'much ado about nothing'? Psychol Med 40:717–729. doi:10.1017/S0033291709991383

Arseneault L, Cannon M, Fisher HL, Polanczyk G, Moffitt TE, Caspi A (2011) Childhood trauma and children's emerging psychotic symptoms: a genetically sensitive longitudinal cohort study. Am J Psychiatry 168:65–72. doi:10.1176/appi.ajp.2010.10040567

Barkataki I, Kumari V, Das M, Taylor P, Sharma T (2006) Volumetric structural brain abnormalities in men with schizophrenia or antisocial personality disorder. Behav Brain Res 169:239–247. doi:10.1016/j.bbr.2006.01.009

Baron-Cohen S (1995) Mindblindness. An essay on autism and theory of mind. Cambridge, MIT Press

Baumeister RF, Heatherton TF (1996) Self-regulation failure: an overview. Psycho Inq 7:1–15. doi:10.1207/s15327965pli0701_1

Beaudoin MN, Hodgins S, Lavoie F (1993) Homicide, schizophrenia and substance abuse or dependency. Can J Psychiatry 38:541–546

Bediou B, Asri F, Brunelin J, Krolak-Salmon, D'Amato T, Saoud M, Tazi I (2007) Emotion recognition and genetic vulnerability to schizophrenia. Br J Psychiatry 191:126–130. doi: 10.1192/bjp.bp.106.028829

Bendall S, Jackson HJ, Hulbert CA, McGorry PD (2008) Childhood trauma and psychotic disorders: a systematic, critical review of the evidence. Schizophr Bull 34:568–579. doi:10.1093/schbul/sbm121

Birbaumer N, Veit R, Lotze M, Erb M, Hermann C, Grodd W, Flor H (2005) Deficient fear conditioning in psychopathy: a functional magnetic resonance imaging study. Arch Gen Psychiatry 62:799–805. doi:10.1001/archpsyc.62.7.799

Bjørkly S (2002a) Psychotic symptoms and violence toward others — a literature review of some preliminary findings. Aggress Violent Beh 7:617–631. doi:10.1016/S1359-1789(01)00049-0

Bjørkly S (2002b) Psychotic symptoms and violence toward others — a literature review of some preliminary findings. Aggress Violent Beh 7:605–615. doi:10.1016/S1359-1789(01)00050-7

Blair RJR (1995) A cognitive developmental approach to morality: investigating the psychopath. Cognition 57:1–29. doi:10.1016/0010-0277(95)00676-P

Blair RJR (2004) The roles of orbital frontal cortex in the modulation of antisocial behavior. Brain Cogn 55:198–208. doi:10.1016/S0278-2626(03)00276-8

Blair RJR (2005) Responding to the emotions of others: dissociating forms of empathy through the study of typical and psychiatric populations. Conscious Cogn 14:698–718. doi:10.1016/j.concog.2005.06.004

Blair RJR (2010) Neuroimaging of psychopathy and antisocial behavior: a targeted review. Curr Psychiatry Rep 12:76–82. doi:10.1007/s11920-009-0086-x

Blair RJR (2012) Considering anger from a cognitive neuroscience perspective. WIREs Cogn Sci 3:65–74. doi:10.1002/wcs.154

Blair RJR, Mitchell D, Peschardt K, Colledge E, Leonard R, Shine J, Murray L, Perrett D (2004) Reduced sensitivity to others' fearful expressions in psychopathic individuals. Pers Indiv Differ 37:1111–1122. doi:10.1016/j.paid.2003.10.008

Blair RJR, Peschardt K, Budhani S, Mitchell D, Pine D (2006a) The development of psychopathy. J Child Psychol Psychiat 47:262–276. doi:10.1111/j.1469-7610.2006.01596.x

Blair KS, Newman C, Mitchell DGV, Richell RA, Leonard A, Morton J, Blair RJR (2006b) Differentiating among prefrontal substrates in psychopathy: neuropsychological test findings. Neuropsychology 20:153–165. doi:10.1037/0894-4105.20.2.153

Bo S, Abu-Akel A, Kongerslev M, Haahr UH, Simonsen E (2011) Risk factors for violence among patients with schizophrenia. Clin Psychol Rev 31(5):711–726. doi:10.1016/j.cpr.2011.03.002

Bobes MA, Ostrosky F, Diaz K, Romero C, Borja K, Santos Y, Valdés-Sosa M (2012) Linkage of functional and structural anomalies in the left amygdala of reactive-aggressive men. Soc Cogn Affect Neurosci. doi:10.1093/scan/nss101

Bornovalova MA, Hicks BM, Iacono WG, McGue M (2010) Familial transmission and heritability of childhood disruptive disorders. Am J Psychiatry 167:1066–1074. doi:10.1176/appi.ajp.2010.09091272

Bowie CR, Reichenberg A, Patterson TL, Heaton RK, Harvey PD (2006) Determinants of real-world functional performance in schizophrenia subjects: correlations with cognition, functional capacity, and symptoms. Am J Psychiatry 163:418–425. doi:10.1176/appi.ajp.163.3.418

Brennan PA, Mednick SA, Hodgins S (2000) Major mental disorders and criminal violence in a Danish birth cohort. Arch Gen Psychiatry 57:494–500. doi:10.1001/archpsyc.57.5.494

Broidy LM, Nagin DS, Tremblay RE, Bates JE, Brame B, Dodge KA, Fergusson D, Horwood JL, Loeber R, Laird R, Lynam DR, Moffitt TE, Pettit GS, Vitaro F (2003) Developmental trajectories of childhood disruptive behaviors and adolescent delinquency: a six-site, cross-national study. Dev Psychol 39:222–245. doi:10.1037/0012-1649.39.2.222

Brothers L, Ring B (1992) A neuroethological framework for the representation of minds. J Cogn Neurosci 4:107–118. doi:10.1162/jocn.1992.4.2.107

Buckholtz JW, Treadway MT, Cowan RL, Woodward ND, Benning SD, Li R, Ansari MS, Baldwin RM, Schwartzman AN, Shelby ES, Smith CE, Cole D, Kessler RM, Zald DH (2010) Mesolimbic dopamine reward system hypersensitivity in individuals with psychopathic traits. Nat Neurosci 13:419–421. doi:10.1038/nn.2510

Burt SA (2009) Rethinking environmental contributions to child and adolescent psychopathology: a meta-analysis of shared environmental influences. Psychol Bull 135:608–637. doi:10.1037/a0015702

Burt SA, McGue M, Krueger RF, Iacono WG (2005) Sources of covariation among the child-externalizing disorders: informant effects and the shared environment. Psychol Med 35:1133–1144. doi:10.1017/S0033291705004770

Cannon M, Clarke MC (2005) Risk for schizophrenia — broadening the concepts, pushing back the boundaries. Schizophrenia Res 79:5–13. doi:10.1016/j.schres.2005.05.027

Cannon M, Jones PB, Murray RM (2002) Obstetric complications and schizophrenia: Historical and meta-analytic review. Am J Psychiatry 159:1080–1092. doi:10.1176/appi.ajp.159.7.1080

Carré JM, Hyde LW, Neumann CS, Viding E, Hariri AR (2012) The neural signatures of distinct psychopathic traits. Social Neuroscience:1–14. doi: 10.1080/17470919.2012.703623

Caspi A, Reichenberg A, Weiser M, Rabinowitz J, Kaplan Z, Knobler H, Davidson-Sagi N, Davidson M (2003) Cognitive performance in schizophrenia patients assessed before and following the first psychotic episode. Schizophrenia Res 65:87–94. doi:10.1016/S0920-9964(03)00056-2

Caspi A, Langley K, Milne B, Moffitt TE, O'Donovan M, Owen MJ, Tomas MP, Poulton R, Rutter M, Taylor A, Williams B, Thapar A (2008) A replicated molecular genetic basis for subtyping antisocial behavior in children with attention-deficit/hyperactivity disorder. Arch Gen Psychiatry 65:203–210. doi:10.1001/archgenpsychiatry.2007.24

Cicchetti D, Rogosch FA, Thibodeau EL (2012) The effects of child maltreatment on early signs of antisocial behavior: genetic moderation by tryptophan hydroxylase, serotonin transporter, and monoamine oxidase A genes. Dev Psychopathol 24:907–928. doi:10.1017/S09545 79412000442

Coccaro EF, Kavoussi RJ, Hauger RL, Cooper TB, Ferris CF (1998) Cerebrospinal fluid vasopressin levels: correlates with aggression and serotonin function in personality-disordered subjects. Arch Gen Psychiatry 55:708–714. doi:10.1001/archpsyc.55.8.708

Coccaro EF, McCloskey MS, Fitzgerald DA, Phan KL (2007) Amygdala and orbitofrontal reactivity to social threat in individuals with impulsive aggression. Biol Psychiatry 62:168–178. doi:10.1016/j.biopsych.2006.08.024

Coccaro EF, Lee R, Kavoussi RJ (2010) Inverse relationship between numbers of 5-HT transporter binding sites and life history of aggression and intermittent explosive disorder. J Psychiatr Res 44:137–142. doi:10.1016/j.jpsychires.2009.07.004

Coccaro EF, Sripada CS, Yanowitch RN, Phan KL (2011) Corticolimbic function in impulsive aggressive behavior. Biol Psychiatry 69:1153–1159. doi:10.1016/j.biopsych.2011.02.032

Coid B, Lewis SW, Reveley AM (1993) A twin study of psychosis and criminality. Br J Psychiatry 162:87–92. doi:10.1192/bjp.162.1.87

Couppis MH, Kennedy CH (2008) The rewarding effect of aggression is reduced by nucleus accumbens dopamine receptor antagonism in mice. Psychopharmacology 197:449–456. doi:10.1007/s00213-007-1054-y

Covell NH, Scalora MJ (2002) Empathic deficits in sexual offenders: an integration of affective, social, and cognitive constructs. Aggress Violent Beh 7:251–270

Crocker AG, Mueser KT, Drake RE, Clark RE, Mchugo GJ, Ackerson TH, Alterman AI (2005) Antisocial personality, psychopathy, and violence in persons with dual disorders: a longitudinal analysis. Crim Justice Behav 32:452–476. doi:10.1177/0093854805276407

Cullen AE, de Brito SA, Gregory SL, Murray RM, Williams SCR, Hodgins S, Laurens KR (2012) Temporal lobe volume abnormalities precede the prodrome: a study of children presenting antecedents of schizophrenia. Schizophr Bull. doi:10.1093/schbul/sbs128

D'Acremont M, van der Linden M (2007) How is impulsivity related to depression in adolescence? Evidence from a French validation of the cognitive emotion regulation questionnaire. J Adolesc 30:271–282. doi:10.1016/j.adolescence.2006.02.007

Dadds MR, Perry Y, Hawes DJ, Merz S, Riddell AC, Haines DJ, Solak E, Abeygunawardane AI (2006) Attention to the eyes and fear-recognition deficits in child psychopathy. Br J Psychiatry 189:280–281. doi:10.1192/bjp.bp.105.018150

Dannlowski U, Stuhrmann A, Beutelmann V, Zwanzger P, Lenzen T, Grotegerd D, Domschke K, Hohoff C, Ohrmann P, Bauer J, Lindner C, Postert C, Konrad C, Arolt V, Heindel W, Suslow T, Kugel H (2012) Limbic scars: long-term consequences of childhood maltreatment revealed by functional and structural magnetic resonance imaging. Biol Psychiatry 71:286–293. doi:10.1016/j.biopsych.2011.10.021

Davidson RJ, Putnam KM, Larson CL (2000) Dysfunction in the neural circuitry of emotion regulation—a possible prelude to violence. Science 289:591–594. doi:10.1126/science.289.5479.591

de Brito SA, Mechelli A, Wilke M, Laurens KR, Jones AP, Barker GJ, Hodgins S, Viding E (2008) Size matters: increased grey matter in boys with conduct problems and callous-unemotional traits. Brain 132:843–852. doi:10.1093/brain/awp011

Dean K, Walsh E, Moran P, Tyrer P, Creed F, Byford S, Burns T, Murray R, Fahy T (2006) Violence in women with psychosis in the community: prospective study. Br J Psychiatry 188:264–270. doi:10.1192/bjp.bp.104.008052

Di Forti M, Iyegbe C, Sallis H, Kolliakou A, Falcone MA, Paparelli A, Sirianni M, La Cascia C, Stilo SA, Marques TR, Handley R, Mondelli V, Dazzan P, Pariante C, David AS, Morgan C, Powell J, Murray RM (2012) Confirmation that the AKT1 (rs2494732) genotype influences the risk of psychosis in cannabis users. Biol Psychiatry 72:811–816. doi:10.1016/j.biopsych.2012.06.020

Dick E (2010) Violent victimisation in a sample of patients with severe mental illness. University of Maastricht, Master's Thesis

Dickson H, Laurens KR, Cullen AE, Hodgins S (2012) Meta-analyses of cognitive and motor function in youth aged 16 years and younger who subsequently develop schizophrenia. Psychol Med 42:743–755. doi:10.1017/S0033291711001693

Dolan MC, Fullam RS (2009) Psychopathy and functional magnetic resonance imaging blood oxygenation level-dependent responses to emotional faces in violent patients with schizophrenia. Biol Psychiatry 66:570–577. doi:10.1016/j.biopsych.2009.03.019

Eisenberg N, Fabes RA (1990) Empathy: conceptualization, measurement, and relation to prosocial behavior. Motiv Emotion 14:131–149. doi:10.1007/BF00991640

Eisenberger NI, Way BM, Taylor SE, Welch WT, Lieberman MD (2007) Understanding genetic risk for aggression: clues from the brain's response to social exclusion. Biol Psychiatry 61:1100–1108. doi:10.1016/j.biopsych.2006.08.007

Ellison-Wright I, Glahn DC, Laird AR, Thelen SM, Bullmore E (2008) The anatomy of first-episode and chronic schizophrenia: an anatomical likelihood estimation meta-analysis. Am J Psychiatry 165:1015–1023. doi:10.1176/appi.ajp.2008.07101562

Erb M, Hodgins S, Freese R, Müller-Isberner R, Jöckel D (2001) Homicide and schizophrenia: maybe treatment does have a preventive effect. Crim Behav Ment Health 11:6–26

Fazel S, Grann M, Carlström E, Lichtenstein P, Långström N (2009) Risk factors for violent crime in schizophrenia: a national cohort study of 13,806 patients. J Clin Psychiatry 70:362–369. doi:10.4088/JCP.08m04274

Finger EC, Marsh AA, Mitchell DG, Reid ME, Sims C, Budhani S, Kosson DS, Chen G, Towbin KE, Leibenluft E, Pine DS, Blair JR (2008) Abnormal ventromedial prefrontal cortex function in children with psychopathic traits during reversal learning. Arch Gen Psychiatry 65:586. doi:10.1001/archpsyc.65.5.586

Frick PJ, Marsee MA (2006) Psychopathy and developmental pathways to antisocial behavior in youth. In: Patrick CJ (ed) Handbook of psychopathy. Guilford, New York, pp 355–374

Frommann N, Brandt M, Schwarze C, Schmidbauer W, Wölwer W (2009) Affect recognition impairments and violence in schizophrenia: a first application of the training of affect recognition (TAR) to offenders suffering from schizophrenia. Eur Arch Psychiatry Clin Neurosci 259:S87. doi:10.1007/s00406-009-0041-1

Fullam R, Dolan M (2006) Emotional information processing in violent patients with schizophrenia: association with psychopathy and symptomatology. Psychiatry Res 141:29–37. doi:10.1016/j.psychres.2005.07.013

Fulwiler C, Ruthazer R (1999) Premorbid risk factors for violence in adult mental illness. Compr Psychiatry 40:96–100. doi:10.1016/S0010-440X(99)90112-8

Fulwiler C, Grossman H, Forbes C, Ruthazer R (1997) Early-onset substance abuse and community violence by outpatients with chronic mental illness. Psychiatr Serv 48:1181–1185

Gatzke-Kopp LM, Beauchaine TP, Shannon KE, Chipman J, Fleming AP, Crowell SE, Liang O, Johnson LC, Aylward E (2009) Neurological correlates of reward responding in adolescents with and without externalizing behavior disorders. J Abnorm Psychol 118:203–213. doi:10.1037/a0014378

George DT, Rawlings RR, Williams WA, Phillips MJ, Fong G, Kerich M, Momenan R, Umhau JC, Hommer D (2004) A select group of perpetrators of domestic violence: evidence of decreased metabolism in the right hypothalamus and reduced relationships between cortical/subcortical brain structures in position emission tomography. Psychiat Res-Neuroim 130:11–25. doi:10.1016/S0925-4927(03)00105-7

Glenn AL, Raine A, Yaralian PS, Yang Y (2010) Increased volume of the striatum in psychopathic individuals. Biol Psychiatry 67:52–58. doi:10.1016/j.biopsych.2009.06.018

Goldman AL, Pezawas L, Mattay VS, Fischl B, Verchinski BA, Zoltick B, Weinberger DR, Meyer-Lindenberg A (2008) Heritability of brain morphology related to schizophrenia: a large-scale automated magnetic resonance imaging segmentation study. Biol Psychiatry 63:475–483. doi:10.1016/j.biopsych.2007.06.006

Gosden NP, Kramp P, Gabrielsen G, Andersen TF, Sestoft D (2005) Violence of young criminals predicts schizophrenia: a 9-year register-based followup of 15- to 19-year-old criminals. Schizophr Bull 31:759–768. doi:10.1093/schbul/sbi015

Green H, McGinnity A, Meltzer H, Ford T, Goodman R (2004) Mental health of children and young people in Great Britain—A survey carried out by the office for national statistics on behalf of the department of health and the Scottish executive. The Licensing Division, Norwich

Gregg TR, Siegel A (2001) Brain structures and neurotansmitters regulating aggression in cats: implications for human aggression. Prog Neuropsychopharmacol Biol Psychiatry 25:91–140. doi:10.1016/S0278-5846(00)00150-0

Gregory S, ffytche D, Simmons A, Kumari V, Howard M, Hodgins S, Blackwood N (2012) The antisocial brain: Psychopathy matters. Arch Gen Psychiatry 69:962–972. doi: 10.1001/archgenpsychiatry.2012.222

Hall JR, Benning SD, Patrick CJ (2004) Criterion-related validity of the three-factor model of psychopathy: personality, behavior, and adaptive functioning. Assessment 11:4–16. doi:10.1177/1073191103261466

Hare RD (2003) Manual for the revised psychopathy checklist, 2nd edn. Multi-Health Systems, Toronto

Hare RD (2006) Psychopathy: a clinical and forensic overview. Psychiat Clin N America 29:709–724. doi:10.1016/j.psc.2006.04.007
Herman JP, Flak J, Jankord R (2008) Chronic stress plasticity in the hypothalamic paraventricular nucleus. Prog Brain Res 170:353–364. doi:10.1016/S0079-6123(08)00429-9
Hicks BM, South SC, DiRago AC, Iacono WG, McGue M (2009) Environmental adversity and increasing genetic risk for externalizing disorders. Arch Gen Psychiatry 66:640–648. doi:10.1001/archgenpsychiatry.2008.554
Ho B, Andreasen NC, Ziebell S, Pierson R, Magnotta V (2011) Long-term antipsychotic treatment and brain volumes: a longitudinal study of first-episode schizophrenia. Arch Gen Psychiatry 68:128–137. doi:10.1001/archgenpsychiatry.2010.199
Hodgins S (1992) Mental disorder, intellectual deficiency, and crime. Arch Gen Psychiatry 49:476–483. doi:10.1001/archpsyc.1992.01820060056009
Hodgins S (1998) Epidemiological investigations of the associations between major mental disorders and crime: methodological limitations and validity of the conclusions. Soc Psychiatry Psychiatr Epidemiol 33:S29–S37
Hodgins S (2004) Criminal and antisocial behaviours and schizophrenia: a neglected topic. In: Gattaz WF, Häfner H (eds) Search for the causes of schizophrenia, vol V. Steinkopff Verlag, Darmstadt, pp 315–341
Hodgins S (2008a) Criminality among persons with severe mental illness. In: Soothill K, Dolan M, Rogers P (eds) Handbook of forensic mental health. Willan Publishing, Devon, pp 400–423
Hodgins S (2008b) Violent behaviour among people with schizophrenia: a framework for investigations of causes, and effective treatment, and prevention. Philos Trans R Soc London B Biol Sci 363:2505–2518. doi:10.1098/rstb 2008.0034
Hodgins S, Côté G (1993) Major mental disorder and antisocial personality disorder: a criminal combination. Bull Am Acad Psychiatry Law 21:155–160
Hodgins S, Janson C (2002) Criminality and violence among the mentally disordered. The Stockholm metropolitan project. Cambridge University Press, Cambridge
Hodgins S, Müller-Isberner R (2004) Preventing crime by people with schizophrenic disorders: the role of psychiatric services. Br J Psychiatry 185:245–250. doi:10.1192/bjp.185.3.245
Hodgins S, Riaz M (2011) Violence and phases of illness: differential risk and predictors. Eur Psychiatry 26:518–524. doi:10.1016/j.eurpsy.2010.09.006
Hodgins S, Côté G, Toupin J (1998). Major mental disorders and crime: an etiological hypothesis. In. Cooke D, Forth A, Hare RD (eds) Psychopathy: theory, research and implications for society. Kluwer Academic Publishers, Dordrecht, p. 231-256
Hodgins S, Kratzer L, McNeil TF (2001) Obstetric complications, parenting, and risk of criminal behavior. Arch Gen Psychiatry 58:746–752. doi:10.1001/archpsyc.58.8.746
Hodgins S, Kratzer L, McNeil TF (2002) Obstetrical complications, parenting practices and risk of criminal behaviour among persons who develop major mental disorders. Acta Psychiatr Scand 105:179–188. doi.10.1034/j.1600-0447.2002.1o109.x
Hodgins S, Hiscoke UL, Freese R (2003) The antecedents of aggressive behavior among men with schizophrenia: a prospective investigation of patients in community treatment. Behav Sci Law 21:523–546. doi:10.1002/bsl.540
Hodgins S, Tiihonen J, Ross D (2005) The consequences of conduct disorder for males who develop schizophrenia: associations with criminality, aggressive behavior, substance use, and psychiatric services. Schizophrenia Res 78:323–335. doi:10.1016/j.schres.2005.05.021
Hodgins S, Alderton J, Cree A, Aboud A, Mak T (2007) Aggressive behaviour, victimisation and crime among severely mentally ill patients requiring hospitalisation. Br J Psychiatry 191:343–350. doi:10.1192/bjp.bp.106.06.029587
Hodgins S, Cree A, Alderton J, Mak T (2008) From conduct disorder to severe mental illness: associations with aggressive behaviour, crime and victimization. Psychol Med 38:975–987. doi:10.1017/S0033291707002164
Hodgins S, Lincoln T, Mak T (2009) Experiences of victimisation and depression are associated with community functioning among men with schizophrenia. Soc Psychiatry Psychiatr Epidemiol 44:448–457. doi:10.1007/s00127-008-0460-8

Hodgins S, Calem M, Shimel R, Williams A, Harleston D, Morgan C, Dazzan P, Fearon P, Morgan K, Lappin J, Zanelli J, Reichenberg A, Jones P (2011) Criminal offending and distinguishing features of offenders among persons experiencing a first episode of psychosis. Early Interv Psychiatry 5:15–23. doi:10.1111/j.1751-7893.2010.00256.x

Honea RA, Meyer-Lindenberg A, Hobbs KB, Pezawas L, Mattay VS, Egan MF, Verchinski B, Passingham RE, Weinberger DR, Callicott JH (2008) Is gray matter volume an intermediate phenotype for schizophrenia? A voxel-based morphometry study of patients with schizophrenia and their healthy siblings. Biol Psychiatry 63:465–474. doi:10.1016/j.biopsych.2007.05.027

Hoptman MJ, Volavka J, Johnson G, Weiss E, Bilder RM, Lim KO (2002) Frontal white matter microstructure, aggression, and impulsivity in men with schizophrenia: a preliminary study. Biol Psychiatry 52:9–14. doi:10.1016/S0006-3223(02)01311-2

Hoptman MJ, Volavka J, Weiss EM, Czobor P, Szeszko PR, Gerig G, Chakos M, Blocher J, Citrome LL, Lindenmayer J, Sheitman B, Lieberman JA, Bilder RM (2005) Quantitative MRI measures of orbitofrontal cortex in patients with chronic schizophrenia or schizoaffective disorder. Psychiat Res-Neuroim 140:133–145. doi:10.1016/j.pscychresns.2005.07.004

Hoptman MJ, Volavka J, Czobor P, Gerig G, Chakos M, Blocher J, Citrome LL, Sheitman B, Lindenmayer J, Lieberman JA, Bilder RM (2006) Aggression and quantitative MRI measures of caudate in patients with chronic schizophrenia or schizoaffective disorder. Neuropsych Clin N 18:509–515. doi:10.1176/appi.neuropsych.18.4.509

Hoptman MJ, D'Angelo D, Catalano D, Mauro CJ, Shehzad ZE, Kelly AMC, Castellanos FX, Javitt DC, Milham MP (2010) Amygdalofrontal functional disconnectivity and aggression in schizophrenia. Schizophr Bull 36:1020–1028. doi:10.1093/schbul/sbp012

Howes OD, Bose SK, Turkheimer F, Valli I, Egerton A, Valmaggia LR, Murray RM, McGuire P (2011) Dopamine synthesis capacity before onset of psychosis: a prospective (18F)-DOPA PET imaging study. Am J Psychiatry 168:1311–1317. doi:10.1176/appi.ajp.2011.11010160

Humphreys M, Johnstone E, Macmillan F (1994) Offending among first episode schizophrenics. J Forensic Psychiatr 5:51–61. doi:10.1080/09585189408410897

Hyatt CJ, Haney-Caron E, Stevens MC (2012) Cortical thickness and folding deficits in conduct-disordered adolescents. Biol Psychiatry 72:207–214. doi:10.1016/j.biopsych.2011.11.017

Javdani S, Sadeh N, Verona E (2011) Expanding our lens: female pathways to antisocial behavior in adolescence and adulthood. Clin Psychol Rev 31(8):1324–1348. doi:10.1016/j.cpr.2011.09.002

Joyal CC, Hallé P, Lapierre D, Hodgins S (2003) Drug abuse and/or dependence and better neuropsychological performance in patients with schizophrenia. Schizophrenia Res 63:297–299. doi:10.1016/S0920-9964(02)00387-0

Joyal C, Putkonen A, Mancini-Marïe A, Hodgins S, Kononen M, Boulay L, Pihlajamaki M, Soininen H, Stip E, Tiihonen J, Aronen H (2007) Violent persons with schizophrenia and comorbid disorders: a functional magnetic resonance imaging study. Schizophrenia Res 91:97–102. doi:10.1016/j.schres.2006.12.014

Kelleher I, Cannon M (2011) Psychotic-like experiences in the general population: characterizing a high-risk group for psychosis. Psychol Med 41:1–6. doi:10.1017/S0033291710001005

Kety SS (1987) The significance of genetic factors in the etiology of schizophrenia: results from the national study of adoptees in Denmark. J Psychiatr Res 21:423–429. doi:10.1016/0022-3956(87)90089-6

Kim-Cohen J, Caspi A, Moffitt TE, Harrington H, Milne BJ, Poulton R (2003) Prior juvenile diagnoses in adults with mental disorder developmental follow-back of a prospective-longitudinal cohort. Arch Gen Psychiatry 60:709–717. doi:10.1001/archpsyc.60.7.709

Kinoshita Y, Shimodera S, Nishida A, Kinoshita K, Watanabe N, Oshima N, Akechi T, Sasaki T, Inoue S, Furukawa TA, Okazaki Y (2011) Psychotic-like experiences are associated with violent behavior in adolescents. Schizophrenia Res 126:245–251. doi:10.1016/j.schres.2010.08.028

Koenigs M, Baskin-Sommers A, Zeier J, Newman JP (2011) Investigating the neural correlates of psychopathy: a critical review. Mol Psychiatry 16:792–799. doi:10.1038/mp.2010.124

Krakowski M, Czobor P (2004) Gender differences in violent behaviors: relationship to clinical symptoms and psychosocial factors. Am J Psychiatry 161:459–465. doi:10.1176/appi.ajp.161.3.459

Krueger RF, Hicks BM, Patrick CJ, Carlson SR, Iacono WG, McGue M (2002) Etiologic connections among substance dependence, antisocial behavior and personality: modeling the externalizing spectrum. J Abnorm Psychol 111:411–424. doi:10.1037/0021-843X.111.3.411

Kumari V, Barkataki I, Goswami S, Flora S, Das M, Taylor P (2009a) Dysfunctional, but not functional, impulsivity is associated with a history of seriously violent behaviour and reduced orbitofrontal and hippocampal volumes in schizophrenia. Psychiat Res-Neuroim 173:39–44. doi:10.1016/j.pscychresns.2008.09.003

Kumari V, Das M, Taylor PJ, Barkataki I, Andrew C, Sumich A, Williams SC, ffytche DH (2009) Neural and behavioural responses to threat in men with a history of serious violence and schizophrenia or antisocial personality disorder. Schizophrenia Res 110:47–58. doi: 10.1016/j.schres.2009.01.009

Kumari V, Gudjonsson G, Raghuvanshi S, Barkataki I, Taylor P, Sumich A, Das K, Kuipers E, Ffytche D, Das M (2012) Reduced thalamic volume in men with antisocial personality disorder or schizophrenia and a history of serious violence and childhood abuse. Eur Psychiatry. doi:10.1016/j.eurpsy.2012.03.002

Large MM, Nielssen O (2011) Violence in first-episode psychosis: a systematic review and meta-analysis. Schizophrenia Res 125(2–3):209–220. doi:10.1016/j.schres.2010.11.026

Laurell J, Belfrage H, Hellström Å (2010) Facets on the psychopathy checklist screening version and instrumental violence in forensic psychiatric patients. Crim Behav. Ment. Health 2):285–294. doi: 10.1002/cbm.779

Laurens K, Hodgins S, Maughan B, Murray R, Rutter M, Taylor E (2007) Community screening for psychotic-like experiences and other putative antecedents of schizophrenia in children aged 9–12 years. Schizophrenia Res 90:130–146. doi:10.1016/j.schres.2006.11.006

Laurens KR, Hodgins S, Mould GL, West SA, Schoenberg PL, Murray RM, Taylor EA (2010) Error-related processing dysfunction in children aged 9 to 12 years presenting putative antecedents of schizophrenia. Biol Psychiatry 67:238–245. doi:10.1016/j.biopsych.2009.07.030

Lee R, Ferris C, van de Kar L, Coccaro EF (2009) Cerebrospinal fluid oxytocin, life history of aggression, and personality disorder. Psychoneuroendocrinology 34:1567–1573. doi:10.1016/j.psyneuen.2009.06.002

Levinson A, Fonagy P (2004) Offending and attachment: the relationship between interpersonal awareness and offending in a prison population with psychiatric. Can J Psychoanal 12:225–251

Lewis MH, Gariépy JL, Gendreau P, Nichols DE, Mailman RB (1994) Social reactivity and D1 dopamine receptors: studies in mice selectively bred for high and low levels of aggression. Neuropsychopharmacology 10):115–122

Lin D, Boyle MP, Dollar P, Lee H, Lein ES, Perona P, Anderson DJ (2011) Functional identification of an aggression locus in the mouse hypothalamus. Nature 470:221–226. doi:10.1038/nature09736

Loeber R, Stouthamer-Loeber M (1998) Development of juvenile aggression and violence: some common misconceptions and controversies. American Psychologist 53(2):242–259

Loeber R, Pardini D, Homish DL, Wei EH, Crawford AM, Farrington DP, Stouthamer-Loeber M, Creemers J, Koehler SA, Rosenfeld R (2005) The prediction of violence and homicide in young men. J Consult Clin Psychol73(6):1074–1088. doi: 10.1037/0022-006X.73.6.1074

Lotze M, Veit R, Anders S, Birbaumer N (2007) Evidence for a different role of the ventral and dorsal medial prefrontal cortex for social reactive aggression: an interactive fMRI study. Neuroimage 34:470–478. doi:10.1016/j.neuroimage.2006.09.028

Malcolm CP, Picchioni MM, DiForti M, Sugranyes G, Cooke E, Joseph C, McQueen G, Paparelli A, Stilo S, O'Connor J, Morgan C, Murray RM, Hodgins S (2011) Pre-morbid conduct disorder symptoms are associated with cannabis use among individuals with a first episode of psychosis. Schizophrenia Res 126:81–86. doi:10.1016/j.schres.2010.11.025

Mathalon DH, Pfefferbaum A, Lim KO, Rosenbloom MJ, Sullivan EV (2003) Compounded brain volume deficits in schizophrenia-alcoholism comorbidity. Arch Gen Psychiatry 60:245–252. doi:10.1001/archpsyc.60.3.245

McCrory E, de Brito SA, Viding E (2010) Research review: the neurobiology and genetics of maltreatment and adversity. J Child Psychol Psychiatry 51:1079–1095. doi:10.1111/j.1469-7610.2010.02271.x

McGowan PO, Sasaki A, D'Alessio AC, Dymov S, Labonté B, Szyf M, Turecki G, Meaney MJ (2009) Epigenetic regulation of the glucocorticoid receptor in human brain associates with childhood abuse. Nat Neurosci 12:342–348. doi:10.1038/nn.2270

McGrath J, Welham J, Scott J, Varghese D, Degenhardt L, Hayatbakhsh MR, Alati R, Gail MW, Bor W, Najman JM (2010) Association between cannabis use and psychosis-related outcomes using sibling pair analysis in a cohort of young adults cannabis use and psychosis-related outcomes. Arch Gen Psychiatry 67:440–447. doi:10.1001/archgenpsychiatry.2010.6

McGrath LM, Mustanski B, Metzger A, Pine DS, Kistner-Griffin E, Cook E, Wakschlag LS (2012) A latent modeling approach to genotype–phenotype relationships: maternal problem behavior clusters, prenatal smoking, and MAOA genotype. Arch Womens Ment Health 15:269–282. doi:10.1007/s00737-012-0286-y

Meyer-Lindenberg A (2011) Neuroimaging and the question of neurodegeneration in schizophrenia. Prog Neurobiol 95:514–516. doi:10.1016/j.pneurobio.2011.07.007

Meyer-Lindenberg A, Buckholtz JW, Kolachana B, Hariri AR, Pezawas L, Blasi G, Wabnitz A, Honea R, Verchinski B, Callicott JH, Egan M, Mattay V, Weinberger DR (2006) Neural mechanisms of genetic risk for impulsivity and violence in humans. Proc Natl Acad Sci USA 103:6269–6274. doi:10.1073/pnas.0511311103

Miller PA, Eisenberg N (1988) The relation of empathy to aggressive and externalizing/antisocial behavior. Psychol Bull 103:324–344. doi:10.1037/0033-2909.103.3.324

Mitchell P (1998) Introduction to theory of mind. Children, autism, and apes. St. Martin's Press, New York

Modestin J, Ammann R (1996) Mental disorder and criminality: male schizophrenia. Schizophr Bull 22:69–82. doi:10.1093/schbul/22.1.69

Moffitt TE (1993) Adolescence-limited and life-course-persistent antisocial behavior: a developmental taxonomy. Psychology Review 100(4):674–701

Moffitt TE, Caspi A (2001) Childhood predictors differentiate life-course persistent and adolescence-limited antisocial pathways among males and females. Dev Psychopathol 13:355–375

Mojtabai R (2006) Psychotic-like experiences and interpersonal violence in the general population. Soc Psychiatry Psychiatr Epidemiol 41:183–190. doi:10.1007/s00127-005-0020-4

Moran P, Hodgins S (2004) The correlates of comorbid antisocial personality disorder in schizophrenia. Schizophr Bull 30:791–802

Morgan C, Fisher H (2006) Environment and schizophrenia: environmental factors in schizophrenia: childhood trauma- acritical review. Schizophr Bull 33:3–10. doi:10.1093/schbul/sbl053

Mueser KT, Rosenberg SD, Drake RE, Miles KM, Wolford G, Vidaver R, Carrieri K (1999) Conduct disorder, antisocial personality disorder and substance use disorders in schizophrenia and major affective disorders. J Stud Alcohol 60:278–284

Mueser KT, Crocker AG, Frisman LB, Drake RE, Covell NH, Essock SM (2006) Conduct disorder and antisocial personality disorder in persons with severe psychiatric and substance use disorders. Schizophr Bull 32:626–636. doi:10.1093/schbul/sbj068

Munkner R, Haastrup S, Joergensen T, Kramp P (2003) The temporal relationship between schizophrenia and crime. Soc Psychiatry Psychiatr Epidemiol 38:347–353. doi:10.1007/s00127-003-0650-3

Munkner R, Haastrup S, Joergensen T, Kramp P (2005) Incipient offending among schizophrenia patients after first contact to the psychiatric hospital system. Eur Psychiatry 20:321–326. doi:10.1016/j.eurpsy.2005.01.008

Munkner R, Haastrup S, Joergensen T, Kramp P (2009) The association between psychopathology of first-episode psychosis patients within the schizophrenia spectrum and previous offending. Nord J Psychiatry 63:124–131. doi:10.1080/08039480802421109

Murray J, Farrington D (2010) Risk factors for conduct disorder and delinquency: key findings from longitudinal studies. Can J Psychiatry 55:633–642

Narayan VM, Narr KL, Kumari V, Woods RP, Thompson PM, Toga AW, Sharma T (2007) Regional cortical thinning in subjects with violent antisocial personality disorder or schizophrenia. Am J Psychiatry 164:1418–1427. doi:10.1176/appi.ajp.2007.06101631

Naudts K, Hodgins S (2006) Neurobiological correlates of violent behavior among persons with schizophrenia. Schizophr Bull 32:562–572. doi:10.1093/schbul/sbj036

Nilsson KW, Sjöberg RL, Damberg M, Leppert J, Öhrvik J, Alm PO, Lindström L, Oreland L (2006) Role of monoamine oxidase A genotype and psychosocial factors in male adolescent criminal activity. Biol Psychiatry 59:121–127. doi:10.1016/j.biopsych.2005.06.024

Nock MK, Kazdin AE, Hiripi EV, KesslerRC (2006) Prevalence, subtypes, and correlates of DSM-IV conduct disorder in the national comorbidity survey replication. Psychol Med 36:699–710. doi: 10.1017/S0033291706007082

Nolan KA, Czobor P, Roy BB, Platt MM, Shope CB, Citrome LL, Volavka J (2003) Characteristics of assaultive behavior among psychiatric inpatients. Psychiatr Serv 54:1012–1016. doi:10.1176/appi.ps.54.7.1012

Odgers CL, Caspi A, Russell MA, Sampson RJ, Arseneault L, Moffitt TE (2012) Supportive parenting mediates neighborhood socioeconomic disparities in children's antisocial behavior from ages 5 to 12. Dev Psychopathol 24:705–721. doi:10.1017/S0954579412000326

Olweus D (1979) Stability of aggressive reaction patterns in males: a review. Psychol Bull 86:852–875. doi:10.1037/0033-2909.86.4.852

Panksepp J (1998) Affective neuroscience: the foundations of human and animal emotions. Oxford University Press, New York

Petitclerc A, Tremblay RE (2009) Childhood disruptive behaviour disorders: review of their origin, development, and prevention. Can J Psychiatry 54:222–231

Polanczyk G, Moffitt TE, Arseneault L, Cannon M, Ambler A, Keefe RSE, Houts R, Odgers CL, Caspi A (2010) Etiological and clinical features of childhood psychotic symptoms: results from a birth cohort childhood psychotic symptoms. Arch Gen Psychiatry 67:328–338. doi:10.1001/archgenpsychiatry.2010.14

Priebe S, Badesconyi A, Fioritti A, Hansson L, Kilian R, Torres-Gonzales F, Turner T, Wiersma D (2005) Reinstitutionalisation in mental health care: comparison of data on service provision from six European countries. BMJ 330:123–126. doi:10.1136/bmj.38296.611215.AE

Puri BK, Counsell SJ, Hamilton G, Bustos MG, Treasaden IH (2008) Brain cell membrane motion-restricted phospholipids in patients with schizophrenia who have seriously and dangerously violently offended. Prog Neuro-Psychoph 32:751–754. doi:10.1016/j.pnpbp.2007.11.025

Raine A, Yaralian PS, Reynolds C, Venables PH, Mednick SA (2002) Spatial but not verbal cognitive deficits at age 3 years in persistently antisocial individuals. Dev Psychopathol 14:25–44

Raine A, Moffitt TE, Caspi A, Loeber R, Stouthamer-Loeber M, Lynam D (2005) Neurocognitive impairments in boys on the life-course persistent antisocial path. J Abnorm Psychol 114:38–49. doi:10.1037/0021-843X.114.1.38

Rasmussen K, Levander S, Sletvold H (1995) Aggressive and non-aggressive schizophrenics: symptom profile and neuropsychological differences. Psychol Crime Law 2:119–129. doi:10.1080/10683169508409770

Read J, van Os J, Morrison AP, Ross CA (2005) Childhood trauma, psychosis and schizophrenia: a literature review with theoretical and clinical implications. Acta Psychiatr Scand 112:330–350. doi:10.1111/j.1600-0447.2005.00634.x

Rice ME, Harris GT (1995) Psychopathy, schizophrenia, alcohol abuse, and violent recidivism. International J Law Psychiatry 18:333–342. doi:10.1016/0160-2527(95)00015-A

Robins LN (1993) Childhood conduct problems, adult psychopathology, and crime. In: Hodgins S (ed) Mental Disorder and Crime. Sage Publications Inc., Newbury Park, pp 173–207

Robins LN, McEvoy JP (1990) Conduct problems as predictors of substance abuse. In: Robins LN, Rutter M (eds) Straight and devious pathways from childhood to adulthood. Cambridge University Press, New York, pp 182–204

Robins LN, Tipp J, Przybeck T (1991) Antisocial personality. In: Robins LN, Regier D (eds) Psychiatric disorders in America: the epidemiologic catchment area study. Macmillan/Free Press, New York, pp 258–290

Rodriguiz RM, Chu R, Caron MG, Wetsel WC (2004) Aberrant responses in social interaction of dopamine transporter knockout mice. Behav Brain Res 148:185–198. doi:10.1016/S0166-4328(03)00187-6

Schanda H, Földes P, Topitz A, Fliedl R, Knecht G (1992) Premorbid adjustment of schizophrenic criminal offenders. Acta Psychiatr Scand 86:121–126. doi:10.1111/j.1600-0447.1992.tb03239.x

Schiffer B, Leygraf N, Muller BW, Scherbaum N, Forsting M, Wiltfang J, Gizewski ER, Hodgins S (2012) Structural brain alterations associated with schizophrenia preceded by conduct disorder: a common and distinct subtype of schizophrenia? Schizophr Bull. doi:10.1093/schbul/sbs115

Schneider F, Gur RC, Koch K, Backes V, Amunts K, Shah NJ, Bilker W, Gur RE, Habel U (2006) Impairment in the specificity of emotion processing in schizophrenia. Am J Psychiatry 163:442–447. doi:10.1176/appi.ajp.163.3.442

Schreier A, Wolke D, Thomas K, Horwood J, Hollis C, Gunnell D, Lewis G, Thompson A, Zammit S, Duffy L, Salvi G, Harrison G (2009) Prospective study of peer victimization in childhood and psychotic symptoms in an nonclinical population at age 12 years. Arch Gen Psychiatry 66:527–536. doi:10.1001/archgenpsychiatry.2009.23

Schug RA, Raine A, Wilcox RR (2007) Psychophysiological and behavioural characteristics of individuals comorbid for antisocial personality disorder and schizophrenia-spectrum personality disorder. Br J Psychiatry 191:408–414. doi:10.1192/bjp.bp.106.034801

Schug RA, Yang Y, Raine A, Han C, Liu J (2010) Structural and psychosocial correlates of birth order anomalies in schizophrenia and homicide. J Nerv Ment Dis 198:870–875. doi:10.1097/NMD.0b013e3181fe7280

Silver E, Arseneault L, Langley J, Caspi A, Moffitt TE (2005) Mental disorder and violent victimization in a total birth cohort. Am J Public Health 95:2015–2021. doi:10.2105/AJPH.2003.021436

Simonoff E, Elander J, Holmshaw J, Pickles A, Murray R, Rutter M (2004) Predictors of antisocial personality: continuities from childhood to adult life. Br J Psychiatry 184:118–127. doi:10.1192/bjp.184.2.118

Sjöberg RL, Nilsson KW, Wargelius H, Leppert J, Lindström L, Oreland L (2007) Adolescent girls and criminal activity: Role of MAOA-LPR genotype and psychosocial factors. Am J Med Genet 144:159–164. doi:10.1002/ajmg.b.30360

Slutske WS, Heath AC, Dinwiddie SH, Madden PAF, Bucholz KK, Dunne MP, Statham DJ, Martin NG (1998) Common genetic risk factors for conduct disorder and alcohol dependence. J Abnorm Psychol 107:363–374. doi:10.1037/0021-843X.107.3.363

Stadler C, Poustka F, Sterzer P (2010) The heterogeneity of disruptive behavior disorders—implications for neurobiological research and treatment. Front Psychiatry 1:21. doi:10.3389/fpsyt.2010.00021

Steadman HJ, Mulvey EP, Monahan J, Robbins PC, Appelbaum PS, Grisso T, Roth LH, Silver E (1998) Violence by people discharged from acute psychiatric inpatient facilities and by others in the same neighborhoods. Arch Gen Psychiatry 55:393–401. doi:10.1001/archpsyc.55.5.393

Steinert T (2002) Prediction of inpatient violence. Acta Psychiatr Scand 106:133–141. doi:10.1034/j.1600-0447.106.s412.29.x

Sterzer P, Stadler C (2009) Neuroimaging of aggressive and violent behaviour in children and adolescents. Front Behav Neurosci 3:35. doi:10.3389/neuro.08.035.2009

Stouthamer-Loeber M, Loeber R, Homish DL, Wei E (2001) Maltreatment of boys and the development of disruptive and delinquent behavior. Dev Psychopathol 13:941–955

Sunak S (2006) Deficient effective experience and violence in schizophrenia. Master's Thesis, King's College London

Swanson JW, Swartz MS, Van Dorn RA, Elbogen EB, Wagner R, Rosenheck RA, Stroup TS, McEvoy JP, Lieberman JA (2006) A national study of violent behavior in persons with schizophrenia. Arch Gen Psychiatry 63:490–499. doi:10.1001/archpsyc.63.5.490

Tangney JP (1991) Moral affect: the good, the bad, and the ugly. J Pers Soc Psychol 61:598–607

Taylor A, Kim-Cohen J (2007) Meta-analysis of gene–environment interactions in developmental psychopathology. Dev Psychopathol 19:1029–1037. doi:10.1017/S095457940700051X

ten Donkelaar HJ, Lammens M, Wesseling P, Thijssen HO, Renier WO (2003) Development and developmental disorders of the human cerebellum. J Neurol 250:1025–1036. doi:10.1007/s00415-003-0199-9

Tengström A, Hodgins S, Kullgren G (2001) Men with schizophrenia who behave violently: the usefulness of an early- versus late-start offender typology. Schizophr Bull 27:205–218

Tengström A, Hodgins S, Grann M, Långström N, Kullgren G (2004) Schizophrenia and criminal offending: the role of psychopathy and substance use disorders. Crim Justice Behav 31:367–391. doi:10.1177/0093854804265173

Teplin LA, McClelland GM, Abram KM, Weiner DA (2005) Crime victimization in adults with severe mental illness comparison with the national crime victimization survey. Arch Gen Psychiatry 62:911–921. doi:10.1001/archpsyc.62.8.911

Tiemeier H, Lenroot RK, Greenstein DK, Tran L, Pierson R, Giedd JN (2010) Cerebellum development during childhood and adolescence: a longitudinal morphometric MRI study. Neuroimage 49:63–70. doi:10.1016/j.neuroimage.2009.08.016

Tiihonen J, Isohanni M, Räsänen P, Koiranen M, Moring J (1997) Specific major mental disorders and criminality: a 26-year prospective study of the 1966 northern Finland birth cohort. Am J Psychiatry 154:840–845

Tiihonen J, Rossi R, Laakso MP, Hodgins S, Testa C, Perez J, Repo-Tiihonen E, Vaurio O, Soininen H, Aronen HJ, Könönen M, Thompson PM, Frisoni GB (2008) Brain anatomy of persistent violent offenders: more rather than less. Psychiat Res-Neuroim 163(3):201–212. doi:10.1016/j.pscychresns.2007.08.012

Tuvblad C, Raine A, Zheng M, Baker LA (2009) Genetic and environmental stability differs in reactive and proactive aggression. Aggressive Behav. 35:437–452. doi:10.1002/ab.20319

Uzefovsky F, Shalev I, Israel S, Knafo A, Ebstein RP (2012) Vasopressin selectively impairs emotion recognition in men. Psychoneuroendocrinology 37:576–580. doi:10.1016/j.psyneuen.2011.07.018

Viding E, Jones AP, Paul JF, Moffitt TE, Plomin R (2008) Heritability of antisocial behaviour at 9: do callous-unemotional traits matter? Dev Sci 11:17–22. doi:10.1111/j.1467-7687.2007.00648.x

Vitacco MJ, Neumann CS, Caldwell MF, Leistico A, van Rybroek GJ (2006) Testing factor models of the psychopathy checklist: youth version and their association with instrumental aggression. J Pers Assess 87:74–83. doi:10.1207/s15327752jpa8701_06

Walker EF, Savoie T, Davis D (1994) Neuromotor precursors of schizophrenia. Schizophr Bull 20:441–451. doi:10.1093/schbul/20.3.441

Wallace C, Mullen PE, Burgess P (2004) Criminal offending in schizophrenia over a 25-year period marked by deinstitutionalization and increasing prevalence of comorbid substance use disorders. Am J Psychiatry 161:716–727. doi:10.1176/appi.ajp.161.4.716

Walsh E, Gilvarry C, Samele C, Harvey K, Manley C, Tyrer P, Creed F, Murray R, Fahy T (2001) Reducing violence in severe mental illness: randomised controlled trial of intensive case management compared with standard care. BMJ 323:1093–1096. doi:10.1136/bmj.323 7321.1093

Walsh E, Moran P, Scott C, McKenzie K, Burns T, Creed F, Tyrer P, Murray RM, Fahy T (2003) Prevalence of violent victimisation in severe mental illness. Br J Psychiatry 183:233–238. doi:10.1192/bjp.183.3.233

Weinberger DR (1987) Implications of normal brain development for the pathogenesis of schizophrenia. Arch Gen Psychiatry 44:660–669. doi:10.1001/archpsyc.1987.01800190080012

Weiss EM, Kohler CG, Nolan KA, Czobor P, Volavka J, Platt MM, Brensinger C, Loughead J, Delazer M, Gur RE, Gur RC (2006) The relationship between history of violent and criminal behavior and recognition of facial expression of emotions in men with schizophrenia and schizoaffective disorder. Aggressive Behav 32:187–194. doi:10.1002/ab.20120

Weissenberger AA, Deli ML, Liow K, Theodore W, Frattali CM, Hernandez D, Zametkin AJ (2001) Aggression and psychiatric comorbidity in children with hypothalamic hamartomas and their unaffected siblings. J Am Acad of Child Adolesc Psychiatry 40:696–703. doi:10.1097/00004583-200106000-00015

Wheeler S, Book A, Costello K (2009) Psychopathic traits and perceptions of victim vulnerability. Crim Justice Behav 36:635–648. doi:10.1177/0093854809333958

Widom CS (1989) The cycle of violence. Science 244:160–166

Young SE, Stallings MC, Corley RP, Krauter KS, Hewitt JK (2000) Genetic and environmental influences on behavioral disinhibition. Am J Med Genet 96(5):684–695

Ziermans TB, Schothorst PF, Schnack HG, Koolschijn PCMP, Kahn RS, van Engeland H, Durston S (2012) Progressive structural brain changes during development of psychosis. Schizophr Bull 38:519–530. doi:10.1093/schbul/sbq113

Psychopathy and Aggression: When Paralimbic Dysfunction Leads to Violence

Nathaniel E. Anderson and Kent A. Kiehl

Abstract Psychopaths can be alarmingly violent, both in the frequency with which they engage in violence and the gratuitous extent of their violent acts. Indeed, one principal utility of the clinical construct of psychopathy is in predicting future violent behavior in criminal offenders. Aggression is a complex construct that intersects psychopathy at many levels. This chapter provides a review of psychopathy as a clinical construct including the most prominent cognitive and neurobiological models, which serve to account for its pathophysiology. We then describe how the brain abnormalities implicated in psychopathy may lead to diverse behavioral outcomes, which can include aggression in its many forms.

Keywords Psychopathy · Violence · Impulsive aggression · Premeditated aggression

Contents

N. E. Anderson (✉)
The Mind Research Network, Albuquerque, NM, USA
e-mail: nanderson@mrn.org

K. A. Kiehl
The Mind Research Network, University of New Mexico,
Albuquerque, NM, USA
e-mail: kkiehl@unm.edu

Curr Topics Behav Neurosci (2014) 17: 369–393
DOI: 10.1007/7854_2013_257

Published Online: 5 December 2013

1 Introduction

Many psychiatric disorders and neuropsychological conditions may be linked to aggressive behavior and violent acts. Among these, psychopathy deserves special attention as this condition is strongly associated with persistent aggressive traits and violent behavior throughout the lifespan. This chapter describes how neural abnormalities associated with psychopaths' emotional dysfunction and personality traits may also be linked to aggressive tendencies in some contexts; however, several important qualifications must be made with regard to this relationship, particularly with respect to how psychopathy and aggression are defined. Psychopaths are often mischaracterized in the popular media, promulgating misconceptions about this psychiatric disorder, which has been well-defined in clinical circles for decades. Furthermore, aggression comes in many forms ranging from drunken bar brawls to calculated homicide, and in some conceptualizations even includes nonviolent actions. In order to achieve a more sophisticated understanding of the relationships between psychopathy and aggression, it will be necessary to review some essential background which has contributed to our contemporary understanding of both constructs. In doing so, this will help to clarify the pathophysiology underlying discrete patterns of aggressive behavior among psychopaths, a population that may indeed be particularly susceptible to violent behavior.

2 Essential Background: Conceptualization of Psychopathy

Descriptions of particularly unscrupulous individuals, who fail to learn from punishment and who habitually violate the rights of others for personal gain, have been well documented throughout history and across cultures (Murphy 1976). However, notions that such behavior may constitute a mental deficiency rather than an evil nature gained traction only relatively recently, around the early nineteenth century. This period marked the infancy of our understanding of abnormal psychology, such that practitioners were only beginning to understand the organic relationship between the brain and individual differences in thoughts and behavior. The clinician Pinel documented the notion of *madness without delirium*, in which individuals retained most outwardly observable faculties such as intelligence and a grasp on reality, while still suffering deficiencies in their character, such as emotions or moral judgment (Pinel 1806). Prichard (1835) extended this notion to a condition he called *moral insanity*, in which a specific and essentially isolated deficit in moral behavior was recognized. These early conceptualizations provided a starting point for our understanding of a variety of disorders, including psychopathy.

As the concept of the *psychopathic personality* gained prominence in the early twentieth century, there remained a need for some descriptive consolidation to unify the construct and bolster its clinical utility. While the construct has continued to evolve, our modern clinical understanding of the psychopath is conceptually founded on Cleckley's (1941) authoritative delineation of 16 core diagnostic traits which characterize the disorder. His case studies paint a vivid picture of the psychopath suffering from an inability to experience empathy, guilt or remorse, and as one who often fails to exercise sound judgment and lacks the basic foundation for socialized, moral behavior. As such, psychopaths show a profound disregard for the rights and well-being of others. At the same time, psychopaths are ordinarily not delusional or psychotic, and they generally perform at normal to high levels of intelligence. This combination of personality traits may be recognized at varying levels among healthy, relatively high functioning members of society; however, the full clinical manifestation of these traits occurs in less than 1 % of the general population (Hare 2003). Individuals who meet clinical criteria for psychopathy tend to fail to achieve normal academic success, frequently lose their jobs or quit without justification, suffer through multiple divorces, and may be socially ostracized due to others' perception of their chronic imprudent choices. As such, psychopathy represents a maladaptive condition severely impairing decision-making, moral judgment, and social functioning.

While a diverse collection of personality traits characterize psychopathy—and indeed define it clinically—contemporary neuroscience research suggests these traits develop from abnormalities in brain structure and function that bias fundamental processes of learning and behavior in measurable ways. As we delineate relationships between psychopathy and aggression, a repeated theme in this chapter will be that certain specific behavioral manifestations of a neuropsychiatric disorder, such as the manifestation of aggressive behavior in psychopathy, are often context-specific and highly variable due to a wide range of developmental influences and etiological factors. A primary focus on neurobiological factors capable of influencing patterns of learning and motivation can provide a more stable reference, and so this chapter will emphasize these features of psychopathy prior to discussions of aggressive outcomes. First, it will be necessary to understand how psychopathy is measured, clinically and for the purposes of empirical research.

2.1 Measuring Psychopathy

Research examining psychopathy and its associated behavioral outcomes requires an appeal to some standard measure of the construct. Most studies have relied on assessment tools specifically developed to account for the range of traits defined by

Cleckley (1941) in his case studies. Of these tools, Hare's Psychopathy Checklist (PCL; Hare 1980), now in a revised form (PCL-R; Hare 1993, 2003), has emerged as the most widely used and well-recognized measure of psychopathy. It is a semi-structured interview paired with a review of institutional files for collateral information, and is designed to be carried out by trained assessors to evaluate the overall presence and severity of 20 items, each graded from zero to two. The recommended clinical cutoff for a diagnosis of psychopathy is 30 out of a maximum 40 points on the scale, although researchers have sometimes used different cutoffs in studies of the underlying biology.

Items on the PCL-R fall into two correlated dimensions, or factors, representing emotional symptoms (Factor 1) and socially deviant behaviors (Factor 2), respectively (Harpur et al. 1988; Hare et al. 1990). This two-factor conceptualization of psychopathy aligns with theoretical treatments of the construct which refer to emotional deficits as the core, precipitating features that drive the development of certain behavioral manifestations including antisocial tendencies (Harpur et al. 1989). There have been recent suggestions for more complex factor solutions for the PCL-R (Cooke and Michie 2001; Hare 2003) nonetheless, each of these alternative solutions have elements which distinguish socially deviant behavior from emotional abnormalities and personality features, and this distinction remains by far the most widely referenced in terms of its correspondence to theoretical constructs of personality as well as etiological and neuropsychological antecedents.

Other measures of psychopathy have been developed for specific purposes such as measuring psychopathic traits in nonincarcerated samples (Levenson et al. 1995; Lilienfeld and Widows 2005) and youth with psychopathic traits (Forth et al. 2003; Frick and Hare 2001). Analyses of these scales generally support the two-factor conceptualization of psychopathy (for a detailed review of the measurement issues in psychopathy, see (Hare and Neumann 2009). Each measure has particular advantages and limitations, and each of these measures has pushed research forward in helpful ways. In addition to refining the clinical construct of psychopathy, the development of the PCL-R and its derivatives has aided the growth of a sound body of empirical literature, including a better understanding of the relationship between psychopathy and aggressive behavior. Indeed, the more accurately one measures a psychological construct (i.e., psychopathy) the more likely one will find stable neurobiological correlates of that construct.

2.2 Psychopathy as a Psychiatric Diagnosis

While the construct of psychopathy was intended to fall under the DSM-IV-TR (APA 2000) categorization of *Antisocial Personality Disorder* (ASPD), it is a much belabored point among researchers and clinicians that it is not well accounted for by the ASPD diagnostic criteria (Cunningham and Reidy 1998; Hare et al. 1991). Early versions of the DSM had utilized a different classification system which included categories such as *sociopathic personality disturbance*; however, this was not specifically synonymous with our modern concept of psychopathy. ASPD was first introduced in DSM-III (APA 1980), which remains in the current version (DSM 5). However, the current diagnostic criteria miss the mark when assessing classic psychopathic symptomatology, as ASPD focuses heavily on criminal behavior and social deviance, while largely ignoring the affective elements considered central to psychopathy (Blackburn 2007). Due to the generality of the symptoms, over 4 % of the general public will meet criteria for ASPD (APA 2000), but less than 1 % meet criteria for psychopathy using the Hare Psychopathy Checklist-Revised (Hare 2003). In forensic samples, where ASPD diagnoses may occur at rates near 80 % or higher, the best estimates for rates of psychopathy are 15–25 % on average (Hare 2003; Hare et al. 1991). Despite the necessary distinction between ASPD and psychopathy as assessed by the Hare PCL-R, antisocial behavior and aggression remain among the most visible characteristics of psychopathy. It must simply be recognized that persistent criminal behavior per se is not sufficient for an accurate diagnosis of psychopathy, which requires the presence of the core emotional deficits. Still, antisocial behavior, but not necessarily criminal behavior, is a critical feature of the disorder.

Longstanding efforts to study nonincarcerated and "successful" psychopaths persist (Babiak and Hare 2006; Gao and Raine 2010; Widom 1977); however, what may seem like a distinct subtype may likely be differentiated as individuals with subclinical levels of a few psychopathic traits. Nevertheless, psychopathic traits, even at subclinical levels, may be related to poor outcomes, including aggressive behaviors. Going forward in our examination of aggression in psychopaths, we should recognize it as a situationally complex behavior which can arise from a more fundamental influence of neurobiology on learning, motivation, and personality traits which characterize psychopaths.

3 Predicting Violent Outcomes

Psychopathy, as measured by the PCL-R, has long been documented as a robust indicator in forensic settings of how likely one is to re-offend (Porter et al. 2001; Salekin et al. 1996), and it is a particularly strong predictor of future violent recidivism (Cornell et al. 1996; Harris et al. 1991; Porter et al. 2009) Indeed, research has indicated that within 1 year of release from prison psychopaths are about 4–6 times more likely to commit another violent crime than are nonpsychopaths (Hemphill et al. 1998). Estimates are fairly consistent across studies: within 10 years after release over 70 % of psychopaths (with a history of violence) commit another violent offense, and 20-year follow-ups indicate that as many as 90 % of psychopaths (with a history of violence) will be re-arrested for violent crimes. In contrast, the average nonpsychopathic violent offender appears to have a risk for violent recidivism around 40 % (Hare et al. 2000; Harris et al. 1991; Hemphill et al. 1998; Rice and Harris 1997). Furthermore, analyses of the actual crimes committed by psychopaths suggest a heightened severity and gratuitous nature to their violent acts (Porter et al. 2003).

These same trends are evident in youth. An accumulating body of literature has provided strong evidence indicating that carefully defined aspects of psychopathic traits, typically referred to as *callous and unemotional traits* when assessed in youth, are apparent at a young age and are persistent across the lifespan (Lynam et al. 2007). These traits are often accounted for by assessment tools designed for behaviorally disordered youth (Forth et al. 2003; Frick and Hare 2001). The combination of persistent behavioral problems (i.e., conduct disorder) and callous, unemotional traits in youth is associated with persistent patterns of severe aggression, and is a particularly strong predictor of future violent offending (Frick et al. 2003; Vitacco and Vincent 2006).

Within forensic samples the PCL-R/PCL-YV reliably predicts violent outcomes, but it should be noted that not all psychopaths are necessarily violent (Cleckley 1976). Violence should not be seen as a unique distinguishing quality of psychopaths, yet among the neural deficits underlying psychopathy, some may contribute to increased risk for violent behavior in various contexts. Examining the wide variety and motivations for aggressive acts and the various neural systems which support them can thus provide us with an informative framework to discuss the specific neural deficits associated with psychopathy.

3.1 Varieties of Aggression

It is easy to recognize variability in aggressive behavior among humans and other animals. Indeed, many have felt the impulse to react violently toward someone who provokes them, and some of us may have even acted upon these impulses—such instincts are not necessarily pathological. It is quite another thing to stalk a potential victim for days, devising a malicious plan to harm someone else for their own benefit. The latter requires a host of violations of ordinary social rules, which are not simply overwhelmed in a momentary fit of rage, but rather, are wholly disregarded in a prolonged plan to achieve some self-serving goal at another's expense. The heterogeneity inherent in violent behavior has prompted the establishment of a variety of categorizations, the most prominent one distinguishing between impulsive and premeditated aggression (McEllistrem 2004).

Impulsive aggression is characterized as a reactive, emotional response to some immediate provocation or frustration. In contrast, *premeditated aggression* results from more complex motivating factors and is largely characterized by predatory actions serving some instrumental purpose. Each of these concepts can be referred to by several alternative terms such as *reactive* versus *proactive* aggression (see elsewhere in this volume), but the conceptual divisions are essentially the same. Factor analyses of reported aggressive acts support the distinction between impulsive and premeditated aggression (Barratt et al. 1999), and some evidence suggests that an individual's aggressive behavior can be characterized as aligning predominantly toward one or the other mode of aggressive tendencies (Stanford et al. 2003). It is also increasingly recognized that aggression need not be strictly defined by physical altercations, but can be complex social actions sometimes referred to as *indirect* or *relational* aggression (Archer and Coyne 2005), and which may include general hostility, bullying, manipulation, and intimidation.

It has sometimes been suggested that psychopaths predominantly commit predatory, instrumental violence (Williamson et al. 1987; Woodworth and Porter 2002) but it is important to recognize that psychopaths also commit impulsive violent acts (Blair 2010) and some have suggested that impulsive risk-taking is a strong motivating factor in psychopaths' aggressive behavior, and is in itself reinforcing (Porter et al. 2000). But this feature is not unique to psychopathy. Impulsive aggression and behavioral disinhibition are prevalent among many psychiatric disorders including post-traumatic stress disorder (Silva et al. 2001), depression (Fava 1998), bipolar disorder (Garno Gunawardane and Goldberg 2008), intermittent explosive disorder (Coccaro 2000), temporal lobe epilepsy (Van Elst et al. 2000), Alzheimers disease (Lai et al. 2003), dementia (Haller et al. 1989), and substance abuse (Boles and Miotto 2003).

It may be more appropriate to consider instrumental forms of aggression a more distinguishing characteristic of psychopathy, particularly among violent offenders (Cornell et al. 1996). Impulsive violence demonstrates a lack of behavioral control consistent with antisocial features of psychopathy, whereas premeditated aggression corresponds to more complex motivational patterns reflecting impaired socialization, lack of empathy, and abnormal moral reasoning. Both impulsive and premeditated violence represent a lack of deliberation or concern for potential consequences, but rather than being limited to acute lapses in behavioral control, for psychopaths this also impinges on long-term planning and goal-oriented behavior. In nonincarcerated samples, conceivably less prone to physical violence, these deficits remain apparent, represented by forms of indirect, relational aggression (Czar et al. 2011; Warren and Clarbour 2009).

As we have noted, complex behavioral outcomes like aggression, violence, and specific forms of criminality can be highly variable and context-specific. As an alternative to examining discrete behaviors directly, it can be more enlightening to focus on personality traits and motivational styles which bias behavior in specific ways and are more proximally related to the neural systems which influence them. As we examine neurobiological models of psychopathy, it will become clear that specific neural deficits differentially influence distinct features of psychopathy; and in turn, these may uniquely account for the occurrence of distinct forms of aggression.

4 Neurobiological Models of Psychopathy

Early recognition of abnormal psychophysiological features in psychopaths have contributed to a large body of research supporting neurobiological models of psychopathy, most of which place a strong emphasis on accounting for a form of emotional dysfunction germane to the construct. Some of the first physiological evidence for these models demonstrated that psychopaths fail to show appropriate autonomic responses in conditioning paradigms (Hare and Quinn 1971; Lykken 1957), and in anticipation of aversive stimuli (Hare 1968, 1982; Hare et al. 1978). This trend has been replicated in many different contexts, as psychopaths show relatively subdued physiological responses to emotionally evocative stimuli including sounds (Verona et al. 2004), emotional words (Williamson et al. 1991), and disturbing images (Herpertz et al. 2001). Generally speaking, the most widely replicated effects in these early studies seem to demonstrate a failure in psychopaths to form associations between specific stimuli and potential harm, which in healthy individuals evokes a physiological cascade of events preparing the body and mind for threatening circumstances (Lang and Davis 2006; Lang et al. 2000).

A prominent extension of these findings came with the recognition that psychopaths fail to demonstrate the ordinary potentiation of startle reflexes under the influence of negative emotional contexts (Patrick et al. 1993). That is, where healthy individuals react with increased startle responses when they are primed by aversive pictures or an otherwise anxiety-provoking state, psychopaths' startle responses remain relatively constant, or even decrease in these circumstances. This rather specific feature of psychopathy has become one of the most consistently replicated physiological features of the disorder, having been documented in a wide range of samples and contexts (Anderson and Stanford 2012; Anderson et al. 2011; Benning et al. 2005; Justus and Finn 2007; Levenston et al. 2000; Patrick 1994; Sutton et al. 2002; Vanman et al. 2003).

The neural architecture supporting the potentiation of startle reflexes has been well described, and we understand that amygdala activation is directly responsible for moderating the magnitude of the startle reflex through its connections with the *caudal pontine reticular* nucleus in the brainstem (Davis 1992; Kim and Davis 1993). The demonstration that affective priming of this startle reflex was impaired in psychopaths provided strong early support for the notion that psychopaths suffered contextual deficits in amygdala activation, and thereby added support to the *Low Fear Hypothesis* of psychopathy (Lykken 1995). This model holds that psychopathy's deficits are due to a specific reduction in the brain's response to potentially threatening features of the environment. This could theoretically account for many of psychopathy's features insomuch as this emotional reaction is an important endogenous signal for us to inhibit certain behaviors and avoid situations which might harm us. Recognizing and/or effectively avoiding potential consequences is certainly an area in which psychopaths show serious deficits; however, several other models ostensibly account for a wider range of deficits associated with psychopathy.

An alternative model emphasizes that our behavior is influenced by the proficiency with which we attend to a broad range of stimuli and allocate cognitive resources appropriately. Incorporating these elements, Newman and colleagues have proposed the *Response Modulation Hypothesis*, which suggests that the specific deficit psychopaths have is in using motivationally relevant cues to shift attention and thereby regulate behavior, but are otherwise able to process emotional stimuli at normal levels (Newman and Lorenz 2003). A subtle consequence of this demonstrated in several empirical studies is that psychopaths are capable of adjusting their behavior appropriately when their primary goal is to do so, or when punishment is the only incentive (Newman and Kosson 1986; Newman et al. 1987, 1990). However, in ordinary circumstances with conflicting rewards and punishments, psychopaths are impaired in shifting their attention to accommodate complex reinforcement contingencies. If one indeed had a specific cognitive deficit of this nature, it is reasonable to suspect that one consequence might be increased incidence of aggression if it satisfies a specific goal, while conflicting, alternative considerations of potential consequences remain unattended or unimportant.

It has been suspected for some time that portions of the prefrontal cortex may also be impaired in psychopaths. Prefrontal injuries have been demonstrated to result in insensitivity to future consequences (Bechara et al. 1994) and also interrupt implementation of advantageous decision-making (Bechara et al. 1997). In fact, some accounts of psychopathy have invoked the *Somatic Marker Hypothesis*, drawing directly from neuropsychological evidence that the orbitofrontal cortex is essential for online evaluation of previously encoded emotional associations (Damasio 1994). The ventromedial/orbitofrontal region of the prefrontal cortex shares extensive reciprocal connections with the amygdala through the uncinate fasciculus. Several reports exist of damage to the orbitofrontal cortex leading to the inception of rash, aggressive behavior accompanied by disinhibited, impulsive traits (Harlow 1993; Meyers et al. 1992; Cato et al. 2004). The consequences of injury to this relatively vulnerable region are conspicuous enough to have spawned the terms *pseudo-psychopathy* or *acquired sociopathy* (Blumer and Benson 1975; Damasio 1994). However, such injuries, particularly those occurring later in life, do not result in the full spectrum of behavioral styles and core personality features of psychopathy. In order to account for traits such as shallow affect and the unique variety of aggressive tendencies typical among psychopaths, including a prevalence of premeditated-instrumental aggression, one must appreciate a wider-reaching, more complex network of brain regions which work together to integrate affective information into several cognitive processes.

With the recent addition of functional neuroimaging to the list of available measures for studying these phenomena, this literature has expanded and allowed for additional specificity in identifying compromised regions of the brain in psychopathy. In reviewing this literature, however, one should understand that examining brain dysfunction using neuroimaging techniques is a delicate process which depends as much upon properly controlling behavioral aspects of a given task as it does on the technical details for adequately modeling brain activity with advanced methods such as functional magnetic resonance imaging (fMRI). To demonstrate an abnormality in brain activity, one must carry out a specific cognitive task and show differentiable patterns of neural activation which are dependent upon a trait or category for which a theoretical basis exists for different functions. This introduces uncertainty at many levels including the appropriateness of trait measurements, efficacy of the task being implemented, and a host of statistical issues which cannot be addressed fully here. Nevertheless, progress is being made, and the accumulating literature has revealed some remarkable consistency in identifying abnormal brain regions in psychopaths (for a full review, see Anderson and Kiehl 2012).

Two prominent, contemporary neurobiological models of psychopathy which have, from their origins, integrated contemporary neuroimaging are proffered by Blair (2006) and Kiehl (2006). These models not only share a number of attributes but also have some important differences. These models implicate components of the limbic system—a contiguous network of brain regions which support emotional processing (including fear) and the integration of affective information into more general behavioral regulation (Papez 1937). Blair's model has emphasized dysfunction in the amygdala resulting in deleterious consequences on reinforcement learning, and attributes some of the functional disturbances in other regions such as the orbitofrontal cortex to hierarchical effects of amygdala dysfunction (Blair 2007). Kiehl's paralimbic dysfunction model integrates evidence for more widely distributed abnormalities in the brains of psychopaths, appearing throughout an extended *paralimbic* network of brain regions, defined by cytoarchitectonic maps as an integrated system (Brodmann 1994), providing a transition from subcortical structures to higher neocortical regions (Kiehl 2006).

The most recent neuroimaging research investigating the concept of psychopathy has largely supported these models, consistently implicating several functional brain regions related to basic emotional processing and higher-order integration of emotion into online cognitive processing. Dysfunction in the amygdala and ventromedial/orbitofrontal cortex is most widely reported in studies of psychopathy, and they likely represent the most robust, easily observed abnormalities in function. Kiehl et al. (2001) were the first to report amygdala dysfunction in criminal psychopaths using fMRI, demonstrating reduced activity there when comparing emotional and nonemotional words. Failure to engage the amygdala and orbitofrontal cortex during tasks which require aversive conditioning (i.e., learning to associate a specific behavior with punishment) has also been consistently associated with psychopathy (Birbaumer et al. 2005; Veit et al. 2002). Reduced activity in these same regions has been associated with psychopathy in a wide variety of other tasks including the *prisoners dilemma* (Rilling et al. 2007), viewing pictures of facial affect (Gordon et al. 2004; Dolan and Fullam 2009), evaluating pictures depicting moral violations (Harenski et al. 2010), and viewing aversive photographic stimuli (Harenski et al. 2009), each of which require a form of emotional processing either directly, or in making task-related decisions.

Beyond the amygdala and orbitofrontal cortex, several other brain areas show evident dysfunction; however, due to the higher-order processing and likely more specialized function of these regions, evident abnormalities there may be more difficult to demonstrate and more highly contextual. In Kiehl et al. (2001) initial report of reduced amygdala function, they also reported reduced activity in the hippocampal formation, parahippocampal gyrus, ventral striatum, and in the anterior and posterior cingulate cortex. In their *prisoner's dilemma* task,

(Rilling et al. 2007) also reported that psychopathy scores were associated with low activity in the anterior cingulate during trials when individuals defected against their partner in crime. Likewise, Birbaumer et al. (2005) and Veit et al. (2002) have demonstrated lower anterior cingulate activity and lower insula activity in psychopaths during aversive conditioning. Abnormally low activity in the right temporal pole of psychopaths has also been reported during an emotion-modulated cognitive task (Müller et al. 2008). Many brain regions rely on contextual emotional information to facilitate specific cognitive processes, and to such a degree that emotional processing areas of the brain are impaired in psychopaths, these diverse domains of brain activity will also show contextual deficits related to reductions in the availability of these systematic responses. Furthermore, these extensive deficits are apparent at a young age with brain imaging findings in youth with psychopathic traits (i.e., conduct disorder combined with callous-unemotional traits) largely mirroring the findings from adults described above (Finger et al. 2008, 2011; Jones et al. 2009; Marsh et al. 2008).

Several reports investigating anatomical features of the brain, independent of function, suggest that psychopaths' brain abnormalities are not limited to patterns of neural activation; rather, morphological brain differences accompanying psychopathic traits mirror the functional deficits noted above in primary and paralimbic regions of the brain. Reduced gray matter volumes have been reported in psychopaths' amygdala and orbitofrontal cortex and cingulate cortex (Boccardi et al. 2011; Yang and Raine 2009; Yang et al. 2010). Others have reported psychopathy-related tissue reductions in the anterior, superior temporal regions (Müller et al. 2008; Yang et al. 2011) and the insula (de Oliveira-Souza et al. 2008). In a large-scale investigation of nearly 300 adult incarcerated participants, psychopathy scores were associated with tissue reductions in the amygdala, orbitofrontal cortex, posterior cingulate, parahippocampal region, and the temporal pole (Ermer et al. 2012), and these diffuse structural differences were largely replicated in a sample of juveniles with psychopathic traits (Ermer et al. 2013).

Given the variety of brain areas implicated in psychopathy and the heterogeneity of some behavioral manifestations of psychopathy, some have preferred to itemize specific features of the disorder, relating them to more discrete neural systems. When this approach is taken, it is most common to adhere to the two-factor conceptualization of psychopathy, differentiating emotional symptoms from behavioral traits and developmental course. Recent attempts to consolidate this literature point out that abnormalities in the structure and function of prefrontal cortical areas are very commonly associated with reactive aggression, impulsivity, and antisocial characteristics (Wahlund and Kristiansson 2009; Yang and Raine 2009; Bufkin and Luttrell 2005), while temporal-limbic abnormalities are more consistently associated with affective, interpersonal facets (Blair 2008; Kiehl 2006). Furthermore, in contrast to prevalent findings that psychopaths present with reduced activity in brain areas related to emotional processing, those whose aggressive acts are primarily characterized as impulsive often show *hyper*activity in the amygdala during emotional primes (Bufkin and Luttrell 2005; Coccaro et al. 2007; Herpertz et al. 2008). The relevance of these accounts is in distinguishing psychopathy from other forms of disinhibitory psychopathology, which further underscores the involvement of widespread paralimbic circuitry, predisposing psychopaths to both reactive aggression and instrumental aggression.

5 Etiologies of Violence in the Psychopath

Diverse etiological factors have been implicated in guiding developmental trajectories in psychopathy, impacting the manifestation of a variety of psychological and behavioral outcomes. This has contributed to descriptions of various subtypes of the construct, a prominent one referring to *primary* and *secondary* psychopaths (Karpman 1941)—a distinction originally intended to differentiate between those with inborn deficits and those who acquired these traits through poor social rearing. The field having progressed beyond a simplistic *nature versus nurture* dichotomy, the term secondary psychopath now generally refers to those with higher levels of anxiety (Lykken 1957; Skeem et al. 2007), more prone to reactionary aggression (Patrick and Zempolich 1998) and impaired prefrontal executive function (Brower and Price 2001; Dolan and Park 2002; Ross et al. 2007). In contrast, primary psychopaths are characterized as less anxious, demonstrating low reactivity to stress and punishment cues (Hare 1982; Verona et al. 2004), normal to high executive function (Dolan 2012; Ross et al. 2007), and whose aggressive tendencies are more typically premeditated (Falkenbach et al. 2008; Patrick and Zempolich 1998).

Among the contemporary models of psychopathy described above, it should be clear that there is a general consensus recognizing functional deficits with deleterious effects on the utility of contextual emotional information in the management of ongoing behavior. This basic deficit impacts essential cognitive functions such as aversive conditioning (Birbaumer et al. 2005), passive avoidance learning (Blair et al. 2004), and ultimately advantageous decision-making when negative consequences should be a relevant consideration (Newman et al. 1987). What remains to be discussed is the extension of these deficits to aggressive behavior. Due to the prevalence of violence among psychopaths, particularly among those who are incarcerated, any model attempting to thoroughly address the development of psychopathic traits and precipitating behavior must make some account for their aggression-prone nature. Furthermore, any such model would benefit from an account of both impulsive and premeditated varieties of aggression. While many have addressed the occurrence of impulsive aggression, relatively fewer models make explicit accounts for the incidence of premeditated aggression.

5.1 Impulsive Aggression

Impulsive aggression is in some ways easier to account for in neurobiological models of behavior. In its essential form, impulsive aggression need not be considered pathological in that rapid defensive and aggressive responses are an adaptive feature of our biology, preserved through evolution. Models for impulsive aggression rely heavily on animal research engaging the mammalian basic threat circuit, which connects the medial hypothalamus and periaqueductal gray matter (Gregg and Siegel 2001). Electrical stimulation of this circuit in animals produces an immediate *defensive rage* response. The amygdala, hippocampus, cingulate gyrus, and prefrontal cortex each send projections to this circuit and can modulate the intensity of this response. In humans, abnormalities in the ventromedial, orbitofrontal cortex have often been associated with unmoderated reactive aggression, conceivably through decreased inhibitory input to this basic threat circuit (Bufkin and Luttrell 2005). Indeed, in contrast to psychopaths' prototypical reduction in primary limbic activity, individuals characterized chiefly by reactive, impulsive aggression demonstrate relative hyperactivity in the amygdala, paired with reduced regulating activity in the ventromedial prefrontal cortex (Meyer-Lindenberg et al. 2006; Coccaro et al. 2007; Raine et al. 1998).

Gene by environment developmental models have been proposed as one etiology for these behavioral tendencies (Buckholtz and Meyer-Lindenberg 2008; Meyer-Lindenberg et al. 2006); however, it remains clear that this frontal executive system is also particularly vulnerable to focal brain injuries, which carry rather informative consequences (Harlow 1993; Meyers et al. 1992; Cato et al. 2004). These injuries have been strongly associated with impulsivity (Berlin et al. 2004) short-sighted decision-making (Bechara et al. 2000), and increases in both verbal and physical aggression (Grafman et al. 1996). These consequences represent a form of behavioral disinhibition which can be seen in psychopaths; however, they do not account for the full spectrum of psychopathic traits, and in terms of aggressive actions, fall more conspicuously under the impulsive-reactive variety. It is reasonable, then, to account for these specific outcomes in psychopathy by appealing to the local impairments recognized in ventromedial regions of the brain by recent neuroimaging studies; however, as described in the previous section, psychopaths' brain abnormalities are not isolated to this region.

5.2 *Instrumental Aggression*

In contrast to the abrupt and ballistic nature of impulsive aggression, instrumental aggression requires planning and enduring cognitive processes which progress uninhibited over longer periods of time. It is not limited by a unitary affective reaction, and thus may be seen as a more profound representation of a failure in governing one's behavior. Premeditated aggression exhibits a futility in what should be a veritable concert of cognitive processes aimed at preventing the behavior including the exercise of such constructs as moral judgment, empathy, and socialization. It is perhaps better represented as an abnormal set of motivational principles, which is missing any impactful consideration for ordinary social rules or the well-being of others. On the other hand, instrumental forms of aggression need not necessarily be considered pathological cither. Given a socially acceptable impetus, such actions may be considered well-founded. What is immediately obvious, however, is that the rightness or wrongness of this action inherently depends on an evaluation of goals.

This form of aggression is less tractable to extant neurobiological models, as its occurrence is more difficult to reduce to a relatively simple brain system. Animal models have demonstrated that differentiable neural circuits from those accounting for impulsive aggression are responsible for this behavior. However, the behavior instigated in these models represents a species-specific form of predation, which again remains a rudimentary analog of the complex behavior we see in humans. The central amygdala connects to ventral periaqueductal gray matter through the lateral hypothalamus, and electrical stimulation of this circuit will instigate stereotyped predation behavior in mammals (Gregg and Siegel 2001). For example, a resting cat will rise, stealthily circle an anesthetized rodent (which it was previously ignoring), and deliver a single calculated strike at the prey. The degree to which this stereotyped behavior effectively represents human predatory aggression is certainly debatable, but the fact that even this simple example of differentiable behavior is governed by a distinct neural circuit is an important consideration. Certainly the motivating factors behind human instrumental aggression can take a more complex form than what we see in the cat. Rather than simply serving an instinct to hunt and eat, human predatory aggression may serve purposes as diverse and complicated as revenge, sexual gratification, monetary reward, interpersonal dominance, or even to achieve a political goal. Ultimately the diversity of such behavior in humans is not succinctly accounted for by any single neurobiological system; however, a more tractable means of considering these influences may be to examine brain deficiencies associated with the chronic disregard for the rights and well-being of others in pursuit of personal goals.

The goal here is not to universally account for all forms of premeditated aggression, but rather to describe what brain systems, that have been demonstrated to be dysfunctional in psychopaths, might influence their increased propensity for premeditated aggression, thus distinguishing their behavior from more common forms of disinhibitory psychopathology. Blair explicitly consolidated neurobiological accounts of psychopathy into a specific model accounting for violent behavior in psychopaths and called it the *violence inhibition mechanism* (VIM) model (Blair 2001). Proposed as an explanation for the occurrence of increased instrumental aggression, VIM described a means by which proper socialization is hindered when ordinary reactions to observing other humans in pain (an aversive unconditioned stimulus) might fail in psychopaths burdened by faulty emotional responses. This model has since been expanded into the *integrated emotional systems* (IES) model (Blair 2006), which accounts for a wider variety of dysfunction related to disrupted subcortical systems in psychopathy extending to expanded paralimbic regions of the brain.

Blair maintains that impulsive, reactive aggression in psychopaths can be largely accounted for by impaired regulation of subcortical and brainstem threat circuitry by the dysfunctional frontal executive system. In contrast, the model proposes that instrumental violence and a great deal of premeditated, antisocial behavior in general are likely elevated in psychopaths due to disruption of critical mechanisms responsible for socialization. In accounting for this disruption in socialization, Blair highlights the involvement of the amygdala in the formation of critical associations, particularly *conditioned stimulus-unconditioned response* associations and *stimulus-reinforcement* associations—processes which facilitate advantageous decision-making through emotional responses. To illustrate, a healthy child finds witnessing the pain of others aversive, and eventually thoughts of acts which cause pain to others become aversive by association. When contextual cues are associated with aversive or rewarding emotional states, a kind of memory trace is established which is capable of biasing future behavior to pursue or avoid similar outcomes when one re-encounters those same contextual cues. And so, a disruption of this type of emotional learning interrupts a great deal of the socialization process as it sabotages the efficacy of direct punishment and associated aversive mental states in guiding the planning of future behavioral patterns.

5.3 Alternative Considerations

As we have already seen, there is certainly a great deal of evidence which suggests that systems governing emotion-related cognition are often dysfunctional in psychopaths, particularly the amygdala. Where Blair's model potentially falls short in accounting for premeditated violence in psychopaths is in the over-specification of one relatively limited system supporting socialization. It is somewhat difficult to rely on neuropsychological evidence to differentiate specific deficits resulting from amygdala damage because isolated, focal damage to this deep tissue is very rare. In contrast to prefrontal brain injuries, reports exploring effects of amygdala lesions point to more subtle and diverse deficits including impaired fear conditioning and emotional episodic memory (Phelps et al. 1998), impaired recognition of facial emotion (Adolphs et al. 1999), and failures in theory of mind reasoning (Stone et al. 2003), which appears to be dependent upon the age at which damage occurs (Shaw et al. 2004). Conspicuously absent from these reports are indications of acquired psychopathic features or increased instrumental aggression.

Socialization is a rather broad term—likely too broad to be monopolized by one anatomical nucleus or even a single neural system. It implies, among other things, an appropriate valuation of one's place among others in society and the appreciation of certain expectations of behavior. Certainly, the formation of stimulus-reinforcement learning is a necessary component of this process (Gao et al. 2009), but many aspects of socialization are also accounted for by behavioral control, delay of gratification, foresight, a balanced appreciation of rewards and consequences, and empathy for other human beings. Another consideration is that socialization depends greatly on the mores and folkways of our communities, biased by one's immediate environment and cultural circumstances. One might suspect, then, that the socialization process can be sabotaged by dysfunction in any of a large number of neural systems, or as Karpman (1941) and others have long suggested, even by a destructive rearing environment.

Dysfunction or abnormal development in a large number of brain regions beyond the amygdala can be detrimental to the socialization process, and potentially instigate behavioral tendencies toward instrumental aggression. When orbitofrontal injuries occur very early in life, the lingering and pervasive behavioral consequences suggest a disruption of the normal socialization processes (Anderson et al. 2000; Anderson and Moore 1995; Taylor and Alden 1997; Williams and Mateer 1992). The anterior cingulate cortex also has important roles in integrating emotion and cognitive responses in error monitoring (Bush et al. 2000) and focal brain damage here may instigate disinhibition, hostility, and difficulty with conflict monitoring and cognitive control (Hornak et al. 2003; di Pellegrino et al. 2007). Anterior portions of the temporal cortex, or temporal pole, have been implicated in highly integrative processes involving emotional and social functions including theory of mind, and damage here can produce a host of symptoms ranging from diminished empathy and hostility to inappropriate behaviors similar to what is seen in Kluver-Bucey syndrome (Olson et al. 2007). So, while the amygdala certainly plays a role in the formation of associations important for socialization, a wider yet highly integrated network of brain areas relies on these associations to effectively implement behavior which we call well-socialized.

Finally, a lingering consideration is, as Cleckley (1941) and others (Patrick 2006) have suggested, violence, per se, is not an essential quality of psychopaths. If we consider psychopathy to be primarily a disorder of personality—even one largely influenced by neurobiological factors—then discrete behavioral consequences such as violence will be a highly idiosyncratic manifestation of how one's cognitive function biases their behavior within a specific environment. A more fundamental quality of psychopaths may simply be that they fail to stop and consider the consequences of their behavior on self and others. Heilbrun (1982) provided evidence that cognitive ability may be a significant moderator of violent outcomes in psychopaths. Furthermore, in examining incarcerated samples we are more likely to observe those whose psychopathic traits have manifested in behavior aligning with more conspicuous forms of crime, readily enforced by police and courts, such as assault and homicide. It is conceivable that psychopaths who benefit from additional resources, or from environments more protected from legal intervention, may be less likely to use physical violence in pursuit of their specific goals. However, these arguments remain largely speculative, and deserve further attention in future research.

Where it seems the extant literature in psychopathy has been particularly informative is in defining aberrant neurocognitive processes that accompany abnormalities in specific brain systems. The subsequent process of extrapolating very specific behavioral consequences such as violence and antisociality from this brain data is more prone to speculation, as higher-order behavior is influenced by such a large number of physiological and environmental variables. While several models exist for the development of psychopathy, many features of these models are ultimately complementary (van Honk and Schutter 2006), and research efforts should continue to explore more precisely the impact recognizable neural deficits have on the way the brain acquires and utilizes information, the way this biases behavior in specific ways, and the varying etiological mechanisms responsible for these neural deficits.

6 Summary and Conclusion

Psychopathy is a disorder that can manifest itself in diverse ways, but which at its core, precipitates from deficits in emotional processing that impair cognitive functions involved in the development of empathy, moral judgment, and sensitivity to future consequences. A common trait among psychopaths is impulsivity and lack of behavioral controls, which may prompt instances of reactive aggression; however, the complex effects of psychopathy's core emotional deficits may also instigate patterns of instrumental aggression, considered by some to distinguish psychopathy from other forms of disinhibitory psychopathology. There remains some debate about whether violence is a common feature of psychopathy per se or if it represents a particular manifestation of psychopathy in an etiological subset of psychopaths; however, available empirical data does not specifically address this, and this suggestion ultimately has little impact on the established recognition of essential neurocognitive deficits which influence the development of psychopathy's definitive personality traits.

Accumulating evidence supports the notion that psychopaths present with abnormal—usually reduced—activity in a closely related network of brain regions responsible for integrating emotional responses into higher cognitive functions. Among the most consistently implicated brain areas are the amygdala, orbitofrontal/ventromedial prefrontal cortex, anterior and posterior cingulate, insula, and anterior–superior temporal cortex. Each of these regions has been demonstrated, through neuroimaging or neuropsychological evidence, to be an integral part of a complex system for utilizing emotional information in behavior management and socialization. The disruption of either the initial encoding of critical associations with emotional responses or the failure of higher-order, integrative cognitive processes accessing these associations could effectively disrupt cognitive processes involved in socialization. Furthermore, these functional abnormalities may result from diverse etiological factors including genetic vulnerabilities, early damage to developing neural systems, and inadequate rearing environments.

To the degree that psychopaths are prone to impulsive, reactive aggression, this is likely related to dysfunction in online regulation of basic threat circuitry and impulse control. High rates of predatory, instrumental aggression, conversely, are more representative of psychopathy's core affective deficits, and likely present as a demonstration of complex failures in socialization, abnormalities in motivational influences, and skewed moral judgments. Psychopaths are characteristically self-centered and generally unable to gauge of the impact of their behavior on others, while retaining the intelligence and observational acuity necessary to effectively manipulate and exploit those who are close to them. In some instances this can take the form of violence.

At the time of this writing, and very likely as you are reading this, public attention has again been drawn to a recent act of brutality. It is, at once, astonishing and familiar. When we encounter these instances of irrational violence, we are forced to remember similar instances that came before it, and we become more sensitive to the aggression that surrounds us every day. Violence and aggression are—it often seems unfortunately—part of our nature, preserved and adapted through evolution. It remains a robust part of our behavior, psychology, and neural architecture. As such, it is also vulnerable to all which can go wrong with the complex physiological systems which govern it. We have made substantial progress in psychology and neuroscience, identifying various forms of aggression and the neural systems which support them, and we continue to make progress understanding both the utility of these complex systems in adapting to our surroundings and the pathological outcomes when things go wrong. On occasions when we are forced to examine the tragic consequences of pathological aggression, we ought to be all the more motivated as students and scientists to understand its occurrence in such a way that we are better equipped to intervene and prevent it in the future.

References

Adolphs R, Tranel D, Hamann S, Young AW, Calder AJ, Phelps EA, Anderson A, Lee GP, Damasio AR (1999) Recognition of facial emotion in nine individuals with bilateral amygdala damage. Neuropsychologia 37(10):1111–1117. doi:10.1016/S0028-3932(99)00039-1

Anderson NE, Kiehl KA (2012) The psychopath magnetized: insights from brain imaging. Trends Cogn Sci 16(1):52–60. doi:10.1016/j.tics.2011.11.008

Anderson NE, Stanford MS (2012) Demonstrating emotional processing differences in psychopathy using affective ERP modulation. Psychophysiology 49(6):792–806. doi:10.1111/j.1469-8986.2012.01369.x

Anderson NE, Stanford MS, Wan L, Young KA (2011) High psychopathic trait females exhibit reduced startle potentiation and increased P3 amplitude. Behav Sci Law 29(5):649–666. doi:10.1002/bsl.998

Anderson SW, Damasio H, Tranel D, Damasio AR (2000) Long-term sequelae of prefrontal cortex damage acquired in early childhood. Dev Neuropsychol 18(3):281–296. doi:10.1207/S1532694202Anderson

Anderson V, Moore C (1995) Age at injury as a predictor of outcome following pediatric head injury: a longitudinal perspective. Child Neuropsychol 1(3):187–202. doi: 10.1080/09297049508400224

Archer J, Coyne SM (2005) An integrated review of indirect, relational, and social aggression. Pers Soc Psychol Rev 9(3):212–230. doi:10.1207/s15327957pspr0903_2

Association AP (1980) Diagnostic and statistical manual of mental disorders, 3rd edn. American Psychiatric Association, Washington, DC, p. 163–170

Association AP (2000) Diagnostic and statistical manual of mental disorders: DSM-IV-TR author. Washington, DC

Babiak P, Hare RD (2006) Snakes in suits: when psychopaths go to work. Harper-Collins, New York

Barratt ES, Stanford MS, Dowdy L, Liebman MJ, Kent TA (1999) Impulsive and premeditated aggression: a factor analysis of self-reported acts. Psychiatry Res 86(2):163–173. doi:10.1016/S0165-1781(99)00024-4

Bechara A, Damasio AR, Damasio H, Anderson SW (1994) Insensitivity to future consequences following damage to human prefrontal cortex. Cognition 50(1):7–15

Bechara A, Damasio H, Tranel D, Damasio AR (1997) Deciding advantageously before knowing the advantageous strategy. Science 275(5304):1293–1295. doi:10.1126/science.275.5304.1293

Bechara A, Tranel D, Damasio H (2000) Characterization of the decision-making deficit of patients with ventromedial prefrontal cortex lesions. Brain J Neurol 123(11):2189–2202

Benning SD, Patrick CJ, Iacono WG (2005) Psychopathy, startle blink modulation, and electrodermal reactivity in twin men. Psychophysiology 42(6):753–762. doi:10.1111/j.1469-8986.2005.00353.x

Berlin H, Rolls E, Kischka U (2004) Impulsivity, time perception, emotion and reinforcement sensitivity in patients with orbitofrontal cortex lesions. Brain J Neurol 127(5):1108–1126. doi:10.1093/brain/awh135

Birbaumer N, Veit R, Lotze M, Erb M, Hermann C, Grodd W, Flor H (2005) Deficient fear conditioning in psychopathy: a functional magnetic resonance imaging study. Arch Gen Psychiatry 62(7):799. doi:10.1001/archpsyc.62.7.799

Blackburn R (2007) Personality disorder and antisocial deviance: comments on the debate on the structure of the Psychopathy Checklist-Revised. J Pers Disord 21(2):142–159. doi:10.1521/pedi.2007.21.2.142

Blair R (2006) The emergence of psychopathy: implications for the neuropsychological approach to developmental disorders. Cognition 101(2):414–442. doi:10.1016/j.cognition.2006.04.005

Blair R (2007) The amygdala and ventromedial prefrontal cortex in morality and psychopathy. Trends cogn sci 11(9):387–392. doi:10.1016/j.tics.2007.07.003

Blair R (2008) The amygdala and ventromedial prefrontal cortex: functional contributions and dysfunction in psychopathy. Philos Trans R Soc B Biol Sci 363(1503):2557–2565. doi:10.1098/rstb 2008.0027

Blair R (2010) Psychopathy, frustration, and reactive aggression: the role of ventromedial prefrontal cortex. Br J Psychol 101(3):383–399. doi:10.1348/000712609X418480

Blair R, Mitchell D, Leonard A, Budhani S, Peschardt K, Newman C (2004) Passive avoidance learning in individuals with psychopathy: modulation by reward but not by punishment. Personality Individ Differ 37(6):1179–1192

Blair RJR (2001) Neurocognitive models of aggression, the antisocial personality disorders, and psychopathy. J Neurol Neurosurg Psychiatry 71(6):727–731

Blumer D, Benson DF (1975) Personality changes with frontal and temporal lobe lesions. In: Benson DF, Blumer D (eds) Psychiatric aspects of neurologic disease, vol 1. Grune and Stratton, New York, pp 151–170

Boccardi M, Frisoni GB, Hare RD, Cavedo E, Najt P, Pievani M, Rasser PE, Laakso MP, Aronen HJ, Repo-Tiihonen E (2011) Cortex and amygdala morphology in psychopathy. Psychiatry Res Neuroimaging 193(2):85–92. doi:10.1016/j.pscychresns.2010.12.013

Boles SM, Miotto K (2003) Substance abuse and violence: A review of the literature. Aggression and Violent Behaviour 8(2):155–174. doi:10.1016/S1359-1789(01)00057-X

Brodmann K (1994) Localisation in the cerebral cortex. Smith-Gordon, London

Brower M, Price B (2001) Neuropsychiatry of frontal lobe dysfunction in violent and criminal behaviour: a critical review. J Neurol Neurosurg Psychiatry 71(6):720–726. doi:10.1136/jnnp.71.6.720

Buckholtz JW, Meyer-Lindenberg A (2008) MAOA and the neurogenetic architecture of human aggression. Trends Neurosci 31(3):120–129. doi:10.1016/j.tins.2007.12.006

Bufkin JL, Luttrell VR (2005) Neuroimaging studies of aggressive and violent behavior current findings and implications for criminology and criminal justice. Trauma Violence Abuse 6(2):176–191. doi:10.1177/1524838005275089

Bush G, Luu P, Posner MI (2000) Cognitive and emotional influences in anterior cingulate cortex. Trends cogn sci 4(6):215–222

Cato MA, Delis DC, Abildskov TJ, Bigler E (2004) Assessing the elusive cognitive deficits associated with ventromedial prefrontal damage: a case of a modern-day Phineas Gage. J Int Neuropsycho Soc 10(3):453–465. doi:10.1017/S1355617704103123

Cleckley H (1941) The mask of sanity; an attempt to reinterpret the so-called psychopathic personality. Mosby, Oxford

Coccaro EF (2000) Intermittent explosive disorder. Curr psychiatry rep 2(1):67–71. doi:10.1007/s11920-000-0045-z

Coccaro EF, McCloskey MS, Fitzgerald DA, Phan KL (2007) Amygdala and orbitofrontal reactivity to social threat in individuals with impulsive aggression. Biol Psychiatry 62(2):168–178. doi:10.1016/j.biopsych.2006.08.024

Cooke DJ, Michie C (2001) Refining the construct of psychopathy: towards a hierarchical model. Psychol Assess 13(2):171. doi:10.1037/1040-3590.13.2.171

Cornell DG, Warren J, Hawk G, Stafford E, Oram G, Pine D (1996) Psychopathy in instrumental and reactive violent offenders. J Consult Clin Psychol 64(4):783

Cunningham MD, Reidy TJ (1998) Antisocial personality disorder and psychopathy: diagnostic dilemmas in classifying patterns of antisocial behavior in sentencing evaluations. Behav Sci law 16(3):333–351

Czar KA, Dahlen ER, Bullock EE, Nicholson BC (2011) Psychopathic personality traits in relational aggression among young adults. Aggressive Behav 37(2):207214. doi:10.1002/Ab.20381

Damasio A (1994) Descartes' error: emotion, reason and the human mind. Putnam, New York

Davis M (1992) The role of the amygdala in fear and anxiety. Annu Rev Neurosci 15(1):353–375. doi:10.1146/annurev.ne.15.030192.002033

de Oliveira-Souza R, Hare RD, Bramati IE, Garrido GJ, Azevedo Ignácio F, Tovar-Moll F, Moll J (2008) Psychopathy as a disorder of the moral brain: fronto-temporo-limbic grey matter reductions demonstrated by voxel-based morphometry. NeuroImage 40(3):1202–1213

di Pellegrino G, Ciaramelli E, Làdavas E (2007) The regulation of cognitive control following rostral anterior cingulate cortex lesion in humans. J Cogn Neurosci 19(2):275–286. doi:10.1162/jocn.2007.19.2.275

Dolan M (2012) The neuropsychology of prefrontal function in antisocial personality disordered offenders with varying degrees of psychopathy. Psychol Med 1(1):1–11. doi:10.1017/S0033291711002686

Dolan M, Park I (2002) The neuropsychology of antisocial personality disorder. Psychol Med 32(3):417–427. doi:10.1017/S0033291702005378

Dolan MC, Fullam RS (2009) Psychopathy and functional magnetic resonance imaging blood oxygenation level-dependent responses to emotional faces in violent patients with schizophrenia. Biol Psychiatry 66(6):570–577. doi:10.1016/j.biopsych.2009.03.019

Ermer E, Cope LM, Nyalakanti PK, Calhoun VD, Kiehl KA (2012) Aberrant paralimbic gray matter in criminal psychopathy. J Abnorm Psychol 121(3):649. doi:10.1037/a0026371

Ermer E, Cope LM, Nyalakanti PK, Calhoun VD, Kiehl KA (2013) Aberrant paralimbic gray matter in incarcerated male adolescents with psychopathic traits. J Am Acad Child Adolesc Psychiatry 52(1):94. doi:10.1016/j.jaac.2012.10.013

Falkenbach D, Poythress N, Creevy C (2008) The exploration of subclinical psychopathic subtypes and the relationship with types of aggression. Personality Individ Differ 44(4):821–832. doi:10.1016/j.paid.2007.10.012

Fava M (1998) Depression with anger attacks. J Clin Psychiatry 59:18-22

Finger EC, Marsh AA, Blair KS, Reid ME, Sims C, Ng P, Pine DS, Blair RJR (2011) Disrupted reinforcement signaling in the orbitofrontal cortex and caudate in youths with conduct disorder or oppositional defiant disorder and a high level of psychopathic traits. Am J Psychiatry 168(2):152–162. doi:10.1176/appi.ajp.2010.10010129

Finger EC, Marsh AA, Mitchell DG, Reid ME, Sims C, Budhani S, Kosson DS, Chen G, Towbin KE, Leibenluft E (2008) Abnormal ventromedial prefrontal cortex function in children with psychopathic traits during reversal learning. Arch Gen Psychiatry 65(5):586. doi:10.1001/archpsyc.65.5.586

Forth AE, Kosson DS, Hare RD (2003) Hare psychopathy checklist: youth version (PCL:YV). MHS, Toronto.
Frick PJ, Cornell AH, Barry CT, Bodin SD, Dane HE (2003) Callous-unemotional traits and conduct problems in the prediction of conduct problem severity, aggression, and self-report of delinquency. J Abnorm Child Psychol 31(4):457–470
Frick PJ, Hare RD (2001) Antisocial process screening device: APSD. Multi-Health Systems, Toronto
Gao Y, Raine A (2010) Successful and unsuccessful psychopaths: a neurobiological model. Behav Sci law 28(2):194–210. doi:10.1002/bsl.924
Gao Y, Raine A, Venables PH, Dawson ME, Mednick SA (2009) Association of poor childhood fear conditioning and adult crime. Am J Psychiatry 167(1):56–60. doi:10.1176/appi.ajp.2009.09040499
Gordon HL, Baird AA, End A (2004) Functional differences among those high and low on a trait measure of psychopathy. Biol Psychiatry 56(7):516–521. doi:10.1016/j.biopsych.2004.06.030
Garno JL, Gunawardane N, Goldberg JF (2008) Predictors of trait aggression in bipolar disorder. Bipolar Disord 10(2):285–292. doi:10.1111/j.1399-5618.2007.00489.x
Grafman J, Schwab K, Warden D, Pridgen A, Brown H, Salazar A (1996) Frontal lobe injuries, violence, and aggression a report of the vietnam head injury study. Neurology 46(5):1231–1238
Gregg TR, Siegel A (2001) Brain structures and neurotansmitters regulating aggression in cats: implications for human aggression. Prog Neuropsychopharmacol Biol Psychiatry 25(1):91–140
Hare RD (1968) Psychopathy, autonomic functioning, and the orienting response. J Abnorm Psychol 73(3p2):1
Hare RD (1980) A research scale for the assessment of psychopathy in criminal populations. Personality Individ Differ 1(2):111–119. doi:10.1016/0191-8869(80)90028-8
Hare RD (1982) Psychopathy and physiological activity during anticipation of an aversive stimulus in a distraction paradigm. Psychophysiology 19(3):266–271. doi:10.1111/j.1469-8986.1982.tb02559.x
Hare RD (2003) Manual for the hare psychopathy checklist, 2nd edn Revised. Multihealth Systems, Toronto
Hare RD, Clark D, Grann M, Thornton D (2000) Psychopathy and the predictive validity of the PCL-R: an international perspective. Behav Sci Law 18(5):623–645
Hare RD, Frazelle J, Cox DN (1978) Psychopathy and physiological responses to threat of an aversive stimulus. Psychophysiology 15(2):165–172
Hare RD, Harpur TJ, Hakstian AR, Forth AE, Hart SD, Newman JP (1990) The revised Psychopathy Checklist: reliability and factor structure. Psychol Assess J Consult Clin Psychol 2(3):338
Hare RD, Hart SD, Harpur TJ (1991) Psychopathy and the DSM-IV criteria for antisocial personality disorder. J Abnorm Psychol 100(3):391
Hare RD, Neumann CS (2009) Psychopathy. In: Millon T (ed) Oxford textbook of psychopathology. vol 12, 2nd edn. Oxford University Press, New York, pp 622–650
Hare RD, Quinn MJ (1971) Psychopathy and autonomic conditioning. J Abnorm Psychol 77(3):223
Harenski CL, Harenski KA, Shane MS, Kiehl KA (2010) Aberrant neural processing of moral violations in criminal psychopaths. J Abnorm Psychol 119(4):863. doi:10.1037/a0020979
Harenski CL, Kim SH, Hamann S (2009) Neuroticism and psychopathy predict brain activation during moral and nonmoral emotion regulation. Cogn Affect Behav Neurosci 9(1):1–15. doi:10.3758/CABN.9.1.1
Harlow J (1993) Classic Text No. 14; recovery from the passage of an iron bar through the head. Hist Psychiatry 4(14):271–273. doi:10.1177/0957154X9300401406
Harpur TJ, Hakstian AR, Hare RD (1988) Factor structure of the Psychopathy Checklist. J Consult Clin Psychol 56(5):741
Harpur TJ, Hare RD, Hakstian AR (1989) Two-factor conceptualization of psychopathy: construct validity and assessment implications. Psychol Assess J Consult Clin Psychol 1(1):6

Harris GT, Rice ME, Cormier CA (1991) Psychopathy and violent recidivism. Law Hum Behav 15(6):625. doi:10.1007/Bf01065856

Heilbrun AB (1982) Cognitive models of criminal violence based upon intelligence and psychopathy levels. J Consult Clin Psychol 50(4):546. doi:10.1037//0022-006x.50.4.546

Haller E, Binder RL, McNiel DE (1989) Violence in geriatric patients with dementia. J Am Acad Psychiatry Law Online 17(2):183–188

Hemphill JF, Hare RD, Wong S (1998) Psychopathy and recidivism: a review. Leg Criminol Psychol 3(1):139–170. doi:10.1111/j.2044-8333.1998.tb00355.x

Herpertz SC, Huebner T, Marx I, Vloet TD, Fink GR, Stoecker T, Jon Shah N, Konrad K, Herpertz-Dahlmann B (2008) Emotional processing in male adolescents with childhood-onset conduct disorder. J Child Psychol Psychiatry 49(7):781–791. doi:10.1111/j.1469-7610.2008.01905.x

Herpertz SC, Werth U, Lukas G, Qunaibi M, Schuerkens A, Kunert HJ, Freese R, Flesch M, Mueller-Isberner R, Osterheider M (2001) Emotion in criminal offenders with psychopathy and borderline personality disorder. Arch Gen Psychiatry 58(8):737

Hornak J, Bramham J, Rolls E, Morris R, O'Doherty J, Bullock P, Polkey C (2003) Changes in emotion after circumscribed surgical lesions of the orbitofrontal and cingulate cortices. Brain J Neurol 126(7):1691–1712. doi:10.1093/brain/awg168

Jones A, Laurens K, Herba C, Barker G, Viding E (2009) Amygdala hypoactivity to fearful faces in boys with conduct problems and callous-unemotional traits. Am J Psychiatry 166(1):95–102. doi:10.1176/appi.ajp.2008.07071050

Justus AN, Finn PR (2007) Startle modulation in non-incarcerated men and women with psychopathic traits. Personality Individ Differ 43(8):2057–2071. doi:10.1016/j.paid.2007.06.020

Karpman B (1941) On the need of separating psychopathy into two distinct clinical types: the symptomatic and the idiopathic. J Crim Psychopathol 3:112–137

Kiehl KA (2006) A cognitive neuroscience perspective on psychopathy: evidence for paralimbic system dysfunction. Psychiatry Res 142(2–3):107–128. doi:10.1016/j.psychres.2005.09.013

Kiehl KA, Smith AM, Hare RD, Mendrek A, Forster BB, Brink J, Liddle PF (2001) Limbic abnormalities in affective processing by criminal psychopaths as revealed by functional magnetic resonance imaging. Biol psychiatry 50(9):677–684. doi:10.1016/S0006-3223(01)01222-7

Kim M, Davis M (1993) Electrolytic lesions of the amygdala block acquisition and expression of fear-potentiated startle even with extensive training but do not prevent reacquisition. Behav Neurosci 107(4):580–595. doi:10.1037/0735-7044.107.4.580

Lai MKP, Tsang SWY, Francis PT, Esiri MM, Keene J, Hope T, Chen CPLH (2003) Reduced serotonin 5-HT1A receptor binding in the temporal cortex correlates with aggressive behavior in Alzheimer disease. Brain Res 974(1):82–87. doi:10.1016/S0006-8993(03)02554-X

Lang PJ, Davis M (2006) Emotion, motivation, and the brain: reflex foundations in animal and human research. Prog Brain Res 156:3–29. doi:10.1016/S0079-6123(06)56001-7

Lang PJ, Davis M, Öhman A (2000) Fear and anxiety: animal models and human cognitive psychophysiology. J Affect Disord 61(3):137–159. doi:10.1016/S0165-0327(00)00343-8

Levenson MR, Kiehl KA, Fitzpatrick CM (1995) Assessing psychopathic attributes in a noninstitutionalized population. J Pers Soc Psychol 68(1):151

Levenston GK, Patrick CJ, Bradley MM, Lang PJ (2000) The psychopath as observer: emotion and attention in picture processing. J Abnorm Psychol 109(3):373. doi:10.1037//0021-843x.109.3.373

Lilienfeld S, Widows M (2005) Professional manual for the psychopathic Personality Inventory-Revised (PPI-R). Psychological Assessment Resources, Lutz

Lykken DT (1957) A study of anxiety in the sociopathic personality. J Abnorm Soc Psychol 55(1):6

Lykken DT (1995) The antisocial personalities. Lawrence Erlbaum, New Jersey

Lynam DR, Caspi A, Moffitt TE, Loeber R, Stouthamer-Loeber M (2007) Longitudinal evidence that psychopathy scores in early adolescence predict adult psychopathy. J Abnorm Psychol 116(1):155. doi:10.1037/0021-843x.116.1.155

Marsh A, Finger E, Mitchell D, Reid M, Sims C, Kosson D, Towbin K, Leibenluft E, Pine D, Blair R (2008) Reduced amygdala response to fearful expressions in children and adolescents with callous-unemotional traits and disruptive behavior disorders. Am J Psychiatry 165(6):712–720. doi:10.1176/appi.ajp.2007.07071145

McEllistrem JE (2004) Affective and predatory violence: a bimodal classification system of human aggression and violence. Aggression Violent Behav 10(1):1–30. doi:10.1016/j.avb.2003.06.002

Meyer-Lindenberg A, Buckholtz JW, Kolachana B, A RH, Pezawas L, Blasi G, Wabnitz A, Honea R, Verchinski B, Callicott JH, Egan M, Mattay V, Weinberger DR (2006) Neural mechanisms of genetic risk for impulsivity and violence in humans. Proc Nat Acad Sci USA 103(16):6269–6274. doi:10.1073/pnas.0511311103

Meyers CA, Berman SA, Scheibel RS, Hayman A (1992) Case report: acquired antisocial personality disorder associated with unilateral left orbital frontal lobe damage. J Psychiatry Neurosci 17(3):121

Müller JL, Sommer M, Döhnel K, Weber T, Schmidt-Wilcke T, Hajak G (2008) Disturbed prefrontal and temporal brain function during emotion and cognition interaction in criminal psychopathy. Behav Sci Law 26(1):131–150. doi:10.1002/bsl.796

Murphy JM (1976) Psychiatric labeling in cross-cultural perspective. Science 191(4231):1019–1028

Newman JP, Kosson DS (1986) Passive avoidance learning in psychopathic and nonpsychopathic offenders. J Abnorm Psychol 95(3):252

Newman JP, Lorenz AR (2003) Response modulation and emotion processing: implications for psychopathy and other dysregulatory psychopathology. In: Davidson RJ, Scherer K, Goldsmith HH (eds) Handbook of affective sciences. Oxford University Press, New York, pp 904–929

Newman JP, Patterson CM, Howland EW, Nichols SL (1990) Passive avoidance in psychopaths: the effects of reward. Personality Individ Differ 11(11):1101–1114. doi:10.1016/0191-8869(90)90021-I

Newman JP, Patterson CM, Kosson DS (1987) Response perseveration in psychopaths. J Abnorm Psychol 96(2):145

Olson IR, Plotzker A, Ezzyat Y (2007) The enigmatic temporal pole: a review of findings on social and emotional processing. Brain J Neurol 130(7):1718–1731. doi:10.1093/brain/awm052

Papez JW (1937) A proposed mechanism of emotion. Arch Neurol Psychiatry 38(4):725

Patrick CJ (1994) Emotion and psychopathy: startling new insights. Psychophysiology 31(4):319–330. doi:10.1111/j.1469-8986.1994.tb02440.x

Patrick CJ (2006) Back to the future: Cleckley as a guide to the next generation of Psychopathy Research. In: Patrick CJ (ed) Handbook of Psychopathy. Guilford Press, New York, pp 605–617

Patrick CJ, Bradley MM, Lang PJ (1993) Emotion in the criminal psychopath: startle reflex modulation. J Abnorm Psychol 102(1):82. doi:10.1037/0021-843X.102.1.82

Patrick CJ, Zempolich KA (1998) Emotion and aggression in the psychopathic personality. Aggression Violent Behav 3(4):303–338. doi:10.1016/S1359-1789(97)00003-7

Phelps EA, LaBar KS, Anderson AK, O'connor KJ, Fulbright RK, Spencer DD (1998) Specifying the contributions of the human amygdala to emotional memory: a case study. Neurocase 4(6):527–540. doi:10.1080/13554799808410645

Pinel P (1806) Section IV: mental derangement distributed into different species: second species of mental derangement. In: Pinel P (ed) A treatise on insantiy. Messers, Cadell and Davies, London, pp 150–156

Porter S, Birt AR, Boer DP (2001) Investigation of the criminal and conditional release profiles of Canadian federal offenders as a function of psychopathy and age. Law Hum Behav 25(6):647. doi:10.1023/A:1012710424821

Porter S, Brinke L, Wilson K (2009) Crime profiles and conditional release performance of psychopathic and non-psychopathic sexual offenders. Legal Criminol Psychol 14(1):109–118. doi:10.1348/135532508x284310

Porter S, Fairweather D, Drugge J, Herve H, Birt A, Boer DP (2000) Profiles of psychopathy in incarcerated sexual offenders. Crim Justice Behav 27(2):216–233. doi:10.1177/0093854800027002005

Porter S, Woodworth M, Earle J, Drugge J, Boer D (2003) Characteristics of sexual homicides committed by psychopathic and nonpsychopathic offenders. Law Hum Behav 27(5):459–470. doi:10.1023/A:1025461421791

Prichard JC (1835) A treatise on insanity and other disorders affecting the mind, vol 1835. Sherwood, Gilbert and Piper, London

Raine A, Meloy JR, Bihrle S, Stoddard J, LaCasse L, Buchsbaum MS (1998) Reduced prefrontal and increased subcortical brain functioning assessed using positron emission tomography in predatory and affective murderers. Behav Sci law 16(3):319–332. doi:10.1002/(Sici)1099-0798(199822)16:3<319::Aid-Bsl311>3.0.Co;2-G

Rice ME, Harris GT (1997) Cross-validation and extension of the Violence Risk Appraisal Guide for child molesters and rapists. Law Hum Behav 21(2):231–241. doi:10.1023/A:1024882430242

Rilling JK, Glenn AL, Jairam MR, Pagnoni G, Goldsmith DR, Elfenbein HA, Lilienfeld SO (2007) Neural correlates of social cooperation and non-cooperation as a function of psychopathy. Biol Psychiatry 61(11):1260–1271. doi:10.1016/j.biopsych.2006.07.021

Ross SR, Benning SD, Adams Z (2007) Symptoms of executive dysfunction are endemic to secondary psychopathy: an examination in criminal offenders and noninstitutionalized young adults. J Pers Disord 21(4):384–399. doi:10.1521/pedi.2007.21.4.384

Salekin RT, Rogers R, Sewell KW (1996) A review and meta-analysis of the Psychopathy Checklist and Psychopathy Checklist-Revised: predictive validity of dangerousness. Clin Psychol Sci Pract 3(3):203-215. doi:10.1111/j.1468-2850.1996.tb00071.x

Shaw P, Lawrence E, Radbourne C, Bramham J, Polkey C, David A (2004) The impact of early and late damage to the human amygdala on 'theory of mind'reasoning. Brain J Neurol 127(7):1535–1548. doi:10.1093/brain/awh168

Silva JA, Derecho DV, Leong GB, Weinstock R, Ferrari MM (2001) A classification of psychological factors leading to violent behavior in posttraumatic stress disorder. J Forensic Sci 46(2):309–316

Skeem J, Johansson P, Andershed H, Kerr M, Louden JE (2007) Two subtypes of psychopathic violent offenders that parallel primary and secondary variants. J Abnorm Psychol 116(2):395. doi:10.1037/0021-843X.116.2.395

Stanford MS, Houston RJ, Mathias CW, Villemarette-Pittman NR, Helfritz LE, Conklin SM (2003) Characterizing aggressive behavior. Assessment 10(2):183–190. doi:10.1177/1073191103010002009

Stone VE, Baron-Cohen S, Calder A, Keane J, Young A (2003) Acquired theory of mind impairments in individuals with bilateral amygdala lesions. Neuropsychologia 41(2):209–220. doi:10.1016/S0028-3932(02)00151-3

Sutton SK, Vitale JE, Newman JP (2002) Emotion among women with psychopathy during picture perception. J Abnorm Psychol 111(4):610. doi:10.1037//0021-843x.111.4.610

Taylor HG, Alden J (1997) Age-related differences in outcomes following childhood brain insults: an introduction and overview. J Int Neuropsychol Soc 3(06):555–567

Van Elst LT, Woermann F, Lemieux L, Thompson P, Trimble M (2000) Affective aggression in patients with temporal lobe epilepsy A quantitative MRI study of the amygdala. Brain : J. Neurol 123(2):234–243. doi:10.1093/brain/123.2.234

van Honk J, Schutter DJLG (2006) Unmasking feigned sanity: a neurobiological model of emotion processing in primary psychopathy. Cogn Neuropsychiatry 11(3):285–306. doi:10.1080/13546800500233728

Vanman EJ, Mejia VY, Dawson ME, Schell AM, Raine A (2003) Modification of the startle reflex in a community sample: do one or two dimensions of psychopathy underlie emotional processing? Personality Individ Differ 35(8):2007–2021. doi:10.1016/S0191-8869(03)00052-7

Veit R, Flor H, Erb M, Hermann C, Lotze M, Grodd W, Birbaumer N (2002) Brain circuits involved in emotional learning in antisocial behavior and social phobia in humans. Neurosci Lett 328(3):233–236. doi:10.1016/S0304-3940(02)00519-0

Verona E, Patrick CJ, Curtin JJ, Bradley MM, Lang PJ (2004) Psychopathy and physiological response to emotionally evocative sounds. J Abnorm Psychol 113(1):99–108. doi:10.1037/0021-843X.113.1.99

Vitacco MJ, Vincent GM (2006) Understanding the downward extension of psychopathy to youth: implications for risk assessment and juvenile justice. Int J Forensic Ment Health 5(1):29–38. doi:10.1080/14999013.2006.10471228

Wahlund K, Kristiansson M (2009) Aggression, psychopathy and brain imaging—Review and future recommendations. Int J Law Psychiatry 32(4):266–271. doi:10.1016/j.ijlp.2009.04.007

Warren GC, Clarbour J (2009) Relationship between psychopathy and indirect aggression use in a noncriminal population. Aggressive Behav 35(5):408–421. doi:10.1002/Ab.20317

Widom CS (1977) A methodology for studying noninstitutionalized psychopaths. J Consult Clin Psychol 45(4):674–683. doi:10.1037/0022-006X.45.4.674

Williams D, Mateer CA (1992) Developmental impact of frontal lobe injury in middle childhood. Brain Cogn 20(1):196–204. doi:10.1016/0278-2626(92)90069-X

Williamson S, Hare RD, Wong S (1987) Violence: criminal psychopaths and their victims. Can J Behav Sci/Revue canadienne des sciences du comportement 19(4):454. doi:10.1037/H0080003

Williamson S, Harpur TJ, Hare RD (1991) Abnormal processing of affective words by Psychopaths. Psychophysiology 28(3):260–273. doi:10.1111/j.1469-8986.1991.tb02192.x

Woodworth M, Porter S (2002) In cold blood: characteristics of criminal homicides as a function of psychopathy. J Abnorm Psychol 111(3):436. doi:10.1037//0021-843x.111.3.436

Yang Y, Raine A (2009) Prefrontal structural and functional brain imaging findings in antisocial, violent, and psychopathic individuals: a meta-analysis. Psychiatry Res 174(2):81. doi:10.1016/j.pscychresns.2009.03.012

Yang Y, Raine A, Colletti P, Toga AW, Narr KL (2010) Morphological alterations in the prefrontal cortex and the amygdala in unsuccessful psychopaths. J Abnorm Psychol 119(3):546. doi:10.1037/a0019611

Yang Y, Raine A, Colletti P, Toga AW, Narr KL (2011) Abnormal structural correlates of response perseveration in individuals with psychopathy. J Neuropsychiatry Clin Neurosci 23(1):107–110. doi:10.1176/appi.neuropsych.23.1.107

Antisocial and Callous Behaviour in Children

Essi Viding, Ana Seara-Cardoso and Eamon J. McCrory

Abstract Antisocial behaviour is one of the most common reasons for a childhood referral to mental health and educational services and represents a substantial public health cost. Callous-unemotional traits can be used to distinguish children who are capable of pre-meditated antisocial behaviour and violence from those whose antisocial behaviour and violence are primarily impulsive and threat reactive. Decades of developmental psychopathology research have shown that children with antisocial behaviour are thus a heterogeneous group and, for interventions to be successful, it is critical that distinct subgroups of children receive services that best match their profile of vulnerabilities and strengths. Recent advances in genetic and brain imaging research in the field have made important contributions to our understanding of the developmental vulnerability that callous-unemotional traits represent. In this chapter, we provide an overview of the current evidence base with regard to genetic and neuroscience findings of callous-unemotional traits and antisocial behaviour with callous-unemotional traits. We also discuss the implications of these findings for prevention and intervention, and finish by outlining what we consider are necessary directions for future research.

Keywords Antisocial behaviour · Callous-unemotional traits · Genetic research · Magnetic resonance imaging research

Contents

E. Viding (✉) · A. Seara-Cardoso · E. J. McCrory
Division of Psychology and Language Sciences, University College London, Gower Street, London WC1 6BT, UK
e-mail: e.viding@ucl.ac.uk

Curr Topics Behav Neurosci (2014) 17: 395–419
DOI: 10.1007/7854_2013_266

Published Online: 20 December 2013

1 Introduction

Antisocial behaviour is one of the most common reasons for a childhood referral to mental health and educational services and represents a substantial public health cost (Scott et al. 2001). We know that children with early onset antisocial behaviour are at risk of developing chronic life-course persistent antisocial problems, as well as several other psychiatric and physical health problems (Kim-Cohen et al. 2003; Odgers et al. 2007). It is also evident from decades of developmental psychopathology research that children with antisocial behaviour are a heterogeneous group and, for interventions to be successful, it is critical that distinct subgroups of children receive services that best match their profile of vulnerabilities and strengths (Frick and Viding 2009).

Callous-unemotional (CU) traits (lack of guilt and empathy, as well as shallow affect) can be used to distinguish between children who are capable of pre-meditated antisocial behaviour and violence (high CU-subtype; AB-HCU) and children whose antisocial behaviour and violence are primarily impulsive and threat reactive (low CU-subtype; AB-LCU). Adults with a combination of CU traits and antisocial behaviour are labelled psychopaths within the criminal justice system. The first extension of the psychopathy concept to children was Bowlby's (1946) description of 'affectionless psychopathy'. It mirrored some of the key features introduced in 1941 by Cleckley (1941) in his seminal work. *The Mask of Sanity* (such as lack of responsiveness to suffering of others), but lay forgotten for a long time until the extension of psychopathy construct to children was proposed again by Frick and colleagues (1994). While it would be entirely inappropriate to suggest that children are psychopaths, it is the case that there is a subset of children with severe conduct problems[1] *and* CU traits that place them at heightened risk for developing adult psychopathy (Lynam et al. 2007). CU traits have been recently introduced as a diagnostic specifier for conduct disorder, termed 'limited prosocial emotions' in the fifth edition of the Diagnostic and Statistical Manual of the American Psychiatric Association (Frick and Moffitt 2010; Sherer and Nickerson 2010).

Longitudinal data show that children with AB-HCU present with a more severe behavioural profile and more long-term problems than children with AB-LCU (Fontaine et al. 2011; Frick and Viding 2009). Furthermore, CU traits have been shown to be associated with overt aggression, delinquency and behavioural dysregulation (Lau and Marsee 2012). Even in the absence of AB, CU traits are associated with poorer outcomes, including risk for developing delinquent behaviours and other types of psychosocial maladjustment (e.g. Barker et al. 2011; Frick et al. 2003; Kumsta et al. 2012; Rowe et al. 2010). Longitudinal data also show that CU traits add to the prediction of serious and persistent criminal behaviour (Pardini and Fite 2010). In short, the affective characteristics of psychopathy (CU traits) can be delineated in children and may be a risk index for later psychopathy, as well as other forms of poor outcome.

Cognitive experimental data suggest that children with AB-HCU are poor at modulating their behaviour in response to punishment in conditioning paradigms (Frick and Viding 2009). In addition, they have difficulties in recognising others' fearful and sad facial expressions and vocal tones (Blair and Viding 2008), show impaired/reduced affective empathy responses to other people distress (De Wied et al. 2012; Schwenck et al. 2012) and report lower levels of empathic concern and sadness in response to other's distress (Pardini and Byrd 2012). These data largely agree with those found in adults with psychopathy, as reviewed elsewhere in this volume. In contrast, AB-LCU is associated with an exaggerated affective response to perceived social threat, such as anger or in some cases even ambiguous, neutral expressions (Dadds et al. 2006; Frick and Viding 2009).

Recent findings suggest that children with AB-HCU may also have an impoverished personal experience of fear and guilt, which could in part explain why they have such difficulty perceiving others' distress (Jones et al. 2010; Marsh et al. 2011b). Children with AB-LCU report comparable experiences of fear and guilt to typically developing children. Interestingly, Jones et al. (2010) found that neither group of children with antisocial behaviour has difficulties in 'mentalising' (perceiving the thoughts and intentions of other people). It is possible therefore that the difficulties that children with AB-HCU exhibit are limited to 'feeling what others feel' and do not extend to difficulties commonly seen in children with autism spectrum disorders, i.e., 'knowing what others think'. This pattern of difficulties and strengths may explain why children with AB-HCU are good at manipulating others to their own advantage, even if such behaviour will cause

[1] Please note that the terms antisocial behaviour and conduct problems will be used interchangeably in this review.

distress to somebody else. Theoretical accounts of AB-HCU propose that normal socialisation is disrupted in these children because they do not form adequate associations between their transgressions and punishment outcome/other people's distress (a form of social 'punishment') (Blair and Viding 2008). By contrast, children with AB-LCU are proposed to form 'hostile attribution biases' and to exhibit aggression as a result of living in unstable and threatening environments (Blair and Viding 2008; Dodge and Crick 1990; Frick and Viding 2009). It also seems to be possible that AB-HCU and AB-LCU children show distinct expectations regarding the consequences of their aggressive behaviour. Recent research has shown that children with AB-HCU are less likely to expect that aggression results in victim suffering, are less concerned about it or about punishment or feelings of remorse, and are more likely to expect that it results in peer dominance. In contrast, children with AB-LCU were more likely to expect that attacking aggressive behaviour towards other people would reduce their aversive behaviour (Pardini and Byrd 2012).

2 Aetiology of CU Traits

A number of twin studies have examined the aetiology of CU traits in children and youth. These studies come from the United States, Sweden and United Kingdom. The samples used in these studies vary in size from moderate (398 twin pairs) to large (3,687 twin pairs), represent different age ranges (7–24 years old) and have used a range of instruments that have relied on both self and other (parent or teacher) ratings. In this chapter, we will concentrate specifically on genetically informative data on CU traits or on antisocial behaviour in the presence of CU traits.

The existing studies have reported remarkably consistent results of moderate to strong heritability for CU traits in children and youth. These studies estimate that 40–78 % of variation in CU traits across the population is due to genetic influences (Bezdjian et al. 2011; Blonigen et al. 2005, 2006; Fontaine et al. 2010; Larsson et al. 2006; Taylor et al. 2003; Viding et al. 2007). Our own data from the Twins Early Development Study (TEDS) has also indicated that at age 7 the group difference between those scoring at the high end for CU traits (top 10 %) and other children is also largely driven by genetic influences ($h_g^2 = 0.67$; Viding et al. 2005). These group heritability estimates appear very similar regardless of whether the CU occurs with ($h_g^2 = 0.80$) or without ($h_g^2 = 0.68$) elevated levels of conduct problems (Larsson et al. 2008).

Twin studies are important for documenting the extent to which environmental factors influence individual differences or group differences in CU traits. Shared environment in twin studies refers to environmental factors that make the members of the twin pair more similar than would be expected by genetic relatedness alone (crudely, this can be inferred if non-identical twin similarity exceeds 50 % of identical twin similarity, as would be expected if only genetic influences were driving twin similarity). Non-shared environment in twin studies refers to environmental factors that make members of the twin pair dissimilar to each other (crudely, this can be inferred if identical twin similarity is less than 100 %). Modest shared environmental influences were detected in only few cases for CU (Fontaine et al. 2010; Viding et al. 2007), but our longitudinal data (described more fully below) indicate that such influences may be particularly important for a handful of girls who have stable and high levels of CU. All studies, including our own, have demonstrated that non-shared environmental influences are particularly important for the development of CU. This does not mean that family or neighbourhood environments are not relevant for development of CU. Rather it suggests that environmental risk factors, including those experienced within the family context, are likely to promote differences between members of the same family (non-shared environment in the twin models). The magnitude of the heritability and environmental estimates for CU traits from child and adolescent samples are in line with previous adult twin data on psychopathic personality traits (Blonigen et al. 2003), as well as other personality dimensions (Bouchard Jr and Loehlin 2001).

2.1 Are There Sex Differences in the Aetiology of CU Traits?

Four studies to date (Bezdjian et al. 2011; Fontaine et al. 2010; Larsson et al. 2007; Viding et al. 2010) have incorporated dizygotic opposite-sex twin pairs in their analyses and formally explored the potential role of qualitative sex differences (i.e., different genes and environments influencing phenotypic variation for males and females). None of these studies reported qualitative sex differences for CU traits. A number of studies have also assessed the possibility of quantitative sex differences (i.e. the same genetic and environmental influences affecting males and females to a different degree). Two studies found little evidence of quantitative sex differences for CU traits (Blonigen et al. 2006; Larsson et al. 2006), but there is also some support for a higher heritability of CU for males (Bezdjian et al. 2011; Fontaine et al. 2010; Viding et al. 2007). For instance, using data from 9462 youths from the Twins Early Development Study (TEDS), Fontaine et al. (2010) found that strong heritability ($h^2 = 0.78$) was observed for boys on a stable high CU trajectory (between 7 and 12 years old). Stable and high levels of CU in girls, however, appeared to be almost entirely driven by shared environmental influences ($c^2 = 0.75$). Replication of this finding is needed given the small number of children who followed the stable and high CU trajectory, and even smaller proportion of this already small group who were females (less than 1 % of the total sample).

2.2 Stability of CU Traits: Genetic and Environmental Contributions

A few twin studies to date have explored the genetic and environmental contributions to the stability of CU traits in childhood and adolescence. Blonigen et al. (2006) focused on two time-points 7 years apart, when the twins were 17- and 24 years old. Their results indicate that the heritability of CU traits remained consistent across time and that 58 % of the stability of CU traits was due to genetic influences. This finding indicates that the stability in CU traits is substantially influenced by genetic factors. Using the TEDS sample, Fontaine and colleagues (2010) reported that stable high trajectory of CU in childhood (between 7 and 12 years) was strongly heritable in boys ($h^2 = 0.78$), but not in girls ($h^2 = 0.00$). This finding suggests that, at least in childhood/early adolescence, genetic influences may drive the stability of high levels of CU traits for boys in particular. Forsman et al. (2008) measured CU traits (as well as impulsivity and grandiosity) and examined genetic and environmental contribution to the stability of these traits between ages 16 and 19. The authors focused on a hierarchical model of psychopathic personality in which a higher order general factor substantially explained the variation in the three psychopathic personality dimensions, both in mid- and late adolescence. Results showed that the observed test-retest correlation of the higher order psychopathic personality factor was high ($r = 0.60$). In addition, as much as 90 % of the test-retest correlation was explained by genetic factors. However, they also found evidence for specific genetic stability in CU. 13 % of the unique genetic effects in the CU dimension at age 19 were shared with the corresponding effects at age 16. Thus, their model provides evidence for aetiologic generality (together with other aspects of psychopathic personality) and aetiologic specificity for the stability of CU traits between mid- and late adolescence. It is of note that, for a subset of children, CU traits are malleable in childhood, either increasing or decreasing with age, rather than remaining persistently high or low (see Fontaine et al. 2010). This is likely to reflect environmental factors interacting with genetic risk to either promote or moderate the development of CU traits, leading to increasing or decreasing CU trajectories, respectively (Fontaine et al. 2010). The challenge for researchers and clinicians is to identify the key environmental factors most influential in this regard, and develop interventions that can promote reduction of CU traits thereby reducing the risk of later maladaptive outcomes.

2.3 Aetiological Overlap Between CU Traits and Antisocial Behaviour

Twin models have also been important in exploring the aetiological overlap between CU traits and antisocial behaviour. Multivariate genetic models can be used to estimate the extent of genetic/environmental correlation, which refers to the degree of overlap between genetic/environmental influences on different traits or behaviours. A few studies to date have demonstrated a modest to moderate genetic correlation between CU traits and antisocial behaviour when co-variation is measured in the whole population (range of $r_g = 0.16–0.57$; Bezdijan et al. 2011; Blonigen et al. 2005; Viding et al. 2007). The genetic overlap may be slightly stronger at the extreme high end of both CU and antisocial behaviour distributions (Viding et al. 2007). Larsson et al. (2007) explored the genetic commonality between three psychopathic personality dimensions (grandiose-manipulative, CU and impulsive-irresponsible dimension) and antisocial behaviour measured at age 13–14 and age 16–17 years. A common genetic factor loaded substantially on both psychopathic personality traits and antisocial behaviour. This was not the case for environmental factors. Forsman et al. (2007) found that externalising behaviour in childhood (age 8–9) was associated with higher levels of psychopathic personality traits in adolescence (age 13–14) among boys but not in girls. Genetic factors were responsible for this association. Another study by Forsman et al. (2010) showed that psychopathic personality in adolescence (16–17) predicted antisocial behaviour in early adulthood (19–20), over and above both concurrent and pre-existing levels of antisocial behaviour. The association between adolescent psychopathic personality and adult antisocial behaviour was mainly explained by genetic effects; a result that can be interpreted as a genetically influenced personality-driven process, where individuals are predisposed to higher risk of involvement in antisocial behaviour because of their psychopathic personality. Finally, a recent study by Bezdjian et al. (2011) demonstrated that CU traits shared genetic influences with both reactive and proactive aggression. The genetic correlation was particularly strong between CU traits and proactive aggression ($r_g = 0.76$).

With regard to environmental influences, modest non-shared environmental correlations have been demonstrated between CU traits and antisocial behaviour/aggression (Bezdijan et al. 2011; Viding et al. 2007). This means that although some child specific environmental factors promote the development of both CU and antisocial behaviour/aggression, the child specific environmental influences for the two constructs are largely independent. We recently conducted a longitudinal monozygotic twin differences study to examine negative parental discipline (e.g., shouting and harsh discipline) as a non-shared environmental factor for CU and antisocial behaviour (Viding et al. 2009). Although negative parental discipline at age 7 had a phenotypic association with both CU traits and antisocial behaviour at age 12, negative parental discipline emerged as a non-shared environmental factor for antisocial behaviour alone. In other words those members of the monozygotic twin pair who received more negative parental discipline at age 7 were also more likely to manifest antisocial behaviours at age 12, even after controlling for baseline differences in the level of antisocial behaviour. This was not true for CU traits and we speculated that the phenotypic association between negative parental discipline and CU traits may reflect the genetic endowment within those families with HCU children, rather than an environmentally driven parenting process that increases risk for CU traits.

2.4 Aetiology of Antisocial Behaviour with and Without CU Traits

Finally, twin studies have also been helpful in exploring the utility of CU traits as a subtyping factor for individuals with antisocial behaviour. Viding et al. (2005) used information from the TEDS sample, to investigate whether the aetiology of teacher-rated antisocial behaviour differs as a function of teacher-rated CU at age 7. The authors separated children with elevated levels of antisocial behaviour (in the top 10 % for the TEDS sample) into two groups based on their CU score (in the top 10 % or not). Antisocial behaviour in children with HCU was under strong genetic influence ($h_g^2 = 0.81$) and no influence of shared environment. In contrast, antisocial behaviour in children without elevated levels of CU showed moderate genetic influence ($h_g^2 = 0.30$) and substantial environmental influence ($c_g^2 = 0.34$, $e_g^2 = 0.26$). Viding et al. (2008) replicated the finding of different heritability estimates for the AB-HCU and AB-LCU groups using the 9-year teacher data from the TEDS study. In addition, they demonstrated that the strong heritability of antisocial behaviour in the AB-HCU group was not driven by co-occurring hyperactivity.

2.5 Molecular Genetic Studies

Despite the substantial literature demonstrating heritable component of CU traits, we know of only a few published molecular genetic studies of child/adolescent CU traits. As no genome-wide significant variants affecting aggression in humans are known (but see below), these concern candidate genes proposed based on their neurochemical effects. The first of these studies was carried out on a relatively small sample of adolescents with ADHD and reported associations between 'emotional dysfunction' scores of psychopathy (CU) and each of the following allelic variants: the val allele of the cathecol-o-methyl-transferase gene; the low-activity allele of monoamine oxidase- A gene (MAOA-L); and the short allele of the serotonin transporter gene (5HTTLPR s) (Fowler et al. 2009). The latter two of these associations were unexpected. Imaging genetic data suggest that MAOA-L and 5HTTLPR s are associated with heightened amygdala activity to emotional stimuli (e.g. Meyer-Lindenberg et al. 2006; Munafò et al. 2008), which is at odds with the reduced amygdala activity to emotional stimuli typically seen in adults with psychopathy and children with AB-HCU (e.g. Birbaumer et al. 2005; Jones et al. 2009; Kiehl et al. 2001; Marsh et al. 2008). It is possible that the findings of Fowler and colleagues are specific to the selected group of adolescents they studied, all of whom had high levels of ADHD symptoms, but relatively low levels of CU traits. A more recent study reported that the long allele of the 5HTTLPR (5HTTLPR l), i.e. the allele conferring low amygdala reactivity, was associated with CU traits in adolescents from low SES backgrounds (Sadeh et al. 2010). This is an extremely interesting finding, as it tentatively suggests that vulnerability to low emotional reactivity may only manifest as high CU traits under disadvantageous socioeconomic conditions. Sadeh et al. (2012) have also reported similar findings in an adult sample of individuals with a forensic/criminal history. Carrying the 5HTTLPR l was associated with the presence of the emotional deficits that characterise the affective factor of psychopathy (i.e. CU traits), however, no moderation by environmental variables was found. Sadeh et al. speculated that this may be due to generally high levels of environmental risk in this particular sample, which would allow for the genotype risk to penetrate and appear as genetic main effect. Sadeh et al. (2012) also found that impulsive and irresponsible lifestyle features of psychopathy were higher among low-activity than high-activity MAO-A carriers.

Two recent studies have also looked at the role of the oxytocin and oxytocin receptor gene variants in relation to CU traits and childhood-onset aggression (Beitchman et al. 2012; Malik et al. 2012). Oxytocin is thought to play an important role in various prosocial behaviours. For example, it has been shown to amplify attachment, human bonding and trust, whilst its dysregulation has been associated with increased aggression (Campbell 2010). Beitchman et al. (2012) genotyped six single-nucleotide polymorphisms (SNPs) in the oxytocin (20p12) and oxytocin receptor genes (3p25) and found that those with the AA genotype of the oxytocin receptor SNP rs237885 had significantly higher CU traits than AC or CC genotype carriers. However, when the same group inspected different oxytocin and oxytocin receptor SNPs, no association with CU traits was found (Malik et al. 2012). Findings with regard to oxytocin genes therefore need replication and refinement.

Finally, new technologies, such as DNA pooling are enabling genome-wide association studies that search for novel single-nucleotide polymorphisms (SNPs), which may be associated with AB-HCU. DNA pooling refers to a genetic screening method that combines DNA from many individuals in a single molecular genetic analysis to generate a representation of allele frequencies. A DNA pool can thus be generated for all cases and all controls and allele frequencies can be compared between these pools. We recently conducted such a study and although no SNPs reached genome-wide significance, there were some potential candidates near neurodevelopmental genes, including ROBO2 (Viding et al. 2010). The association with ROBO2 has been recently replicated in an independent sample, although it appears that the association concerns antisocial behaviour in general, rather than CU specifically (Dadds et al. in press).

Findings from genetic studies could be fruitfully incorporated into imaging genetic investigations of psychopathy. We have recently used twin design to document that aberrant structural brain development in AB-HCU reflects genetic, rather than environmental vulnerability to this behavioural outcome (Rijsdijsk et al. 2010). Investigations with specific genotypes are still pending.

2.6 Summary: Genetic Research

Numerous twin studies from different laboratories suggest that both individual and group differences in CU traits are moderately to strongly heritable. Child specific (non-shared) environmental factors are also important in accounting for individual and group differences in CU traits. However, environmental factors that make children growing up in the same family similar to each other (termed shared environmental factors in twin models) do not typically play a role in individual and group differences in CU traits. Girls with stable high CU traits represent a possible exception, as shared environmental factors appear to be important for the development of CU traits in this group.

Twin studies can go beyond answering questions about relative importance of heritable and environmental influences. To date, such studies have advanced our knowledge about CU traits in several important ways. First, they have demonstrated that stability of CU traits is typically driven by genetic influences (particularly in boys), but also that there is substantial environmental influence that may serve to increase or moderate levels of CU over time, thereby contributing to an increasing or decreasing trajectory of CU trait development. Third, a monozygotic twin differences data suggest that negative parental practices do not act as a child specific environmental risk factor for CU, but may instead reflect genetic vulnerability within families. However, aspects of positive parenting are yet to be investigated within this framework and may be a promising environmental modulator of CU traits. Fourth, a number of studies have shown that CU and antisocial behaviour share a degree of genetic risk and (to a modest extent) child specific environmental risk factors. Finally, antisocial behaviour in the presence (but not in the absence) of CU traits appears strongly heritable, suggesting that CU traits are a useful subtyping index for children with disruptive behaviours.

Research into specific polymorphisms that increase risk of CU traits or AB-HCU is still in its infancy and only a handful of studies have been conducted to date. It is too early to draw firm conclusions from the existing, meagre evidence base, although there are promising leads with regard to genotypes that predispose to low emotional reactivity, as well as those that may affect early neurodevelopment.

3 Neural Correlates of CU Traits

3.1 Functional Magnetic Resonance Imaging Studies

Children with AB-HCU share an affective profile with adult psychopaths showing reduced sensitivity to visual or vocal displays of others' distress and poor modulation of behaviour in response to punishment (Blair and Viding 2008). In line with the behavioural and experimental neuropsychology data, fMRI findings for children with AB-HCU also indicate functional deficits consistent with low emotional responsiveness to others' distress and poor ability to learn from reinforcement information. Aberrant neural functioning (as compared with typically developing children or children with ADHD) has been observed for children with AB-HCU in the amygdala, ventromedial prefrontal cortex (vmPFC), orbitofrontal cortex (OFC), anterior insula and caudate; brain areas that are involved in processing basic emotional salience, reinforcement learning and emotion regulation.

Recent studies have reported reduced amygdala activity in children with AB-HCU as compared with typically developing children or children with ADHD (Jones et al. 2009; Marsh et al. 2008, 2011a; Sebastian et al. 2012; Viding et al. 2012; White et al. 2012). Children with AB-HCU show reduced amygdala response to others' distress (e.g. Jones et al. 2009; Marsh et al. 2008; White et al. 2012) and reduced functional coupling between amygdala and OFC when viewing fearful facial expressions (Marsh et al. 2008). Reduced amygdala function in children with AB-HCU also seems to extend to more complex forms of social cognition, such as categorisation of legal and illegal behaviours in a moral judgment task (Marsh et al. 2011a) or affective theory of mind judgements (Sebastian et al. 2012). Furthermore, CU traits and conduct problems also seem to present differential contributions to amygdala activity in children with AB (Sebastian et al. 2012; Viding et al. 2012). Viding, Sebastian et al. (2012) have demonstrated a differential amygdala activity pattern to preattentively presented fear across the spectrum of CU traits in children with AB. In this study, boys with AB-HCU compared with AB-LCU presented significantly lower amygdala activity to backwardly masked fearful versus calm faces, whilst amygdala activity level in typical developing boys was intermediate between those of the two AB groups. These findings not only indicate that reduced amygdala activation to salient stimuli in children with AB-HCU encompasses even early stages of information processing, but also suggest an affective processing deficit specific to this group. White et al. (2012) have also demonstrated that reduced amygdala reactivity in response to fearful faces was associated with CU, but not with other traits commonly associated with AB such as impulsivity. A recent study from our group (Sebastian et al. 2012) has shown that the unique variance associated with CU traits was related to decreased amygdala activity, while unique variance associated with conduct problems was associated with increased amygdala activity to emotionally salient social scenes. These results clearly suggest that reduced amygdala activation is characteristic of the subgroup of children with conduct problems and HCU rather than of children with conduct problems more generally.

Abnormal vmPFC and OFC response to punishment and reward in adolescents with AB-HCU has also been reported (Finger et al. 2008, 2011). In one study, participants had to choose the 'correct' stimulus from a pair of items. From time to time, the reinforcement associations reversed and a previously rewarded stimulus became unrewarded, while the previously unrewarded stimuli became rewarded. Finger et al. (2008) reported that typically developing children and children with ADHD showed a reduction in vmPFC activity following an unexpected punishment. Such reduction in vmPFC activity has been shown to co-occur with prediction error (Mitchell 2011). In contrast, youth with AB-HCU did not show this reduction in vmPFC activity. In another study, using a passive avoidance paradigm where participants had to learn which stimuli were 'good' (rewarded) and which were 'bad' (unrewarded), Finger et al. (2011) also demonstrated that children with AB-HCU showed less OFC and caudate responsiveness to early stimulus-reinforcement exposure, and less OFC responsiveness to rewards. These neural differences are likely to index compromised sensitivity to early reinforcement information in the OFC and caudate and compromised sensitivity to reward outcome information in the OFC in adolescents with AB-HCU. These fMRI findings in AB-HCU are in line with those typically reported in studies of adult psychopaths (e.g. Birbaumer et al. 2005; Kiehl et al. 2001) and suggest functional neural bases for why individuals with AB-HCU appear unaffected by other people's distress and often make and repeat disadvantageous decisions.

3.2 Structural Magnetic Resonance Imaging Studies

To date, only two studies have reported structural MRI correlates of AB-HCU in children. De Brito et al. (2009, 2011) found that, compared with typically developing boys, boys with AB-HCU had increased grey matter concentration (GMC), in several brain areas implicated in decision-making, moral processing and self-reflection. These included OFC, insula, anterior cingulate cortex (ACC), posterior cingulate cortex (PCC) and superior temporal cortex. Subsequently, De Brito et al. (2011) showed that, compared with typically developing boys, those with AB-HCU exhibited *decreased* white matter concentration (WMC) in a subset of the brain areas where increased GMC was found for boys with AB-HCU, including ACC and superior temporal cortex. These findings indicate that children with AB-HCU are characterised by atypical neural structure in many of the same areas where grey and white matter abnormalities have also been reported in adults with psychopathy (Gao et al. 2009; Yang and Raine 2009). However, somewhat puzzlingly, the direction of the effect (at least for grey matter) is different in the child, as compared with adult studies. It is of note that De Brito et al. (2009, 2011) studied children who were between 10 and 13 years of age. Recent brain imaging evidence in normative samples suggests that grey matter decreases and white matter increases in several of the brain areas implicated in AB-HCU during this period of early adolescence (Gogtay et al. 2004). This is contrary to the pattern observed for the 10–13-year-old boys with AB-HCU, possibly indicative of aberrant brain maturation for this group in early adolescence. More recently, Sumich et al. (2012) have found that AB-HCU boys do not show the typical maturational pattern of decline in the event-related potential component N200 amplitude in midline frontal and temporal electrodes, further evidencing a possible atypical maturation for this group of boys. These data are not necessarily at odds with the findings from adult studies, which clearly represent a very different developmental stage. Distinct developmental disorders can follow markedly different patterns of structural brain development (Shaw et al. 2010) and future longitudinal studies should probe the exact developmental pattern characteristic of AB-HCU.

A number of functional and structural MRI studies have focused on children with antisocial behaviour without subtyping on CU traits (e.g. see reviews of Sterzer and Stadler 2009 and Rubia 2011; see also Passamonti et al. 2010; Rubia et al. 2009; Fairchild et al. 2011), but these are difficult to interpret in the context of AB-HCU as the relative composition of those individuals who are high versus low on CU traits is unclear.

3.3 Summary: Magnetic Resonance Imaging Research

The sparse magnetic resonance imaging evidence base suggests that AB-HCU is associated with atypical patterns of brain structure and function, particularly in the areas critical for affective processing, affective decision-making and moral emotions. The findings are broadly in line with those reported in studies of adult psychopaths and suggest neural bases for the types of traits and behaviours associated with AB-HCU. Specifically, children with AB-HCU show lower amygdala reactivity to fearful faces than typically developing children or children with ADHD. They also show abnormal activity in vmPFC, OFC and caudate; these are areas associated with prediction error and monitoring of reward outcomes during affective decision-making and reversal learning.

4 Integrating Genetic and Neuroimaging Findings

Genetic vulnerability to HCU may contribute to some of the neural vulnerabilities characteristic of AB-HCU and integration of genetic and neuroimaging approaches may yield novel information important for understanding the development of psychopathy. We recently conducted a twin study that investigated whether GMC on those areas that differentiated children with AB-HCU from typically developing children were heritable (Rijsdijsk et al. 2010). We also investigated whether common genetic influences were important for GMC and AB-HCU status and whether such common genetic influences were responsible for the phenotypic association between higher GMC and AB-HCU. Of the brain areas that showed group differences and heritable influences, left PCC, right dACC shared a moderate degree of genetic influences with AB-HCU. These common genetic influences were entirely responsible for the phenotypic association between GMC and AB-HCU. These findings provide preliminary evidence that left PCC and right dACC may constitute intermediate phenotypes for the development of psychopathy. Both left PCC and right dACC are involved in empathy for pain, moral judgments and self-referential thinking (including judgments and obligations), and as such represent logical intermediate phenotype markers for the development psychopathy. It is premature to speculate regarding the putative mechanism by which genes could increase GMC and in turn lead to increased levels of AB-HCU. However, future imaging genetic studies could investigate the role of specific neurodevelopmental genes in explaining GMC differences between AB-HCU and typically developing children. It is also of interest to investigate genetic contributions to functional brain differences using both twin and candidate gene imaging approaches. For example, it would be important to explore the role of 5HTLPPR in modulating neural activity to emotional stimuli in children with high levels of CU traits and different levels of social adversity.

5 Treatment Implications

Conceptually, children with AB-HCU would be hypothesised to respond differentially to treatment (as compared with AB-LCU) given their distinct pattern of aetiological and neurocognitive vulnerability. For example, we have considered evidence that these children have genetic vulnerability to antisocial behaviour and show functional and structural neural abnormalities that may predispose them to deviant development of emotion processing, reinforcement learning and empathy. Only relatively recently have studies begun to examine how treatment outcome might vary in relation to a child's level of CU traits, and whether the efficacy of treatment strategies varies depending on level of CU traits.

Several parenting studies by Mark Dadds, David Hawes and colleagues have provided preliminary evidence for two conclusions regarding the nature of CU traits in preadolescent children. *Firstly, children with high levels of CU traits appear to respond less well to some typical parenting interventions.* Following a 10-week standardised parenting programme for children aged 3–8 years of age with conduct problems it was found that high CU traits uniquely predicted poor response to treatment, even after controlling for family characteristics (Hawes and Dadds 2005). Observational data relating to the parent–child interactions at home indicated that HCU children differed only in their response to the 'time-out' procedure. This suggests differential responsiveness to a traditional treatment component, consistent with the evidence from the experimental literature that these children may be relatively punishment insensitive (Frick and Viding 2009). *Second, high levels of CU traits appear malleable in a subset of children.* In a subsequent study, Hawes and Dadds (2007) examined the malleability and stability of CU traits in this sample including 6-month follow-up. They found that CU scores in a subset of the sample dropped significantly following treatment. While this change may reflect problems of measuring CU traits accurately (e.g. parents over-reporting these traits at presentation), it seems likely that the effect is consistent with a genuine malleability of CU traits in some children. This finding is in line with that reported for older preadolescent children (Pardini et al. 2007). A sample of 120 aggressive children in the fifth grade were followed over a 1-year period; those exposed to lower levels of physical punishment and reporting greater levels of parental warmth and involvement showed decreases in CU traits over time. *Third, while coercive parenting seems to particularly exacerbate conduct problems in boys with low levels of CU traits, parental warmth appears to particularly ameliorate conduct problems in boys with high levels of CU traits* (Pasalich et al. 2011). In this study of 95 clinic referred boys aged 2–12 years, Pasalich et al. used direct observations of family interaction and a Five-Minute Speech Sample (rather than questionnaires) to measure coercive parenting and parental warmth respectively. These findings add empirical weight to existing evidence that parental warmth and involvement is associated with changes in levels of CU traits over time (Pardini et al. 2007). In particular, they suggest that the affective quality of the parent–child relationship (as opposed to discipline) is more important for the socialisation of under-aroused children who have blunted emotional responses by virtue of their high CU traits (Kochanska 1997; Pasalich et al. 2011). How to effectively promote the affective quality of the parent–child relationship should be at the forefront of studies focusing on improvement of treatment efficacy.

These findings are at least suggestive that parenting practices, which are amenable to change through intervention, may influence levels of CU traits over time. However, the nature of any relationship between parenting and CU traits is likely to be complex and may involve bidirectional influences. A recent study by Hawes et al. (2011) examined the relationship between CU traits and parenting in a large community sample of children aged between 3 and 10 years of age. CU traits predicted change in relation to inconsistent discipline, corporal punishment and parental involvement with the latter two factors moderated by child age and sex. It is possible, therefore, that to some degree CU traits may serve to elicit escalating levels of harsh and inconsistent discipline by parents; however, it remains possible that child driven effects on negative parenting are primarily attributable to conduct problems, and not CU traits (Larsson et al. 2008; Pasalich et al. 2011).

A separate line of research that may have treatment implications has investigated the atypical pattern of eye gaze and eye contact in children with high CU traits. As outlined earlier in this paper, children with high levels of CU traits have deficits in fear recognition and do not automatically orient to eye-region of the face (Dadds et al. 2006; Frick and Viding 2009; Sylvers et al. 2011). However, when these children were asked to 'look at the eyes' of the stimulus faces, the recognition deficits disappeared (Dadds et al. 2006; Dadds and Rhodes 2008). It has also been shown that children, aged between 5 and 15 years of age with high CU traits show consistent impairments in their eye contact with their parents (Dadds et al. 2011). This finding has been replicated in a younger sample in a task requiring a 'loving' interaction between the child and their mother (Dadds et al. 2011). Understanding deficits in these domains is potentially critical to intervention in two ways. First, deficits in basic social interaction processes may be relevant in informing targets for parenting and child-based interventions. Second, given that most psychologically based intervention relies on establishing an effective therapeutic alliance it will be important to reconsider traditional approaches to establishing rapport when engaging therapeutically with children with high CU traits. In a recent study, Dadds et al. (Dadds et al. 2012) report the findings of randomised control trial of emotion recognition training (ERT) versus treatment as usual in a group of children (n = 195) referred for a range of behavioural and emotional problems. While ERT did not have an effect on the group as a whole, children with higher levels of CU traits (even those without a diagnosis of conduct disorder) showed more improvement in affective empathy and reduced conduct problems in response to ERT. Strikingly, ERT served to reduce level of conduct problems compared to baseline in children with high CU traits, while TAU (treatment-as-usual) was associated with an increase in conduct problems in this group. These preliminary findings serve to underscore the heterogeneous nature of treatment response in children presenting with conduct problems and provides a promising basis to develop tailored treatment strategies for children with higher levels of CU traits.

Increasingly, behavioural interventions are also being delivered in school settings (Viding et al. 2011). Given that children with AB-HCU often come from families characterised by multiple difficulties (where parents are also likely to have genetic and neurocognitive vulnerabilities), school settings may provide an important context for delivering consistent intervention. One of the most commonly studied problem behaviours in school setting is bullying, and both CU traits and a combination of CU traits and antisocial behaviour are associated with increased rates of bullying behaviour (e.g. Crapanzano et al. 2011; Muñoz et al. 2011; Viding et al. 2009). Neuroscience findings (e.g. Finger et al. 2011; Marsh et al. 2008; Sebastian et al. 2012) indicate there is a growing case for developing tailored approaches to reduce bullying behaviour in children with AB-HCU because they are unlikely to be well served by the most commonly implemented approaches (Smith et al. 2004). For example, educative approaches often aim to elicit empathy in the bully and focus on the distress they cause other children as a means to engender in them a motivation to change. The second commonly used intervention approach, exemplified by 'zero-tolerance' policies, is essentially punitive and involves exclusion from school and other high-level disciplinary sanctions. The research we have reviewed above indicates that neither of the above approaches is likely to work as effectively in children with callous-unemotional conduct problems because they have difficulties empathising and are less responsive to punishment. A more successful approach could involve, for example, the establishment of a system of rewards for behaviour incompatible with bullying; in parallel, there would be a need to ensure that rewards for bullying behaviour (e.g. gaining peer dominance, status and goods) were minimised, for example through close supervision by adults or peer mentors. Currently, however, schools rarely implement systemic and peer-supported approaches to bullying (Sherer and Nickerson 2010).

Genetic and neuroscience research is helping to inform a model of developmental vulnerability to adult psychopathy. Although this research suggests that children with high CU traits or AB-HCU are genetically and neurocognitively vulnerable, existing research also strongly endorses that CU traits are malleable during childhood. Longitudinal, genetically informative data suggest that environmental influences account for a substantial proportion of variance in CU traits and are also important for change in these traits. Phenotypic data has demonstrated that positive parenting and parental involvement can reduce CU traits over time (Hawes et al. 2011). These data are in line with the notion that despite genetic risk for AB-HCU (or later psychopathy), there are no genes that directly code for psychopathic behaviour. Genes code for proteins that influence characteristics such as neurocognitive vulnerabilities that may in turn increase risk for developing psychopathy. This risk may only manifest itself under unfavourable environmental circumstances and genetic variants implicated in CU and AB-HCU are likely to confer advantages, as well as disadvantages, depending on the environmental context.

A particularly important implication from the neuroscience research is that interventions for children with AB-HCU may need to be tailored such that they take into account their distinct pattern of neurocognitive vulnerability. Specifically, it may be fruitful to avoid punishment-oriented or explicit empathy induction strategies. Preliminary evidence from behavioural and treatment studies suggests that enhancing positive parenting and parental involvement, as well as applying consistent rewards may represent promising foci for future treatment research.

6 Future Agenda

6.1 Aetiology of CU Traits

Genetic research is likely to advance greatly in the coming decade, including novel epigenetic approaches that may help us uncover mechanisms of gene environment interaction or studies of rare copy number variants that may affect smaller subsets of individuals at risk for developing psychopathy. It is useful to consider the following pointers with regard to current, but in particular future molecular genetic investigations. *First, there are no genes for psychopathy.* Genes code for proteins that influence characteristics such as neurocognitive vulnerabilities that may in turn increase risk for developing psychopathy. *Second, genetic risk for psychopathy may only manifest itself under unfavourable environmental circumstances* (e.g. Sadeh et al. 2010) and genetic variants implicated in CU and AB-HCU are likely to include several common polymorphisms that confer advantages, as well as disadvantages, depending on the environmental context. *Third, we know that the neurocognitive vulnerabilities associated with psychopathy (or risk for development of psychopathy) are at least partially distinct from those associated with antisocial behaviour in general.* This suggests that the risk alleles for psychopathy may not be the same as risk alleles as those for antisocial behaviour in the absence of CU traits. (See Glenn (2011) for a review of this in relation to 5-HTTLPR and

Viding and Jones (2008) for a more general proposal of differential genetic vulnerability in AB-HCU and AB-LCU). The study by Sadeh et al. (2010) is in line with this possibility, as it demonstrated not only that the long allele of the 5-HTTLPR predisposed individuals to CU traits in low SES environments, but also that the short allele of the same gene predisposed individuals to impulsive antisocial behaviour. Buckholtz and Meyer-Lindenberg (2008) have also speculated that MAOA-L allele, which has received a lot of attention as a risk allele for antisocial behaviour (see also chapter by Leucht et al.), may predispose to threat reactive and impulsive, rather than psychopathic antisocial behaviour. The MAOA-L genotype is associated with a pattern of hyper-reactivity of emotion processing areas of the brain (Meyer-Lindenberg et al. 2006), which is in direct contrast to the pattern typically reported for individuals with psychopathy (e.g. Birbaumer et al. 2005; Kiehl et al. 2001). It is interesting to note that some studies have reported increased vulnerability to antisocial behaviour in the presence of the MAOA-H allele (e.g. Manuck et al. 2000). These may reflect false positive findings, but could also reflect the relative composition of HCU and LCU individuals in different studies.

6.2 Neural Correlates of CU Traits

Unfortunately only one of the neuroimaging studies to date have involved an explicit comparison of children with AB-HCU and AB-LCU (Viding et al. 2012) and only one other study has examined the independent contributions of AB and CU to neural activity (Sebastian et al. 2012). Such studies are clearly needed in the future.

In addition, extant studies have only utilised a limited number of paradigms. To date these have included: simple gender decision task when viewing facial stimuli of emotional and neutral content; passive avoidance learning; reversal learning, affective and non-affective theory of mind cartoons, and categorisation of legal and illegal behaviours. In the future, it would be of interest to investigate neural responses in children with AB-HCU to stimuli related to empathy, morality and emotion regulation. It would also be of interest to assess functional and structural brain development longitudinally, including the identification of possible ‘brain biomarkers’ that might predict future behavioural outcomes. Combining imaging methodologies, with genotyping and study of environmental risk is similarly likely to prove informative, particularly as we try to understand multifinality of outcomes for children with AB-HCU. Although this group of children are at an increased risk of developing psychopathy, not all of them do so. By combining different levels of analyses longitudinally, we will be able to develop an integrated model of AB and its relation to CU traits that can better inform approaches to prevention and intervention.

6.3 Development of Psychopathy: Where do we go from Here?

Integrating information across multiple levels of analyses and combining different methodologies within a single study, while keeping in mind multiple possible developmental pathways to antisocial behaviour (equifinality) and different possible outcomes following childhood CU traits (multifinality) (Cicchetti and Rogosch 1996) is important if the field of developmental psychopathology in general is to advance. In relation to psychopathy, this approach is already bearing fruit (Blair and Viding 2008; Frick and Viding 2009). The current evidence base suggests that AB is associated with a number of different developmental trajectories, consistent with the notion of equifinality. Different aetiological pathways can lead to high levels of antisocial behaviour; as we have seen children with AB-HCU appear to be the most genetically vulnerable to persistent antisocial behavioural problems and, in some cases, psychopathy. However, the current data also clearly indicate that there are no genes for antisocial behaviour or psychopathy, not even in this group with a stronger genetic predisposition. Rather, variation in genes is likely to code for variation in information/affective processing styles. The information processing style of children with AB-HCU is characterised by low emotional reactivity to others' distress and difficulty in learning from sanctions. We have also reviewed neuroimaging evidence that this group of children show related atypical patterns of neural function and structure. Yet, these patterns are unlikely to be fixed and determinate. Multifinality characterises outcomes for children with AB-HCU as with all children showing antisocial behaviour. For example, twin studies suggest that environmental factors can influence both the level of CU traits and of antisocial behaviour. Data from clinic and community studies is consistent with this notion, and preliminary data indicate that children with AB-HCU may be particularly responsive to warm parenting practices and reward-based strategies.

In the light of the fact that neurobiological research on CU traits and AB-HCU is in its infancy, several important questions remain outstanding. *First, what genes are involved in vulnerability to AB-HCU?* In the coming years, we are likely to discover that some of the polymorphisms we thought were important may simply represent false positive findings and others, which at first sight appeared less intuitive, may represent true genetic risk. *Second, are genetic or neural biomarkers predictive of long-term outcome and treatment response?* Because our diagnostics systems are based on behavioural criteria, we can comfortably predict that there will not be a single basis to any given disorder (genetic or otherwise); but we may be able to isolate specific biomarkers that can provide clues to developmental risk. *Finally, how do both genetic and environmental risk factors pertinent for psychopathy manifest at the neural level across development?* Currently, most research has been cross-sectional in nature. We need a much better picture of how atypical patterns of neural function develop over time in children with AB-HCU if we are to tease apart which effects reflect developmental immaturity or delay, and which reflect an abnormal pattern of development.

Addressing these questions effectively represents a challenge to basic science research, but further advances in this field will help us better understand how risk factors in childhood may lead to the development of psychopathy in adulthood. We believe that a better delineation of (even putative) causal mechanisms has the potential to inform more effective approaches to prevention and treatment for at-risk children.

Acknowledgements Essi Viding and Eamon McCrory received grant support from the Economic and Social Research Council (RES-062-23- 2202) when writing this chapter. Ana Seara-Cardoso received Ph.D. funding from the Portuguese Foundation for Science and Technology (Fundação para a Ciência e Tecnologia; SFRH/BD/60279/2009).

References

Barker ED, Oliver BR, Viding E, Salekin RT, Maughan B (2011) The impact of prenatal maternal risk, fearless temperament and early parenting on adolescent callous-unemotional traits: a 14-year longitudinal investigation. J Child Psychol Psychiatry 52(8):878–888

Beitchman JH, Zai CC, Muir K, Berall L, Nowrouzi B, Choi E, Kennedy JL (2012) Childhood aggression, callous-unemotional traits and oxytocin genes. Eur Child Adolesc Psychiatry 21(3):125–132

Bezdjian S, Raine A, Baker LA, Lynam DR (2011) Psychopathic personality in children: genetic and environmental contributions. Psychol Med 41(3):589–600

Birbaumer N, Veit R, Lotze M, Erb M, Hermann C, Grodd W, Flor H (2005) Deficient fear conditioning in psychopathy: a functional magnetic resonance imaging study. Arch Gen Psychiatry 62(7):799–805

Blair RJR, Viding E (2008) Psychopathy. Rutter's Child Adolesc Psychiatry :852–863

Blonigen DM, Carlson SR, Krueger RF, Patrick CJ (2003) A twin study of self-reported psychopathic personality traits. Personality Individ Differ 35(1):179–197

Blonigen DM, Hicks BM, Krueger RF, Patrick CJ, Iacono WG (2005) Psychopathic personality traits: heritability and genetic overlap with internalizing and externalizing psychopathology. Psychol Med 35(5):637–648

Blonigen DM, Hicks BM, Krueger RF, Patrick CJ, Iacono WG (2006) Continuity and change in psychopathic traits as measured via normal-range personality: a longitudinal-biometric study. J Abnorm Psychol 115(1):85–95

Bouchard TJ Jr, Loehlin JC (2001) Genes, evolution, and personality. Behav Genet 31(3):243–273

Bowlby J (1946) Forty-four juvenile thieves: their home life. Bailliere, Tindall & Cox, London

Buckholtz JW, Meyer-Lindenberg A (2008) MAOA and the neurogenetic architecture of human aggression. Trends Neurosci 31(3):120–129

Campbell A (2010) Oxytocin and human social behavior. Pers Soc Psychol Rev 14(3):281–295

Cicchetti D, Rogosch FA (1996) Equifinality and multifinality in developmental psychopathology. Dev Psychopathol 8(4):597–600

Cleckley H (1941) The mask of insanity, 1st edn. Mosby, St Louis

Crapanzano AM, Frick PJ, Childs K, Terranova AM (2011) Gender differences in the assessment, stability, and correlates to bullying roles in middle school children. Behav Sci Law 29(5):677–694

Dadds MR, Cauchi AJ, Wimalaweera S, Hawes DJ, Brennan J (2012) Outcomes, moderators, and mediators of empathic-emotion recognition training for complex conduct problems in childhood. Psychiatry Res 199(3):201–207

Dadds MR, Jambrak J, Pasalich D, Hawes DJ, Brennan J (2011) Impaired attention to the eyes of attachment figures and the developmental origins of psychopathy. J Child Psychol Psychiatry 52(3):238–245
Dadds MR, Moul C, Cauchi A, Hawes DJ, Brennan J (in press) Replication of a ROBO2 polymorphism associated with conduct problems but not psychopathic tendencies in children. Psychiatric Genetics
Dadds MR, Perry Y, Hawes DJ, Merz S, Riddell AC, Haines DJ, Solak E, Abeygunawardane AI (2006) Attention to the eyes and fear-recognition deficits in child psychopathy. Br J Psychiatry 189:280–281
Dadds MR, Rhodes T (2008) Aggression in young children with concurrent callous-unemotional traits: can the neurosciences inform progress and innovation in treatment approaches? Philos Trans R Soc B: Biol Sci 363(1503):2567–2576
De Brito SA, Mechelli A, Wilke M, Laurens KR, Jones AP, Barker GJ, Hodgins S, Viding E (2009) Size matters: increased grey matter in boys with conduct problems and callousunemotional traits. Brain 132(4):843–852
De Brito SA, McCrory EJ, Mechelli A, Wilke M, Jones AP, Hodgins S, Viding E (2011) Small, but not perfectly formed: decreased white matter concentration in boys with psychopathic tendencies. Mol Psychiatry 16:476–477
De Wied M, Van Boxtel A, Matthys W, Meeus W (2012) Verbal, facial and autonomic responses to empathy-eliciting film clips by disruptive male adolescents with high versus low callous-unemotional traits. J Abnorm Child Psychol 40(2):211–223
Dodge KA, Crick NR (1990) Social information-processing bases of aggressive behavior in children. Pers Soc Psychol Bull 16(1):8–22
Fairchild G, Passamonti L, Hurford G, Hagan CC, Von Dem Hagen EAH, Van Goozen SHM, Goodyer IM, Calder AJ (2011) Brain structure abnormalities in early-onset and adolescent-onset conduct disorder. Am J Psychiatry 168(6):624–633
Finger EC, Marsh AA, Blair KS, Reid ME, Sims C, Ng P, Pine DS, Blair RJR (2011) Disrupted reinforcement signaling in the orbitofrontal cortex and caudate in youths with conduct disorder or oppositional defiant disorder and a high level of psychopathic traits. Am J Psychiatry 168(2):152–162
Finger EC, Marsh AA, Mitchell DG, Reid ME, Sims C, Budhani S, Kosson DS, Chen G, Towbin KE, Leibenluft E, Pine DS, Blair JR (2008) Abnormal ventromedial prefrontal cortex function in children with psychopathic traits during reversal learning. Arch Gen Psychiatry 65(5):586–594
Fontaine NMG, McCrory EJP, Boivin M, Moffitt TE, Viding E (2011) Predictors and outcomes of joint trajectories of callous-unemotional traits and conduct problems in childhood. J Abnorm Psychol 120(3):730–742
Fontaine NMG, Rijsdijk FV, McCrory EJP, Viding E (2010) Etiology of different developmental trajectories of callous-unemotional traits. J Am Acad Child Adolesc Psychiatry 49(7):656–664
Forsman M, Larrson H, Andershed H, Lichtenstein P (2007) Persistent disruptive childhood behavior and psychopathic personality in adolescence: a twin study. Br J Dev Psychol 25: 383–398
Forsman M, Lichtenstein P, Andershed H, Larsson H (2008) Genetic effects explain the stability of psychopathic personality from mid- to late adolescence. J Abnorm Psychol 117(3):606–617
Forsman M, Lichtenstein P, Andershed H, Larsson H (2010) A longitudinal twin study of the direction of effects between psychopathic personality and antisocial behaviour. J Child Psychol Psychiatry 51(1):39–47
Fowler T, Langley K, Rice F, Van Den Bree MBM, Ross K, Wilkinson LS, Owen MJ, O'Donovan MC, Thapar A (2009) Psychopathy trait scores in adolescents with childhood ADHD: the contribution of genotypes affecting MAOA, 5HTT and COMT activity. Psychiatr Genet 19(6):312–319
Frick PJ, Cornell AH, Barry CT, Bodin SD, Dane HE (2003) Callous-unemotional traits and conduct problems in the prediction of conduct problem severity, aggression, and self-report of delinquency. J Abnorm Child Psychol 31(4):457–470

Frick PJ, Lilienfeld SO, Ellis M, Loney B, Silverthorn P (1999) The association between anxiety and psychopathy dimensions in children. J Abnorm Child Psychol 27(5):383–392

Frick PJ, Moffitt TE (2010) A proposal to the DSM-V childhood disorders and the ADHD and disruptive behavior disorders work groups to include a specifier to the diagnosis of conduct disorder based on the presence of callous-unemotional traits. American Psychiatric Association, Washington, pp 1–36

Frick PJ, O'Brien BS, Wootton JM, McBurnett K (1994) Psychopathy and conduct problems in children. J Abnorm Psychol 103(4):700–707

Frick PJ, Viding E (2009) Antisocial behavior from a developmental psychopathology perspective. Dev Psychopathol 21(4):1111–1131

Gao Y, Glenn AL, Schug RA, Yang Y, Raine A (2009) The neurobiology of psychopathy: a neurodevelopmental perspective. Can J Psychiatry 54(12):813–823

Glenn AL (2011) The other allele: exploring the long allele of the serotonin transporter gene as a potential risk factor for psychopathy: a review of the parallels in findings. Neurosci Biobehav Rev 35(3):612–620

Gogtay N, Giedd JN, Lusk L, Hayashi KM, Greenstein D, Vaituziz AC, Nugent TF III, Herman DH, Clasen LS, Toga AW, Rapoport JL, Thompson PM (2004) Dynamic mapping of human cortical development during childhood through early adulthood. Proc Natl Acad Sci 101:8174–8179

Hawes DJ, Dadds MR (2005) The treatment of conduct problems in children with callous-unemotional traits. J Consult Clin Psychol 73(4):737–741

Hawes DJ, Dadds MR (2007) Stability and malleability of callous-unemotional traits during treatment for childhood conduct problems. J Clin Child Adolesc Psychol 36(3):347–355

Hawes DJ, Dadds MR, Frost ADJ, Hasking PA (2011) Do childhood callous-unemotional traits drive change in parenting practices? J Clin Child Adolesc Psychol 40(4):507–518

Hicks BM, Patrick CJ (2006) Psychopathy and negative emotionality: analyses of suppressor effects reveal distinct relations with emotional distress, fearfulness, and anger-hostility. J Abnorm Psychol 115(2):276–287

Jones AP, Happé FG, Gilbert F, Burnett S, Viding E (2010) Feeling, caring, knowing: different types of empathy deficit in boys with psychopathic tendencies and autism spectrum disorder. J Child Psychol Psychiatry 51(11):1188–1197

Jones AP, Laurens KR, Herba CM, Barker GJ, Viding E (2009) Amygdala hypoactivity to fearful faces in boys with conduct problems and callous-unemotional traits. Am J Psychiatry 166(1): 95–102

Kiehl KA, Smith AM, Hare RD, Mendrek A, Forster BB, Brink J, Liddle PF (2001) Limbic abnormalities in affective processing by criminal psychopaths as revealed by functional magnetic resonance imaging. Biol Psychiatry 50(9):677–684

Kim-Cohen J, Caspi A, Moffitt TE, Harrington H, Milne BJ, Poulton R (2003) Prior juvenile diagnoses in adults with mental disorder: developmental follow-back of a prospective-longitudinal cohort. Arch Gen Psychiatry 60(7):709–717

Kochanska G (1997) Mutually responsive orientation between mothers and their young children: implications for early socialization. Child Dev 68(1):94–112

Kumsta R, Sonuga-Barke E, Rutter M (2012) Adolescent callous-unemotional traits and conduct disorder in adoptees exposed to severe early deprivation. Br J Psychiatry 200(3):197–201

Larsson H, Andershed H, Lichtenstein P (2006) A genetic factor explains most of the variation in the psychopathic personality. J Abnorm Psychol 115(2):221–230

Larsson H, Tuvblad C, Rijsdijk FV, Andershed H, Grann M, Lichtenstein P (2007) A common genetic factor explains the association between psychopathic personality and antisocial behavior. Psychol Med 37(1):15–26

Larsson H, Viding E, Plomin R (2008) Callous-unemotional traits and antisocial behavior: genetic, environmental, and early parenting characteristics. Crim Justice Behav 35(2): 197–211

Lau KSL, Marsee MA (2012) Exploring narcissism, psychopathy, and machiavellianism in youth: examination of associations with antisocial behavior and aggression. J Child Family Stud:1–13

Lynam DR, Derefinko KJ, Caspi A, Loeber R, Stouthamer-Loeber M (2007) The content validity of juvenile psychopathy: an empirical examination. Psychol Assess 19(3):363–367

Malik AI, Zai CC, Abu Z, Nowrouzi B, Beitchman JH (2012) The role of oxytocin and oxytocin receptor gene variants in childhood-onset aggression. Genes Brain Behav 11(5):545–551

Manuck SB, Flory JD, Ferrell RE, Mann JJ, Muldoon MF (2000) A regulatory polymorphism of the monoamine oxidase-A gene may be associated with variability in aggression, impulsivity, and central nervous system serotonergic responsivity. Psychiatry Res 95(1):9–23

Marsh AA, Finger EC, Fowler KA, Jurkowitz ITN, Schechter JC, Yu HH, Pine DS, Blair RJR (2011a) Reduced amygdala-orbitofrontal connectivity during moral judgments in youths with disruptive behavior disorders and psychopathic traits. Psychiatry Res Neuroimaging 194(3):279–286

Marsh AA, Finger EC, Mitchell DGV, Reid ME, Sims C, Kosson DS, Towbin KE, Leibenluft E, Pine DS, Blair RJR (2008) Reduced amygdala response to fearful expressions in children and adolescents with callous-unemotional traits and disruptive behavior disorders. Am J Psychiatry 165(6):712–720

Marsh AA, Finger EC, Schechter JC, Jurkowitz ITN, Reid ME, Blair RJR (2011b) Adolescents with psychopathic traits report reductions in physiological responses to fear. J Child Psychol Psychiatry 52(8):834–841

Meyer-Lindenberg A, Buckholtz JW, Kolachana B, Hariri AR, Pezawas L, Blasi G, Wabnitz A, Honea R, Verchinski B, Callicott JH, Egan M, Mattay V, Weinberger DR (2006) Neural mechanisms of genetic risk for impulsivity and violence in humans. Proc Natl Acad Sci USA 103(16):6269–6274

Mitchell DGV (2011) The nexus between decision making and emotion regulation: a review of convergent neurocognitive substrates. Behav Brain Res 217(1):215–231

Munafò MR, Brown SM, Hariri AR (2008) Serotonin transporter (5-HTTLPR) genotype and amygdala activation: a meta-analysis. Biol Psychiatry 63(9):852–857

Muñoz LC, Qualter P, Padgett G (2011) Empathy and bullying: exploring the influence of callous-unemotional traits. Child Psychiatry Hum Dev 42(2):183–196

Odgers CL, Caspi A, Broadbent JM, Dickson N, Hancox RJ, Harrington H, Poulton R, Sears MR, Thomson WM, Moffitt TE (2007) Prediction of differential adult health burden by conduct problem subtypes in males. Arch Gen Psychiatry 64(4):476–484

Pardini DA, Byrd AL (2012) Perceptions of aggressive conflicts and others' distress in children with callous-unemotional traits: 'I'll show you who's boss, even if you suffer and I get in trouble'. J Child Psychol Psychiatry 53(3):283–291

Pardini DA, Fite PJ (2010) Symptoms of conduct disorder, oppositional defiant disorder, attention-deficit/hyperactivity disorder, and callous-unemotional traits as unique predictors of psychosocial maladjustment in boys: advancing an evidence base for DSM-V. J Am Acad Child Adolesc Psychiatry 49(11):1134–1144

Pardini DA, Lochman JE, Powell N (2007) The development of callous-unemotional traits and antisocial behavior in children: are there shared and/or unique predictors? J Clin Child Adolesc Psychol 36(3):319–333

Pasalich DS, Dadds MR, Hawes DJ, Brennan J (2011) Do callous-unemotional traits moderate the relative importance of parental coercion versus warmth in child conduct problems? An observational study. J Child Psychol Psychiatry 52(12):1308–1315

Passamonti L, Fairchild G, Goodyer IM, Hurford G, Hagan CC, Rowe JB, Calder AJ (2010) Neural abnormalities in early-onset and adolescence-onset conduct disorder. Arch Gen Psychiatry 67(7):729–738

Rijsdijsk FV, Viding E, De Brito S, Forgiarini M, Mechelli A, Jones AP, McCrory E (2010) Heritable variations in gray matter concentration as a potential endophenotype for psychopathic traits. Arch Gen Psychiatry 67(4):406–413

Rowe R, Costello EJ, Angold A, Copeland WE, Maughan B (2010) Developmental pathways in oppositional defiant disorder and conduct disorder. J Abnorm Psychol 119(4):726–738

Rubia K (2011) "Cool" inferior frontostriatal dysfunction in attention-deficit/hyperactivity disorder versus "hot" ventromedial orbitofrontal-limbic dysfunction in conduct disorder: a review. Biol Psychiatry 69(12):69–87

Rubia K, Smith AB, Halari R, Matsukura F, Mohammad M, Taylor E, Brammer MJ (2009) Disorder-specific dissociation of orbitofrontal dysfunction in boys with pure conduct disorder during reward and ventrolateral prefrontal dysfunction in boys with pure ADHD during sustained attention. Am J Psychiatry 166(1):83–94

Sadeh N, Javdani S, Jackson JJ, Reynolds EK, Potenza MN, Gelernter J, Lejuez CW, Verona E (2010) Serotonin transporter gene associations with psychopathic traits in youth vary as a function of socioeconomic resources. J Abnorm Psychol 119(3):604–609

Sadeh N, Javdani S, Verona E (2012) Analysis of monoaminergic genes, childhood abuse, and dimensions of psychopathy. J Abnorm Psychol. Advance online publication

Scott S, Knapp M, Henderson J, Maughan B (2001) Financial cost of social exclusion: follow up study of antisocial children into adulthood. BMJ 323:1–5

Schwenck C, Mergenthaler J, Keller K, Zech J, Salehi S, Taurines R, Romanos M, Schecklmann M, Schneider W, Warnke A, Freitag CM (2012) Empathy in children with autism and conduct disorder: group-specific profiles and developmental aspects. J Child Psychol Psychiatry 53(6):651–659

Sebastian CL, McCrory EJP, Cecil CAM, Lockwood PL, De Brito SA, Fontaine NMG, Viding E (2012) Neural responses to affective and cognitive theory of mind in children with conduct problems and varying levels of callous-unemotional traits. Arch Gen Psychiatry 69(8):814–822

Shaw P, Gogtay N, Rapoport J (2010) Childhood psychiatric disorders as anomalies in neurodevelopmental trajectories. Hum Brain Mapp 31(6):917–925

Sherer YC, Nickerson AB (2010) Anti-bullying practices in American schools: perspectives of school psychologists. Psychol Sch 47(3):217–229

Smith PK, Pepler D, Rigby K (2004) Bullying in schools: how successful can interventions be? Cambridge University Press, Cambridge

Sterzer P, Stadler C (2009) Neuroimaging of aggressive and violent behaviour in children and adolescents. Front Behav Neurosci 3:35

Sumich A, Sarkar S, Hermens DF, Kelesidi K, Taylor E, Rubia K (2012) Electrophysiological correlates of CU traits show abnormal regressive maturation in adolescents with conduct problems. Pers Individ Differ 53(7):862–867

Sylvers PD, Brennan PA, Lilienfeld SO (2011) Psychopathic traits and preattentive threat processing in children: a novel test of the fearlessness hypothesis. Psychol Sci 22(10): 1280–1287

Taylor J, Loney BR, Bobadilla L, Iacono WG, McGue M (2003) Genetic and environmental influences on psychopathy trait dimensions in a community sample of male twins. J Abnorm Child Psychol 31(6):633–645

Viding E, Blair RJR, Moffitt TE, Plomin R (2005) Evidence for substantial genetic risk for psychopathy in 7-years-olds. J Child Psychol Psychiatry 46(6):592–597

Viding E, Fontaine NMG, Oliver BR, Plomin R (2009) Negative parental discipline, conduct problems and callous-unemotional traits: monozygotic twin differences study. Br J Psychiatry 195(5):414–419

Viding E, Frick PJ, Plomin R (2007) Aetiology of the relationship between callous-unemotional traits and conduct problems in childhood. Br J Psychiatry 49:33–38

Viding E, Hanscombe KB, Curtis CJC, Davis OSP, Meaburn EL, Plomin R (2010) In search of genes associated with risk for psychopathic tendencies in children: a two-stage genome-wide association study of pooled DNA. J Child Psychol Psychiatry 51(7):780–788

Viding E, Jones AP (2008) Cognition to genes via the brain in the study of conduct disorder. Q J Exp Psychol 61(1):171–181

Viding E, Jones AP, Frick PJ, Moffitt TE, Plomin R (2008) Heritability of antisocial behaviour at 9: do callous-unemotional traits matter? Dev Sci 11(1):17–22

Viding E, McCrory EJ, Blakemore SJ, Frederickson N (2011) Behavioural problems and bullying at school: can cognitive neuroscience shed new light on an old problem? Trends Cognit Sci 15(7):289–291

Viding E, Sebastian CL, Dadds MR, Lockwood PL, Cecil CAM, De Brito SA, McCrory EJ (2012) Amygdala response to pre-attentive masked fear in children with conduct problems: the role of callous-unemotional traits. Am J Psychiatry 169(10):1109–1116

White SF, Marsh AA, Fowler KA, Schechter JC, Adalio C, Pope K, Sinclair S, Pine DS, Blair RJR (2012) Reduced amygdala response in youths with disruptive behavior disorders and psychopathic traits: decreased emotional response versus increased top-down attention to nonemotional features. Am J Psychiatry 169(7):750–758

Yang Y, Raine A (2009) Prefrontal structural and functional brain imaging findings in antisocial, violent, and psychopathic individuals: a meta-analysis. Psychiatry Res Neuroimaging 174(2):81–88

Aggression in Children and Adolescents

Arlette Buchmann, Sarah Hohmann, Daniel Brandeis, Tobias Banaschewski and Luise Poustka

Abstract Aggressive behaviour is a common phenomenon during childhood and adolescence, but at the same time it is an important associated feature of many psychiatric disorders during this age period. Persistent aggression is related to a variety of negative outcomes in adulthood, including low socioeconomic status and unemployment, criminal behaviour and social isolation. The great heterogeneity of aggressive behaviour still hampers our understanding of causal mechanisms. Still, over the past years, the identification of specific subtypes of aggression has opened possibilities for new and individualized treatment approaches. This article provides information on different subtypes of aggression in children and adolescents, on individual differences that contribute to aggression during development and on possible underlying processes related to aggressive behaviour in young people. Current treatment approaches as well as new emerging treatment possibilities are discussed.

Keywords Aggression · Children · Adolescents · Disruptive behavior · Conduct disorder · Callous unemotional traits

Contents

A. Buchmann · S. Hohmann · D. Brandeis · T. Banaschewski · L. Poustka (✉)
Medical Faculty Mannheim, Department of Child and Adolescent Psychiatry and Psychotherapy, Central Institute of Mental Health, Heidelberg University, J5, 68159 Mannheim, Germany
e-mail: Luise.Poustka@zi-mannheim.de

D. Brandeis
Department of Child and Adolescent Psychiatry and Zurich Center for Integrative Human Physiology, University of Zurich, Zurich, Switzerland

Curr Topics Behav Neurosci (2014) 17: 421–442
DOI: 10.1007/7854_2013_261

Published Online: 24 December 2013

Aggression, usually defined as a behavioural style aiming at deliberately harming other people or objects, is a common phenomenon during early childhood (Nagin and Tremblay 1999). In certain situations, aggression is considered as an adaptive function, but becomes abnormal and maladaptive, when rules limiting it are seriously and frequently broken (Haller et al. 2005). Maladaptive aggression in children is associated with negative impact on social development as well as a variety of negative outcomes (Maughan and Rutter 2001).

Children showing no physical aggression like fighting, kicking and pushing before the age of 3 years is rare; in preschool children, physical aggression has been described as usually linked to frustration, anger, impulsivity and unmet needs, but without intent to hurt others (Wakschlag et al. 2010). While the prevalence of certain forms of aggression (e.g. physical fighting) has been shown to decrease during development (Lahey et al. 1998; Loeber and Hay 1997), the number of non-aggressive and covert conduct problems seems to increase from middle childhood to adolescence (Achenbach et al. 1991; Loeber and Farrington 1998). Childhood onset-disruptive symptoms are associated with an increased risk of continuing and enhanced aggressive behaviour and an overall lower psychosocial adjustment. The beginning before adolescence (age 10 childhood onset) is commonly characterised by more severe aggression and a higher risk for criminal outcomes in adulthood as compared to adolescent onset. Onset after age 16 is rare, and shows a considerably lower risk for a persisting course.

In the majority of children, aggressiveness peaks during the first 3 years of life and then subsequently decreases due to an increase of cognitive and self-control competencies (Connor 2002; Tremblay 2010). In some children, however, aggressive and disruptive behaviours become a persistent and repetitive pattern and are associated with significant impairment and diagnostic criteria of a mental disorder. As there is a considerable overlap of a child's disruptive behaviour and normative misbehaviours, the importance of an age-related specification of disruptive symptoms in early childhood has repeatedly been emphasised. Concerning pathological or age inappropriate aggressive/violent behaviour, there is currently no separate diagnostic category. In child and adolescent psychiatry, aggressive behaviours are usually subsumed under the diagnostic entity of conduct disorders (CD). As epidemiological studies demonstrate, between 5 and 10 % of children in the age range 8–16 years show significant persistent oppositional, disruptive or aggressive behaviour problems (Hill 2002). Boys outnumber girls across different ages 3–4 times. Increased severity and persistence of disruptive symptoms is considered as a precursor of antisocial personality disorder and is strongly associated with long-term negative outcomes such as violence, dropping out of school, unemployment, drug abuse, criminality and social isolation.

1 Subtypes of Aggression in Children and Adolescents

Due to the extensive symptom heterogeneity, the differentiation of several subtypes of aggression has been broadly proposed. In order to obtain a better understanding of the correlates and biological underpinnings of aggressive behaviour, this approach also is promising for the development of optimised individual treatment options.

1.1 Reactive Versus Proactive Aggression

Amongst various concepts of aggression, two major types of aggression have been proposed: (i) reactive or impulsive aggression, defined as an angry, defensive response to threat, frustration or provocation and (ii) proactive or instrumental aggression, defined as a deliberate, coercive behaviour controlled by external reinforcements and used to obtain a desired goal (Crick and Dodge 1996).

The distinction between reactive and proactive aggression is closely associated with the question of the motive or intent behind the aggressive act. Reactive aggression refers to the response to a threatening or frustrating event. It is associated with impulsivity, high arousal and deficient affective control but is not goal directed. According to Dodge et al. (1997) reactive aggression in children may result from an increased tendency to perceive a threatening intent in their counterparts. In contrast, proactive aggression is described as planned and driven by the anticipation of reward (Stadler et al. 2010). In addition, evidence suggests different correlates of these two forms of aggression. While purposeful aggression seems to be linked to CD and antisocial personality disorder, reactive aggression also appears as a common symptom of many other disorders, such as depression and anxiety disorders. Longitudinal twin studies have indicated that the higher stability of proactive aggression over time is largely genetically mediated. In contrast, reactive aggression which has been found to decrease over time was reported to be mainly due to genetic and nonshared environmental influences (Tuvblad et al. 2009).

1.2 Psychopathy and Callous–Unemotional Traits

The concept of psychopathy offers another possibility to differentiate types of aggression and was therefore extended to childhood and adolescence. Proactive or instrumental aggression has often been related to callous–unemotional (CU) traits. This concept comprises the lack of remorse or guilt, lack of empathy and shallow or deficient affect and was suggested as an indicator of psychopathy in adolescents (Raine et al. 2006). Although these traits designate chronic antisocial behaviour and a more severe form of aggression, it has to be emphasised that only a very small percentage of adolescents with CD show characteristics of psychopathy as adults (Frick 2009).

Elevated CU traits have been reported to be highly heritable with only little influence of shared environmental factors. This stands in contrast to the substantial importance of environmental factors for conduct problems without elevated levels of CU. In a recent study, Kumsta et al. (2012) provided evidence that CU traits may also develop as a result of severe early deprivation. However, in their sample of Romanian adoptees with a history of exceptionally severe institutional deprivation, high CU traits were not significantly associated with either CD or ODD, further supporting the distinction of CD and CU traits. A specification of the CD diagnosis concerning CU traits was included in the fifth revision of the DSM due to the clinical relevance and reduced treatment benefit associated with this facet of personality.

2 Aggression in Child and Adolescent Psychiatric Disorders

2.1 Oppositional Defiant and Conduct Disorders

ODD is a recurrent pattern of negativistic, defiant, disobedient and hostile behaviour toward authority figures, which leads to impairment, and CD is a repetitive and persistent pattern of behaviour in which the basic rights of others and major age-appropriate societal norms or rules are violated (American Psychiatric Association 2000). CD may be diagnosed when a child seriously misbehaves with aggressive (or non-aggressive) behaviours against people, animals or property that may be characterised as belligerent, destructive, threatening, physically cruel, deceitful, disobedient or dishonest. This may include stealing, intentional injury and forced sexual activity. The diagnoses of oppositional defiant disorder (ODD) and conduct disorder (CD) follow the ICD-10 criteria (World Health Organization 1993) and DSM-IV-TR (American Psychiatric Association 2000). The principal difference is that, in DSM-IV-TR, ODD and CD are separated, whereas in ICD-10 the criterion set of items required for CD is very close to what would be obtained by combining the DSM-IV-TR ODD and CD items. It is still not clear how valid the distinction is between ODD and CD, because items in each category are clearly age related (Angold and Costello 2001; Hill 2002).

The DSM-IV-TR distinguishes conduct problems based on age of onset (aged 10 or younger vs. 11 or older) of first CD symptoms is supported by various studies. Early onset is a strong predictor of persistence through childhood to adulthood. Early onset antisocial behaviour that persists into adulthood seems to be rather different from later onset syndromes in which the manifestations are mainly evident in adolescence. Children with ODD/CD with early-onset 'life course persistent group' have lower IQ, more attentional and impulsivity problems, poorer scores on neuropsychological deficits, greater peer difficulties and they are more likely to come from adverse family circumstances (Hinshaw et al. 1993; Lahey et al. 1998; Moffitt and Caspi 2001; Rutter et al. 1998). According to ICD-10, conduct disorders are distinguished based on whether dissocial or aggressive behaviour (and not merely oppositional, defiant, disruptive behaviour) is entirely, or almost entirely, confined to the home and/or to interactions with members of the nuclear family or immediate household (conduct disorder confined to the family context), whether or not there is significant pervasive abnormality in the individual's relationships with other children and integration into peer groups (socialised vs. unsocialised conduct disorder) and whether there is a comorbidity with hyperkinetic disorder (hyperkinetic conduct disorder) or emotional disorders (mixed disorders of conduct and emotions).

Evidence suggests that ODD and CD reflect the same genetic liability (Eaves et al. 2000). Furthermore, there is evidence for a subtyping of ODD/CD according to the distinction between overt (confrontational) and covert (concealing) conduct disorder (Achenbach 1991; Fergusson et al. 1994). Particularly, aggression and persistent physical fighting are of prognostic importance for later delinquency (Loeber and Farrington 1998; Loeber and Stouthamer-Loeber 1998; Tremblay et al. 1991; Vitiello and Stoff 1997). Proactive aggression appears more important for later maladjustment than reactive aggression (Dodge 1991; Vitaro et al. 1998). However, compared to children showing proactive aggression, reactive children are more likely to have been physically abused and to have poor peer relationships, to have shown aggression from an earlier age and to suffer from AD/HD (Dodge et al. 1997).

Generally, prognosis of CD is regarded as poor (Lahey et al. 1997). In adolescence and adulthood, children with conduct problems have an increased risk for antisocial behaviours and other psychiatric disorders (Feehan et al. 1994; Offord et al. 1992; Robins 1991). Loeber et al. (1997, 1998) have distinguished three different developmental sequences to conduct disorder and delinquency: (1) an overt pathway, characterised by minor aggression followed by early physical fighting and proceeding to more serious violence; (2) a covert pathway before age 15, starting from minor covert behaviours (lying and stealing) to more serious forms of property damage (fire setting or vandalism); and (3) an authority conflict pathway before age 12, in which oppositional and defiant behaviours and authority avoidance are prominent (truancy, running away, staying out late at night). These pathways have been identified in several independent samples (Elliott 1994; LeBlanc 1997; Loeber et al. 1999; Tolan and Gorman-Smith 1998). However, most individuals in clinic samples follow various pathways (Loeber and Coie 2001).

According to Loeber et al. (2000), there is a heterotypic developmental continuity (changing manifestations of the same disorder) in ODD, CD and antisocial personality, with AD/HD influencing the developmental progression from less serious to more serious manifestations of CD (Burke et al. 2002). However, CD is often preceded and predicted by persistent ODD symptoms (Lahey et al. 1998) and in the absence of ODD ADHD seems to be only a weak predictor of a later onset of CD (Biederman et al. 1996; Loeber et al. 1995; Satterfield and Schell 1997), for reviews see (Hinshaw 1994; Jensen et al. 1997; Loeber and Keenan 1994) and of antisocial behaviour in adulthood (Farrington et al. 1990; Loeber et al. 1995; Magnusson and Bergmann 1990). Generally speaking, the age of onset of first symptoms of conduct disorder has been continually described as a strong predictor of persisting aggressive behaviour through childhood to adulthood. However, the validity of early onset versus late onset cases was questioned by Loeber et al. (2000). Indeed, research on developmental pathways of onset of physical aggression suggests that onset of chronic physical aggression after early childhood seems to be rather the exception (Barker et al. 2007; NICHD Early Child Care Research Network 2004; van Lier et al. 2009). While neurodevelopmental impairments of the child in interaction with psychosocial adversity are well-known risk factors for early onset-disruptive problems, high levels of adverse family circumstances and their interaction are particularly associated with persistent antisocial behaviour (Moffitt and Caspi 2001; Pitzer et al. 2010).

2.2 Attention Deficit Hyperactivity Disorder (ADHD)

Children with ADHD tend to engage in high levels of aggression, and it has been discussed whether aggressive behaviour is part of the psychopathology of ADHD itself rather than a result of comorbid CD (King and Waschbusch 2010). The co-occurrence of ADHD with oppositional defiant and/or conduct disorders (ODD/CD) is the most prevalent comorbidity and poses a significant clinical and public health problem. ODD and/or CD are present in approximately 40–70 % of children with AD/HD. Among children with ODD and/or CD, 40–60 % are estimated to have AD/HD, which is a higher rate than expected by chance. Odds ratios for AD/HD in the presence of conduct disorder are about ten (Angold et al. 1999). Due to the high comorbidity of hyperactivity (ADHD) and conduct disorder (Patterson et al. 2000) the question arose whether one problem behaviour can be differentiated from another. It has been suggested that the higher level of aggressive behaviour in children suffering from ADHD is independent of comorbid conduct problems and may be one expression of increased levels of impulsivity. Emotional impulsivity (or emotional dysregulation) as the reduced control over flush of anger and the tendency to act without forethought, even if that act is harmful to others is associated with an increased risk for aggression in children with ADHD. Both Barkley (2010) and Martel (2009) suggested that emotional dysregulation may be one important factor leading to high rates of aggression in children with ADHD; high comorbidity may reflect the difficulty that the same behaviours can be attributed to different syndromes or represent an increased disease severity. As another explanation, comorbidity was discussed as a separate condition. Factor analytic approaches support the distinction of hyperactive and antisocial problem behaviour. Although hyperactivity in childhood was shown to be associated with adolescent antisocial behaviour, the reversed relationship of pure antisocial behaviour predicting later hyperactivity could not be confirmed. Further support for the distinction of both problems comes from the familial association study implemented by (Faraone et al. 1997). They reported elevated rates of CD and ASPD among relatives of individuals suffering from ADHD and CD, whereas this was not the case in patients with ADHD + ODD or ADHD alone. Patterson et al. (2000) assumed that the correlation of both disorders reflect different points in the progression of a shared underlying process. In their study of families living in a high crime density region, the authors were able to show that the presence of disrupted parental discipline practices accounted for the correlation of hyperactivity and antisocial behaviour in 10-year-old boys. These results suggest disrupted discipline as a causal mechanism underlying the shift from hyperactivity to antisocial behaviour.

2.3 Substance Use Disorders

Together with mood disorders, substance use disorders form one of the most consistently implicated comorbidities in CD with a prevalence of approximately 60 %. Elevated rates of aggressive behaviour have repeatedly been shown to predict both the initiation as well as continuation of substance use in children and adolescents (Swadi 1999). According to self-ratings as well as teacher-ratings, aggression at age 9 was reported to predict substance use disorders at age 21 (Reinherz et al. 2000). Thus, symptoms of CD seem to precede the onset of substance abuse, while childhood ADHD is thought to be less relevant in the prediction of alcohol and drug problems (Mannuzza et al. 1998). Poor behavioural regulation, characterised by deficits in the planning and execution of goal-directed behaviour, was proposed to be a common underlying factor for both antisocial behaviour and substance abuse in adolescence (Cadoret et al. 1995). As childhood disruptive behaviour is a strong predictor of alcohol and drug abuse in adolescence and adulthood, strategies for effective interventions and prevention need to include approaches to reduce aggression and antisocial behaviours (Clark et al. 2002).

2.4 Depression

An impressive body of research highlights the prevalence of comorbidity between CD and depression in children and adolescents. Greene et al. (2002) recently reported on a rate of over 50 % of clinically referred children with CD meeting also criteria for major depression and also in epidemiological samples comorbidity of both disorders is higher than would be expected by chance (Wolff and Ollendick 2006). Several hypotheses exist in order to explain the frequent co-occurence of mood disorders and CD; whether both disorders share common underlying risk factors or one disorder causes the other one by influencing developmental trajectories and increasing the risk for affected individuals, is to date not satisfactorily clarified; related to the association of aggression and depression, the idea of a 'masked depression' suggested that children with less inhibition or behavioural control may externalise their experienced feelings of sadness or irritability, and will falsely appear rather disruptive than depressed; another proposal claimed that cognitive impairments related to depression like increased distractibility and irritability may prevent affected children from thinking properly through negative consequences of their actions and thus increase the risk for disruptive and aggressive behaviour. However, most studies, although not all, report that aggression and conduct problems precede the onset of depressive symptoms (Biederman et al. 1995; Nock et al. 2006). The co-occurence of disruptive behaviour with depression is highly meaningful for its impact on degree of impairment and level of psychopathology in affected youth. Adolescents with disruptive disorder and comorbid depressive disorder were reported to display greater risk for suicide attempts and poorer global functioning, compared to those with disruptive disorder alone (Lewinsohn et al. 1995). Moreover, studies investigating treatment efficacy in depressed adolescents with comorbid CD show poor treatment outcomes in this group for both behavioural treatment as well as anti-depressant medication (Hughes et al. 1990; Rohde et al. 2001). As in substance abuse, interventions specific for this population are needed.

2.5 Disruptive Mood Dysregulation Disorder

Disruptive mood dysregulation disorder (DMDD) is a new proposed diagnosis for DSM-5 which describes a population of children with severe and age inappropriate temper outbursts, such as in the form of verbal rages or physical aggression towards people or property as well as persistently irritable or angry mood. This newly incorporated category into the classification of mental disorders was designed for better characterisation of those children that were frequently misdiagnosed as having a bipolar disorder or 'difficult ADHD'. These patients often posed a problem regarding diagnostic classification and also, inevitably, intervention. The core feature of DMDD is chronic irritability, a common phenomenon in child and adolescent psychiatry, which is a mood symptom referring to easy annoyance and touchiness that can manifest in anger and temper outburst (Stringaris 2011). The incorporation of DMDD into DSM-5 has been an issue of much debate, as for example the specificity of the new disorder was questioned (Axelson et al. 2011). Alternative concepts, like characterising irritability as a dimension rather than as a category, have been proposed (Mikita and Stringaris 2013). DMDD was already reported to be not definitely distinguishable from ODD and CD. However, rationale for the introduction of DMDD was to provide a 'diagnostic home' for children displaying severe non-episodic irritability (Stringaris 2011).

2.6 Autism Spectrum Disorders

Autism spectrum disorders (ASD) are heterogeneous neurodevelopmental disorders characterised by impairments in social interaction and communication as well as restricted, repetitive and stereotyped patterns of behaviour (American Psychiatric Association 2000). Children and adolescents with ASD show significantly elevated rates of self-injurious and aggressive behaviour compared to healthy controls, at least when associated with comorbid mental retardation (Green et al. 2000; McClintock et al. 2003; Tsakanikos et al. 2007). In preschoolers with ASD, externalising and aggressive behaviours were reported to be associated with lower non-verbal IQ, decreased expressive language abilities and lower levels of adaptive functioning (Hartley et al. 2008). In a sample of 1,380 children and adolescents between 4 and 17 years, 35 % of the participants displayed aggressive acts against caregivers or other persons, as reported by their parents (Kanne and Mazurek 2011). These rates were markedly higher than those reported by previous studies in ASD or intellectual disability (Emerson et al. 2001; Hartley et al. 2008; Holden and Gitlesen 2006; Matson and Rivet 2008), suggesting that aggression in ASD is an even more serious problem than previously assumed. Also, motives behind aggressive behaviour seem to differ between children with ASD and non-ASD samples: while typically developing children engage in aggressive acts primarily for social goals, children with ASD seem to display disruptive or aggressive behaviour in order to prevent interference with repetitive behaviours (Reese et al. 2005). Age at onset of aggression in ASD seems to occur typically in early childhood, and may place a significant impairment to later development, as aggressive behaviour is associated with increased risk for, e.g. mood disorders in ASD (Ming et al. 2008) as well as with later medication with anti-psychotics and use of psychiatric services in adulthood (Tsakanikos et al. 2007). Also, individuals with ASD may be overrepresented among violent offenders in adulthood (Soderstrom et al. 2004).

3 Risk factors and Correlates for Aggression and Conduct Disorder

At the individual level, reduced autonomic reactivity is the most frequently replicated correlate of antisocial behaviour in older children and adolescents, similar to findings in adults. The assumption that reduced hypothalamic pituitary adrenal

(HPA) axis activity is a significant characteristic of antisocial behaviour in children was confirmed in a meta-analysis by Hawes et al. (2009). Nevertheless, their results also indicate that decreased cortisol levels are particularly associated with proactive aggression, whereas aggressive children with low levels of CU traits rather show HPA-axis hyperactivity. In accordance with findings in adult samples, reduced responsiveness of the amygdala when exposed to affective stimuli seems likewise to be a particular characteristic for the CU forms of aggressive behaviour in children and adolescents. Hyporeactivity of the amygdala is discussed as marker of impaired functioning of neural networks which are involved in emotion processing in CD (Sterzer et al. 2005), particularly when combined with higher CU traits (Sebastian et al. 2012). In addition, evidence showing relative deactivation of ACC during emotional exposure in aggressive children was suggested to represent an impairment of cognitive functions which are important for emotion regulation (Sterzer et al. 2005). Both impaired recognition of emotional stimuli and deficits in the cognitive control of emotional behaviour may interfere with social learning processes and the inhibition of aggressive behaviour. Also, poor IQ and deficits of executive functioning in children seem to constitute predisposing factors for conduct problems due to decreased self-control and problem-solving competencies (Lynam and Henry 2001). Difficulties in understanding own feelings and signs of distress in others may encourage the use of aggressive behaviours in children with low verbal performance.

3.1 Early Risk Factors

Pre- and perinatal influences, such as maternal smoking and alcohol use during pregnancy and birth complications are well-known risk factors in association with conduct disorder. However, evidence indicates that obstetric factors do not themselves determine the development of aggressive behaviour problems but constitute an increased vulnerability, in particular for the influence of an adverse family environment (Raine 2002). Prenatal environmental risk factors may predispose children for aggressive behaviour problems through their impact on gene expression as suggested by recent epigenetic studies (Tremblay 2010), among other mechanisms. A wide range of adverse family factors, including poverty and parents suffering from psychiatric disorders are strongly related to early onset disrupted behaviour (Pitzer et al. 2010). There is a strong correlation between prenatal risk factors and birth complications with psychosocial adversity and the challenge to disentangle their independent impact is not negligible. Confounding of these factors may result in biased multiple-risk groups and therefore contribute to inconsistent findings (Laucht et al. 2000).

Apart from executive deficits, the observation of biased information processing in aggressive children was assumed to result from repeated experience of physical punishment and neglect. For example, aggressive children were found to show a greater tendency to attribute hostile intentions to others in ambiguous social situations. This increased probability to perceive threat may account at least in part for the propensity to act in an aggressive manner. Moreover, parental modelling, e.g. the process of learning to effectively use aggression to exert control, is probably involved in the transmission of aggressive behaviour within the family. However, it has to be noted that childhood maltreatment was not only repeatedly found to increase the risk for aggression and conduct problems but is associated with a variety of psychiatric disorders.

3.2 *Influence of Parenting*

The most important factor for the development of social skills in children is parental discipline. It is a well replicated finding that children's aggression is promoted and maintained by inconsistent and inconsequent parenting practices characterised by the use of punishment and physical or emotional aggression in combination with low positive involvement. Confusing environmental experiences may interfere with correct information processing and the development of an effective regulation of motivational systems, which can lead to behavioural disturbances. On the other hand, negative parenting can be initiated by characteristics of a child such as a 'difficult temperament'. Concerning aggression, the interaction of the infant's temperament and parenting practices was described in the coercive family process model by Patterson (1982). According to this model, the parent and child interact in a way that negative or aversive behaviour of the child is negatively reinforced by the parents' behaviour. As an example, if the child breaks a rule, parents may initially insist on their demands to which the child responds by yelling and hitting. Finally, parents are likely to give in, which negatively reinforces the child's coercive behaviours. The parents' behaviour is also reinforced in this process by removal of the child's aversive behaviour. As a consequence, the probability of disobedience and escalation towards each other increases. Both, the parents and the child are reinforced to use aversive behaviour in the future. It was repeatedly found that parents of children with CD show higher rates of antisocial behaviour as well as hostile parenting practices. This may indicate coercive parenting to represent a correlate of genetic factors involved in the transmission of aggressive behaviours. On the other hand, there is also evidence for a meditational role of coercive parenting for the association between parental antisocial behaviour and the offspring's conduct problems (Bank et al. 1993).

In addition, the impact of inconsistent parenting was found to vary according to individual characteristics of the child. While a strong effect of parenting was found among children with low CU traits, parenting quality seems to be less important for the development of aggressive behaviour in children scoring high on these traits.

3.3 Genetic Risk Factors

This is in line with adoption studies showing that genetic risk is significantly modified by the specific features of the rearing environment. In a Swedish adoption study (Bohmann 1996), adoptees whose biological parents were not antisocial and were reared in low-risk families had a risk of adult criminality of 3 %. In the presence of a high-risk family it was 6 %; among those with criminal biological parents raised in low-risk families the rate was 12 %. In the presence of both biological and environmental risk the rate was 40 %.

There are in the meantime a number of investigations reporting significant interaction of environmental characteristics and specific genes in the prediction of aggressive behaviour. For instance, those genes functionally involved in serotonergic neurotransmission (MAO-A; 5-HTTLPR see chapter by M. Laucht and Buckholtz and Meyer-Lindenberg) were found to moderate the susceptibility to aggression in maltreated children. Weder et al. (2009) reported increased susceptibility to aggression among MAO-A carriers in children exposed to maltreatment. However, this study also demonstrated that genetic differences become irrelevant above a certain level of adversity.

4 Implications for Clinical Diagnostics and Treatment

As the development of aggressive behaviour in children and adolescent is influenced by many different etiological factors, there are as many different treatment options to be considered. As mentioned above, aggressive behaviour should not to be treated like an independent psychiatric condition but may rather be a phenomenon within many common psychiatric disorders (e.g. ADHD, CD, ASD, mood disorders or substance abuse). This implicates the necessity for a broad psychiatric examination of every patient with aggressive behaviour problems, also in consideration of possible psychiatric comorbidities.

4.1 Diagnostic Instruments

Questionnaires on a broad range of symptoms such as, e.g. the Child Behavior Checklist (CBCL, Achenbach 1991) or the Strengths and Difficulties questionnaire (SDQ, Goodman 1997; Goodman 1999) can be useful in this context, particularly since the CBCL does also differentiate between aggressive and dissocial behaviour. Moreover, the examiner should detect any abnormal psychosocial circumstances in the surroundings of the particular patient, as experiences like, e.g. abuse or neglect can crucially affect the progress of the treatment.

Besides the assessment of general psychopathology, it is important to distinguish between different kinds of disruptive behaviour disorders (CD/ODD) and to detect the extent of aggressive behaviour by means of specific instruments (Knapp et al. 2012). There are several evaluated and standardised measures available with which practitioners are recommended to assess treatment effects and outcomes, such as, e.g. the children's aggression scales (teachers and parent versions, Halperin et al. 2002, 2003) or the impulsive/remediated aggression scale (Kronenberger et al. 2007) which are to be completed by parents, teachers or other attachment figures of the particular child. Besides the registration of aggressive symptoms at baseline these scales can serve as measures for monitoring effects of, e.g. pharmacological treatment. Furthermore, according to recently published guidelines on management of maladaptive aggression (Knapp et al. 2012), assessments should include information about parenting styles and general developmental conditions as those could influence the decision on treatment alternatives. As the reactive type of aggression seems to differ from the proactive type concerning the underlying aetiology and treatment response, it seems reasonable to assess which type of aggressive behaviour does predominantly exist in a child. For this purpose several scales have been designed, amongst others the Dodge and Coie Scale (Dodge and Coie 1987, a teacher's report) or the scale by Raine et al. (2006), which is a self-report measure consisting of ten items (for more information on possible instruments see Hubbard et al. 2010). On the basis of the results of a comprehensive and accurate diagnostic investigation, the practitioner should develop an individual treatment plan for the affected child, taking into account possible underlying psychiatric conditions, comorbidities and psychosocial circumstances. To achieve consent and further compliance for the planned treatment, the practitioner should provide the child and its family with general information about the specific disorder or condition, possible treatment options and their availability. These psychoeducational interventions should be tailored to the particular needs of the respective family, e.g. information material should be provided in their native language and be comprehensible to non-professionals.

4.2 Behavioural Interventions

Taking into account the severe and long-lasting impact of aggressive and disruptive behaviour on further development of children and adolescents, effective treatment options are essential. When choosing a treatment option, several aspects should be included in the considerations: Motivation of the child or young person and his/her family, psychosocial circumstances and availability of treatment programmes. In most cases, a multimodal approach including parenting programmes, therapeutic group treatments and individualised therapy on a CBT basis as well as pharmacological or psychosocial interventions seem to be the most promising. Psychotherapeutic interventions mainly focus on problem-solving skills, anger management and hostile attribution biases. Within the recent US-guideline on treatment for children with maladaptive aggressive behaviour (Scotto Rosato et al. 2012) as well as in expert recommendations focussing on identical issues (Pappadopulos et al. 2011) parenting programmes as, e.g. the ‘incredible years’ (Eyberg et al. 2008; Reid et al. 2004) programme or ‘Triple P’ (Thomas and Zimmer-Gembeck 2007) are recommended as first-line treatment options, especially in young children. Most of these programmes include trainings of parents and/or teachers in effective application of positive reinforcement strategies to achieve a decrease of unwanted behaviour. Another focus lies on the reduction of negative interaction patterns within families and measures to improve intrafamiliar communication structures. One such intensive treatment programme for older children and adolescents and their families is, e.g. multisystemic therapy (Littell et al. 2005) which proved effective in reducing aggressive problem behaviours and enhanced better family and peer relations. Besides these family-based interventions, group-based trainings for children and adolescents turned out to be another promising approach for treating aggressive behaviour problems. Key aspects of these programmes are usually practice of social- and problem-solving skills, acquisition of self-management strategies and an improvement of self-control. Several so-called ‘summer camp’ approaches for children and adolescents with ADHD (Pelham and Hoza 1996) and/or disruptive behaviour problems (Grasmann and Stadler 2011), by which an intensification of treatment can be accomplished, have proven to be effective in reducing externalising behaviour problems and could represent an interesting alternative to conventional outpatient treatment options on a weekly or bi-weekly basis. Especially in older children, individualised CBT-based therapeutic interventions have a substantial empirical support (Scotto Rosato et al. 2012) in managing anger and aggressive behaviour.

The majority of the existing programmes predominantly address the reactive/impulsive type of aggression; until now, there are only few evidence=based treatment options available concerning psychotherapeutic or pharmacological treatment of children showing proactive aggression or callous–unemotional traits. Due to its strong genetic influence, proactive aggression shows a high stability over time and seems to be far more difficult to treat than reactive/impulsive aggression. Therapeutic approaches mainly focus on an improvement of empathic abilities and emotional responsiveness as well as the enhancement of moral development (Stadler et al. 2013). Specific programmes comprise trainings for emotion recognition and perspective taking. In a recent study, Dadds et al. (2012) reported on the effectiveness of an emotion recognition skills training on reducing aggressive behaviour in children with callous unemotional traits.

Findings from long-term studies suggest a negative outcome especially for aggressive behaviour with early onset. For this reason, intervention programmes aiming at the early identification and treatment of high-risk groups are promising. The ‘Enriched Environment’ programme (Raine et al. 2003) addresses variables like negative parenting style and psychosocial adversity in 3-year olds and showed good long-term effects in the prevention of conduct disorders, antisocial behaviours and schizotypal personality. Other approaches focus on early regulation problems, improvement of parent–child interactions and attachment and were shown to be able to enhance normalisation of neuroendocrinological stress-parameters in children (Bakermans-Kranenburg et al. 2008; Dozier et al. 2008).

4.3 *Psychopharmacological Interventions in Aggressive Behaviour*

Psychopharmacological interventions can form a crucial component in the treatment plan for children with aggressive behaviour problems. Clearly, psychopharmacology should not be applied on a monotherapeutical basis. During the past years, a trend towards prescription of psychopharmacological substances for the treatment of disruptive behaviour problems in children and adolescents was reported (Martin and Leslie 2003; Olfson et al. 2006). At the same time, they observed a general decrease in the intensity of in- and outpatient mental health services due to a shortness of resources and an increase in cost pressure on health care systems.

If behavioural and psychosocial interventions alone do not lead to sufficient success in the treatment of aggressive behaviour, additional psychopharmacological treatment should be considered. Scotto-Rosato and colleagues (2012) reviewed 29 RCTs that included direct measures for aggression and of which 28 were placebo-controlled. They found the atypical antipsychotic risperidone to have the best effects on aggressive behaviour (mean ES 0.72, 10 studies), closely followed by stimulants (mean ES = 0.6, 6 studies) and mood stabilisers (best evidence for lithium) (mean ES 0.42, 6 studies). Due to more severe side effects of antipsychotic medication like weight gain, sedation and metabolic side effects, stimulants or mood stabilisers were recommended as first-line options despite the better evidence for the efficacy of atypical antipsychotics in treating aggressive behaviour. Especially in ADHD with associated maladaptive aggression, published guidelines (Pappadopulos et al. 2011) recommend treating of underlying ADHD symptoms first, before targeting additional disruptive symptoms with other medication but stimulants. However, these recommendations are obviously not always followed by clinicians treating ‘difficult’ children and must be viewed with circumspection because, as discussed above, aggression cuts across a variety of disorders which will respond differently to psychopharmacological agents. In line with this, a naturalistic study on a large sample of children showing maladaptive aggression with various diagnoses demonstrated that clinicians increasingly seem to be inclined to skip stimulant medication and use in the first step antipsychotics in order to reduce aggression, despite the presence of ADHD symptoms (Penzner et al. 2009).

In general, the medication classes described above predominantly affect symptoms of reactive aggression like, e.g. impulsivity and are less successful in individuals with predominantly proactive aggressive behaviour. In children and adolescents with a psychiatric condition such as ADHD or mood disorders, practitioners should choose a medication according to current treatment-recommendations for this specific disorder. Accordingly, in patients with ADHD and aggressive behaviour problems, stimulants would be the medication of choice. Stimulants (Nevels et al. 2010) were also found to be effective concerning the treatment of aggressive behaviour in children and adolescents with CD and/or ODD without ADHD symptoms; Lithium proved to be superior compared to placebo in terms of decreasing inpatient aggression over 4 weeks of treatment (Deshmukh et al. 2010). However, authors reported more serious side effects like nausea and an increase in urinary frequency in subjects receiving lithium when compared to the placebo group. In the case of persistence of severe aggressive symptoms despite an adequate trial of treatment of the underlying condition, health care professionals should consider adding an atypical antipsychotic. Typical antipsychotics are not recommended for use as a continuous pharmacological treatment in aggressive children and adolescents, mainly due to extrapyramidal side effects.

5 Conclusions

Aggressive and disruptive behaviour is common among children and adolescents, and is part of the symptomatology not only in conduct disorders, but also in, e.g. mood disorders, ADHD, substance use disorders or/and pervasive developmental disorders. Persistent aggression is associated with a variety of negative outcomes in adulthood, including low socioeconomic status and unemployment, criminal behaviour and social isolation; thus, aggressive behaviour accounts for a substantial proportion of health burden in adulthood. Research efforts have demonstrated interactions of biological mechanisms with psychosocial factors influencing development and persistence of aggressive behaviour in children and adolescents. Still, the great heterogeneity of disruptive/aggressive behaviour might still hamper our understanding of causal mechanisms, and categorical diagnoses like CD may only insufficiently picture specific basic psychopathological dimensions. Current classification systems do not allow a distinction between different subtypes of aggression. A first step for a better differentiation of aggressive behaviour in DSM-5 is the incorporation of a callous-unemotional subtype to CD, based on evidence indicating these characteristics delineate a particularly malignant form of antisocial behaviour (Frick and Moffitt 2010). Especially, biological research has opened possibilities of identifying specific characteristics of aggressive behaviour over the past years, which in turn can stimulate the development of new and individualised treatment approaches.

References

Achenbach TM (1991) Manual for the child behaviour checklist/4-18 and 1991 profile. Department of Psychiatry, University of Vermont, Burlington

Achenbach TM, Howell CT, Quay HC et al (1991) National survey of problems and competencies among four- to sixteen-year-olds: parents' reports for normative and clinical samples. Monogr Soc Res Child Dev 56:1–131

American Psychiatric Association (2000) Diagnostic and statistical manual of mental disorders 4th edition text revision (DSM-IV-TR). American Psychiatric Association, Washington DC

Angold A, Costello EJ (2001) The epidemiology of disorders of conduct: nosological issues and comorbidity. In: Hill J, Maughan B (eds) Conduct disorders in childhood and adolescence. Cambridge University Press, Cambridge

Angold A, Costello EJ, Erkanli A (1999) Comorbidity. J Child Psychol Psychiatry 40:57–87

Axelson DA, Birmaher B, Findling RL et al (2011) Concerns regarding the inclusion of temper dysregulation disorder with dysphoria in the diagnostic and statistical manual of mental disorders (Fifth Edition). J Clin Psychiatry 72:1257–1262

Bakermans-Kranenburg MJ, van Ijzendoorn MH, Mesman J et al (2008) Effects of an attachment-based intervention on daily cortisol moderated by dopamine receptor D4: a randomized control trial on 1 to 3-year-olds screened for externalizing behaviour. Dev Psychopathol 20:805–820

Bank L, Forgatch MS, Patterson GR et al (1993) Parenting practices of single mothers: mediator of negative contextual factors. J Marriage Fam 55:371–384

Barker ED, Seguin JR, White HR et al (2007) Developmental trajectories of male physical violence and theft: relations to neurocognitive performance. Arch Gen Psychiatry 64:592–599

Barkley RA (2010) Deficient emotional self-regulation: a core component of attention deficit/ hyperactivity disorder. JARD 1:5–37

Biederman J, Faraone S, Mick E et al (1995) Psychiatric comorbidity among referred juveniles with major depression: fact or artifact? J Am Acad Child Adolesc Psychiatry 34:579–590

Biederman J, Faraone S, Milberger S et al (1996) Predictors of persistence and remission of ADHD into adolescence: results from a four-year prospective follow-up study. J Am Acad Child Adolesc Psychiatry 35:343–351

Bohmann M (1996) Predisposition to criminality: Swedish adoption studies in retrospect. In: Bock GR, Goode JA (eds) Genetics of criminal and antisocial behavior. Wiley, Chichester, pp 99–114

Burke JD, Loeber R, Birmaher B (2002) Oppositional defiant disorder and conduct disorder: a review of the past 10 years, part II. J Am Acad Child Adolesc Psychiatry 41:1275–1293

Cadoret RJ, Yates WR, Troughton E et al (1995) Adoption study demonstrating two genetic pathways to drug abuse. Arch Gen Psychiatry 52:42–52

Clark DB, Vanyukov M, Cornelius J (2002) Childhood antisocial behavior and adolescent alcohol use disorders. Alcohol Res Health 26:109–115

Connor DF (2002) Aggression and antisocial behavior in children and adolescents: research and treatment. Guilford Press, New York

Crick NR, Dodge KA (1996) Social information-processing mechanisms in reactive and proactive aggression. Child Dev 67:993–1002

Dadds MR, Cauchi AJ, Wimalaweera S et al (2012) Outcomes, moderators, and mediators of empathic-emotion recognition training for complex conduct problems in childhood. Psychiatry Res 199:201–207

Deshmukh P, Kulkarni G, Barzman D (2010) Recommendations for pharmacological management of inpatient aggression in children and adolescents. Psychiatry (Edgmont) 7:32–40

Dodge KA (1991) The structure and function of reactive and proactive aggression. In: Pepler DJ, Rubin KH (eds) The development and treatment of childhood aggression. Erlbaum, Hilldale, pp 201–218

Dodge KA, Coie JD (1987) Social-information-processing factors in reactive and proactive aggression in children's peer groups. J Pers Soc Psychol 53:1146–1158

Dodge KA, Lochman JE, Harnish JD et al (1997) Reactive and proactive aggression in school children and psychiatrically impaired chronically assaultive youth. J Abnorm Psychol 106:37–51

Dozier M, Peloso E, Lewis E et al (2008) Effects of an attachment-based intervention on the cortisol production of infants and toddlers in foster care. Dev Psychopathol 20:845–859

Eaves L, Rutter M, Silberg JL et al (2000) Genetic and environmental causes of covariation in interview assessments of disruptive behavior in child and adolescent twins. Behav Genet 30:321–334

Elliott DS (1994) Serious violent offenders: onset, developmental course, and termination–the American society of criminology 1993 presidential address. Criminology 32:1–21

Emerson E, Kiernan C, Alborz A et al (2001) The prevalence of challenging behaviors: a total population study. Res Dev Disabil 22:77–93

Eyberg SM, Nelson MM, Boggs SR (2008) Evidence-based psychosocial treatments for children and adolescents with disruptive behavior. J Clin Child Adolesc Psychol 37:215–237

Faraone SV, Biederman J, Jetton JG et al (1997) Attention deficit disorder and conduct disorder: longitudinal evidence for a familial subtype. Psychol Med 27:291–300

Farrington DP, Loeber R, van Kammen WB (1990) Long-term criminal outcomes of hyperactivity-impulsivity-attention deficit and conduct problems in childhood. In: Robins LN, Rutter M (eds) Straight and devious pathways to adulthood. Cambridge University Press, Cambridge, pp 62–81

Feehan M, McGee R, Raja SN et al (1994) DSM-III-R disorders in New Zealand 18-year-olds. Aust N Z J Psychiatry 28:87–99

Fergusson DM, Horwood LJ, Lynskey MT (1994) Structure of DSM-III-R criteria for disruptive childhood behaviors: confirmatory factor models. J Am Acad Child Adolesc Psychiatry 33:1145–1155

Frick PJ (2009) Extending the construct of psychopathy to youth: implications for understanding, diagnosing, and treating antisocial children and adolescents. Can J Psychiatry 54:803–812

Frick PJ, Moffitt TE (2010) A proposal to the DSM-V childhood disorders and the ADHD and disruptive behavior disorders work groups to include a specifier to the diagnosis of conduct disorder based on the presence of callous-unemotional traits. http://www.dsm5.org/Proposed%20Revision%20Attachments/Proposal%20for%20Callous%20and%20Unemotional%20Specifier%20of%20Conduct%20Disorder.pdf

Goodman R (1997) The strengths and difficulties questionnaire: a research note. J Child Psychol Psychiatry 38:581–586

Goodman R (1999) The extended version of the strengths and difficulties questionnaire as a guide to child psychiatric caseness and consequent burden. J Child Psychol Psychiatry 40:791–799

Grasmann D, Stadler C (2011) VIA–an intensive therapeutic treatment program for conduct disorders. Z Kinder Jugendpsychiatr Psychother 39:23–30

Green J, Gilchrist A, Burton D et al (2000) Social and psychiatric functioning in adolescents with Asperger syndrome compared with conduct disorder. J Autism Dev Disord 30:279–293

Greene RW, Biederman J, Zerwas S et al (2002) Psychiatric comorbidity, family dysfunction, and social impairment in referred youth with oppositional defiant disorder. Am J Psychiatry 159:1214–1224

Haller J, Mikics E, Halasz J et al (2005) Mechanisms differentiating normal from abnormal aggression: glucocorticoids and serotonin. Eur J Pharmacol 526:89–100

Halperin JM, McKay KE, Grayson RH et al (2003) Reliability, validity, and preliminary normative data for the children's aggression scale-teacher version. J Am Acad Child Adolesc Psychiatry 42:965–971

Halperin JM, McKay KE, Newcorn JH (2002) Development, reliability, and validity of the children's aggression scale-parent version. J Am Acad Child Adolesc Psychiatry 41:245–252

Hartley SL, Sikora DM, McCoy R (2008) Prevalence and risk factors of maladaptive behaviour in young children with autistic disorder. J Intellect Disabil Res 52:819–829

Hawes DJ, Brennan J, Dadds MR (2009) Cortisol, callous-unemotional traits, and pathways to antisocial behavior. Curr Opin Psychiatry 22:357–362
Hill J (2002) Biological, psychological and social processes in the conduct disorders. J Child Psychol Psychiatry 43:133–164
Hinshaw S (1994) Conduct disorder in childhood: conceptualization, diagnosis, comorbidity, and risk status for antisocial functioning in adulthood. In: Fowles DC, Sutker P, Goodman SH (eds) Progress in experimental personality and psychopathology research. Springer, New York, pp 3–44
Hinshaw SP, Lahey BB, Hart EL (1993) Issues of taxonomy and comorbidity in the development of conduct disorder. Dev Psychopathol 5:31–49
Holden B, Gitlesen JP (2006) A total population study of challenging behaviour in the county of Hedmark, Norway: prevalence, and risk markers. Res Dev Disabil 27:456–465
Hubbard JA, McAuliffe MD, Morrow MT et al (2010) Reactive and proactive aggression in childhood and adolescence: precursors, outcomes, processes, experiences, and measurement. J Pers 78:95–118
Hughes CW, Preskorn SH, Weller E et al (1990) The effect of concomitant disorders in childhood depression on predicting treatment response. Psychopharmacol Bull 26:235–238
Jensen PS, Martin D, Cantwell DP (1997) Comorbidity in ADHD: implications to research, practice, and DSM-IV. J Am Acad Child Adolesc Psychiatry 36:1065–1079
Kanne SM, Mazurek MO (2011) Aggression in children and adolescents with ASD: prevalence and risk factors. J Autism Dev Disord 41:926–937
King S, Waschbusch DA (2010) Aggression in children with attention-deficit/hyperactivity disorder. Expert Rev Neurother 10:1581–1594
Knapp P, Chait A, Pappadopulos E et al (2012) Treatment of maladaptive aggression in youth: CERT guidelines I. Engagement, assessment, and management. Pediatrics 129:e1562–e1576
Kronenberger WG, Giauque AL, Dunn DW (2007) Development and validation of the outburst monitoring scale for children and adolescents. J Child Adolesc Psychopharmacol 17:511–526
Kumsta R, Sonuga-Barke E, Rutter M (2012) Adolescent callous-unemotional traits and conduct disorder in adoptees exposed to severe early deprivation. Br J Psychiatry 200:197–201
Lahey BB, Loeber R, Quay HC, Frick PJ, Grimm J (1997) Oppositional defiant disorder and conduct disorder. In: Widiger TA, Frances AJ, Pincus HA, Ross R, First MB, Davis W (eds) DSM-IV sourcebook, vol 3. American Psychiatric Associaion, Washington, DC, pp 189–209
Lahey BB, Loeber R, Quay HC et al (1998) Validity of DSM IV subtypes of conduct disorder based on age of onset. J Am Acad Child Adolesc Psychiatry 37:435–442
Laucht M, Esser G, Baving L et al (2000) Behavioral sequelae of perinatal insults and early family adversity at 8 years of age. J Am Acad Child Adolesc Psychiatry 39:1229–1237
LeBlanc M (1997) Identification of potential juvenile offenders. Eur J Crim Policy Res 5:9–32
Lewinsohn PM, Rohde P, Seeley JR (1995) Adolescent psychopathology: III. The clinical consequences of comorbidity. J Am Acad Child Adolesc Psychiatry 34:510–519
Littell JH, Popa M, Forsythe B (2005) Multisystemic therapy for social, emotional, and behavioral problems in youth aged 10–17. Cochrane Database Syst Rev CD004797
Loeber R, Farrington DP (1998) Serious and violent juvenile offenders: risk factors and sucessful interventions. Sage, Thousand Oaks
Loeber R, Farrington DP, Stouthamer-Loeber M, van Kammen WB (1998) Antisocial behavior and mental health problems: explanatory factors in childhood and adolescence. Erlbaum, Mahwah
Loeber R, Keenan K, Zhang Q (1997) Boys' experimentation and persistence in developmental pathways towards serious delinquency. J Child Fam Stud 6:321–357
Loeber R, Burke JD, Lahey BB et al (2000) Oppositional defiant and conduct disorder: a review of the past 10 years, part I. J Am Acad Child Adolesc Psychiatry 39:1468–1484
Loeber R, Coie J (2001) Continuities and discontinuities of development, with particular emphasis on emotional and cognitive components of disruptive behaviour. In: Hill J, Maughan B (eds) Conduct disorders in childhood and adolescence. Cambridge University Press, Cambridge

Loeber R, Green SM, Keenan K et al (1995) Which boys will fare worse? Early predictors of the onset of conduct disorder in a six-year longitudinal study. J Am Acad Child Adolesc Psychiatry 34:499–509

Loeber R, Hay D (1997) Key issues in the development of aggression and violence from childhood to early adulthood. Annu Rev Psychol 48:371–410

Loeber R, Keenan K (1994) The interaction between conduct disorder and its comorbid conditions: effects of age and gender. Clin Psychol Rev 14:497–523

Loeber R, Stouthamer-Loeber M (1998) Development of juvenile aggression and violence. Some common misconceptions and controversies. Am Psychol 53:242–259

Loeber R, Wei E, Stouthamer-Loeber M et al (1999) Behavioral antecedents to serious and violent juvenile offending: joint analyses from the Denver youth survey, Pittsburgh youth study, and the Rochester development study. Stud Crime Prev 8:245–263

Lynam DR, Henry V (2001) The role of neuropsychological deficits in conduct disorders. In: Hill J, Maughan B (eds) Conduct disorders in childhood and adolescence. Cambridge University Press, Cambridge

Magnusson D, Bergmann LR (1990) A pattern approach to the study of pathways from childhood to adulthood. In: Robins LN, Rutter M (eds) Straight and devious pathways from childhood to adulthood. Cambridge University Press, Cambridge, pp 101–115

Mannuzza S, Klein RG, Bessler A et al (1998) Adult psychiatric status of hyperactive boys grown up. Am J Psychiatry 155:493–498

Martel MM (2009) Research review: a new perspective on attention-deficit/hyperactivity disorder: emotion dysregulation and trait models. J Child Psychol Psychiatry 50:1042–1051

Martin A, Leslie D (2003) Psychiatric inpatient, outpatient, and medication utilization and costs among privately insured youths, 1997–2000. Am J Psychiatry 160:757–764

Matson JL, Rivet TT (2008) The effect of severity of autism and PDD-NOS symptoms on challenging behaviors in adults with intellectual disabilities. J Dev Phys Disabil 20:41–51

Maughan B, Rutter M (2001) Antisocial children grown up. In: Hill J, Maughan B (eds) Conduct disorders in childhood and adolescence. Cambridge University Press, Cambridge, pp 507–552

McClintock K, Hall S, Oliver C (2003) Risk markers associated with challenging behaviours in people with intellectual disabilities: a meta-analytic study. J Intellect Disabil Res 47:404–416

Mikita N, Stringaris A (2013) Mood dysregulation. Eur Child Adolesc Psychiatry 22(Suppl 1):S11–S16

Ming X, Brimacombe M, Chaaban J et al (2008) Autism spectrum disorders: concurrent clinical disorder. J Child Neurol 23:6–13

Moffitt TE, Caspi A (2001) Childhood predictors differentiate life-course persistent and adolescence-limited antisocial pathways among males and females. Dev Psychopathol 13:355–375

Nagin D, Tremblay RE (1999) Trajectories of boys' physical aggression, opposition, and hyperactivity on the path to physically violent and nonviolent juvenile delinquency. Child Dev 70:1181–1196

Nevels RM, Dehon EE, Alexander K et al (2010) Psychopharmacology of aggression in children and adolescents with primary neuropsychiatric disorders: a review of current and potentially promising treatment options. Exp Clin Psychopharmacol 18:184–201

NICHD Early Child Care Research Network (2004) Trajectories of physical aggression from toddlerhood to middle school: Predictors, correlates, and outcomes. SRCD Monogr 69:1–146

Nock MK, Kazdin AE, Hiripi E et al (2006) Prevalence, subtypes, and correlates of DSM-IV conduct disorder in the National Comorbidity Survey Replication. Psychol Med 36:699–710

Offord DR, Boyle MH, Racine YA et al (1992) Outcome, prognosis, and risk in a longitudinal follow-up study. J Am Acad Child Adolesc Psychiatry 31:916–923

Olfson M, Blanco C, Liu L et al (2006) National trends in the outpatient treatment of children and adolescents with antipsychotic drugs. Arch Gen Psychiatry 63:679–685

Pappadopulos E, Rosato NS, Correll CU et al (2011) Experts' recommendations for treating maladaptive aggression in youth. J Child Adolesc Psychopharmacol 21:505–515

Patterson GR (1982) A social learning approach. In: Patterson GR et al (eds) Coercive familiy processes, vol 3. Castalia, Eugene
Patterson GR, Degarmo DS, Knutson N (2000) Hyperactive and antisocial behaviors: comorbid or two points in the same process? Dev Psychopathol 12:91–106
Pelham WE Jr, Hoza B (1996) Intensive treatment: a summer treatment program for children with ADHD. In: Hibbs ED, Jensen PS (eds) Psychosocial treatments for child and adolescent disorders: Empirically based strategies for clinical practice. American Psychological Association, Washington, pp 311–340
Penzner JB, Dudas M, Saito E et al (2009) Lack of effect of stimulant combination with second-generation antipsychotics on weight gain, metabolic changes, prolactin levels, and sedation in youth with clinically relevant aggression or oppositionality. J Child Adolesc Psychopharmacol 19:563–573
Pitzer M, Esser G, Schmidt MH et al (2010) Early predictors of antisocial developmental pathways among boys and girls. Acta Psychiatr Scand 121:52–64
Raine A (2002) Biosocial studies of antisocial and violent behavior in children and adults: a review. J Abnorm Child Psychol 30:311–326
Raine A, Dodge K, Loeber R et al (2006) The reactive-proactive aggression questionnaire: differential correlates of reactive and proactive aggression in adolescent boys. Aggress Behav 32:159–171
Raine A, Mellingen K, Liu J et al (2003) Effects of environmental enrichment at ages 3–5 years on schizotypal personality and antisocial behavior at ages 17 and 23 years. Am J Psychiatry 106:1627–1635
Reese RM, Richman DM, Belmont JM et al (2005) Functional characteristics of disruptive behavior in developmentally disabled children with and without autism. J Autism Dev Disord 35:419–428
Reid MJ, Webster-Stratton C, Baydar N (2004) Halting the development of conduct problems in head start children: the effects of parent training. J Clin Child Adolesc Psychol 33:279–291
Reinherz HZ, Giaconia RM, Hauf AM et al (2000) General and specific childhood risk factors for depression and drug disorders by early adulthood. J Am Acad Child Adolesc Psychiatry 39:223–231
Robins LN (1991) Conduct disorder. J Child Psychol Psychiatry 32:193–212
Rohde P, Clarke GN, Lewinsohn PM et al (2001) Impact of comorbidity on a cognitive-behavioral group treatment for adolescent depression. J Am Acad Child Adolesc Psychiatry 40:795–802
Rutter M, Giller H, Hagell A (1998) Antisocial behaviour by young people. Cambridge University Press, Cambridge
Satterfield JH, Schell A (1997) A prospective study of hyperactive boys with conduct problems and normal boys: adolescent and adult criminality. J Am Acad Child Adolesc Psychiatry 36:1726–1735
Scotto Rosato N, Correll CU, Pappadopulos E et al (2012) Treatment of maladaptive aggression in youth: CERT guidelines II. Treatments and ongoing management. Pediatrics 129:e1577–e1586
Sebastian CL, McCrory EJ, Cecil CA et al (2012) Neural responses to affective and cognitive theory of mind in children with conduct problems and varying levels of callous-unemotional traits. Arch Gen Psychiatry 69:814–822
Soderstrom H, Sjodin AK, Carlstedt A et al (2004) Adult psychopathic personality with childhood-onset hyperactivity and conduct disorder: a central problem constellation in forensic psychiatry. Psychiatry Res 121:271–280
Stadler C, Kröger A, Grasmann D, Sterzer P (2013) Störungen des Sozialverhaltens, Dissozialität und Delinquenz. In: Lehmkuhl G, Poustka F, Holtmann M, Steiner H (eds) Lehrbuch der Kinder- und Jugendpsychiatrie. Hogrefe, Göttingen, pp 1094–1126
Stadler C, Poustka F, Sterzer P (2010) The heterogeneity of disruptive behavior disorders—implications for neurobiological research and treatment. Front Psychiatry 1:21

Sterzer P, Stadler C, Krebs A et al (2005) Abnormal neural responses to emotional visual stimuli in adolescents with conduct disorder. Biol Psychiatry 57:7–15

Stringaris A (2011) Irritability in children and adolescents: a challenge for DSM-5. Eur Child Adolesc Psychiatry 20:61–66

Swadi H (1999) Individual risk factors for adolescent substance use. Drug Alcohol Depend 55:209–224

Thomas R, Zimmer-Gembeck MJ (2007) Behavioral outcomes of Parent-Child Interaction Therapy and Triple P-Positive Parenting Program: a review and meta-analysis. J Abnorm Child Psychol 35:475–495

Tolan PH, Gorman-Smith D (1998) Development of serious and violent offending careers. In: Loeber R, Farrington DP (eds) Serious and violent juvenile offenders: risk factors and sucessful intervention. Sage, Thousand Oaks, pp 68–87

Tremblay RE (2010) Developmental origins of disruptive behaviour problems: the 'original sin' hypothesis, epigenetics and their consequences for prevention. J Child Psychol Psychiatry 51:341–367

Tremblay RE, Loeber R, Gagnon C et al (1991) Disruptive boys with stable and unstable high fighting behavior patterns during junior elementary school. J Abnorm Child Psychol 19:285–300

Tsakanikos E, Costello H, Holt G et al (2007) Behaviour management problems as predictors of psychotropic medication and use of psychiatric services in adults with autism. J Autism Dev Disord 37:1080–1085

Tuvblad C, Raine A, Zheng M et al (2009) Genetic and environmental stability differs in reactive and proactive aggression. Aggress Behav 35:437–452

van Lier PA, Vitaro F, Barker ED et al (2009) Developmental links between trajectories of physical violence, vandalism, theft, and alcohol-drug use from childhood to adolescence. J Abnorm Child Psychol 37:481–492

Vitaro F, Gendreau PL, Tremblay RE et al (1998) Reactive and proactive aggression differentially predict later conduct problems. J Child Psychol Psychiatry 39:377–385

Vitiello B, Stoff DM (1997) Subtypes of aggression and their relevance to child psychiatry. J Am Acad Child Adolesc Psychiatry 36:307–315

Wakschlag LS, Tolan PH, Leventhal BL (2010) Research review: 'Ain't Misbehavin'': towards a developmentally-specified nosology for preschool disruptive behavior. J Child Psychol Psychiatry 51:3–22

Weder N, Yang BZ, Douglas-Palumberi H et al (2009) MAOA genotype, maltreatment, and aggressive behavior: the changing impact of genotype at varying levels of trauma. Biol Psychiatry 65:417–424

Wolff JC, Ollendick TH (2006) The comorbidity of conduct problems and depression in childhood and adolescence. Clin Child Fam Psychol Rev 9:201–220

World Health Organization (1993) The ICD-10 classification of mental and behavioural disorders: diagnostic criteria for research. WHO, Geneva

Translational Clinical Neuroscience Perspectives on the Cognitive and Neurobiological Mechanisms Underlying Alcohol-Related Aggression

Anne Beck, Adrienne J. Heinz and Andreas Heinz

Abstract Alcohol-related violence, a longstanding, serious, and pervasive social problem, has provided researchers from diverse disciplines with a model to study individual differences in aggressive and violent behavior. Of course, not all alcohol consumers will become aggressive after drinking and similarly, not all individuals with alcohol use disorders will exhibit such untoward behavior. Rather, the relationship is best conceptualized as complex and indirect and is influenced by a constellation of social, cognitive, and biological factors that differ greatly from one person to the next. Animal experiments and human studies have elucidated how these mechanisms and processes explain (i.e., mediate) the relation between acute and chronic alcohol consumption and aggressive behavior. Further, the rich body of literature on alcohol-related aggression has allowed for identification of several potential high-yield targets for clinical intervention, e.g., cognitive training for executive dysfunction; psychopharmacology targeting affect and threat perception, which may also generalize to other psychiatric conditions characterized by aggressive behavior. Here we aim to integrate pertinent findings, derived from different methodological approaches and theoretical models, which explain heterogeneity in aggressive responses to alcohol. A translational platform is provided, highlighting common factors linking alcohol and aggression that warrant further, interdisciplinary study in order to reduce the devastating social impact of this phenomenon.

Keywords Alcohol · Violence · Aggression · Cognitive · Executive functioning · Neuroscience · Serotonin · Impulsivity

---

A. Beck · A. Heinz (✉)
Department of Psychiatry and Psychotherapy, Campus Charité Mitte,
Charité - Universitätsmedizin Berlin, Chariteplatz 1,
10117 Berlin, Germany
e-mail: andreas.heinz@charite.de

A. J. Heinz
Center for Health Care Evaluation, VA Palo Alto Health Care System
and Department of Psychiatry and Behavioral Sciences,
Stanford University School of Medicine, Menlo Park, CA, USA

Curr Topics Behav Neurosci (2014) 17: 443–474
DOI: 10.1007/7854_2013_258

Published Online: 14 December 2013

Contents

1 Introduction

Worldwide, alcohol has been implicated in approximately one-half of violent crimes (Murdoch et al. 1990; Pernanen 1993) and in the United States, approximately 3 million violent crimes occur each year in which victims perceive the offender to have been under the influence of alcohol (Greenfeld 1998). Of note, alcohol has been identified as a major contributing factor in 32 % of all homicides in the United States (Smith et al. 1999); in Germany, 31.8 % of violent acts in 2011, e.g., aggravated assaults, involved alcohol intoxication (Bundeskriminalamt 2011). Moreover, alcohol use has been implicated in approximately one-half of all sexual assaults (Collins and Messerschmidt 1993), and the US National Crime Victimization Survey reported that (based on victims' perceptions) the perpetrator had been drinking in 26.8 % of rapes and sexual assaults (NCVS 2006). In terms of intimate partner violence, a national study among a subgroup of couples in the United States revealed that 30–40 % of men and 27–34 % of women were drinking at the time they perpetrated violence against their partner (Caetano et al. 2001). Finally, nearly 40 % of child victimizers reported having been drinking at the time they perpetrated the crime (Greenfeld 1996). With such high prevalence rates, the economic burden associated with alcohol-related crimes is high and highlights the urgent need for intervention. For instance, a study in New South Wales, Australia reported that expected cost *per* alcohol-related incident of disorderly conduct, property damage, assault and sexual offence (in 2006 dollar values) was $501, $1,166, $3,982, and $5,976, respectively (Byrnes et al. 2012).

Epidemiological and survey data clearly demonstrate that aggressive and violent behavior is more common among individuals with hazardous alcohol use and alcohol use disorders (AUD). Incidence of violent behavior in various samples of male alcohol-dependent patients has been estimated as high as 50 % (Jaffe et al. 1988; Mayfield 1976; Nicol et al. 1973; Schuckit and Russell 1984), and compared to individuals without AUD, those with an AUD are at significantly higher risk (up to fivefold) for involvement in violence (Arseneault et al. 2000; Coid et al. 2006; Swanson et al. 1990). In terms of hazardous drinking, a meta-analytic study showed that compared to individuals with very low or moderate alcohol use, those who got drunk at least once a year were almost twice as likely to engage in violence (Lipsey et al. 1997). In addition, a cross-sectional British study showed that hazardous drinking was associated with a sixfold increase in risk for reporting victim and perpetrator injury when intoxicated (Coid et al. 2006). Finally, a longitudinal study with adolescents reported that increases in alcohol abuse and dependence symptoms were significantly associated with increased rates of violent crime (Fergusson and Horwood 2000).

Reports in the literature on temporal patterns of alcohol-related violence among heavy and problem drinkers demonstrate that aggression is more likely to occur after heavier, *acute* alcohol intake. Two studies with arrestees and felons reported that the acute effects of drinking (drinking in events related to the arrest) were positively associated with committing a violent offense (Collins and Schlenger 1988; Wiley and Weisner 1995). Interestingly though, the chronic effects of alcohol (i.e., drinking patterns in the past 12 months) did not discriminate between violent and nonviolent charges. In line with this, compared to days with nonviolent partner conflicts, alcoholic men consumed significantly more alcohol in the 12 h prior to violent partner conflict (Murphy et al. 2005). Finally, a New Zealand study of alcohol-dependent individuals reported that violence was best accounted for by alcohol and substance use prior to the offence (Arseneault et al. 2000). Alcohol's acute effects are known to facilitate aggressive behavior more so than its chronic effects, e.g., Chermack and Blow 2002; Fals-Stewart 2003, and as such, it may be concluded that higher quantity and frequency of (acute) alcohol consumption is the decisive factor and contributes to the high prevalence of violent behavior observed in AUD populations.

Studies of alcohol-related aggression conducted among treatment populations also suggest that heavier chronic alcohol use patterns contribute to the incidence and severity of violent acts. For instance, treatment-seeking alcohol-dependent men, as compared to healthy controls, were four times more likely to exhibit violence toward their female partners, and partner violence decreased 1 year post-treatment only among patients who did not relapse to drinking (O'Farrell 2003). In a similar sample, general patterns of alcohol use as well as alcohol use on the conflict day were related to severity of partner violence; again, acute alcohol use levels were significantly higher on days involving interpersonal conflict (Chermack and Blow 2002). Collectively, these data illustrate the complexity of alcohol-related aggression and highlight the importance of clinical conceptualizations that consider the host of putative cognitive and biological mechanisms impacted by problem drinking, which in turn confer risk for aggressive and violent behavior.

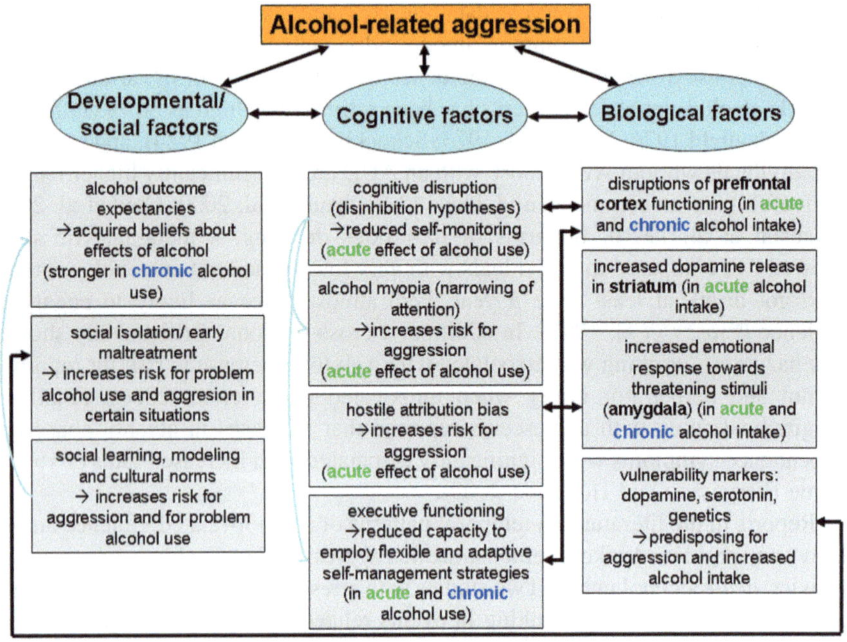

Fig. 1 Alcohol-related aggression as a multi-causal phenomenon

2 Social-Cognitive Mechanisms and Models

One of the more empirically scrutinized ways in which alcohol is believed to promote aggressive behavior is via its acute effects on cognitive functioning. In addition, a number of stable cognitive and social-cognitive factors have been shown to increase vulnerability to aggressive behavior under alcohol (Chermack and Giancola 1997). Among them are executive functioning, e.g., inhibition, error-tracking, information processing, attention, individual differences in alcohol outcome expectancies such as "alcohol will make me feistier", and social learning experiences, e.g., repeated exposure to family or peers who become aggressive following alcohol consumption (see Fig. 1). In this section, under the rubric of empirically supported cognitive theories and models, we discuss the role of cognitive and social-cognitive factors implicated in acute alcohol-induced aggression.

2.1 Theories Invoking Cognitive Disruption

The indirect-cause model of alcohol-related aggression espouses that alcohol compromises multiple domains of functioning and thereby increases the probability of engaging in aggressive behavior (Graham 1980). The "disinhibition hypothesis" (Graham 1980) posits that the pharmacological properties of alcohol disrupt the functioning of brain centers implicated in maintaining inhibitory control over more primal behavior. That is, alcohol is supposed to facilitate aggression through disruption of cognitive processes critical for self-regulation, e.g., attention, inhibition, information processing, decision making. Several more nuanced explanations have since expanded upon this broad conceptualization and describe more specific yet *indirect* pathways by which alcohol's effects on cognition contribute to aggressive behavior.

Alcohol consumption is presumed to cause a narrowing of attention, a phenomenon termed alcohol myopia (Steele and Josephs 1990). By way of analogy, disinhibition can be equated to viewing a scene through a distorted or blurry camera lens, whereas myopia is akin to using a camera's zoom function to view a small part of the scene with extreme clarity at the expense of the broader picture. Several models maintain that when alcohol decreases inhibition or narrows the perceptual field, the probability of emitting disinhibited behavior, e.g., aggression, increases. This is because alcohol reduces the cognitive resources available to process information and as such, proximal, instigating cues in the immediate context, e.g., being bumped into in a bar, can be awarded far more salience than external, inhibitory cues, e.g., considering that it was accidental or considering distal consequences of responding aggressively (Taylor and Leonard 1983).

Importantly though, competition for attention between instigating and inhibitory cues (i.e., inhibition conflict) is necessary for alcohol to bias one toward an excessive, potentially aggressive response (Steele and Southwick 1985). Within this framework, alcohol is thought to reduce (high) inhibition conflict, thus attention is biased towards only the most provocative and salient cues, i.e., attention-allocation (Giancola et al. 2010). Hence the decision-making process between two potential responses, e.g., telling a bar patron to "please be careful" versus shoving him, is strongly influenced by instigating cues and less so by conflict-mitigating information (Ito et al. 1996). The first laboratory examination of the attention-allocation model adapted for aggression showed that under moderate cognitive load (i.e., holding four elements in sequential order in memory) and in the presence of low-provocation cues, those who had ingested alcohol administered lower intensity shocks to a fictitious competitor (i.e., lower aggression) than sober participants, (i.e., effects of alcohol on aggression were attenuated (Giancola and Corman 2007). Interestingly, among intoxicated participants, those under small or large cognitive load, e.g., high or low inhibition conflict, demonstrated significantly more aggression compared to those under moderate load (Giancola and Corman 2007). These data indicate that aggressive responding under alcohol is governed by the amount of cognitive space available for processing inhibitory cues, and that both high and low (but not moderate) demands increase aggression.

Other approaches have considered how alcohol's effects on the processing of social/interpersonal information may promote aggressive behavior (Sayette et al. 1993). For instance, alcohol may increase the likelihood of making erroneous and hostile attributions about another person's actions, which would be (objectively) considered ambiguous in intent from an objective standpoint, a phenomenon termed hostile attribution bias (HAB; Nasby et al. 1980). Within a social context, e.g., a bar, under the influence of alcohol, the ambiguous actions of others such as a frown may signal an uncertain level of threat (Taylor et al. 1976). In such circumstances, an individual may misinterpret ambiguous interpersonal cues, resulting in HAB and thereby increasing the probability of emitting an aggressive response. Of note, alcohol tends to explain more variance in aggressive behavior under conditions of low provocation rather than high provocation, e.g., threat of physical assault, because high-threat situations are generally more likely to engender aggressive responding, independent of the effects of alcohol (Ito et al. 1996; Giancola et al. 2002).

Alcohol-related disinhibition is also been shown to impact intrapersonal self-monitoring processes central to self-regulation. The self-awareness hypothesis stipulates that alcohol promotes aggressive behavior by disrupting cognitive encoding of self-relevant information (e.g., tone, posture, body language) that is vital to sustain self-awareness (Hull 1981). In support of this theory, one study reported that intoxicated individuals who could see themselves in a mirror or on video camera (i.e., high self-awareness condition) administered fewer shocks to others in a game than individuals in the low self-awareness condition (Bailey et al. 1983). Increasing self-focused attention may reduce the alcohol-induced gap between internal states and external standards of comportment (Ito et al. 1996).

2.2 *Theories Invoking Social Learning, Memory Structures, and Alcohol Outcome Expectancies*

Alcohol outcome expectancies (AOE) are the beliefs and predictions held by an individual about the effects that alcohol consumption will have on their subjective experience (mood), behavior and performance (Goldman et al.1987). AOEs are hypothesized to be stored in memory and organized within one's alcohol schema (i.e., theoretical cognitive structure or template for the effects of alcohol; Goldman et al. 1999). The strength and accessibility of these expectancies is typically quantified using self-report questionnaires, e.g., Fromme et al. 1993, and cognitive tasks, e.g., Bartholow and Heinz 2006; Stacy et al. 1994, respectively. Conceptual models invoking AOEs indicate that the anticipation of or the belief that alcohol has been consumed functions to activate memories and expectancies for alcohol, which in turn influence behavior under alcohol (for review see Goldman et al. 1999b). Of note, even in the absence of alcohol, but in the presence of alcohol cues, individuals with stronger expectations for alcohol to facilitate aggression rated the ambiguous behavior of a fictional character as significantly more hostile (Bartholow and Heinz 2006).

Principles of semantic networking theory have facilitated our understanding of how the alcohol expectancy memory network is organized and the putative processes by which it is activated (Rather et al. 1992). When alcohol-related stimuli are encountered, (e.g., beer bottle, party setting, alcohol concepts), they are represented as nodes of information linked together in memory on the basis of intrinsic meaning and learning, and they are supposedly stimulated in the form of a "spreading activation" (Collins and Loftus 1975). Activation of these alcohol concepts in memory (including expectancies) then occurs as a function of their semantic proximity to the nature of the alcohol-related stimuli encountered, e.g., champagne glasses versus beer-bong. Modeling of alcohol-related memory structures and processes with Euclidean distance-based algorithms suggests that heavy drinkers' memory networks for alcohol expectancies are more tightly configured than those of light drinkers (Rather and Goldman 1994). For instance, light drinkers associate drinking with sedative effects, whereas individuals with heavier drinking patterns tend to expect more positive and arousing effects of alcohol (Rather et al. 1992). Heavy drinkers also demonstrate stronger associations in memory between alcohol and positive outcomes than with negative outcomes (Leigh and Stacy 1998; Weingardt et al. 1996). The alcohol memory network also includes aggressive expectancies and action tendencies, though the number and strength of these connections is likely determined by individual learning and experience with alcohol-related aggression, (Goldman 1999).

AOEs have been documented in children of pre-school age with no prior experience of personal alcohol consumption (Zucker et al. 1995), and it is surmised that social learning and modeling contributes to their development (Bandura 1977). Indeed, outcome expectancies for alcohol are gradually shaped over time by both direct use of alcohol and by vicarious social learning experiences, e.g. modeling of alcohol use by family and peers or media messages. These constructs are supposed to hold a reciprocally deterministic relationship with drinking behaviors, such that the positive and negative outcomes of using alcohol reinforce expectations and, in turn, expectancies influence the experience of outcomes (Maisto et al. 1999; Leigh and Stacy 1998). Further, as experience with alcohol mounts, as in the case of chronic drinkers, the connections between expectancies and outcomes in memory strengthen and can be accessed with greater efficiency (Leigh and Stacy 1998; Stacy et al. 1994; Weingardt et al. 1996). Once these connections in memory solidify, explicit awareness of outcome expectancies declines and upon activation, they may drive behavior in an automatic or habitual manner (see Tiffany 1990). AOEs have demonstrated powerful predictive validity concerning a wide range of drinking outcomes (Goldman et al. 1999b; Jones et al. 2001) as well as treatment outcomes in AUD populations (Brown 1985) and appear to exercise a significant influence over drinking behavior.

2.3 Executive Functioning

Executive functioning (EF) comprises several subsets of cognitive abilities, including but not limited to strategic planning, problem solving, cognitive flexibility, and attentional capacity, (Luria (1980); Posner and Petersen (1990)), which contribute to the planning, initiation and regulation of goal-directed behavior.

Evidence from clinical populations suggests that individual differences in EF may have a distal and an indirect influence on general alcohol use and drinking outcomes (Giancola et al. 2002; Giancola 2007). Indeed, alcohol-dependent individuals, when compared to healthy controls, demonstrate impairments across a range of executive functioning abilities, e.g., Bates et al. 2002; Giancola and Moss 1998; Loeber et al. 2010; Ratti et al. 2002; Stavro et al. 2012; Sullivan et al. 2000). A comprehensive review of the PET, fMRI and neuropsychological data concluded that alcoholism is characterized by varying degrees of impaired frontal lobe function (Moselhy et al. 2001) and compensatory activations during frontal lobe-related working memory tasks, which predict relapse (Charlet et al. 2013). Importantly, deficits in EF are also indicated in suboptimal decision making and risky (impulsive) behavior, a behavioral profile commonly observed in AUD, e.g., aggressive behavior or lack of consideration for consequences. For example, individuals with alcohol dependence tend to discount larger future rewards over smaller immediate rewards (i.e., delay discounting) at a higher rate than healthy controls (Mackillop et al. 2011). Given evidence that the relation between alcohol and aggression is stronger among individuals with lower EF (Giancola 2004), alcohol-dependent patients may be more prone to respond aggressively under alcohol due to the chronic effects of alcohol on EF. In recognition that alcohol-induced impairment of EF may be associated with alcohol-related aggression and also that individual differences in EF can moderate the strength of this relationship, Giancola (2000) put forth an integrated EF framework of alcohol-related aggression. He postulated that four key abilities which inhibit impulsive aggressive actions - attending to and appraising situational information, taking the perspective of others, considering consequences of one's actions, defusing a hostile situation - are impaired by alcohol. According to this framework, alcohol promotes impulsive aggression indirectly by disrupting EF (mediating effect), and that this relationship is strongest among individuals with lower baseline EF (moderating effect). Of note, studies including imaging or electrophysiological methods are needed to verify whether the acute effects of alcohol indeed indirectly promote aggression by compromising brain areas indicated in executive control, such as the prefrontal cortex.

3 Alcohol and Aggressive Behavior: How Alcohol Affects the Brain

Neurobiological research further elucidates the complex relations between alcohol and aggression. Animal studies have shown that the acute consumption of alcohol stimulates release of monoaminergic neurotransmitters, including dopamine and serotonin, within the ventral and dorsal striatum (Diana et al. 1992; Mathews et al. 2006; Puglisiallegra et al. 1991). In addition, alcohol simultaneously inhibits cortical excitability by stimulating the release of GABA (Gamma-aminobutyric acid), and blocks glutamatergic neurotransmission, e.g. via interference with a glycine binding site that modulates NMDA [N-Methyl-D-Aspartat] receptor function, and thus induces mainly sedative effects (DiChiara et al. 1996; Tsai et al. 1995; Heinz et al. 2009). As a result, the effects of alcohol on behavior are largely dose-dependent. For example in rodents, low doses of alcohol facilitate aggression in a subset of animals, whereas higher doses induce sedation and motor disorganization (Murphy et al. 2005; Takahashi et al. 2010a; Weerts et al. 1993). In this section, we will discuss different neurobiological mechanisms by which alcohol can affect brain circuits indicated in executive functioning, behavioral control, and emotion processing (see Fig. 1).

3.1 The Role of Prefrontal and Amygdalar Functioning

In human studies, the acute effects of alcohol have been primary investigated with respect to impairments of executive functions, which are associated with the prefrontal cortex (PFC) (Chermack and Giancola 1997): disruptions of PFC functioning, as observed in lesion studies, are associated with behavioral changes including problems in initiating, shifting and stopping behavior, anticipating, planning and sequencing deficits, impaired abstract reasoning as well as apathy, emotional instability, and aggression (Adolphs 2009; Peterson et al. 1990). For example, previous research showed that up to 70 % of patients with traumatic brain injuries developed high irritability and aggression, which causes significant distress to their families (McKinlay et al. 1981). Interestingly, alcohol intake has been observed in placebo-controlled studies to induce comparable impairments in PFC functioning, including attentional processes (Schreckenberger et al. 2004), planning (Peterson et al. 1990), response inhibition and flexibility (Easdon and Vogel-Sprott 2000; Finn et al. 1999; Mulvihill et al. 1997; Schweizer et al. 2006) as well as set-shifting (Guillot et al. 2010; Lyvers and Maltzman 1991). Additionally, consumption of alcohol in moderate doses has been shown to induce significant impairments in the ability to detect errors and adjust performance accordingly in a classical flanker task (Ridderinkhof et al. 2002). These impairments were indexed as abnormal amplitudes in brain electrical activity associated with the mediofrontal and anterior cigulate cortex (ACC). Such deficits in executive control might contribute to thoughtless, inappropriate or excessive behavior that could lead to increased vulnerability of engaging in aggressive or even violent responses when confronted with provocations (Hawkins and Trobst 2000).

Importantly, several subregions of the prefrontal cortex, e.g., the mentioned ACC and ventromedial PFC, are directly connected with limbic areas, such as the amygdala via (predominantly inhibitory) projections (Barbas 2000; Amaral and Price 1984), and are thus associated with the regulation of emotional behavior (Ochsner and Gross 2005). Impaired amygdala function is associated with different forms of aggressive behavior. For instance, instrumental or "cold" aggression is likely associated with reduced amygdala reactivity, whereas reactive or so-called "hot" aggression appears to be associated with heightened activation of the amygdala (Birbaumer et al. 2005; Coccaro et al. 2007; Jones et al. 2009; Lee et al. 2008;

Marsh et al. 2008; Sterzer et al. 2005; Sterzer and Stadler 2009). Functional neuroimaging studies among aggressive individuals, e.g., subjects diagnosed with intermittent explosive disorder (a psychiatric disorder characterized by reactive aggressive behavior) suggest that impaired activation of medial prefrontal structures together with heightened amygdala responses might be a key explanatory mechanism of reactive aggression in emotionally provocative social situations (Coccaro et al. 2007; Lee et al. 2008; Sterzer et al. 2005). Thus, these data support the hypothesis that the prefrontal cortex plays an important role in the regulation of aggressive behavior (Davidson et al. 2000).

Interestingly, the above described effects of acute alcohol intake on the brain also resemble aspects of how stress influences neurotransmitter balance in the brain. Indeed stress, as well as alcohol, induces dopamine release in the striatum and in fronto-cortical circuits (Dazzi et al. 2002; Diana et al. 1992; Engleman et al. 2006; Puglisiallegra et al. 1991). In rodent studies, employing social isolation or defeat as a stress factor, stress has been observed to disrupt functions in both the amygdala and the prefrontal cortex as well as to interfere with dopaminergic modulation of these areas: uncontrollable episodes of social defeat stress produced tolerance to opiate analgesia and behavioral sensitization to challenges with amphetamine and cocaine, and these effects persisted for months; the authors suggested that the disrupted monoamine neurotransmitter balance within mesocorticolimbic brain structures contributes to increased drug intake among animals exposed to stress (Miczek et al. 2004). Similar social stress factors (i.e., social exclusion stress) have been associated with increased striatal dopamine concentrations (modulating levels of D2 receptor binding) in nonhuman primates; moreover, the alterations in D2 receptor levels correlated with increased intake of dopaminergic drugs of abuse (Nader and Czoty 2005). Of note here, elevated dopamine levels in the PFC and ventral striatum (including the nucleus accumbens) were associated with the initiation of aggressive attacks and threatening behavior in rodents (de Almeida et al. 2005). Other animal studies indicate that both alcohol and stress (i.e., conditioned fear) induce dopamine release in the amygdala (McBride 2002; Yokoyama et al. 2005). In humans, dopamine synthesis capacity in the amygdala (measured with 6-[(18)F]fluoro-L-DOPA positron emission tomography) correlated with functional activation of the amygdala and the anterior cingulate cortex elicited by aversive pictures (Kienast et al. 2008). To summarize, acute alcohol intake disrupts executive functions, including prefrontal control of emotional behavior. In addition, alcohol increases (dopaminergic) activation of the striatum and sensitizes responses in the amygdala. Given that both brain areas are implied in the processing of social conflicts and related emotions, alcohol's effect on these brain areas may indirectly promote aggressive behavior.

Figure 1 illustrates the developmental, social, cognitive and biological factors identified in the literature to increase risk for alcohol-related aggression and violence. Alcohol outcome expectancies are acquired beliefs about the effects of alcohol on emotion, cognition and behavior; individuals with heavier alcohol use demonstrate more robust expectancies (i.e., for alcohol to promote aggression). Social learning experiences, derived from parental modeling as well as peer and cultural norms, help to shape expectancies for the effects of alcohol on behavior. Early maltreatment in animal models and among humans is associated with increased risk for problematic alcohol use as well as increased DA release in the striatum (i.e., disrupted reward processing, overvaluation of alcohol-related stimuli; dysfunctional reward-associated learning; impulsive reward-seeking). The acute effects of alcohol disrupt cognitive processes in a number of ways and function to increase disinhibition (i.e., reduced self-monitoring), alcohol myopia (i.e., narrowing of attention; attention awarded to instigating social cues at the expense of mitigating social cues), hostile attribution bias (i.e., increased threat perception; perceive potentially ambiguous cues as threatening). The latter might be caused by disinhibited amygdala function (and therefore increased emotional responses towards aversive and threatening stimuli), caused by decreased prefrontal inhibitory projections.

Executive functioning (e.g., planning, inhibition, judgment, mental flexibility) is disrupted by the acute effects of alcohol and is impaired among many individuals with alcohol use disorder. In addition, lower executive functioning is a risk factor for aggressive and violent behavior, independent of alcohol ostensibly because it reflects reduced capacity to employ flexible and adaptive self-management strategies. Acute and chronic [problematic] alcohol use also disrupts cortical processes in the prefrontal cortex where executive functions are localized. Such disruptions have been shown to contribute to sub-optimal and impulsive decision making, as well as impaired flexible set-shifting – a cognitive capacity which helps individuals adapt to constant changes in the environment.

4 Human and Animal Research: Vulnerability Factors and Explanatory Theories

Research in animals and in humans emphasizes that only a subset of individuals becomes aggressive after having consumed alcohol. Accordingly, a number of individual factors have been shown to increase vulnerability to engaging in violent behavior following acute alcohol intake. Among them are reduced executive functioning and alcohol outcome expectancies (see Section II, Chermack and Taylor 1995; Giancola 2004; Ito et al. 1996; Rohsenow and Bachorowski 1984), gender, i.e., a higher risk for males (Giancola et al. 2009; Giancola and Zeichner 1995), personality features such as trait anger (Parrott and Zeichner 2002), high irritability (Godlaski and Giancola 2009), sensation seeking (Cheong and Nagoshi 1999) and lack of empathy (Giancola 2003), maladaptive motives for drinking, e.g., drinking to cope (Mihic et al. 2009) as well as social approval of aggression as an acceptable social interaction (Levinson and Giancola 2011). Moreover, situational aspects of interpersonal aggression have to be taken into account such as provocation (Giancola et al. 2002, Taylor et al. 1979), threat (Taylor et al. 1976) and social pressure (Taylor and Sears 1988). This section reviews key individual difference vulnerability factors that confer risk for problematic alcohol use and aggressive behavior (see Fig. 1).

4.1 *Neurobiological Vulnerability Markers: Dopamine, Serotonin, and Genetics*

4.1.1 Rodent Models

Studies in rodents show that increased levels of dopamine in the prefrontal cortex and ventral striatum (de Almeida et al. 2005) and an associated low availability of striatal dopamine D2 receptors (Dalley et al. 2007) are associated with impulsivity and the initiation of aggressive behavior. However, individual differences in dopamine function, e.g., dopamine release measured via microdialysis, have not been directly associated with levels of aggressive behavior (as indexed by aggressive acts toward an intruder in the animals' cage) (van Erp and Miczek 2007). Variations in serotonergic neurotransmission, on the other hand, do seem to differentiate between rodents with and without alcohol-heightened aggression. Chiavegatto et al. (2010) observed a reduced serotonin receptor expression for nearly all receptor subtypes (except for 5–HT3 receptors) in the prefrontal cortex of mice exhibiting alcohol-related aggression. Another study investigating in vivo changes in serotonin levels during aggressive behavior observed a sustained decrease in serotonin in the frontal cortex to 80 % of baseline levels during and after the confrontation. This finding supports the hypotheses that cortical serotonin deficits mediate an increased propensity for violent behavior, including the relation between alcohol and aggression (van Erp and Miczek 2000).

Different serotonin receptors and the localization of these receptors and their interaction with other neurotransmitter systems appear to play specific roles with respect to their effects on aggressive behavior (Chiavegatto et al. 2010). For example, application of a GABA-B agonist appears to increase aggressive behavior when it is applied in the dorsal raphe area, where it activates serotonergic neurons. This effect can be blocked by a 5-HT1A agonist, i.e. by the activation of auto-receptors that inhibit serotonergic projections (Takahashi et al. 2010a, b). Increased aggression following GABA-B agonist application in the dorsal raphe is independent of alcohol intake (Takahashi et al. 2010a, b), On the other hand, the application of a 5-HT1B agonist increases aggression in rodents with a history of alcohol self-administration when this application is localized in the medial prefrontal cortex, while *systemic* application of a 5-HT1B agonist appears to reduce aggression in mice with a history of alcohol self-administration (Fish et al. 2008; Takahashi et al. 2010b). In the medial prefrontal cortex of mice with alcohol related aggression 5-HT1B auto-receptor stimulation blocks local 5HT release and can reduce local GABAergic inhibition and may thus facilitate aggressive behavior manifestations (Cai et al. 2002; Fink and Gothert 2007; Faccidomo et al. 2008; Fish et al. 2008). These findings suggest that impaired serotonergic neurotransmission in the medial prefrontal cortex contributes to alcohol-associated aggression. Altogether, such observations support the hypothesis that alcohol consumption interacts with various serotonin receptors to modulate aggression following alcohol intake particularly via interactions with PFC functioning (Chiavegatto et al. 2010; Takahashi et al. 2010b).

Specific effects of different serotonin receptors with respect to alcohol-associated aggression may also depend upon the interaction with GABAergic neurons. For example, agonists of 5-HT1A and 5-HT1B receptors have been shown to inhibit GABAergic neurons when applied in the ventral tegmental area, amygdala or hippocampus; in turn, inhibition of GABAergic interneurons appears to facilitate striatal dopamine release (Cai et al. 2002; Fink and Gothert 2007). Conversely, agonists of 5-HT2A and 5-HT3 receptors increase GABA release in the frontal cortex and hippocampus of rats (Cai et al. 2002; Fink and Gothert 2007). Given that GABAergic neurotransmission per se plays a major role in the context of alcohol consumption (alcohol stimulates GABA release as mentioned in the introduction) as well as aggression (administration of GABA-B agonists increase aggressive behavior in rodents), its interplay with the serotonergic neurotransmitter system is crucial for the understanding of alcohol-associated aggression.

4.1.2 Primate Models

Animal experiments in nonhuman primates have helped to confirm the critical role of serotonergic neurotransmission in heightened vulnerability toward alcohol-related aggression. Moreover, these studies emphasize the importance of the interaction between neurobiology and environmental influences. Studies in rhesus monkeys showed that early social isolation is a massive stressor (Clarke et al. 1996; Jones et al. 1992, Higley et al. 1996a, b) and induced a long-term, trait-like reduction in central serotonergic neurotransmission (Barr et al. 2004; Higley et al. 1996a, b), which in turn was found to be associated with increased excessive alcohol intake as well as impulsive aggression. Monkeys who were separated from their mothers directly after birth had a trait-like reduction of the serotonin metabolite 5-hydroxyindoleacetic acid (5–HIAA) in the cerebrospinal fluid (CFS) compared to their mother-reared companions, which persisted into adulthood (Higley et al. 1996a, b). Imaging studies went one step further and showed that adult monkeys who had experienced early isolation from their mothers and who had low levels of 5–HIAA levels in the CSF displayed an increased availability of brainstem (raphe) serotonin transporters (5–HTTs) (Heinz et al. 1998).

Interestingly, further studies revealed that a reduction of 5–HIAA concentrations was only found in monkeys carrying at least one "short" copy of the SLC6A4 (5–hydroxytryptamine transporter gene)-linked polymorphic region (5-HTTLPR). This short (s) copy of the 5HTTLPR is associated with decreased transporter expression and function in primates and in humans (Bennett et al. 2002; Lesch et al. 1996; Reimold et al. 2007). This s-allele was also associated with elevated levels of aggressiveness in male nonhuman primates, and levels of aggression were especially heightened among monkeys who were also exposed to early social isolation stress (Schwandt et al. 2010). Monkeys carrying such a short copy of the 5-HTTLPR displayed elevated 5-HTT availability in the brainstem (raphe) area following social isolation such authors concluded that can availability could be due to reductions in serotonin turnover and hence reduced competition between endogenous serotonin and the (displaceable) radioligand used to bind to serotonin transporters (Heinz et al. 1998). This hypothesis was supported by further studies, in which serotonin content was depleted and later restored: endogenous serotonin restoration measurably displaced serotonin transporter radioligand [123I]beta-CIT binding in the brainstem (raphe area). Moreover, the higher the previous 5-HIAA levels in the cerebrospinal fluid (CFS), the larger the displacement of the radioligand β-CIT from brainstem (raphe) serotonin transporters following restoration of normal serotonin levels (Heinz et al. 2004a). These findings confirm that monkeys with higher cerebrospinal fluid 5-HIAA levels display increased extracellular serotonin concentrations, which in vivo compete with radioligand binding to serotonin transporters (Heinz et al. 1998, 2004a). Monkeys that displayed a trait-like reduction in CSF 5-HAA levels and who showed negative correlation between CSF 5-HIAA levels and serotonin transporters (i.e., lower 5-HIAA levels associated with higher serotonin transporter availability) following social isolation (Heinz et al. 1998, 2004a) thus appear to have a reduced overall in central serotonin turnover.

When radioligands are used that are not displaced by endogenous serotonin, the absolute density of transporters can be measured in vivo. Such studies have revealed that rhesus monkeys with reduced CSF 5-HIAA levels following social isolation stress, also display a reduction in absolute 5-HTT density in the brainstem, thalamus and striatum (Ichise et al. 2006; Hummerich et al. 2004). These findings suggest that early social isolation stress severely impairs serotonergic neurotransmission, causing a long-lasting, trait-like reduction in serotonin turnover rates and also a reduction in absolute serotonin transporter density. Importantly, these alterations in serotonin turnover and transporter density were correlated with self-initiated aggression, increased anxiety and impaired social competence (i.e., lack of reciprocal behaviors such as grooming) (Heinz et al. 1998). Also, studies in rhesus monkeys confirmed observations in rodents with respect to the interplay of serotonergic neurotransmission and GABAergic activation in the prefrontal cortex: animals with a low level of serotonin turnover, measured as low concentrations of CSF 5-HIAA, where less sensitive to stimulation of GABA-A receptors in the prefrontal cortex (Doudet et al. 1995; Heinz et al. 1998, 2001).

These observations suggest that the inhibitory effects of prefrontal GABA-A receptor stimulation are reduced in rhesus monkeys with low serotonin turnover following social isolation stress. This finding is particularly interesting with respect to alcohol-associated aggression: alcohol is a GABA-A receptor stimulant, and hence the inhibitory and sedative effects of alcohol appear to be reduced in rhesus monkeys exposed to early social isolation stress, who also display impaired serotonergic neurotransmission. Behaviorally, these monkeys were less sedated and ataxic following acute alcohol intake. Importantly, a low level of alcohol-associated sedation is a partially heritable factor, which predisposes to excessive alcohol intake in humans (Hinckers et al. 2006; Schuckit 1994). Individuals showing such a low level of acute alcohol responses can consume more alcohol before they are strongly sedated, and it has also been argued that humans with a low level of response to alcohol intake lack an internal warning sign to signal that too much ethanol is being consumed (Heinz et al. 2011). In accordance with these ideas, it was observed that rhesus monkeys with low serotonin turnover (low CSF 5-HIAA levels) and associated increase in radioligand binding to brainstem serotonin transporters also display a low response to acute alcohol and high levels of alcohol intake. Moreover, the amount of alcohol intake in a free-choice paradigm was directly and positively correlated with the relative availability of brainstem serotonin transporters (Heinz et al. 2003). In humans, a genetic disposition toward a high availability of brainstem 5-HTT was also associated with a low acute response to alcohol and increased consumption in adolescents (Hinckers et al. 2006).

These experiments suggest that both genetic and environmental factors can interact with serotonergic neurotransmission and hence with alcohol intake. Indeed, a relatively high availability of serotonin transporters (with respect to turnover rates) may either result from early social isolation experiences or be directly caused by the genetic constitution of the serotonin transporter per se (Heinz et al. 2000). In both situations, the high availability of serotonin transporters in the brainstem (raphe) area can affect serotonin release in different projection sites, with blunted release in the medial prefrontal cortex known to interact with GABAergic inhibition and hence alcohol-associated sedation (Doudet et al. 1995; Heinz et al. 1998, 2001; Faccidomo et al. 2008; Fish et al. 2008). Studies in nonhuman primates confirm that social stress factors can exert direct impact upon serotonin transporter availability and serotonin neurotransmission, for example in the prefrontal cortex, where it modifies local alcohol effects and facilitates both excessive alcohol intake as well as aggressive behavior (Clarke et al. 1996; Doudet et al. 1995; Heinz et al. 1998, 2001; Higley et al. 1991, 1993). Finally, evidence for the critical role of social stress factors is further strengthened by results of a twin study (Johnson et al. 1996) showing that poor home placement and other stress factors during childhood were associated with increased alcohol intake later on.

4.2 Human Studies: Stress, Genetic Effects on Serotonergic Neurotransmission, and Aggression

As discussed in detail elsewhere in this volume, the role of serotonergic neurotransmission in the pathophysiology of aggressive behavior in humans is supported by genetic studies and may be mediated by serotonin effects on the processing of aversive, threatening stimuli. Indeed, the 5-HTTLPR has been associated with increased amygdala activation in response to aversive pictures: subjects carrying the short allele (that is also associated with lower serotonin transporter expression and function) display higher levels of functional activation of the amygdala when exposed to aversive versus neutral stimuli (Heinz et al. 2005; Pezawas et al. 2005). These subjects were also at higher risk for depression and other negative mood states following exposure to traumatic life events (Caspi et al. 2002; Karg et al. 2011; but see Risch et al. 2009). The environmental effects on serotonergic neurotransmission may be specific to certain neurodevelopmental periods. For example, another genetic variation that affects serotonergic neurotransmission, a single mutation in the gene that encodes monoamine oxidase A (MAO-A), has been associated with aggressive behavior when subjects were exposed to trauma during their early development (Brunner et al. 1993, Kim-Cohen et al. 2006).

Of note, MAO-A levels are modulated by tobacco use and hence not surprisingly, a PET study observed no direct correlation between MAO-A genotype and expression in vivo (Freedman et al. 2005). Rather, MAO-A genotype may interact with neurodevelopmental processes including the structure of serotonergic innervations independent of MAO-A availability in adults. Also with respect to 5-HTTLPR, tobacco use interfered with the impact of the serotonin transporter genotype on in vivo transporter expression: short-allele carriers only displayed reduced transporter levels when they did not smoke (Kobiella et al. 2011).

Several studies suggest that the MAO-A genotype is correlated with alcohol intake and negative mood states (Saraceno et al. 2009). As with the 5-HTTLPR, the same genotype that interacts with human aggression also appears to affect alcohol intake and the risk for mood disorders. As hypothesized, this MAO-A genotype was associated with functional activation in the amygdala during implicit processing of angry and fearful faces and in the cingulate cortex during perceived social rejection (Eisenberger et al. 2007). MAO-A genotype also appears to affect the interaction between the amygdala and the ventromedial prefrontal cortex (vmPFC) such that carriers of the low-expressing genetic variant exhibited dysregulated amygdala activation and increased functional coupling with vmPFC. However, this effect was limited to men and not found in women (Buckholtz et al. 2008). A comparable role of the 5-HTTLPR has been observed with respect to functional coupling between amygdala activation elicited by aversive stimuli and the medial prefrontal cortex (Heinz et al. 2005). Furthermore, both the MAO genotype and the serotonin transporter genotype (5-HTTLPR) appear to increase aggressive behavior when humans have been exposed to early social adversity (Caspi et al. 2002; Kim-Cohen et al. 2006; Reif et al. 2007).

In all, these data suggest that a specific interaction between developmentally early social stress factors and genetic variations affecting serotonin reuptake and metabolism appear to play a key role in the predisposition toward affect dysregulation and aggressive behavior. These studies suggest that both the MAO and 5-HTT genotype increase sensitivity to social rejection and enhance limbic activation, particularly amygdala activation, when confronted with threatening environmental stimuli (Eisenberger et al. 2007; Heinz et al. 2005). However, these effects do not necessarily appear to be dependent on current expression of MAO-A or serotonin transporter levels (Freedman et al. 2005; Kobiella et al. 2011). Rather, serotonin transporter genotype directly interacted with amygdala volume and indirectly with functional activation of the amygdala, which was itself dependent upon amygdala volume. This observation supports the hypothesis that variation in genes affecting serotonergic neurotransmission can directly impact volume and thus also functional response of neuronal structures, e.g., in the amygdala, when individuals are confronted with threatening stimuli. Altogether, animal experiments and human studies show that both environmental and genetic factors regulate the expression and sensitivity of serotonin receptors and transporters as well as serotonin metabolism, which in turn modulate the risk to develop aggressive behavior. The manifestation of aggressive behavior and its neurobiological correlates appears to be specifically associated with developmental insults such as early social isolation stress, adversity or maltreatment. These observations highlight the importance of early social and therapeutic support for children exposed to adversity and maltreatment.

5 Chronic Alcohol Intake and Aggression in Humans

Chronic alcohol consumption as well as alcohol dependence impact a number of neural systems and processes that may also increase the risk for alcohol-related aggression. Chronic alcohol intake exerts a substantial impact on monoaminergic, specifically serotonergic, neurotransmission, which is known to play a major role in aggressive behavior. Further, chronic alcohol and nicotine consumption is associated with a reduction in serotonin transporter availability in the brainstem as well as other changes in monoaminergic, GABAergic and glutamatergic neurotransmission. Interestingly, this reduction in serotonin transporter level reverses during detoxification from alcohol, but only among nonsmokers (Cosgrove et al. 2009; Heinz et al. 2000).

There are several lines of evidence to suggest that the association between chronic alcohol intake and aggressive behavior is mediated by changes in serotonergic functioning (see Fig. 1). First, alcohol-dependent patients with early onset of the disease and a greater genetic disposition toward alcohol reduced in turnover nuerotransmission as evidenced by low serotonin metabolites (5-HIAA) in the cerebrospinal fluid and by increased levels of impulsive aggression and anxiety (Bohman et al. 1987; Cloninger 1987, Virkkunen et al. 1994). In a similar vein, human twin and adoption studies demonstrate an increased risk for both alcohol dependence and violent behavior, while they failed to show a significant heritability for violent behavior independent of alcohol use disorders (Bohman et al. 1982; Carey 1996; Johnson et al. 1996). These studies suggest that there is a common hereditary factor, which increases the risk for both alcohol dependence and increased aggressive behavior, and which may impact on 5-HT neurotransmission.

5.1 *Perception of Threat and Anxiety*

As described in Section III, limbic brain areas such as the amygdala exercise a prominent role in the regulation of emotional behavior, e.g., related to the perception of threat or fear. It has been suggested that heightened levels of aggression as well as alcohol intake are due to a failure to learn from aversive outcomes

(Cloninger 1987; Gray 1982), and that this failure is due to serotonin dysfunction, which impairs behavioral inhibition (Gray 1982). However, rather than being unable to experience punishment, increased attention to threatening stimuli may promote aggression in subjects with chronic alcohol intake. Indeed, serotonergic dysfunction was associated with negative mood states and violent suicide attempts in humans suffering from affective and addictive disorders (Heinz et al. 2001; Mann et al. 1996). Accordingly, impaired serotonergic neurotransmission may facilitate processing of aversive and threatening environmental stimuli, thus predisposing an individual to react more strongly to threatening environmental cues (Heinz et al. 2011). Moreover, it is known that 5–HTTLPR short (s) allele carriers (the genetic variant linked with decreased transporter expression and function) display an increased amygdala response when confronted with affectively negative stimuli (Heinz et al. 2005; Pezawas et al. 2005) (see Fig. 1). Such genetic effects on serotonin transporter expression may be additionally heightened in alcohol-dependent patients, who show reduced 5–HTT availability in the raphe area, depending on their 5–HTT genotype (Heinz et al. 1998, 2000).

The individual may then, depending on circumstances and previous experiences, react with either anxious withdrawal or impulsive aggression toward the threatening stimuli (Heinz et al. 2001). Excessive alcohol intake may represent a failed attempt to cope with heightened presence of threatening environmental cues and associated negative mood states and - due to alcohol-related neurotoxicity - further impair serontonin function. In this context, acute as well as chronic reductions in central and prefrontal serotonin availability can reduce inhibitory functions and at the same time increase behavioral rigidity, eventually contributing to inflexible response patterns toward socially threatening contexts and thus facilitating aggressive acts (Clarke et al. 2004; Chamberlain et al. 2006). A double-blind, randomized, and placebo-controlled study showed that fluoxetine, a selective serotonin reuptake-inhibitor that blocks serotonin transporters and increases synaptic 5-HT and reduces anger and aggression in a group of alcohol-dependent perpetrators of domestic violence (George et al. 2011). Nevertheless, such data are only correlational and prospective data are needed to determine whether chronic alcohol intake or excessive alcohol use predispose to aggression via serotonergic dysfunction. Such data would also facilitate identification of other factors beyond serotonergic neurotransmission that link excessive alcohol use and aggression.

In summary, there is growing evidence that changes in serotonergic neurotransmission, potentially arising from genetic factors or early socially adverse experiences (Heinz et al. 1998, 2000), can predispose to a low intoxicating effect of acute alcohol intake, which in turn facilitates excessive alcohol intake in adolescents in addition to impulsive aggression (Hinckers et al. 2006; Schuckit et al. 1999). Studies investigating the interaction between changes in monoaminergic neurotransmitter systems and brain activation associated with threatening cues, e.g., presented during alcohol administration, could address our current knowledge gap and help to further our understanding of the neurobiology of alcohol-related aggression.

5.2 *Neuroadaptations in Chronic Alcohol Consumption: A Link to Executive Functions, Impulsivity and Impulsive Aggressiveness*

An important characteristic of impulsivity in alcohol dependence (similar to other forms of drug addiction) is dysfunctional reward anticipation such that immediate rewards are preferred over delayed ones even though the latter, delayed reward would promise to be larger (Moeller et al. 2001). This phenomenon, termed delay discounting, is considered to be an important element of impulsivity and might be associated with early onset of alcohol dependence (preference of immediate "alcohol reward") and a number of social problems (Bechara 2005; Heinz et al. 2001). From a clinical perspective, addicted individuals often struggle to reject drugs on the basis of long-term outcomes e.g., quit smoking to improve health later in life. Such problems with consideration of future consequences, e.g., poor risk-forecasting, can lead to a "myopic" or narrow appreciation for the impact on current decisions on future well-being. Indeed, numerous studies have suggested that alcohol-dependent patients are more impulsive when compared to the general nonaddicted population (Dom et al. 2006; Rubio et al. 2008), and they display a higher rate of impulse control disorders including aggressive-impulsive behavior than observed among healthy control subjects (Lejoyeux et al. 1998; Virkkunen et al. 1994).

Impaired reward expectation is associated, at the neurobiological level, with alteration of the dopaminergically innervated ventral striatum, which in turn, is supposed to contribute to impulsiveness and impulsive aggression. In a rodent study, low striatal dopamine D2 receptor availability was associated with heightened impulsivity (Dalley et al. 2007). Although this reduced dopamine D2 receptor availability could be a heritable trait, animal as well as human studies have shown that chronic alcohol consumption also reduces D2 receptor availability and sensitivity (Heinz et al. 1996, 2004b; Rommelspacher et al. 1992; Volkow et al. 1996). These alterations in dopaminergic neurotransmission can functionally interfere with the expectation of rewards mediated by the dopaminergically innervated striatum. Indeed, a fMRI study in alcohol-dependent patients observed an association between impaired ventral striatal activity during the processing of reward-indicating cues and enhanced impulsivity (Beck et al. 2009). It was suggested that impulsive subjects might have problems tracking reward expectation over various periods of time, rendering them prone toward discounting rewards further out on the temporal horizon (Hariri et al. 2006).

The ventral striatum is a core region of the brain reward system and reduced sensitivity in this region could lead to increased reward seeking behavior, especially for immediate rewards like alcohol. Increased propensity for reward seeking can thus predispose individuals to involvement in risky situations including aggressive interactions (Corte and Sommers 2005). Given that impulsivity is a robust correlate of aggressive behavior (Edwards et al. 2003; McCloskey et al. 2008), this personality trait could be a mediating factor between (acute and chronic) alcohol intake and aggression. Apart from delay discounting, further aspects of impulsivity include problems with response inhibition as well as decision making and (rapidly) changing intentions and goals (Barratt 1982). Acute alcohol has been shown to increase these cognitive and behavioral facets of impulsivity (Dougherty et al. 2008; Marczinski et al. 2007).

In addition to dopamine, serotonergic dysfunction can contribute to impulsive behavior via impairment of response inhibition (Cloninger 1987; Higley and Linnoila 1997). That is, the effect of chronic alcohol consumption on serotonergic neurotransmission may further augment alcohol-associated impulsiveness and aggressive behavior. For instance, serotonin and dopamine turnover rates acutely increase after chronic alcohol consumption in alcohol-preferring animals, which suggests that low dopamine and serotonin turnover rates of animals predisposing to excessive alcohol intake are counteracted by alcohol-induced dopamine and serotonin release; however, chronic alcohol intake impairs both neurotransmitter systems (Heinz et al. 2001; Lemarquand et al. 1994). Indeed, in rodents with alcohol-induced aggression, there was a reduced serotonin release in the prefrontal cortex following application of a 5-HT1B agonist (Faccidomo et al. 2008). This finding suggests that chronic alcohol consumption can interact with serotonergic functioning in the prefrontal cortex among individuals with a disposition toward alcohol-associated aggression. This hypothesized interaction may promote aberrant activation in prefrontal cortex and thus increases the likelihood of impaired behavioral control.

Indeed, as described in Section III, various studies in animals and humans indicate that serotonergic dysfunction interferes with different functions of the PFC, e.g., response inhibition, behavioral control as well as flexible adaptation in social situations involving confrontation or hostility (Chamberlain et al. 2006; Clarke et al. 2004). Correspondingly, a study among impulsive-aggressive individuals showed that PFC activation (including orbitofrontal, ventral medial frontal and cingulate cortex) is blunted compared with control subjects after a serotonergic challenge with fenfluramine, indicating impaired serotonergic modulation of the frontal cortex in some subjects with heightened aggression (Siever et al. 1999). In the context of chronic alcohol consumption, similar executive functions are impaired in alcohol-dependent patients, as indexed by both behavioral (Giancola and Moselhy 1998; Loeber et al. 2010; Ratti et al. 2002; Sullivan et al. 2000) and neurobiological measures (Moselhy et al. 2001). Given that the association between alcohol and aggression seems to be most prominent in individuals with impaired executive functions (Giancola 2004), we suggest that alcohol-dependent persons—with at least a substantial part of their executive impairment due to chronic alcohol consumption—might be more vulnerable to react aggressively under alcohol. However, there are no studies to date testing this hypothesis directly.

6 Future Directions

Much remains to be studied concerning alcohol-related aggression and more specifically why some individuals aggress under alcohol whereas others do not. For instance, there is some evidence suggesting that trait impulsivity (Cunradi and Ames 2011; Lejuez et al. 2010) and low distress tolerance (Anestis et al. 2011; Daughters et al. 2008; Wray et al. 2012) help to explicate the relation between alcohol and aggression. That is, individuals who struggle to inhibit behavior or delay gratification and who have problems tolerating uncomfortable emotional states may be at greater risk for alcohol-related violence. Future research should examine whether impulsive behavior is associated with increased threat perception (Heinz et al. 2005) and whether the neural correlates of these multifaceted traits determine whether targeted interventions addressing these risk factors reduce the incidence and social costs of alcohol-related aggression.

Experimental data indicate that the effects of alcohol-induced aggression are more pronounced in men than women, e.g., Bushman 1997; Bushman, and Cooper 1990; Giancola et al. 2002). Therefore it may not be surprising that the preponderance of research on aggression among individuals with AUD is characterized by an underrepresentation of women. Researchers have called for increased focus on alcohol-related aggression among women, (e.g., Collins et al. 2007), as women do perpetrate a significant amount of alcohol-related violence, and in some circumstances, more so than males, e.g., Wells et al. 2007, 2008. Of particular importance, alcohol abuse by parents has deleterious effects on children, e.g., Maffli 2001. Data from a national survey of 2,760 families with children revealed that maternal alcohol use (but not drug use) was significantly associated with physical abuse toward children (Berger 2005). Finally, the co-occurrence of alcohol-related victimization and perpetration in the *same* encounter is more unique to women (Parks et al. 2008). As such, women (as compared to men) are more susceptible to completing a cycle of alcohol-related violence.

Neuroscience-based computerized cognitive training interventions have already been shown to drive enduring plastic changes in cortical processes in a variety of clinical populations (Vinogradov et al. 2012). More recently they have shown promising potential to help cognitively rehabilitate individuals with AUD in the context of traditional treatment, (e.g., Fals-Stewart and Lam 2010; Houben et al. 2011; Rupp et al. 2012). Improvements in cognitive functioning may increase patients' capacity to absorb clinical material from existing, empirically supported treatments, and help them to effectively recall and utilize more adaptive coping skills in provocative circumstances (i.e., to inhibit the urge to aggress). Additionally, the tendency for individuals with AUD to discount future rewards (avoiding injury, incarceration) in favor of immediate rewards (revenge, drinking) may also be reduced with cognitive training, e.g., Bickel et al. 2011. Further research is needed to determine whether aggressive behavior in AUD populations may be reduced with cognitive training by strengthening executive functions critical to self-regulation, and whether neurobiological factors can help to identify subgroups of patients who profit most from certain training modules.

Acknowledgments Support for this project was provided in part by VA Health Services Research and Development Funds for Dr. Adrienne Heinz.This work was also supported by the German Research Foundation (DFG: grants DFG FOR 16/17, HE 2597/14-1; Excellence Cluster EXC 257) and by the Bundesministerium für Bildung und Forschung (grants 01GQ0411 and 01QG87164 and Nationales Genomforschungsnetz grant 01 GS 08 159) for Prof. Andreas Heinz.

References

Adolphs R (2009) The social brain: neural basis of social knowledge. Annu Rev Psychol 60:693–716

Amaral DG, Price JL (1984) Amygdalo-cortical projections in the monkey (macaca-fascicularis). J Comp Neurol 230:465–496

Anestis MD, Lavender JM, Marshall-Berenz EC, Gratz KL, Tull MT, Joiner TE (2011) Evaluating Distress Tolerance Measures: Interrelations and Associations with Impulsive Behaviors. Cognit Ther Res 36:593–602

Arseneault L, Moffitt TE, Caspi A, Taylor PJ, Silva PA (2000) Mental disorders and violence in a total birth cohort: results from the Dunedin Study. Arch Gen Psychiatry 57(10):979–986

Bailey D, Leonard K, Cranston J, Taylor S (1983) Effects if alcohol and self-awareness on human physical aggression. Pers Soc Psychol Bull 9:289–295

Bandura A (1977) Social learning theory. Prentice-Hall, Englewood Cliffs

Barbas H (2000) Connections underlying the synthesis of cognition, memory, and emotion in primate prefrontal cortices. Brain Res Bull 52:319–330

Barr CS, Newman TK, Lindell S, Shannon C, Champoux M, Lesch KP, Suomi SJ, Goldman D, Higley JD (2004) Interaction between serotonin transporter gene variation and rearing condition in alcohol preference and consumption in female primates. Arch Gen Psychiatry 61:1146–1152

Barratt ES (1982) Impulsivity, behavioral dyscontrol, and conscious awareness. Behav Mot Contr Psychiatr Disord S04:217

Bates ME, Bowden SC, Barry D (2002) Neurocognitive impairment associated with alcohol use disorders: Implications for treatment. Exp Clin Psychopharmacol 10:193–212

Bartholow BD, Heinz A (2006) Alcohol and aggression without consumption: alcohol cues, aggressive thoughts, and hostile perception bias. Psychol Sci 17(1):30–37

Beck A, Schlagenhauf F, Wüstenberg T, Hein J, Kienast T, Kahnt T, Schmack K, Hägele C, Knutson B, Heinz A, Wrase J (2009) Ventral striatal activation during reward anticipation correlates with impulsivity in alcoholics. Biol Psychiatry 66:734–742

Bennett AJ, Lesch KP, Heils A, Long JC, Lorenz JG, Shoaf SE, Champoux M, Suomi SJ, Linnoila MV, Higley JD (2002) Early experience and serotonin transporter gene variation interact to influence primate CNS function. Mol Psychiatry 7:118–122

Bechara A (2005) Decision making, impulse control and loss of willpower to resist drugs: a neurocognitive perspective. Nature Neurosci 8:1458–1463

Berger L (2005) Income, family characteristics, and physical violence toward children. Child Abuse Negl 29:107–133

Bickel WK, Yi R, Landes RD, Hill PF, Baxter C (2011) Remember the future: working memory training decreases delay discounting among stimulant addicts. Biol Psychiatry 69(3):260–265

Birbaumer N et al (2005) Deficient fear conditioning in psychopathy: a functional magnetic resonance imaging study. Arch Gen Psychiatry 62:799–805

Bohman M, Cloninger CR, Sigvardsson S, Vonknorring AL (1982) Predisposition to petty criminality in Swedish adoptees. I. Genetic and environmental heterogeneity. Arch Gen Psychiatry 39:1233–1241

Bohman M, Cloninger R, Sigvardsson S, Vonknorring AL (1987) The genetics of alcoholisms and related disorders. J Psychiatr Res 21:447–452

Brown SA (1985) Reinforcement expectancies and alcoholism treatment outcome after a one year follow-up. J Stud Alcohol Drugs 46:304–308

Brunner HG, Nelen M, Breakefield XO, Ropers HH, Vanoost BA (1993) Abnormal behavior associated with a point mutation in the structural gene for monoamine oxidase A. Science 262:578–580

Buckholtz JW, Callicott JH, Kolachana B, Hariri AR, Goldberg TE, Genderson M, Egan MF, Mattay VS, Weinberger DR, Meyer-Lindenberg A (2008) Genetic variation in MAOA modulates ventromedial prefrontal circuitry mediating individual differences in human personality. Mol Psychiatry 13:313–324

Bundeskriminalamt (2011) Polizeiliche Kriminalstatistik (PKS) 2011 - IMK-Kurzbericht

Bushman BJ, Cooper HM (1990) Effects of alcohol on human-aggression: an integrative research review. Psychol Bull 107(3):341–354

Bushman BJ (1997) Effects of alcohol on human aggression. In: Galanter M (ed) Recent developments in alcoholism. Plenum Press, New York, pp 227–244

Byrnes JM, Doran CM, Shakeshaft AP (2012) Cost per incident of alcohol-related crime in New South Wales. Drug alcohol rev 31(7):854–860

Caetano R, Schafer J, Cunradi CB (2001) Alcohol-related intimate partner violence among white, black, and hispanic couples in the united states. Alcohol Res Health 25:58–65

Cai X, Flores-Hernandez J, Feng J, Yan Z (2002) Activity-dependent bidirectional regulation of GABA(A) receptor channels by the 5–HT4 receptor-mediated signalling in rat prefrontal cortical pyramidal neurons. J Physiol 540:743–759

Carey G (1996) Aggression and violence: genetic, neurobiological, and biosocial perspectives. Stoff DM, Cairns RB (eds) Lawrence Earlbaum Associates, New Jersey, pp 3–21

Caspi A, McClay J, Moffitt TE, Mill J, Martin J, Craig IW, Taylor A, Poulton R (2002) Role of genotype in the cycle of violence in maltreated children. Science 297:851–854

Chamberlain SR, Müller U, Blackwell AD, Clark L, Robbins TW, Sahakian BJ (2006) Neurochemical modulation of response inhibition and probabilistic learning in humans. Science 311:861–863

Charlet K, Beck A, Jorde A, Wimmer L, Vollstädt-Klein S, Gallinat J, Walter H, Kiefer F, Heinz A. (2013) Increased neural activity during high working memory load predicts low relapse risk in alcohol dependence. Addict Biol. doi: 10.1111/adb.12103 [Epub ahead of print]

Cheong J, Nagoshi CT (1999) Effects of sensation seeking, instruction set, and alcohol/placebo administration on aggressive behavior. Alcohol 17:81–86

Chermack ST, Blow FC (2002) Violence among individuals in substance abuse treatment: the role of alcohol and cocaine consumption. Drug Alcohol Depend 66(1):29–37

Chermack ST, Giancola PR (1997) The relation between alcohol and aggression: an integrated biopsychosocial conceptualization. Clin Psycho Rev 17:621–649

Chermack ST, Taylor SP (1995) Alcohol and human physical aggression: pharmacological, versus expectancy effects. J Stud Alcohol 56:449–456

Chiavegatto S, Quadros IMH, Ambar G, Miczek KA (2010) Individual vulnerability to escalated aggressive behavior by a low dose of alcohol: decreased serotonin receptor mRNA in the prefrontal cortex of male mice. Genes Brain Behav 9:110–119

Clarke AS, Hedeker DR, Ebert MH, Schmidt DE, McKinney WT, Kraemer GW (1996) Rearing experience and biogenic amine activity in infant rhesus monkeys. Biol Psychiatry 40:338–352

Coccaro EF, McCloskey MS, Fitzgerald DA, Phan KL (2007) Amygdala and orbitofrontal reactivity to social threat in individuals with impulsive aggression. Biol Psychiatry 62:168–178

Coid J, Yang M, Roberts A, Ullrich S, Moran P, Bebbington P, Brugha T et al (2006) Violence and psychiatric morbidity in the national household population of Britain: public health implications. Br J Psychiatry 189(1):12–19

Collins AM, Loftus EF (1975) A spreading activation theory of semantic processing. Psychol Rev 82:407–428

Collins JJ, Schlenger WE (1988) Acute and chronic effects of alcohol use on violence. J Stud Alcohol Drugs 49(06):516

Collins JJ, Messerschmidt PM (1993) Epidemiology of alcohol-related violence. Alcohol Health Res World 17:93–100

Collins RL, Quigley B, Leonard KE (2007) Women's physical aggression in bars: an event-based examination of precipitants and predictors of severity. Aggressive Behav 33(4):304–313

Corte CM, Sommers MS (2005) Alcohol and risky behaviors. Annu Rev Nurs Res 23:327–360

Cosgrove KP, Krantzler E, Frohlich EB, Stiklus S, Pittman B, Tamagnan GD, Baldwin RM, Bois F, Seibyl JP, Krystal JH, O'Malley SS, Staley JK (2009) Dopamine and serotonin transporter availability during acute alcohol withdrawal: effects of comorbid tobacco smoking. Neuropsychopharmacology 34:2218–2226

Clarke HF, Dalley JW, Crofts HS, Robbins TW, Roberts AC (2004) Cognitive inflexibility after prefrontal serotonin depletion. Science 304:878–880

Cloninger CR (1987) Neurogenetic adaptive-mechanisms in alcoholism. Science 236:410–416

Cunradi CB, Ames GM (2011) The Relationship of alcohol problems to the risk for unidirectional and bidirectional intimate partner violence among a sample of blue-collar couples. Violence Vict 26(2):147–158

Dalley JW, Fryer TD, Brichard L, Robinson ES, Theobald DE, Lääne K, Peña Y, Murphy ER, Shah Y, Probst K, Abakumova I, Aigbirhio FI, Richards HK, Hong Y, Baron JC, Everitt BJ, Robbins TW (2007) Nucleus Accumbens D2/3 receptors predict trait impulsivity and cocaine reinforcement. Science 315:1267–1270

Daughters SB, Sargeant MN, Bornovalova MA, Gratz KL, Lejuez CW (2008) The relationship between distress tolerance and antisocial personality disorder among male inner-city treatment seeking substance users. J Pers Disord 22(5):509–524

Davidson RJ, Putnam KM, Larson CL (2000) Dysfunction in the neural circuitry of emotion regulation — a possible prelude to violence. Science 289:591–594

Dazzi L, Serra M, Vacca G, Ladu S, Latrofa A., Trapani G, Biggio G (2002) Depletion of cortical allopregnanolone potentiates stress-induced increase in cortical dopamine output. Brain research 932(1):135–139

de Almeida RMM, Ferrari PF, Parmigiani S, Miczek KA (2005) Escalated aggressive behavior: dopamine, serotonin and GABA. Eur J Pharmacol 526:51–64

Diana M, Gessa GL, Rossetti ZL (1992) Lack of tolerance to ethanol-induced stimulation of mesolimbic dopamine system. Alcohol Alcohol 27:329–333

Di Chiara G, Acquas E, Tanda G (1996) Ethanol as a neurochemical surrogate of conventional reinforcers: the dopamine-opioid link. Alcohol 13:13–17

Dom G, D'Haene P, Hulstijn W, Sabbe B (2006) Impulsivity in abstinent early- and late-onset alcoholics: differences in self-report measures and a discounting task. Addiction 101:50–59

Doudet D, Hommer D, Higley JD, Andreason PJ, Moneman R, Suomi SJ, Linnoila M (1995) Cerebral glucose-metabolism, Csf 5–hiaa levels, and aggressive behavior in rhesus monkeys. Am J Psychiatry 152:1782–1787

Dougherty DM, Marsh-Richard DM, Hatzis ES, Nouvion SO, Mathias CW (2008) A test of alcohol dose effects on multiple behavioral measures of impulsivity. Drug Alcohol Depend 96:111–120

Easdon CM, Vogel-Sprott M (2000) Alcohol and behavioral control: impaired response inhibition and flexibility in social drinkers. Exp Clin Psychopharmacol 8:387–394

Edwards DW, Scott CL, Yarvis RM, Paizis CL, Panizzon MS (2003) Impulsiveness, impulsive aggression, personality disorder, and spousal violence. Violence Vict 18:3–14

Eisenberger NI, Way BM, Taylor SE, Welch WT, Lieberman MD (2007) Understanding genetic risk for aggression: clues from the brain's response to social exclusion. Biol Psychiatry 61:1100–1108

Engleman, EA, Ingraham, CM, McBride, WJ, Lumeng, L, Murphy, JM (2006) Extracellular dopamine levels are lower in the medial prefrontal cortex of alcohol-preferring rats compared to Wistar rats. Alcohol 38:5–12

Faccidomo S, Bannai M, Miczek KA (2008) Escalated aggression after alcohol drinking in male mice: dorsal raphe and prefrontal cortex serotonin and 5–HT1B receptors. Neuropsychopharmacology 33:2888–2899

Fals-Stewart W (2003) The occurrence of partner physical aggression on days of alcohol consumption: a longitudinal diary study. J Consult Clin Psychol 71(1):41–52

Fals-Stewart W, Lam WKK (2010) Computer-assisted cognitive rehabilitation for the treatment of patients with substance use disorders: a randomized clinical trial. Exp Clin Psychopharmacol 18(1):87–98

Fergusson DM, Horwood LJ (2000) Alcohol abuse and crime: a fixed-effects regression analysis. Addiction 95(10):1525–1536

Fink KB, Gothert M (2007) 5–HT receptor regulation of neurotransmitter release. Pharmacol Rev 59:360–417

Finn PR, Justus A, Mazas C, Steinmetz JE (1999) Working memory, executive processes and the effects of alcohol on Go/No-Go learning: testing a model of behavioral regulation and impulsivity. Psychopharmacology 146:465–472

Fish EW, McKenzie-Quirk SD, Bannai M and Miczek KA (2008) 5–HT1B receptor inhibition of alcohol-heightened aggression in mice: comparison to drinking and running. Psychopharmacology 197:145–156

Freedman NMT, Mishani E, Krausz Y, Weininger J, Lester H, Blaugrund E, Ehrlich D, Chisin R (2005) In vivo measurement of brain monoamine oxidase B occupancy by rasagiline, using C–11–L-deprenyl and PET. J Nucl Med 46:1618–1624

Fromme K, Stroot EA, Kaplan D (1993) Comprehensive effects of alcohol: development and psychometric assessment of a new expectancy questionnaire. Psychol Assess 5:19–26

George DT, Phillips MJ, Lifshitz M, Lionetti TA, Spero DE, Ghassemzedeh N, Doty L, Umhau JC, Rawlings RR (2011) Fluoxetine treatment of alcoholic perpetrators of domestic violence: a 12-week, double-blind, randomized, placebo-controlled intervention study. J Clin Psychiatry 72:60–65

Giancola PR, Zeichner A (1995) An investigation of gender differences in alcohol-related aggression. J Stud Alcohol 56:573–579

Giancola PR, Moss HB (1998) Executive cognitive functioning in alcohol use disorders. In M. Galanter (Ed.), Recent developments in alcoholism. The Consequences of Alcoholism: Medial Neuropsychiatric Economic Cross-Cultural pp. 227–251. New York: Plenum Press

Giancola PR (2000) Executive functioning: A conceptual framework for alcohol-related aggression. Experimental and Clinical Psychopharmacology 8(4):576–597

Giancola PR, Helton EL, Osbourne AB, Terry MK, Fuss AM, Westerfield JA (2002) The effects of alcohol on provocation on aggressive behavior in men and women. J Stud Alcohol 63:64–73

Giancola PR (2003) The moderating effects of dispositional empathy on alcohol-related aggression in men and women. J Abnorm Psychol 112:275–281

Giancola PR (2004) Executive Functioning and Alcohol-Related Aggression. J Abnorm Psychol 113(4):541–555

Giancola PR (2007) The underlying role of aggressivity in the relationship between executive function and alcohol consumption. Addict Behav 32:765–783

Giancola PR, Corman MD (2007) Alcohol and aggression - a test of the attention-allocation model. Psychol Sci 18(7):649–655

Giancola PR, Levinson CA, Corman MD, Godlaski AJ, Morris DH, Phillips JP, Holt JC (2009) Men and women, alcohol and aggression. Exp. Clin. Psychopharmacol. 17:154–164

Giancola PR, Josephs RA, Parrott DJ, Duke AA (2010) Alcohol myopia revisited: clarifying aggression and other acts of disinhibition through a distorted lens. Perspect Psychological Sci 5:265–278

Godlaski AJ, Giancola PR (2009) Executive functioning, irritability, and alcohol-related aggression. Psychol Addict Behav 23:391–403

Goldman MS, Darkes J, Del Boca FK (1999a) Expectancy mediation of biopsychosocial risk for alcohol use and alcoholism. In: Kirch L (ed) How expectancies shape experience. American Psychological Association, Washington, pp 233–262

Goldman MS, Del Boca FK, Darkes J (1999) Psychological theories of drinking and alcoholism. Leonard KE, Blane HT (eds.), Guilford Press, New York, pp 203–246

Goldman MS, Brown SA, Christiansen BA (1987) Expectancy theory: thinking about drinking. In: Blane HT, Leonard KE (eds) Psychological theories of drinking and alcoholism. Guilford Press, New York, pp 181–226

Goldman MS (1999) Expectancy operation: cognitive–neural models and architectures. In: Kirsch I (ed) How expectancies shape experience. American Psychological Association, Washington, pp 41–63

Graham K (1980) Theories of intoxicated aggression. J Behav Sci 12:141–158

Gray JA (1982) The neuropsychology of anxiety. An inquiry into the function of the septo-hippocampal system. Oxford University Press, New York

Greenfeld LA (1996) Child victimizers: violent offenders and their victims: US Department of Justice, Bureau of Justice Statistics, Office of Juvenile Justices and Delinquency Prevention

Greenfeld LA (1998) Alcohol and crime: an analysis of national data on the prevalence of alcohol involvement in crime: US Department of Justice

Guillot CR, Fanning JR, Bullock JS, McCloskey MS, Berman ME (2010) Effects of alcohol on tests of executive functioning in men and women: a dose response examination. Exp Clin Psychopharmacol 18:409–417

Hariri AR, Brown SM, Williamson DE, Flory JD, de Wit H, Manuck SB (2006) Preference for immediate over delayed rewards is associated with magnitude of ventral striatal activity. J Neurosci 26:13213–13217

Hawkins KA, Trobst KK (2000) Frontal lobe dysfunction and aggression: conceptual issues and research findings. Aggress Violent Behav 5:147–157

Heinz A, Dufeu P, Kuhn S, Dettling M, Gräf K, Kürten I, Rommelspacher H, Schmidt LG (1996) Psychopathological and behavioral correlates of dopaminergic sensitivity in alcohol-dependent patients. Arch Gen Psychiatry 53:1123–1128

Heinz A, Higley JD, Gorey JG, Saunders RC, Jones DW, Hommer D, Zajicek K, Suomi SJ, Lesch KP, Weinberger DR, Linnoila M (1998) In vivo association between alcohol intoxication, aggression, and serotonin transporter availability in nonhuman primates. Am J Psychiatry 155:1023–1028

Heinz A, Jones DW, Mazzanti C, Goldman D, Ragan P, Hommer D, Linnoila M, Weinberger DR (2000) A relationship between serotonin transporter genotype and in vivo protein expression and alcohol neurotoxicity. Biol Psychiatry 47:643–649

Heinz A, Mann K, Weinberger DR, Goldman D (2001) Serotonergic dysfunction, negative mood states, and response to alcohol. Alcohol Clin Exp Res 25:487–495

Heinz A, Jones DW, Gorey JG, Bennet A, Suomi SJ, Weinberger DR, Higley JD (2003) Serotonin transporter availability correlates with alcohol intake in non-human primates. Mol Psychiatry 8:231–234

Heinz A, Jones DW, Zajicek K, Gorey JG, Juckel G, Higley JD, Weinberger DR (2004a) Depletion and restoration of endogenous monoamines affects beta-CIT binding to serotonin but not dopamine transporters in non-human primates. J Neural Transm 68:29–38

Heinz A, Siessmeier T, Wrase J, Hermann D, Klein S, Grüsser SM, Flor H, Braus DF, Buchholz HG, Gründer G, Schreckenberger M, Smolka MN, Rösch F, Mann K, Bartenstein P (2004b) Correlation between dopamine D(2) receptors in the ventral striatum and central processing of alcohol cues and craving. Am J Psychiatry 161:1783–1789

Heinz A, Braus DF, Smolka MN, Wrase J, Puls I, Hermann D, Klein S, Grüsser SM, Flor H, Schumann G, Mann K, Büchel C (2005) Amygdala-prefrontal coupling depends on a genetic variation of the serotonin transporter. Nat Neurosci 8:20–21

Heinz A, Beck A, Grusser SM, Grace AA, Wrase J (2009) Identifying the neural circuitry of alcohol craving and relapse vulnerability. Addict Biol 14:108–118

Heinz AJ, Beck A, Meyer-Lindenberg A, Sterzer P, Heinz A (2011) Cognitive and neurobiological mechanisms of alcohol-related aggression. Nat Rev Neurosci 12(7):400–413

Higley JD, Suomi SJ, Linnoila M (1991) Csf monoamine metabolite concentrations vary according to age, rearing, and sex, and are influenced by the stressor of social separation in rhesus monkeys. Psychopharmacology 103:551–556

Higley JD, Thompson WW, Champoux M, Goldman D, Hasert MF, Kraemer GW, Scanlan JM, Suomi SJ, Linnoila M (1993) Paternal and maternal genetic and environmental contributions to cerebrospinal-fluid monoamine metabolites in rhesus monkeys (Macaca mulatta). Arch Gen Psychiatry 50:615–623

Higley JD, Suomi SJ, Linnoila M (1996a) A nonhuman primate model of type II excessive alcohol consumption? Part 1. Low cerebrospinal fluid 5–hydroxyindoleacetic acid concentrations and diminished social competence correlate with excessive alcohol consumption. Alcohol Clin Exp Res 20:629–642

Higley JD, Suomi SJ, Linnoila M (1996b) A nonhuman primate model of type II alcoholism? Part 2. Diminished social competence and excessive aggression correlates with low cerebrospinal fluid 5-hydroxyindoleacetic acid concentrations. Alcohol Clin Exp Res 20:643–650

Higley JD, Linnoila M (1997) Low central nervous system serotonergic activity is traitlike and correlates with impulsive behavior. A nonhuman primate model investigating genetic and environmental influences on neurotransmission. Ann NY Acad Sci 836:39–56

Hinckers AS, Laucht M, Schmidt MH, Mann KF, Schumann G, Schuckit MA, Heinz A et al (2006) Low level of response to alcohol as associated with serotonin transporter genotype and high alcohol intake in adolescents. Biol Psychiatry 60:282–287

Houben K, Wiers RW, Jansen A (2011) Getting a grip on drinking behavior: training working memory to reduce alcohol abuse. Psychol Sci 22(7):968–975

Hull J (1981) A self-awareness model of the causes and effects of alcohol consumption. J Abnorm Psychol 90:586–600

Hummerich R, Reischl G, Ehrlichmann W, Machulla HJ, Heinz A, Schloss P (2004) DASB—in vitro binding characteristics on human recombinant monoamine transporters with regard to its potential as positron emission tomography (PET) tracer. J Neurochem 90:1218–1226

Ichise M, Vines DC, Gura T, Anderson GM, Suomi SJ, Higley JD, Innis RB (2006) Effects of early life stress on [C-11] DASB positron emission tomography imaging of serotonin transporters in adolescent peer- and mother-reared rhesus monkeys. J Neurosci 26:4638–4643

Ito TA, Miller N, Pollock VE (1996) Alcohol and aggression: a meta-analysis on the moderating effects of inhibitory cues, triggering events, and self-focused attention. Psychol Bull 120(1):60–82

Jaffe JH, Babor TF, Fishbein DH (1988) Alcoholics, aggression and antisocial personality. J Stud Alcohol 3(49):211–218

Johnson EO, vandenBree MBM, Pickens RW (1996) Subtypes of alcohol-dependent men: a typology based on relative genetic and environmental loading. Alcohol Clin Exp Res 20:1472–1480

Jones GH, Hernandez TD, Kendall DA, Marsden CA, Robbins TW (1992) Dopaminergic and serotonergic function following isolation rearing in rats: study of behavioral responses and postmortem and in vivo neurochemistry. Pharmacol Biochem Behav 43:17–35

Jones BT, Corbin W, Fromme K (2001) A review of expectancy theory and alcohol consumption. Addiction 96:57–72

Jones AP, Laurens KR, Herba CM, Barker GJ, Viding E (2009) Amygdala hypoactivity to fearful faces in boys with conduct problems and callous-unemotional traits. Am J Psychiatry 166:95–102

Karg K, Burmeister M, Shedden K, Sen S (2011) The serotonin transporter promoter variant (5-HTTLPR), stress, and depression meta-analysis revisited: evidence of genetic moderation. Arch Gen Psychiatry 68(5):444–454

Kienast T, Hariri AR, Schlagenhauf F, Wrase J, Sterzer P, Buchholz HG, Smolka MN, Gründer G, Cumming P, Kumakura Y, Bartenstein P, Dolan RJ, Heinz A (2008) Dopamine in amygdala gates limbic processing of aversive stimuli in humans. Nat Neurosci 11:1381–1382

Kim-Cohen J, Caspi A, Taylor A, Williams B, Newcombe R, Craig IW, Moffitt TE (2006) MAOA, maltreatment, and gene–environment interaction predicting children's mental health: new evidence and a meta-analysis. Mol Psychiatry 11:903–913

Kobiella A, Reimold M, Ulshöfer DE, Ikonomidou VN, Vollmert C, Vollstädt-Klein S, Rietschel M, Reischl G, Heinz A, Smolka MN (2011) How the serotonin transporter 5-HTTLPR polymorphism influences amygdala function: the roles of in vivo serotonin transporter expression and amygdala structure. Transl Psychiatry 30(1):e37

Lee TMC, Chan SC, Raine A (2008) Strong limbic and weak frontal activation to aggressive stimuli in spouse abusers. Mol Psychiatry 13:655–656

Leigh BC, Stacy AW (1998) Individual differences in memory associations involving the positive and negative outcomes of alcohol use. Psychol Addict Behav 12(1):39–46

Lejoyeux M, Feuche N, Loi S, Solomon J, Ades J (1998) Impulse-control disorders in alcoholics are related to sensation seeking and not to impulsivity. Psychiatry Res 81:149–155

Lejuez CW, Magidson JF, Mitchell SH, Sinha R, Stevens MC, de Wit H (2010) Behavioral and biological indicators of impulsivity in the development of alcohol use, problems, and disorders. Alcohol Clin Exp Res 34(8):1334–1345

Lemarquand D, Pihl RO, Benkelfat C (1994) Serotonin and alcohol intake, abuse, and dependence: findings of animal studies. Biol Psychiatry 36:395–421

Lesch KP, Bengel D, Heils A, Sabol SZ, Greenberg BD, Petri S, Benjamin J, Müller CR, Hamer DH, Murphy DL (1996) Association of anxiety-related traits with a polymorphism in the serotonin transporter gene regulatory region. Science 274:1527–1531

Levinson CA, Giancola PR (2011) Beliefs about aggression moderate alcohol's effects on aggression. Exp Clin Psychopharmacol 19:64–74

Lipsey MW, Wilson DB, Cohen MA, Derzon JH (1997) Is there a causal relationship between alcohol use and violence? In: Galanter M (ed) Alcohol and violence: epidemiology, neurobiology, psychology, family issues, 13th edn. Plenum Press, New York, pp 245–278

Loeber S, Duka T, Marquez HW, Nakovics H, Heinz A, Mann K, Flor H (2010) Effects of repeated withdrawal from alcohol on recovery of cognitive impairment under abstinence and rate of relapse. Alcohol Alcohol 45(6):541–547

Luria A (1980) Higher Cortical functions in Man. Basic Books, New York

Lyvers MF, Maltzman I (1991) Selective effects of alcohol on wisconsin card sorting test-performance. Br J Addict 86:399–407

MacKillop J, Amlung MT, Few LR, Ray LA, Sweet LH, Munafò MR (2011) Delayed reward discounting and addictive behavior: a meta-analysis. Psychopharmacology 216(3):305–321

Maffli E (2001) Problem drinking and relatives. In: Klingemann H, Gmel G (eds) Mapping the social consequences of alcohol consumption. Kluwer Academic Publishers, Dordrecht, pp 79–91

Maisto SA, Carey KB, Bradizza CM (1999) Social learning theory. In: Leonard KE, Blane HT (eds) Psychological theories of drinking and alcoholism, 2nd edn. Guilford Press, New York, pp 106–163

Mann JJ, Malone KM, Sweeney JA, Brown RP, Linnoila M, Stanley B, Stanley M (1996) Attempted suicide characteristics and cerebrospinal fluid amine metabolites in depressed inpatients. Neuropsychopharmacology 15:576–586

Marczinski CA, Combs SW, Fillmore MT (2007) Increased sensitivity to the disinhibiting effects of alcohol in binge drinkers. Psychol Addict Behav 21:346–354

Marsh AA et al (2008) Reduced amygdala response to fearful expressions in children and adolescents with callous-unemotional traits and disruptive behavior disorders. Am J Psychiatry 165:712–720

Mathews TA, John CE, Lapa GB, Budygin EA, Jones SR (2006) No role of the dopamine transporter in acute ethanol effects on striatal dopamine dynamics. Synapse 60:288–294
Mayfield D (1976) Alcoholism, alcohol, intoxication and assaultive behavior. Diseases of the nervous system 37(5):288–291
McBride WJ (2002) Central nucleus of the amygdala and the effects of alcohol and alcohol-drinking behavior in rodents. Pharmacol Biochem Behav 71:509–515
McCloskey MS, Lee R, Berman ME, Noblett KL, Coccaro EF (2008) The relationship between impulsive verbal aggression and intermittent explosive disorder. Aggress Behav 34:51–60
McKinlay WW, Brooks DN, Bond MR, Martinage DP, Marshall MM (1981) The short-term outcome of severe blunt head injury as reported by relatives of the injured person. J Neurol Neurosurg Psychiatry 44:527–533
Miczek KA, Covington HE, Nikulina EA, Hammer RP (2004) Aggression and defeat: persistent effects on cocaine self-administration and gene expression in peptidergic and aminergic mesocorticolimbic circuits. Neurosci Biobehav Rev 27:787–802
Mihic L, Wells S, Graham K, Tremblay PF, Demers A (2009) Situational and respondent-level motives for drinking and alcohol-related aggression: a multilevel analysis of drinking events in a sample of Canadian University students. Addict Behav 34:264–269
Moeller FG, Barratt ES, Dougherty DM, Schmitz JM, Swann AC (2001) Psychiatric aspects of impulsivity. Am J Psychiatry 158:1783–1793
Moselhy HF, Georgiou G, Kahn A (2001) Frontal lobe changes in alcoholism: a review of the literature. Alcohol Alcohol 36(5):357
Mulvihill LE, Skilling TA, VogelSprott M (1997) Alcohol and the ability to inhibit behavior in men and women. J Stud Alcohol 58:600–605
Murdoch D, Pihl RO, Ross D (1990) Alcohol and crimes of violence: present issues. Int J Addict 25(9):1065–1081
Murphy CM, Winters J, O'Farrell TJ, Fals-Stewart W, Murphy M (2005) Alcohol consumption and intimate partner violence by alcoholic men: comparing violent and nonviolent conflicts. Psychol Addict Behav 19(1):35–42
Nader MA, Czoty PW (2005) PET imaging of dopamine D2 receptors in monkey models of cocaine abuse: genetic predisposition versus environmental modulation. Am J Psychiatry 162:1473–1482
Nasby W, Hayden B, Depaulo BM (1980) Attributional bias among aggressive boys to interpret ambiguous social stimuli as displays of social hostility. J Abnorm Psychol 89:459–468
NCVS (2006) National crime victimization survey. In US Department of Justice, Bureau of Justice Assistance
Nicol AR, Gunn JC, Gristwood J, Foggitt RH, Watson JP (1973) The relationship of alcoholism to violent behaviour resulting in long-term imprisonment. Br J Psychiatry 123(572):47
Ochsner KN, Gross JJ (2005) The cognitive control of emotion. Trends Cogn Sci 9:242–249
O'Farrell TJ, Fals-Stewart W, Murphy M, & Murphy CM (2003) Partner violence before and after individually based alcoholism treatment for male alcoholic patients. J Consult Clin Psychol 71:92–102
Parks KA, Hsieh YP, Bradizza CM, Romosz AM (2008) Factors influencing the temporal relationship between alcohol consumption and experiences with aggression among college women. Psychol Addict Behav 22(2):210–218
Parrott DJ, Zeichner A (2002) Effects of alcohol and trait anger on physical aggression in men. J Stud Alcohol 63:196–204
Pernanen K (1993) Causal attributions in the explanation of alcohol-related accidents. Addiction 88:897–906
Peterson JB, Rothfleisch J, Zelazo PD, Pihl RO (1990) Acute alcohol-intoxication and cognitive-functioning. J Stud Alcohol 51:114–122
Pezawas L, Meyer-Lindenberg A, Drabant EM, Verchinski BA, Munoz KE, Kolachana BS, Egan MF, Mattay VS, Hariri AR, Weinberger DR (2005) 5–HTTLPR polymorphism impacts human cingulate-amygdala interactions: a genetic susceptibility mechanism for depression. Nat Neurosci 8:828–834

Posner MI, Petersen SE (1990) The attention system of the human brain. Annu Rev Neurosci 13:25–42

Puglisiallegra S, Imperato A, Angelucci L, Cabib S (1991) Acute stress induces time-dependent responses in dopamine mesolimbic system. Brain Res 554:217–222

Rather BC, Goldman MS, Roehrich L, Brannick M (1992) Empirical modeling of an alcohol expectancy memory network using multidimensional-scaling. J Abnorm Psychol 101(1):174–183

Rather BC, Goldman MS (1994) Drinking-related differences in the memory organization of alcohol expectancies. Exp Clin Psychopharmacol 2(2):167–183

Ratti MT, Bo P, Giardini A, Soragna D (2002) Chronic alcoholism and the frontal lobe: which executive functions are impaired? Acta Neurol Scand 105:276–281

Reif A, Rösler M, Freitag CM, Schneider M, Eujen A, Kissling C, Wenzler D, Jacob CP, Retz-Junginger P, Thome J, Lesch KP, Retz W (2007) Nature and nurture predispose to violent behavior: serotonergic genes and adverse childhood environment. Neuropsychopharmacology 32:2375–2383

Reimold M, Smolka MN, Zimmer A, Batra A, Knobel A, Solbach C, Mundt A, Smoltczyk HU, Goldman D, Mann K, Reischl G, Machulla HJ, Bares R, Heinz A (2007) Reduced availability of serotonin transporters in obsessive-compulsive disorder correlates with symptom severity - a [11C]DASB PET study. J Neural Transm 114(12):1603–1609. Epub 2007 Aug 22

Ridderinkhof KR, de Vlugt Y, Bramlage A, Spaan M, Elton M, Snel J, Band GP (2002) Alcohol consumption impairs detection of performance errors in mediofrontal cortex. Science 298:2209–2211

Risch N, Herrell R, Lehner T, Liang KY, Eaves L, Hoh J, Griem A, Kovacs M, Ott J, Merikangas KR (2009) Interaction between the serotonin transporter gene (5-HTTLPR), stressful life events, and risk of depression: a meta-analysis. JAMA 301(23):2462–2471

Rohsenow DJ, Bachorowski JA (1984) Effects of alcohol and expectancies on verbal aggression in men and women. J Abnorm Psychol 93:418–432

Rommelspacher H, Raeder C, Kaulen P, Bruning G (1992) Adaptive changes of dopamine-D(2) receptors in rat brain following ethanol withdrawal: a quantitative autoradiographic investigation. Alcohol 9:355–362

Rubio G, Jiménez M, Rodríguez-Jiménez R, Martínez I, Avila C, Ferre F, Jiménez-Arriero MA, Ponce G, Palomo T (2008) The role of behavioral impulsivity in the development of alcohol dependence: a 4–year follow-up study. Alcohol Clin Exp Res 32:1681–1687

Rupp CI, Kemmler G, Kurz M, Hinterhuber H, Fleischhacker WW (2012) Cognitive remediation therapy during treatment for alcohol dependence. J Stud Alcohol Drugs 73:625–634

Saraceno L, Munafo M, Heron J, Craddock N, van den Bree MBM (2009) Genetic and non-genetic influences on the development of co-occurring alcohol problem use and internalizing symptomatology in adolescence: a review. Addiction 104:1100–1121

Sayette M, Wilson G, Elias M (1993) Alcohol and aggression: a social information processing analysis. J Stud Alcohol 54:399–407

Schreckenberger M et al (2004) Acute alcohol effects on neuronal and attentional processing: striatal reward system and inhibitory sensory interactions under acute ethanol challenge. Neuropsychopharmacology 29:1527–1537

Schuckit MA, Russell JW (1984) An evaluation of primary alcoholics with histories of violence. J Clin Psychiatry 45:3–6

Schuckit MA (1994) Low level of response to alcohol as a predictor of future alcoholism. Am J Psychiatry 151:184–189

Schuckit MA, Mazzanti C, Smith TL, Ahmed U, Radel M, Iwata N, Goldman D (1999) Selective genotyping for the role of 5-HT2A, 5-HT2C, and GABA alpha 6 receptors and the serotonin transporter in the level of response to alcohol: a pilot study. Biol Psychiatry 45(5):647–651

Schwandt ML, Lindell SG, Sjöberg RL, Chisholm KL, Higley JD, Suomi SJ, Heilig M, Barr CS (2010) Gene–environment interactions and response to social intrusion in male and female rhesus macaques. Biol Psychiatry 67:323–330

Schweizer TA et al (2006) Neuropsychological profile of acute alcohol intoxication during ascending and descending blood alcohol concentrations. Neuropsychopharmacology 31:1301–1309

Siever LJ, Buchsbaum MS, New AS, Spiegel-Cohen J, Wei T, Hazlett EA, Sevin E, Nunn M, Mitropoulou V (1999) d, l-fenfluramine response in impulsive personality disorder assessed with [F–18] fluorodeoxyglucose positron emission tomography. Neuropsychopharmacology 20:413–423

Smith GS, Branas CC, Miller TR (1999) Fatal nontraffic injuries involving alcohol: a metaanalysis. Ann Emerg Med 33:659–668

Stacy AW, Leigh BC, Weingardt KR (1994) Memory accessibility and association of alcohol use and its positive outcomes. Exp Clin Psychopharmacol 2:269–282

Stavro K, Pelletier J, Potvin S (2012) Widespread and sustained cognitive deficits in alcoholism: a meta-analysis. Addict Biol. doi:10.1111/j.1369-1600.2011.00418.x

Steele C, Southwick L (1985) Alcohol and social behavior: the psychology of drunken excess. J Pers Soc Psychol 48:18–34

Steele C, Josephs R (1990) Alcohol myopia: its prized and dangerous effects. Am Psychol 45:921–933

Sterzer P, Stadler C, Krebs A, Kleinschmidt A, Poustka F (2005) Abnormal neural responses to emotional visual stimuli in adolescents with conduct disorder. Biol Psychiatry 57:7–15

Sterzer P, Stadler C (2009) Neuroimaging of aggressive and violent behaviour in children and adolescents. Front Behav Neurosci 3:35

Sullivan EV, Rosenbloom MJ, Pfefferbaum A (2000) Pattern of motor and cognitive deficits in detoxified alcoholic men. Alcohol Clin Exp Res 24:611–621

Swanson JW, Holzer III CE, Ganju VK, Jono RT (1990). Violence and psychiatric disorder in the community: evidence from the Epidemiologic Catchment Area surveys. Hosp Community Psychiatry 41:761–770

Takahashi A, Kwa C, Debold JF, Miczek KA (2010a) GABAA receptors in the dorsal raphe, nucleus of mice: escalation of aggression after alcohol consumption. Psychopharmacology 211:467–477

Takahashi A, Shimamoto A, Boyson CO, Debold JF, Miczek KA (2010b) GABA(B) receptor modulation of serotonin neurons in the dorsal raphe nucleus and escalation of aggression in mice. J Neurosci 30:11771–11780

Taylor SP, Gammon CB, Capasso DR (1976) Aggression as a function of the interaction between of alcohol and threat. J Pers Soc Psychol 34:938–941

Taylor SP, Schmutte GT, Leonard KE, Cranston JW (1979) The effects of alcohol and extreme provocation on the use of a highly noxious electric shock. Motiv Emot 3:73–81

Taylor S, Leonard K (1983) Alcohol and human physical aggression. In: Geen R, Donnerstein E (eds) Aggression: theoretical and empirical reviews, vol 2. Academic Press, New York, pp 77–101

Taylor SP, Sears JD (1988) The effects of alcohol and persuasive social pressure on human physical aggression. Aggress Behav 14:237–243

Tiffany ST (1990) A cognitive model of drug urges and drug-use behavior: role of automatic and nonautomatic processes. Psychol Rev 97:147–168

Tsai GC, Gastfriend DR, Coyle JT (1995) The glutamatergic basis of human alcoholism. Am J Psychiatry 152:332–340

van Erp AMM, Miczek KA (2000) Aggressive behavior, increased accumbal dopamine, and decreased cortical serotonin in rats. J Neurosci 20:9320–9325

van Erp AMM, Miczek KA (2007) Increased accumbal dopamine during daily alcohol consumption and subsequent aggressive behavior in rats. Psychopharmacology 191:679–688

Vinogradov S, Fisher M, de Villers-Sidani E (2012) Cognitive training for impaired neural systems in neuropsychiatric illness. Neuropsychopharmacology 37(1):43–76

Virkkunen M, Kallio E, Rawlings R, Tokola R, Poland RE, Guidotti A, Nemeroff C, Bissette G, Kalogeras K, Karonen SL (1994) Personality profiles and state aggressiveness in finnish alcoholic, violent offenders, fire setters, and healthy volunteers. Arch Gen Psychiatry 51:28–33

Volkow ND, Wang GJ, Fowler JS, Logan J, Hitzemann R, Ding YS, Pappas N, Shea C, Piscani K (1996) Decreases in dopamine receptors but not in dopamine transporters in alcoholics. Alcohol Clin Exp Res 20:1594–1598

Weerts EM, Tornatzky W, Miczek KA (1993) Prevention of the pro-aggressive effects of alcohol in rats and squirrel-monkeys by benzodiazepine receptor antagonists. Psychopharmacology 111:144–152

Weingardt KR, Stacy AW, Leigh BC (1996) Automatic activation of alcohol concepts in response to positive outcomes of alcohol use. Alcohol Clin Exp Res 20(1):25–30

Wells S, Speechley M, Koval JJ, Graham K (2007) Gender differences in the relationship between heavy episodic drinking, social roles, and alcohol-related aggression in a us sample of late adolescent and young adult drinkers. Am J Drug Alcohol Abuse 33(1):21–29

Wells S, Mihic L, Tremblay PF, Graham K, Demers A (2008) Where, with whom, and how much alcohol is consumed on drinking events involving aggression? Event-level associations in a Canadian national survey of university students. Alcohol Clin Exp Res 32(3):522–533

Wiley JA, Weisner C (1995) Drinking in violent and nonviolent events leading to arrest: evidence from a survey of arrestees. J Crim Justice 23(5):461–476

Wray TB, Simons JS, Dvorak RD, Gaher RM (2012) Trait-based affective processes in alcohol-involved "risk behaviors". Addict Behav 37(11):1230–1239

Yokoyama M et al (2005) Amygdalic levels of doparnine and serotonin rise upon exposure to conditioned fear stress without elevation of glutamate. Neurosci Lett 379:37–41

Zucker RA, Kincaid SB, Fitzgerald HE, Bingham CR (1995) Alcohol schema acquisition in preschoolers: differences between children of alcoholics and children of nonalcoholics. Alcohol Clin Exp Res 19(4):1011–1017

Index

Curr Topics Behav Neurosci (2014) 17: 475–479
DOI: 10.1007/978-3-662-44281-4

MIX
Papier aus verantwortungsvollen Quellen
Paper from responsible sources
FSC® C105338

If you have any concerns about our products,
you can contact us on
ProductSafety@springernature.com

In case Publisher is established outside the EU,
the EU authorized representative is:
Springer Nature Customer Service Center GmbH
Europaplatz 3, 69115 Heidelberg, Germany

Printed by Libri Plureos GmbH
in Hamburg, Germany